International Conference on Mathematics, Science Education (ICMScE 2017)

Strengthening math and science education to Promote ASEAN Community

Journal of Physics: Conference Series Volume 895

Bandung, Indonesia
24 May 2017

ISBN: 978-1-5108-4895-5
ISSN: 1742-6588

Printed from e-media with permission by:

Curran Associates, Inc.
57 Morehouse Lane
Red Hook, NY 12571

Some format issues inherent in the e-media version may also appear in this print version.

Printed by Curran Associates, Inc. (2017)

For permission requests, please contact the Institute of Physics
at the address below.

Institute of Physics
Dirac House, Temple Back
Bristol BS1 6BE UK

Phone: 44 1 17 929 7481
Fax: 44 1 17 920 0979

techtracking@iop.org

Additional copies of this publication are available from:

Curran Associates, Inc.
57 Morehouse Lane
Red Hook, NY 12571 USA
Phone: 845-758-0400
Fax: 845-758-2633
Email: curran@proceedings.com
Web: www.proceedings.com

TABLE OF CONTENTS

Assessing High School Students' Pro-Environmental Behaviour

N Hidayah[1*] **and R R Agustin**[2]

[1]Sekolah Menengah Atas Negeri 1 Lembang, Dinas Pendidikan Provinsi Jawa Barat, Jl. Maribaya No. 68, Lembang, Bandung Barat, Indonesia
[2]International Program of Science Education, Universitas Pendidikan Indonesia, Jl. Dr. Setiabudi No. 229, Bandung 40154, Indonesia

[*]nurul.hidayah@jabarprov.go.id

Abstract. This paper aims to reveal students' pro-environmental behavior in a High School. Self-reported behavior assessment was administered in this study involving students with age range 15 to 18 years. Pro-environmental behavior in this study comprises six domains. Those are recycling, waste avoidance, consumerism, energy conservation, mobility and transportation, and vicarious conservation behavior. Pro-environmental behavior (PEB) of science class students was compared to behavior of non-science class students. Effect of students' grade level and extracurricular activity on the behavior was evaluated. Study revealed that science could improve students' PEB. It is because environmental topics are covered in science class. Student's involvement in extracurricular activity may enhance PEB as well. In conclusion, students' PEB is influenced by class program (science or non-science) but it is not influenced by time length in learning science. This finding could be consider by science educator in choosing strategy to enhance student's pro-environmental behaviour.

1. Introduction

It was agreed that social, economy and environment should compromise in order to build sustainable development [1].According to this, environment is an aspect should be considered in all aspects of live. Nowadays, environment faces many problem including pollution, water scarcity, species distinction and also global warming [2,3]. Those deterioration result from human behavior[4].

Education is believed as a crucial agent in changing behavior to become more environmental friendly. In regard to environment, education should cover not only knowledge but also attitude and behavior [5]. Most of Education systems including in Indonesia establish primary school to secondary school as compulsory education. Thus, assessing secondary school students' pro-environmental behavior can be an indicator of Environmental Education success. It is also expected a feedback on environmental education practice in high school. In order to evaluate environmental literacy including pro-environmental behavior of high school students, many researches were conducted. It was revealed that demographic factor influence pro-environmental behaviour of high school students in Israel and Colcata, India [6,7]. There is also relationship between environmental knowledge and attitude of high school students who has enrolled environmental education class in Texas, USA [8]. Survey on high school students' knowledge and behavior toward environment which focused on 3R in West Malaysia revealed that urban student has better understanding rather than rural student [9]. Meanwhile, effect of schooling system into pro-environmental behavior of adolescent and adult was conducted in Thailand [10]. In Indonesian context, surveys to explore university students' knowledge, attitude and behaviour toward environment were conducted [11]. Perception, knowledge and attitude of Community in Jakarta toward environment were also explored [12]. Survey on believe, attitude and intention of High School student toward environment was also conducted [13]. However, study on high school students' pro-environmental behavior is still needed.

Environmental education is set in educational system through three approaches; separation, infusion and integration. In Indonesian High School Environmental education can be taught as separated subject through local curriculum, infused environmental topics into subjects such as Biology and Geography or integrated into subjects such as literacy or anthropology. Thus, students among different program (Science, Language and Social) may exposure to environmental issues throughout subjects they learn. Students also involve in extracurricular activities in their school.

Based on the background above, this study aims to reveal pro-environmental behavior of High School students related to their grade level (class), study program they choose and involvement in extracurricular activities. To achieve this aim, survey based on self-assessment was conducted.

2. Research Method

This quantitative research was conducted through survey design. This design is a procedure to administer a survey to describe the attitudes, opinions, behaviour or characteristics of population [14]. Population of this research is tenth and eleventh grade students of a High School in West Bandung, Indonesia. Sample of this study is 231 students were randomly chosen.

To assess pro-environmental behaviour, questionnaire based on self-reported behaviour was used. The instrument was adapted from Kaiser, Oerke and Bogner [15] which comprises six domains and 40 items. Two items regarding season condition were excluded and five items related to cultural context were added. Each item was pre-tested to test the validity and reliability. Thirty four out of forty three items is valid. The instrument is reliable with Cronbachs' Alpha reliability coefficient 0,888. The questionnaire used a Likert response scale from 1 (never) to 5(always). The questionnaire also included a set question of student's background namely class, program, and extracurricular involvement. Students were classified according to grade or school level into two groups; tenth grade and eleventh grade. Students were categorized into Social, Language and Science class according to program they choose. Students were divided into two group based on their involvement in extracurricular activities (yes or no).

The ordinal data from questionnaire were changed into interval number through Method of Successive Interval (MSI) prior to statistics analysis. Data were analysed using statistics software determining means and standard deviation through descriptive statistics. Multivariate Analysis of Variance (MANOVA) was conducted to examine relationship between grade, program, involvement in extracurricular activities and pro-environmental behaviour (PEB) generally and six domains of PEB.

3. Result and Discussion

3.1. Effect of School Program on Pro-environmental Behavior

According to multivariate test, students' PEB is affected by program ($p=0.007$) and involvement in extracurricular activities ($p=0.026$) but is not affected by their grade ($p=0.076$). Bonferroni Post Hoc Test indicated students' program (Language, Science and Social) significantly affected energy conservation, waste avoidance and recycling behaviour. Means of those behaviours between three programs are shown on figure 1 and 2. Bonferroni test showed that there is a significance differences between behaviours of science students and behaviours of social students, meanwhile the differences is not significant for language and social either language and science students.

Figure 1 shows the comparison of pro-environmental behaviour among students from different study program namely Language, Science and Social. According to the figure science students tent to have higher score than social students in terms of general pro-environmental behaviour. Meanwhile, language students tent to have in between score of the behaviour. This finding is contrary to study in Kolkata, India which there is no difference between behaviour of Science and Commerce student and PEB of Science students are lower than PEB of Art Students[7]. It is expected that Science Students have higher PEB than others due to their exposure to environmental issue in some Science Subjects.

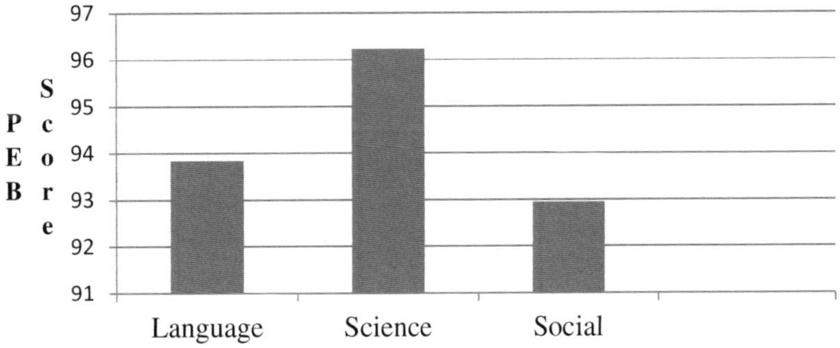

Figure 1. Pro-environmental Behaviour Score of Students from Different Study Program

Post-Hoc Test about relationship between study program and six pro-environmental behaviour domains indicated a significant difference on energy conservation, waste avoidance and recycling behaviour. Means score of those score are showed on figure 2.

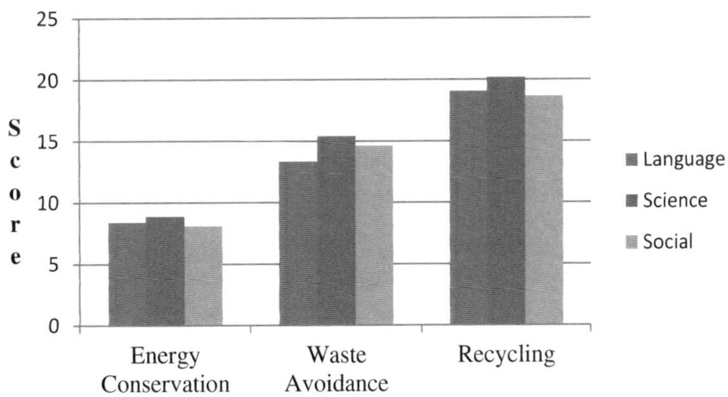

Figure 2. Behaviours Score of Students from Different Study Program

Figure 2 shows that science students tent to have the highest score on energy conservation, waste avoidance and recycling behaviour. This is because science students may accept more information about those topics. However, there is no significant difference on other three domains namely transportation and mobility, consumerism, and mediated conservation behaviour. Behaviours regarding mobility and transportation of student from Language, Science and Social Class are almost the same. It is because behaviour such as walking and choosing bicycle or public transportation rather than motorcycle is presumably influenced by complex factors. Geographical condition and poor management of public transport are among probable factors.

3.2. Effect of Students' Grade Level on Pro-environmental Behavior

In term of students' grade level, it is indicated that students' grade level (class) has no significant effect on pro-environmental behavior (PEB) and six domains of PEB. Among general PEB and PEB domains, energy conservation is the only domain affected by students' grade level as shown in the table 1.

Table 1. p Value of Effect of Students' Grade Level on Pro-environmental Behaviours

Behaviour	p Value
General PEB	0.311
Energy Conservation	0.013
Transportation and Mobility	0.979
Waste Avoidance	0.541
Recycling	0.568
Consumerism	0.355
Vicarious Conservation Behaviour	0.974

When pro-environmental behaviours between tenth and eleventh grade students were compared, there is no statistically significant difference even tenth grade science student and eleventh grade science student. Meanwhile tenth grade science students in this study have not learnt Ecosystem and Pollution concept formally in the classroom whether the eleventh grade have experience learning on those concepts. This finding challenges teachers to think how to teach student environmental concepts in order to encourage pro-environmental behaviour. Approaches, method and media should be considered. There is an alternative to conduct more active learning in Science classroom as these approaches could promote students ecological affinity [13].

3.3. Effect of Extracurricular Activities on Pro-environmental Behavior

Students' involvement in extracurricular activities has a significant effect on pro-environmental behavior. Extracurricular activities have significant effect on almost all behavior except vicarious conservation behavior (p=0.186). Student who involve in extracurricular activities incline to have higher environmental behavior than student who does not involve in this activities. Extracurricular activities often provide student broad horizon about environment. They also often give students opportunities to build tight bonding with nature and engage in some environmental projects.

This study did not explore types of extracurricular activities students join whether it relates to art (e.g. music, dance, and theatre), religion (e.g. Youth Moslem Association), sports (e.g. basketball, volleyball and football team), literacy (e.g. Library Lover), Science (e.g. Science Club) or health and nature (e.g. red cross and nature lover). Thus further study regarding this need to be conducted.

4. Conclusion

This study tried to picture students' pro-environmental behaviour of a High School. The findings cannot be generalized into all high school in Indonesia at national or regional level. Further studies to explore the behaviour within larger population still have to be conducted. This study revealed that pro-environmental behaviour relates to program student chose and extracurricular activities student joined but does not relate to students' grade level. Since there is no difference between pro-environmental behaviour of tenth grade students and eleventh grade student it is highly recommended to establish researches about strategy, approach or model in order to affect students' environmental behaviour in infused and integrated environmental education curriculum.

5. References

[1] United Nation Organization 2005 *2005 World Summit Outcome* (New York: UN General Assembly)

[2] Steinfeld H, Gerber P, Wassenaar T, Castel V, Rosales M and Haan C 2006 *Livestock's Long Shadow : Environmental Issues and Option* (FAO)

[3] Najam A, Runnalls D and Halle M 2007 *Environment and globalization: Five propositions* (Manitoba:iisd) p 4

[4] Steg L and Vlek C 2007 Human behaviour and environmental sustainability problems, driving forces and research topics *J. Soc. Issues* **63** 1-19

[5] UNESCO/UNEP 1977 *Intergov. Conf. on Environmental Education* (Tbilisi : UNEP)

[6] Negev M, Sagy G, Garb Y, Salzberg A and Tal A 2008 Evaluating the environmental literacy of Israeli elementay and high school students *J. Env. Education* **39** 3-20

[7] Sengupta M, Das J and Maji P 2010 Environmental awareness and environmental related behaviour of twelfth grade students in Kolkata effects of stream and gender *J. Anwesa* 1-8

[8] Bradley JC, Waliczek TM and Zajicek JM 1999 Relationship between environmental knowledge and environmental attitude of high school students *J. Env.Education* **30** 17-21

[9] Embong A M, Hashim H M, Malik N A and Iskandar S 2013 Assessing the understanding and practices of 3R's among secondary school students *Int. Conf. on Environment and Electrical Engineering (Wroclaw)* (IEEE) p 1-6

[10] ChankrajangAderhold J, Davydov V Yu, Fedler F, Klausing H, Mistele D, Rotter T, Semchinova O, Stemmer J and Graul J 2001 *J. Ecol. Economics* **131** 434-48

[11] Ma'ruf, Surya S and Dessy P 2016 Knowledge , attitudes and behavior of university students towards environmental issues in Indonesia *J. Sains Humanika* **8** 81–8.

[12] Bradley J C, Waliczek T M and Zajicek J M 1999 Relationship between environmental knowledge and environmental attitude of high school students *J. of Env. Education* **30** 17–21

[13] Kusmawan U, O 'toole J M, Reynolds R and Bourke S 2009 Beliefs, attitudes, intentions and locality: the impact of different teaching approaches on the ecological affinity of Indonesian secondary school students *Int. Res. in Geo. and Env. Education* **18** 157–169

[14] Creswel J W 2012 *Educational Research : Planning, Conducting and Evaluating Quantitative and Qualitative Research* ed 4 (New York : Pearson) p 375-92

[15] Kaiser, F. G., Oerke, B., & Bogner, F. X. (2007). Behavior-based environmental attitude: Development of an instrument for adolescents. *Journal of Environmental Psychology*, *27*(3), 242–251. https://doi.org/10.1016/j.jenvp.2007.06.004

Student Needs to Practicum Guidance in Physiology of Animals Based on Guided Inquiry

R Widiana*, S Susanti and D Susanti
Program Studi Pendidikan Biologi, STKIP PGRI Sumatera Barat, Jl. Gunung Pangilun, Padang 25173, Indonesia

*rinaroesdi68@gmail.com

Abstract. The achievement of the subject of animal physiology requires that the students actively and creatively find their knowledge independently in understanding the concepts, theories, physiological processes, decompose, assemble, compare and modify physiological processes in relation to the fluctuation of environmental factors through practicum activities. The achievement of this lesson has not been fully realized because the learning resources used can't guide, direct and make the independent students achieve their learning achievement and the practical handbook used has not been able to lead the students active and creative in finding their own knowledge. The practical handbook used so far consists only of the introduction of materials, work steps and questions. For that, we need to develop guided inquiry guide based on the needs of students. Objectives this study produces a practical handbook that fits the needs of the students. The research was done by using 4-D models and limited to define stage that is student requirement analysis. Data obtained from the questionnaire and analysed descriptively. The questionnaire obtained an average of 88.16%. So the needs of students will guide guided inquiry based inquiry both to be developed.

1. Introduction

Learning in university not only provides the materials or concepts but must provide a learning experience that allows creative development and student independence in learning. One of the student learning activities that demand creativity and independence students is a practical activity. One of the subjects that are practiced in Biology Education program is Animal Physiology.

Animal physiology is one of the compulsory subjects in the study program of Biology Education. This subject discusses the physiological processes in vertebrate animal bodies. The learning achievement of the subject requires the active and creative students to find their knowledge independently to understanding the concepts, theories, physiological processes, decompose, composing, comparing and modifying physiological processes in relation to the fluctuation of environmental factors through practical activities. One of the factors that can support the creativity and independence of students in the implementation of the practical activities is the availability of adequate laboratory guidance.

The practicum guidance is a form of teaching material. A practice guide is a guideline for the implementation of a practice that contains procedures for preparation, implementation, and data analysis. The practicum guidance is compiled and written by a person or group of teaching staff who handles the practical by following scientific writing rules [1]. The components of the practicum guidance according to Rohman and Sofan [2] are a) Introduction, b) Objectives, c) Basic theories, d) Tools and materials, e) procedures, f) Data/results of observations, g) Discussion and h) Questions and conclusions.

Based on observations of the students and lecturers of Biology Education Program found that Practicum guidance used have not been able to lead active and creative students in finding their own knowledge so that the achievement of learning from the implementation of the lab can not be achieved. The guidance of the existing practicum is that the learning activities are still focused on the lecturers, causing several problems, including 1) the students are unable to relate the physiological concepts in theory to the laboratory activities; 2) the laboratory activities have not emphasized the scientific activities in problem-solving and have not used Approach, 3) practice topics are limited to the type of experiments that can be done only [3], 4) guide provides detailed instructions so that students lose the benefits of the experiment, 5) the instructions are very rigid and does not allow the flexibility of students in creative so that student creativity is hampered [4-7] and 6) the student is unable to communicate his idea effectively in making practice report. For example, students are unable to attribute variations of the rate of respiratory processes with nerve work and fluctuations in animal environmental factors.

What is the solution to this problem? Therefore, it is necessary to develop a practicum guidance that is able to lead students actively, creatively and independently in building their knowledge and forming active, productive and conducive educational interaction between students and lecturers. A student-centered and active approach that uses inquiry-based laboratory activity has been shown to produce more skills related to independent and critical thinking and enhances students' curiosity and motivation than any other traditional approach [8-11] Before doing the development of teaching materials, to get the teaching materials that match the characteristics and needs of students need to be done in advance of student analysis. Based on the above, the definition of this research has been done by the student's analysis with the aim to get the design model for the development of practical handbook according to the characteristics and the needs of the students.

2. Experimental Method

This research uses Four-D development model developed by Thiagarajan. This research has been done until define stage and the activity done at this stage is student analysis. This research is a development research using 4D model (define, design, develop and disseminate). But the stage that is carried out only to the define stage is to analyze the needs of students Respondents consisted of 45 students of the 5th semester who have followed animal physiology practicum, responded taken from 5 classes with 3 lecturers, each sample class was taken 25%.The data in this study were obtained from the questionnaire instrument of modified student needs from Sophia [11] The data obtained were processed by descriptive analysis using modified formulas from Riduwan [12] as follows.

Value Requirement = (Score obtained / Score maximum) X 100%

The level of achievement of needs category using Purwanto classification [13] in the following table.

Table 1. Categories of Analysis needs of students

Level of Achievement (%)	Category
90-100	Very Good
80-89	Good
65-79	Enough
55-64	Not Good
0-55	Very Bad

3. Result and Discussion

Based on the results of interviews that have been conducted by researchers with lecturers of Animal Physiology lecturers obtained information that during this practical handbook has been using practical guides and practical activities have been able to run well but the guides available not yet able to lead students to be creative, active and independent in finding Concepts and build knowledge and form an active, productive and conducive educational interaction between students and lecturers. The result of student requirement analysis based on practical handbook through questioner distribution to students is shown in Table 2.

Table 2. Results of Student Needs Analysis Based On Animal Physiology Practical Handbook

No	Questio	Student Answers	Percentage (%)
1	Do you already have an animal physiology guide	Yes	100
2	Do you like the activities of Animal physiology practical?	Yes	100
3	Is the work done in accordance with the theories learned in the classroom?	Yes	100
4	Does the animal physiology guide you use to contain the basic theories about the material to be practiced?	Yes	83,33
5	Does the lab workbook you use have images that support the theoretical basis of the material being practiced?	Yes	58,33
6	Are your practical handbook used in order to do practical work?	Yes	100
7	Is the language used in practical handbook easy to understand?	Yes	91,66
8	Does the practical handbook you use have column sheets for observation results?	Yes	100
9	Does the practical handbook you use have a column for assessment?	Yes	100
10	Is the work order in your practical handbook easy to understand?	Yes	91,66
11	Is the practical handbook you use to help explain abstract learning concepts?	Yes	83,33
12	Does your practical have applied the following scientific ways: formulating problems, formulating hypotheses, collecting data, testing hypothesis formulating hypotheses and drawing conclusions?	Yes	58,33
13	Do you have prior knowledge/read lines before practicing?	Yes	91,66
14	Can you conclude the results of the lab?	Yes	100
15	Can you understand the practice of formulating hypotheses according to the Guided Inquiry steps of identifying mapping the scope of the problem, planning and predicting outcomes, investigating data collection, interpreting data and developing conclusions and reflecting?	Yes	75
16	Is there a conducive atmosphere at the time (fun)?	Yes	83,33
17	Does the laboratory on your campus have a laboratory/technician?	Yes	100
18	Is at the time of practical all students participate / active role in practical?	Yes	58,33
19	Is it possible to increase your knowledge after practical?	Yes	100

Based on Table 1 it can be seen that 100% of students answered already have a practical handbook, 100% like practical animal physiology, 100% practical has been carried out according to theory, 83.33% guidance guide already contains the required theory, 58,33% Which is required, 100% guides contain discipline of practice, 91.66% of the language used is easy to understand, 100% guides have observation columns, 100% has assessment columns, 91.66% of working order is easy to understand, 83.33 % Guidance is able to help explain the concept of abstract learning, 58.33% practical activities have applied for scientific work, 91.66% of students have preliminary knowledge before the lab, 100% of students can conclude the lab results, 75% of students are able to understand the practical according to Step guided inquiry, 83.33% the atmosphere is conducive laboratory, 100% laboratory already has a labour /technician, only 58.33% of students who have active participation in the lab and 100% practical can increase knowledge.

The result of the analysis from the student's answer shown that the students of Biology Education that program need practicum guidance that can help the students to understand the abstract concept and stimulate the active participation of the students in the practical implementation, is the practical handbook which has the following characteristics: 1) In accordance with the achievement of learning 2) systematic, 3) contains the required theories, 4) can help explain the concepts of learning that are abstract, 5) can guide and train students in applying scientific work, Implementation of the inquiry approach in the lab can increase involvement in scientific work. Students complete a physiological research project from developing individual hypotheses to preparing journal-formatted reports. Skills associated with formulas and hypothesis testing, experimental design, and data analysis, interpretation, and presentation are required when students complete their reports [14]. 6) can understand the material lab According to guided inquiry step and 7) can motivate students to actively participate in practical activities, Application of inquiry approach in the implementation of animal physiology practice can improve LO achievement of students' material understanding from 57% to 92%, LO hi design [1], and to increased student motivation and commitment to the course [16]. Regular course commitment also encourages cooperation among students and increases the likelihood of peer guidance.

Based on this result, it is necessary to develop guidance of Animal Physiology based on guided inquiry approach. The inquiry approach is the right approach because the inquiry approach has several steps which are suitable for the practical activity. The steps are 1) orientation, 2) formulate the problem, 3) formulate the hypothesis, 4) collect data, 5) test the hypothesis and 6) formulate the conclusion. In addition, the inquiry approach can train learners in developing thinking skills through questions [17].

The inquiry consists of two levels: open inquiry and guided inquiry. Departing from the analysis that active participation and student independence in the implementation of the lab is still low, the inquiry approach chosen in this guidance development is guided inquiry approach. In a guided inquiry approach, learners seek and find with the help of teachers as facilitators [18]. Guided inquiry model can provide learners with the opportunity to learn how to find facts, concepts, and principles through thinking and scientific skills [19], as well as improve positive attitude and high-level thinking [20]. In addition, guided inquiry model can also attract the attention of learners to do the practice independently and mentored.

The result of student's analysis also showed that 91,66% of students have knowledge and readiness before doing practically. This indicates that the students already have experience with the material that practical before. This condition supports the use of guided inquiry approaches in the development of this Physiology Animal practical handbook. Guided inquiry based lab activities are laboratory activities that encourage learners to build their cognitive knowledge. In addition, through practical activities learners will gain experience directly, the results of the experience can be either new knowledge or verification of prior knowledge. Knowledge is the result of two stages of learning that is assimilation and accommodation that occurs in practical activities. Assimilation occurs when new information (the result of practicum activities) matches the prior knowledge. If learners get new information that is not in accordance with previous knowledge, then learners experience the stage of accommodation. At this stage, there is an imbalance of knowledge so that learners will accommodate (find out and find solutions to the problem) of that knowledge to fit the knowledge possessed previously [21].

4. Conclusion

From the student's analysis, it can be concluded that the students need guidance of Physiology of Animals based on guided inquiry in Biology Education Program. Based on these results, the research can be continued at the design and development stage of Practicum guides based on guided inquiry approach.

Acknowledgments

The authors express their immense gratitude to KEMENRISTEK DIKTI who has funded all research and publication of the results of this research.

References

[1] Meyhandoko, A.2013 Development of Contextual Practical Guidance with Utilization of Local Environmental Condition in Pollution Material Learning at SMAN 2 Rembang (Semarang: Department of Biology FMPA Semarang State University) p 45
[2] Rohman and A. Sofan 2013 Strategy and Design of Learning System Development (Jakarta: Pustaka Karya Achievement) p 97
[3] National Research Council. 2000. Inquiry and the National Science Education Standards: a Guide for Teaching and Learning (Washington, DC: National Academy)
[4] National Research Council. 2002. Learning and Understanding: Improving Advanced Study of Mathematics and Science in U. S. High Schools. (Washington, DC: National Academy)
[5] Saunders DK, Sievert LM. Providing StudentsT he Opportunity to Think Critically and Creatively Through Student Designed Laboratory Exercises. Bioscience 28: 9 –15, 2002.
[6] Schroeder C., Scott TP, Tolson H, Huang TY, Lee YH.2007. A Metaanalysis of National Research: effects of Teaching Strategies on Student Achievement in Science in The United States. J Res Sci Teach 44: 1436 –1460.
[7] Springer I, Stanne MF, Donova, SS. 1997. Effects of Small-Group Learning on Undergraduates in Science, Mathematics, Engineering, and Technology: a Meta-Analysis. Madison, WI: National Institute for Science Education and University of Wisconsin-Madison, Research Monograph 11
[8] Myers MJ, Burgess AB. 2003. Inquiry-Based Laboratory Course Improves Students' Ability to Design Experiments and Interpret Data. Adv Physiol Ed 27: 26 –33, 2003
[9] National Research Council. 1999. How People Learn: Brain, Mind, Experience, and School. Committee on Developments in The Science of Learning. (Washington, DC: National Academy)
[10] VanderArk SD, Ely D. 1992. Biochemical and Galvanic Skin Responses to Music Stimuli by College Students in Biology and Music. Percept Mot Skills 74:1079 –1090.
[11] Sophia, A 2017. Pengembangan Penuntun Praktikum Taksonomi Invertebrata Berbasis Pendekatan Saintifik untuk Mahasiswa (Padang: Universitas Negeri Padang).
[12] Riduwan 2007 Belajar Mudah Penelitian untuk Guru-karyawan dan Peneliti Pemula (Bandung: Alfabeta)
[13] Purwanto, N 2004. Prinsip-Prinsip dan Teknik Evaluasi Pengajaran (Bandung: Remaja Rosdakarya)
[14] River, D.B. 2002. Using A Course-Long Theme for Inquiry Based Laboratories In a Comparative Physiology Course. Advanced In Physiology Education. 26(4): 317-326
[15] Casotti, G., L.R. Danner, and M. T. Knabb. 2008. Successful implementation of inquiry-based physiology laboratories in undergraduate major and nonmajor courses. Advanced In Physiology Education. 32: 286-296
[16] National Research Council. From Analysis to Action: Undergraduate Education in Science, Mathematics, Engineering, and Technology. Report of a Convocation. Washington, DC:

Center for Science, Mathematics, and Engineering Education, National Research Council, 1996.
[17] Massialis BG. 1985. Discovery and Inquiry-based Programs. In: The International Encyclopedia of Education, edited by Husen T, Postlethwaite TN. Oxford: Pergamon, p. 1415–1418.
[18] Sulistina, O. 2010 Pengembangan Pembelajaran Kimia dengan Model Inkuiri Terbimbing (Malang: Jurusan Kimia FMIPA Universitas Negeri Malang) p 7
[19] Dimyati dan Mudjiono 2006 Belajar dan Pembelajaran (Jakarta: Rineka Cipta) p 173
[20] Kurniati danWahyuningrum, D. 2011. Pengembangan Pembelajaran Inkuiri Terbimbing di
[21] SMA/MA Melalui Penyusunan Modul Praktikum Isolasi dan Identifikasi Senyawa
[22] dalam Daun Tanaman Mint (Menthacordifolia Opiz). Prosding Simposium
[23] Nasional Inovasi Pembelajaran dan Sains. Bandung, 22-23 Juni 2011.
[24] Rustaman, N Y. 2005 Perkembangan Penelitian Pembelajaran Berbasis Inkuiri dalam Pendidikan Sains. Makalah disajikan pada Seminar Nasional II Himpunan Ikatan Sarjana dan Pemerhati

IOP Conf. Series: Journal of Physics: Conf. Series **895** (2017) 012006 doi:10.1088/1742-6596/895/1/012006

Student Analysis of Handout Development based on Guided Discovery Method in Process Evaluation and Learning Outcomes of Biology

S Nerita*, A Maizeli and A Afza

[1]Program Studi Pendidikan Biologi, Sekolah Tinggi Keguruan dan Ilmu Pendidikan Persatuan Guru Republik Indonesia Sumatera Barat, Jl. Gunung Pangilun, Padang 25173, Indonesia

*siskabio@gmail.com

Abstract. Process Evaluation and Learning Outcomes of Biology subjects discusses the evaluation process in learning and application of designed and processed learning outcomes. Some problems found during this subject was the student difficult to understand the subject and the subject unavailability of learning resources that can guide and make students independent study. So, it necessary to develop a learning resource that can make active students to think and to make decisions with the guidance of the lecturer. The purpose of this study is to produce handout based on guided discovery method that match the needs of students. The research was done by using 4-D models and limited to define phase that is student requirement analysis. Data obtained from the questionnaire and analyzed descriptively. The results showed that the average requirement of students was 91,43%. Can be concluded that students need a handout based on guided discovery method in the learning process.

1. Introduction

Process Evaluation and Learning Outcomes of Biology subjects are one of the compulsory subjects in Biology Education of Sekolah Tinggi Keguruan dan Ilmu Pendidikan Persatuan Guru Republik Indonesia Padang. This subject discusses the evaluation process in learning and application of designed and processed learning outcomes. Students as prospective educator should be able to master this subject, So they can analyze and evaluate their student's learning outcomes appropriately.

Some problems found during the evaluation of the learning process, the student to understand material and the subject unavailability of learning resources that can guide and make students independent study. One of them is in the design of knowledge scoring instruments. In addition to the design of the instrument required an understanding of biological materials. For example on the organizational structure of life. Before designing questions, students must understand the content of the material, such as cells, tissues, organs, organ systems and organisms. After that, then students design the question based on the cognitive level of Bloom's Taxonomy from C1 level to C6 level. This also applies to other biological materials. To design questions, students need help in concept discovery. The solution to solve the problem is to apply guided discovery methods.

Learning with guided discovery method hoping the students are really active learners to find their own materials to learn [1]. Learning based guided discovery method is one of the innovative learning models that can provide active learning conditions to students. This lesson emphasizes the process of solving problems faced scientifically. Students not only just take notes and memorize the material, but students are actively thinking and can finally make conclusions with the guidance of lecturers [2]. Students learn better when they are led through activities that allow them to discover concepts for themselves (guided discovery) than from the presentation of concepts by an instructor [3].

Learning with guided discovery method can be converted into the handout based on guided discovery method. So required student analysis in developing handout based on guided discovery method. Needs analysis is an important step in terms of developing curriculum and syllabus for different courses. It is necessary to become familiar with students' objectives, attitudes, learning habits and expectations of the course in order to have an appropriate and suitable syllabus and course design [4]. Needs analysis should be carried out as frequently as possible [5] Teachers have to promote meaningful learning with different tools to respond to student need [6]. The purpose of this study is to produce *handout* based on guided discovery method that suitable for the students.

2. Experimental Method

This research used 4-D development model [7]. 4-D consists of 4 phase, define, design, develop, and disseminate. This research has done until define stage, that was student analysed. Data collection used questionnaires of student needs analysis. The subjects of this researched were 20 students of Biology Education Sekolah Tinggi Keguruan dan Ilmu Pendidikan Persatuan Guru Republik Indonesia Sumatera Barat. They are 20 until 21 years old which have taken Evaluation Process and Learning Outcomes of Biology subject. Data was taken on 12 January 2017.

3. Result and Discussion

The results of student needs analysis can be seen in Table 1 below.

Table1. The results of student needs analysis

No	Question	Choice of Student Answers	Percentage %
1	Do you like learning by using teaching materials?	Yes	100
2	Why are you using teaching materials?	Because the teaching materials help to study independently	90,48
3	Do you have difficulty in learning process without teaching materials?	Yes	100
4	What kinds of teaching materials (print) do you know?	*Handout*	95,24
5	Do you like the handout as a learning resource in process evaluation and learning outcomes of biology?	Yes	100
6	How is the presentation of the material you expect from a handout?	Guided (directed) to find concepts	71,43
7	What style of language do you like about a handout?	Standard, but not rigid	85,71
8	Do you think it needs to develop handouts for learning process?	Need	100
9	Do you know about guided discovery method?	Yes	71,43
10	Do you think it should be developed	Yes	100

IOP Conf. Series: Journal of Physics: Conf. Series **895** (2017) 012006 doi:10.1088/1742-6596/895/1/012006

	handout based on guided discovery method in Process Evaluation and Learning Outcomes of Biology?

Table 1 showed that 100% students like learning to use teaching materials. 90,48% students stated that teaching materials can help self-study. 100% students stated that they have difficulty in lectures without using teaching materials. 95,24% students know that handout is one type of teaching materials. 100% students like handouts as learning resources in Process Evaluation and Learning Outcomes of Biology. 71,43% students stated that they like the presentation of handouts in evaluation lectures by being guided (directed) to find concepts. 85,71% students liked the style of the language in the standard but not rigid. 100% students agree that handouts need to be developed for lectures. 71,43% student know about guided discovery methods. 100% students stated should be developed handout based on guided discovery method in process evaluation and learning outcomes of biology.

Based on the data, can be seen that students like learning by using teaching materials. The teaching materials are all materials are arranged systematically which displays the complete figure of competence that will be mastered by learners and used in the learning process with the purpose of planning and review of the implementation of learning [8]. Teaching materials are preferred by students because it is easier to understand the material. With the teaching materials, lecturers do not have to speak in class, but as facilitators for students. The lecturer's task is not to provide the knowledge that students should memorize, but to prepare a learning environment that enables students to active build their knowledge. Presence teaching materials can facilitate students to learn independently [9].

The questionnaire analysis also showed that students know and like handout as teaching materials. Handout is written materials prepared by a teacher to enrich the knowledge of learners [10]. With provided handouts, students do not need to write fast without focus on the concepts. Therefore they have more time to listen and focus on the educational content [11]. Students like handouts because the grammar is simpler so it is easy to understand. Furthermore, the preparation of handouts is also in accordance with face-to-face meetings in the semester learning plan, so students are not confused about the depth and breadth of the material and practical in its use.

Questionnaires that have been disseminated also indicate that the student knows about guided discovery methods and needs to be handout development based guided discovery method. Guided discovery method is a method that can guide and direct students to learn independently and find their own knowledge. In this lesson, students are encouraged to learn largely through active engagement with concepts and principles, and teachers encourage students to have experience and experiment that allows them to find principles for themselves. Therefore, it is hoped that with this guided discovery method can improve the learning outcomes of students [12]. Learning using guided discovery methods can improve student learning outcomes [13]. Guided discovery method was more effective than the conventional method in promoting chemistry students' achievement in chemistry [14].

Using active learning approaches like guided discovery learning compared to the traditional methods is more effective for the improvement of problem-solving, decision making, critical thinking, creative thinking and interpersonal skills [15]. Higher order thinking skill can not be developed directly and instantly [16]. This indicates that the guided discovery method influences the students' critical thinking skills.

This guided discovery method can be converted into teaching materials, that is handout based on guided discovery method. In accordance with the results of the questionnaire analysis, that students agreed to do handout development based on guided discovery method. Based on this, it can be seen that the students need the material in the form of handouts based on guided discovery method.

4. Conclusion

The conclusion of this study is that students need a handout based on guided discovery method in the learning process.

Acknowledgments
Our thanks to Kementerian Riset, Teknologi dan Pendidikan Tinggi who have funded this research.

References

[1] Suherman E 2003 Common TextBook Contemporary Learning Math Strategy (Bandung: JICA Universitas Pendidikan Indonesia) p 212
[2] Widdihartono R 2004 *Model-model Pembelajaran Matematika SMP* (Departemen Pendidikan Nasional) p 5
[3] Achera L J, R R Belecina and M D Garvida 2015 The Effect of Group Guided Discovery Approach on the performance of Students in Geometry *International Journal of Multidisciplinary Research and Modern Education (IJMRME)* p 339
[4] Moiinvaziri M 2014 Students' Voice: A Needs Analysis of University General English Coursein Iran *Journal of Language Studies* p 57
[5] Todea L and R Demarcsek 2016 Needs Analysis for Language Course Design. A Case Study for Engineering and Bussines Students IOP Conf. Series: Materials Science and Engineering 200 (2017) 012064 p 9
[6] Herold J F 2014 A Cognitive Analysis of Students' Activity : An Example in Mathematics *Australian Journal of Teacher Education* p 151
[7] Thiagarajan S, Dorothy S S, and Melvyn I S 1974 *Instructional Development for Training Teacher of Exeptional Children* (Indiana: Indiana University) p 6-9
[8] Prastowo A 2012 Panduan Kreatif Membuat Bahan Ajar Inovatif: Menciptakan Metode Pembelajaran yang Menarik dan Menyenangkan (Jogjakarta: Diva Press) p 180
[9] Dimas A, Cari, Suparmi, Sarwanto, J Handhika 2016 Profil Analisis Kebutuhan Bahan Ajar Mahasiswa Materi Dinamika Gerak pada Mata Kuliah Fisika Dasar *Prosiding Seminar Nasional Fisika* p 42
[10] Majid A 2008 Perencanaan Pembelajaran Mengembangkan Standar Kompetensi Guru (Bandung: Remaja Rosdakarya) p 172
[11] Avval F Z, L Jarahi, K Ghazvini and M Youssefi Distribution of Handouts in Undergraduate Class to Create More Effective Educational Environment International *Journal of Education and Research* p 5
[12] Nurhadi 2004 Pembelajaran Kontekstual dan implementasinya pada KBK (Malang: UM Press) p 122
[13] Juweni, Sumadji, T C Wulandari 2016 Penerapan Metode Penemuan Terbimbing untuk Meningkatkan Hasil Belajar Siswa *Prosiding Seminar Nasional Matematika 2016 Universitas Kanjuruhan Malang* p 177
[14] Akani O 2007 Effect of Guided Discovery Method of Instruction And Students' Achievement in Chemistry at the Secondary School Level in Nigeria p 6226
[15] Makoolati N, M Amini, H Raisi, Sh. Yazdani, AV. Razeghi The Effectiveness of Guided Discovery Learning on the Learning and Satisfaction of Nursing Students *Hormozgan Medical Journal* p 494
[16] Fianti, F L Najwa, S Linuwih 2016 Development of Open Ended Problems for Measuring The Higher-Order-Thinking-Skills of High School Students on Global Warming Phenomenon *IOP Conf. Series: Journal of Physics: Conf. Series* 824 (2017) 012008 p1

IOP Conf. Series: Journal of Physics: Conf. Series **895** (2017) 012007 doi:10.1088/1742-6596/895/1/012007

Mathematical Understanding of the Underprivileged Students through GeoGebra

A Amam[1,3*], A T Fatimah [1,3], W Hartono[2] and A Effendi [1,3]
[1]Departemen Pendidikan Matematika, Universitas Galuh, Jl. R.E Martadinata No. 150 Ciamis 46274 Jawa Barat, Indonesia
[2]Departemen Pendidikan Matematika, Universitas Swadaya Gunung Djati, Perjuangan No. 01 Kota Cirebon 45134 Jawa Barat, Indonesia
[3]Departemen Pendidikan Matematika, Sekolah Pasca Sarjana Universitas Pendidikan Indonesia, Jl. Dr. Setiabudi No. 229, Bandung 40154, Indonesia

*amam@unigal.ac.id

Abstract. A student's mathematical understanding in high school from poor families in the district of Ciamis is still low. After reviews the various literature and earlier research, consequently, researchers convince that learning mathematics with GeoGebra can help students improve for the better understanding. Our long-term goal of this research is to support the implementation of new curriculum, namely ICT-based learning mathematics. Another goal is to give a basic mastery skill regarding mathematics software to students from underprivileged families. Moreover, the specific objective of this study is to examine the students' mathematical understanding from underprivileged families after the implementation of learning with GeoGebra. We use a quantitative comparative research method to determine differences in the mathematical understanding of students' from underprivileged families before and after mathematics learning with GeoGebra. Accordingly, the students of senior high school from underprivileged family in Baregbeg, Ciamis district, are the population of this study. This research is using purposive sampling. The instrument is in the form of a test question, which is the test of mathematical understanding. Research results show that the mathematical understanding students' from underprivileged families after the mathematics learning with GeoGebra becomes better than before. The novelty of this research is that students understand the material of trigonometry through the use of modules, aided by GeoGebra in learning activities. Thus, the understanding has an impact on improving students' mathematical understanding. Students also master the use of GeoGebra Software. Implementing these two things will be very useful for the next lesson.

1. Introduction

Students must master the ability of mathematical understanding which is the purpose of learning mathematics [1]. Mathematical understanding is a key to solve mathematical problems. The ability to solve mathematical problems is important. Recently research even observes the gesture of students in solving mathematical problems [2]. Most students with underprivileged backgrounds have poor mathematical understanding skills. National exam results corroborate the claim. Allegedly, It is because of their limitations on access to technology. Therefore, we need a technology that can help the learning process. GeoGebra software is one tool that can help students in understanding the concept of mathematics. Especially students who need visualization help. Because this software can construct the problem becomes more real. So that will facilitate students in understanding the problem and solve it more accurately.

According to Indonesian constitution article 31, education is the right of every citizen in Indonesia. Consequently, the government had to organize education for its citizens within the framework of the intellectual life of the nation with having faith, devotion, and noble character. The intelligent nation automatically improves the welfare of the nation. Moreover, to examine the extent of welfare, there are eight fields that should be the focus of attention which are population, health and nutrition, education, employment, extent and patterns of consumption, housing and the environment, poverty, and other social becomes a reference in an effort to improve the quality of life [3].

The implementation of education must reach all the people. Education is not only enjoyed by wealthy citizens. Accordingly, the government should seek education equally for all its citizen, ensure that every citizen has a minimum of elementary and secondary education. Moreover, education also does not know city or remote, rich or poor, prosperous or underprivileged family. All student should get the same opportunity in obtaining education let alone in math lessons. As Schnell and Preediger [4] "equity and opportunity are required to study especially for underachieving students and the opportunity to gain access to basic mathematics. This is particularly important because of the role of mathematics as a gatekeeper for higher education (secondary and tertiary), especially in the fields of science, technology, engineering, and mathematics (STEM) whose job prospects are good in many countries".

Poor and underprivileged families still feel the lack of access to quality education. The National Family Planning Coordinating Board in Indonesia classifies the level of family welfare, including the underprivileged family classification. Underprivileged families are families who have not been able to meet basic needs (basic need) at a minimum, such as the need for spiritual, food, clothing, board, health and family planning. Moreover, underprivileged families have the main characteristics, which are unable to fulfil family needs which include, first, basic needs. Second, to reach health services. Third, the floor part of the house is the ground, fourth, lack of education.

Guarantees to the poor are listed in the 1945 Constitution article 34. Consequently, the state should develop social security for all Indonesians and empower the poor to make their lives prosperous and dignified. One way to achieve it is with education. For a long time, the mathematics education discourse about equity and the discourse about mathematics enrichment for students of high potential were conducted rather separately: On the one hand, the discourse on equity and access mainly referred to offering access to a minimum of basic mathematics, without systematically taking into account that there might be high potentials among the underprivileged students. Only recently, the underprivileged access to tertiary education came in view [5].

Quality education can certainly encourage the achievement of welfare and enhancement of the dignity of a nation. Efforts to realize the achievement of quality education is to develop technology-based education. Education without balanced technology development will lead to miss with other nations. Technological development efforts for education are absolutely necessary. In schools education, the most important aspect is how to provide effective and meaningful learning so as to impact students' ability in understanding the concepts. The use of technology should develop continuing. Mathematical learning using computer technology continues to develop. Consequently, students can use it in everyday life on various problem-solving. We expect the use of technology in mathematics learning can also improve students' mathematical abilities.

The students' mathematical understanding ability is still not satisfactory. The results of Wahyudin [6] provide claims about it. In detail, Wahyudin found five weaknesses that exist in the students, one of which is the students lack the ability to understand and explore the basic concepts (axiomatic, definitions, rules and theorems) mathematics-related subjects that become the focus of attention.

Seeing the influence of GeoGebra in mathematics learning on a mathematical understanding for students of underprivileged families on the trigonometric material is the focus of this research. We provide opportunities for students from underprivileged families to use GeoGebra in the learning process of mathematics. GeoGebra is a free dynamic software for teaching mathematics and learning that provide geometry and algebraic feature in a fully-fledged software environment [7]. There is a several consideration to choose the software for a learning process. First, GeoGebra is easy to use for students and teachers because it has a fairly complete tool. Second, GeoGebra provides visualization in various representations to facilitate mathematical understanding. Third, GeoGebra can compute more than ordinary calculators.

The use of empirical investigations with GeoGebra contributes to student-centered learning by creating a discussion environment and it has been revealed in many studies that dynamic mathematics software used in learning environments in recent years has affected students positively. It was determined that GeoGebra, a dynamic mathematics software, had positive effects on students' motivation and increased students' interest towards the course [8]. The main reason for the implementation of the system GeoGebra into the mathematics education is the creation of a dynamic and interactive digital environment stimulating active learning [9].

The problem formulation in this research is: Is the mathematical understanding of underprivileged students after getting mathematics learning with GeoGebra assisted better than before obtaining mathematics learning with GeoGebra assisted?

2. Experimental Method

The method used in this research is an experimental method with experimental quasi one group pretest-posttest design as for research model which conducted is laboratory research using computer assisted GeoGebra software. This study was conducted on 10 high school students with underprivileged family background and had a very low initial mathematical ability. The research procedure conducted is as follows: first, Selection of students with background ability of low mathematical understanding and come from underprivileged families. Second, pretest the ability of mathematical understanding. Third, conditioned students to learn mathematics on the subject of trigonometry using computer with GeoGebra software. We carry out the learning process in 4 meetings. Every meeting is 60 minutes. The trigonometric subjects in this study are the comparison of trigonometry on right triangles (first meeting), angle and quadrant (second meeting), trigonometric functions (third meeting), and related angles at the fourth meeting. Students learn by using module guide and GeoGebra software on the computer. Module containing materials and worksheets that guide students to complete blank fields of questions. The GeoGebra software is helping to answer the questions. Thus, in addition to students understanding the trigonometric material as a concept, students can also visualize trigonometric material through GeoGebra software. Fourth, post-test of mathematical understanding ability. Fifth, data analysis using normality test of pretest-posttest data, homogeneity test with Levene and paired sample t-test one right side.

3. Result and Discussion

3.1 Results

The researchers collected data in the form of a test score of mathematical understanding ability before and after the action. To know the ability of students understanding of underprivileged family after obtaining learning with GeoGebra assisted better than before, hence used t-test by first doing a prerequisite test that is normality and homogeneity test. The analysis of mathematical comprehension test scores of underprivileged family students is presented in Table 1

Table 1. The Results of Statistical tests of Mathematical Understanding Tests

Mathematical understanding	Sig. Test For Normality	Sig. Homogeneity Test	Mean	Sig. 2-tailed Test Difference Of Two Mean
Pretest	0.423	0.781	24.03125	0.00
Posttest	0.900		77.0625	

Table 1 shows the results of research data analysis using the help of SPSS software. Both data are a normal distribution and homogeneous. Then, we apply the dependent t test one right side.
The learning process of trigonometry with GeoGebra-assisted, for four meetings, is seen in the activity of figure 1 and figure 2.

Figure 1. Students understand the module

Figure 2. Results of student work using GeoGebra

Figure 1 shows students understanding the trigonometric material in a computer lab using a module. The module used contains activities that require students to practice using GeoGebra software. Figure 2 shows one of the students' works done by GeoGebra where the student visualizes the function f (x) = sin (x) by inputting the function to the GeoGebra software by first setting the x-axis and y-axis on the property. Finally, Figure 3 shows some student activity results in the module through GeoGebra software.

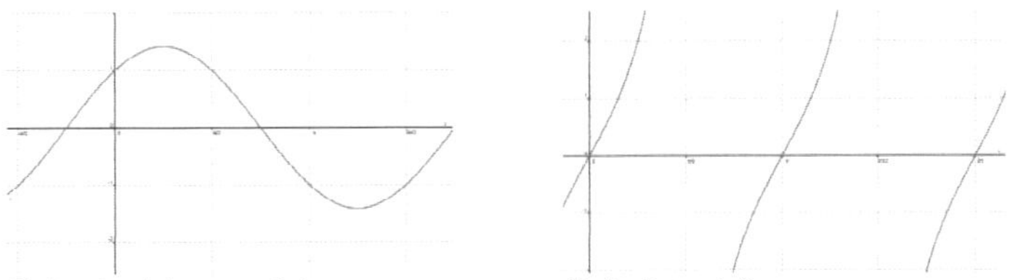

f(x)=sin (x) + cos(x) f(x)=2 tan(x)

Figure 3. Output function through Geogebra

A more detailed explanation of the research is in Table 2. The result of dependent t test on pre-test and the post-test result shows that post-test result of students' comprehension ability is better than pre-test result.

Table 2. The Results of the Test Statistic Analysis of Mathematical Understanding Tests

Mathematical understanding	Test For Normality	Homogeneity Test	The conclusion of the Mean Differences Test
Pretest	Normal Distribution	Homogen	The average ability of mathematical understanding students of underprivileged family after obtaining mathematical learning with GeoGebra-assisted better than before obtaining mathematics learning with GeoGebra assisted
Posttest	Normal Distribution		

3.2 Discussion

The ability of students' mathematical understanding after obtaining mathematical learning with GeoGebra-assisted is better than before learning. The exposure of the results of the study shows that fact. This fact can not be separated from the success of the learning process.

We carry out the learning process in the mathematics laboratory of Mathematics Education Studies Program of Galuh University which has complete enough facilities to support the learning process. Each student uses a computer with GeoGebra software in it. According to Hohenwarter and Fuchs [5], GeoGebra is very useful as a medium of mathematics learning with various activities as follows: 1) as a media of demonstration and visualisation; 2) as a construction tool; 3) as a tool of the invention process.

We started learning with the introduction of GeoGebra software that is on the look of GeoGebra. The appearance of GeoGebra consists of 1) Menu (located at the top), menu consists of File, Change, Display, Options, Tools, Window, and Help; 2) Tool Bar (located on the second row, contains icons/symbols); 3) The Left Wind (consisting of free objects and bound objects, is where algebraic forms are displayed); 4) Right Window (where graph is displayed); 5) Input Section (located at bottom left); 6) Symbol Section (contains a list of symbols); 7) Function Section (contains a list of functions); 8) Help section (contains a list of help); 9) Command section (contains a list of commands).

The use of GeoGebra software makes it easier for teachers as facilitators to understand the concept of trigonometry in students. The systematically organized "Trigonometry with GeoGebra" module also help the learning process. The module directs students to understand trigonometric concepts visually. Visualization helps the construction of knowledge so that the discovery and understanding of trigonometric concepts become easier.

The usefulness of GeoGebra is to visualize trigonometric concepts, i.e. angles and quadrants, trigonometric ratios, trigonometric functions, and related angles. Visualization of such concepts is to construct and discover the definitions and properties of trigonometry. The diverse visualization of triangular triangles can understand the concept of trigonometric comparison. Visualization and construction of quadrant concepts, boundary angles, and special angles are to obtain the values of the special angles and the values of the series. Moreover, the visualization and construction of trigonometric comparison values of special and other angles are very easy with a graph of trigonometric functions.

The ease of use of GeoGebra also increases students' enthusiasm and confidence, whereas new students are first acquainted with GeoGebra software. Students come from an underprivileged family with no ICT facilities. Therefore, students are given the opportunity to use ICT that they do not necessarily enjoy at home. The hope is to open up their insights so that in the future they can develop their potential with confidence. So they can increase the dignity and the degree of the family into a prosperous family.

4. Conclusion

Based on the results of data processing, analysis, findings, and discussion, there are conclusions as follows: 1) Mathematics learning with GeoGebra-assisted gives effect to the result of mathematical understanding ability test of underprivileged students; 2) This lesson motivates underprivileged students to learn and understand mathematical concepts. Some of the recommendations of this research activity are: 1) Need further coaching so that underprivileged students can optimise ability to utilise GeoGebra software in learning mathematics; 2) Continue to develop the use of GeoGebra software in mathematics learning so that math learning is easy to understand.

Acknowledgments

Our gratitude goes to Mrs. Siti as the mathematics teacher of SMAN 1 Baregbeg Ciamis who has helped prepare the research object and Dr. Awang kustiawan as the head of LPPM Unigal who has helped provide research funding.

IOP Conf. Series: Journal of Physics: Conf. Series **895** (2017) 012007 doi:10.1088/1742-6596/895/1/012007

References

[1] NCTM. (2000). *Principles and standards for school mathematics*. Reston, VA: National Council of Teachers of Mathematics.

[2] Harisman, Y., Noto, M. S., Bakar, M. T., & Amam, A. (2017, February). The Different Patterns of Gesture between Genders in Mathematical Problem Solving of Geometry. In *Journal of Physics: Conference Series* (Vol. 812, No. 1, p. 012039). IOP Publishing.

[3] BPS. (2013). Indikator kesejahteraan rakyat. *Badan Pusat Statistik Provinsi Aceh.*

[4] Schnell, S., & Prediger, S. (2017). *Eurasia Journal of Mathematics, Science & Technology Education*, 131.

[5] Pateman, N. A. & Lim, C. S. (2013). The Politics of Equity and Access in Teaching and Learning Mathematics. In M. A. Clements, A. J. Bishop, C. Keitel, J. Kilpatrick, & F. K. S. Leung (Eds.), *Third International Handbook of Mathematics Education* pp. 243–263. New York: Springer.

[6] Wahyudin. (1999). Kemampuan Guru Matematika, Calon Guru Matematika dan Siswa dalam Mata Pelajaran Matematika.Disertasi PPS UPI.Tidak diterbitkan.

[7] Hohenwarter, M., and Fuchs, K. (2004, July). Combination of dynamic geometry, algebra and calculus in the software system GeoGebra. In *Computer Algebra Systems and Dynamic Geometry Systems in Mathematics Teaching Conference*.

[8] Zengin, Y. (2017). *International Journal of Mathematical Education in Science and Technology*, pp. 1-21.

[9] Lukáč, S., & Sekerák, J. (2015, November). Implementation of feedback in digital learning environment with a focus on system GeoGebra. In *Emerging eLearning Technologies and Applications (ICETA), 2015 13th International Conference on* pp. 1-7. IEEE.

IOP Conf. Series: Journal of Physics: Conf. Series **895** (2017) 012009 doi:10.1088/1742-6596/895/1/012009

Designing Interactive Electronic Module in Chemistry Lessons

F S Irwansyah[1*], I Lubab[1], I Farida[1] and M A Ramdhani[2]
[1]Program Studi Pendidikan Kimia, UIN Sunan Gunung Djati Bandung, Jl. A. H. Nasution No. 105, Indonesia
[2]Program Studi Informatika, Fakultas Saintek UIN Sunan Gunung Djati Bandung, Jl. A. H. Nasution No. 105, Indonesia

[*]ferli@uinsgd.ac.id

Abstract. This research aims to design electronic module (e-module) oriented to the development of students' chemical literacy on the solution colligative properties material. This research undergoes some stages including concept analysis, discourse analysis, storyboard design, design development, product packaging, validation, and feasibility test. Overall, this research undertakes three main stages, namely, Define (in the form of preliminary studies); Design (designing e-module); Develop (including validation and model trial). The concept presentation and visualization used in this e-module is oriented to chemical literacy skills. The presentation order carries aspects of scientific context, process, content, and attitude. Chemists and multi media experts have done the validation to test the initial quality of the products and give a feedback for the product improvement. The feasibility test results stated that the content presentation and display are valid and feasible to be used with the value of 85.77% and 87.94%. These values indicate that this e-module oriented to students' chemical literacy skills for the solution colligative properties material is feasible to be used.

1. Introduction

As a part of science, Chemistry is directly related to its applications in everyday life. By studying chemistry, students are expected to learn about themselves and the environment, thus be able to apply the knowledge in daily life and in solving the problem as the further prospect of studying chemistry [1]. Chemical literacy skill is defined as the level of individual ability to have scientific knowledge and using it to think critically and creatively, and to analyze scientific phenomenon based on scientific evidence happening in daily life. Chemical literacy skill consists of four levels, namely 1) nominal literacy, capable of recognizing core concepts; 2) functional literacy, capable of determining core concepts; 3) conceptual literacy, capable of using chemical concept to identify phenomena happening in daily life; and 4) multidimensional literacy, capable of linking the concept and scientific inquiry to a decision-making process based on social point of view.

The implementation of information and technology develop rapidly in every aspect of life [2, 3]. Implementation of IT in a field of learning is believed to be able to improve student's learning ability [4]. Learning process by implementing IT will provide the learning materials using various media which will affect the student's learning achievement in a positive way [5], learning using multimedia has a positive effect on students' learning outcomes [6].

A module specifically supporting students' chemical literacy development has not been developed yet. Generally, the module and learning in chemistry lessons emphasize more on less applicative content aspects. One of the materials in high school chemistry lessons, relevant to be developed into a module for chemical literacy development, is colligative properties of solutions. This material is a concept presenting properties and concepts based on principles and processes. Moreover, the learning of solution colligative properties is closely related to daily life, for example, the process of melting snow with salt. To make the learning process more significant, it is required a content presentation integrated to aspects of the scientific process, context and attitude. However, the learning of solution colligative properties emphasizes more on symbolic level and chemical calculation problem solving [7].

The colligative nature of the solution is a chemical matter that belongs to a type of concept that states the nature and concepts based on principles and processes. This material of colligative nature is closely related to daily life such as the melting process of snow by using salt. Consequently, it requires aspects of the scientific process, context, and attitude to make the learning process become meaningful. Nevertheless, the materials of colligative nature of solution usually emphasize on the level of symbolic and problem solving which focuses on the scientific content of the materials [7]. Science actually has a strategic function because it can be used to develop the students' potential and ability both in cognitive aspect (science content), psychomotor aspect (science process), and effective aspect (science attitude) [8]. Therefore, it is necessary to develop a module able to facilitate students' chemical literacy skills. Accordingly, an electronic module (e-module) is a good development choice since the conventional one (printed module) is less interactive and has a static or monotone pictorial display while e-module can interactively present materials featured by multimedia such as videos, animations, simulations, and question with a direct feedback [9]. Using multimedia can minimize the number of intrinsic and extrinsic cognitive burden so that it can facilitate the externalization of information thinking, memorizing and processing. Learning content becomes more dynamic, effective and enjoyable. The use of multimedia is considered capable of improving the level of learning outcome [5].

This article carries the research study and the development aiming to create a product 'e-module' or electronic module oriented to students' chemical literacy skill concerning the topic of colligative properties of solutions

2. Methodology

The making process of e-module of solution's colligative nature uses design-based research method [10] [11]. Overall, this research undertakes three main stages, namely, Define, in the form of preliminary studies; Design, designing e-module; Develop including validation and model trial. Instruments used in this research consist of e-module storyboard and questionnaire. The storyboard is arranged as a guideline on e-module production. The production of e-module should be in line with the flowchart in order to make it easy to understand and perform and follow by user thoroughly and significantly. Criteria used in this e-module development are interaction model and functionality [4]. Flowchart and storyboard are made separately for each topic. Before e-module is made, the storyboard is reviewed first by experts. The questionnaire is used to determine the feasibility of e- module. The feasibility test is done by expert validators, consisting of 1 multimedia expert and 2 study field experts. Aspects asked in the questionnaire are the opinions of experts about the quality and aspects of the material present in the e- module. Furthermore, limited trials were conducted to 20 students to know the opinions of users regarding the ease to master the concept and use e- module oriented to students' chemical literacy skills.

3. Result and Discussion

E-Module presents materials of colligative properties of solutions, including four scientific aspects i.e. scientific context, process, content, and attitude. The development model of e-module consists of three stages: 1) presenting authentic problems (scientific context) in daily life in the form of videos or related pictures; 2) using the scientific context to develop the process aspect of the targeted skill by presenting the process of how an event occurs that is related to the context; 3) developing aspects of scientific attitude by giving some problems in terms of social, economic and environmental matters to make students able to argue scientifically. By using these ways, the scientific oriented of the e-module characteristic can be fully pursued [12]. This e-module also implies that technology should be highly adaptable and applicable by ignoring environmental effects [13]. Based on the analysis result of the concept and discourse of each topic, the learning objectives of and students' chemical literacy skill developed are set. The analysis result becomes a guideline to determine and identify main elements to be displayed in e-module. The following is the resume of the analysis result of the objectives and chemical literacy developed.

Table 1. Learning Objectives and Chemical Literacy on Solution Colligative Properties Material

Sub-Materials and Learning Objectives	**Chemical Literacy Developed**	
Saturated Steam Pressure Reduction of Solutions (ΔP): Describing the phenomenon of steam pressure reduction of solutions	Process : Context : Content : Attitude :	Concluding Aquarium illustration videos and questions related to evaporation Material explanation Growing high curiosity
Freezing Point Depression of Solutions (ΔT_f): Analyzing the objectives of using salt to melt snow based on the concept of freezing point depression of solutions	Process : Context : Content : Attitude :	Predicting Videos about melting snow with salt and questions relating to chemical literacy Animations about freezing point of solvents and solutions with explanation Growing high curiosity
Boiling Point Elevation of Solutions (ΔT_f): Analyzing cooking process based on the concept of boiling point elevation of solution	Process : Context : Content : Attitude :	Concluding Videos about cooking water and instant noodle, and questions relating to chemical Literacy as well. Animations about boiling point of solvents and solutions with explanation Considering an efficient cooking technique
Osmotic Pressure of Solutions (π): Analyzing the process of sea water purification based on the concept of reverse osmosis principle.	Process : Context : Content : Attitude :	Concluding videos about sea water purification questions relating to chemical literacy Animations about reverse osmosis with explanation Being aware of the importance of clean water supposed to be used wisely

The development of *E-module* at each topic follows the storyboard arranged [14]. The main pages of e-module for the topic of colligative properties of solutions consist of:

a. Home: Menu, containing links and materials profiles listed in the electronic module, which user can choose to learn. This display is similar to the table of contents of books.
b. Instructions for use: Instructions about how to use *e-module.*
c. Learning Objectives: chemical literacy indicators, consisting of 10 indicators relating to scientific process and attitude skills.
d. Material pages carry the concept of colligative properties of solutions based on chemical literacy indicators i.e. students' scientific process, context, content, and attitude.
e. On pages carrying process, aspect is animation, video or pictures relating to materials presented.
f. On pages carrying context, aspect is animation, video or pictures relating to materials presented.
g. Learning evaluation pages contain ten question, each, about colligative properties of solutions. The question refers to targeted learning objectives. The student must answer the questions one by one because if they skip even just one, they will not be able to continue or go to the next question automatically. After answering all questions, they will get a feedback in the form of a score and a note whether they are a success or not.
h. Evaluation pages of scientific attitude contain 16 questions about the scientific attitude toward phenomena happening in daily life. Those questions are related to phenomena happening in daily life. Once they answer all questions, they will get a feedback in the form of a note whether they have already had scientific attitude or not

Materials of nature colligative of the solution shown in four menus, which consist of four different themes namely aquarium, snow melting, cooking process, and osmotic reverse. Each theme explains sub concept of Saturated Steam Pressure Reduction of Solutions (ΔP), Freezing Point Depression of Solutions (ΔTf), Boiling Point Elevation of Solutions (ΔTf), and Osmotic Pressure of Solutions (π). The materials also are shown in four aspects namely scientific context, scientific process, scientific content, and scientific attitude that oriented to chemical literacy. One part that belongs to scientific context is sea water purification by using the technology of osmotic reverse and shown in the video. The visualization of scientific context aspects from osmotic reverse theme can be seen in figure 2.

Figure 1. Title Part

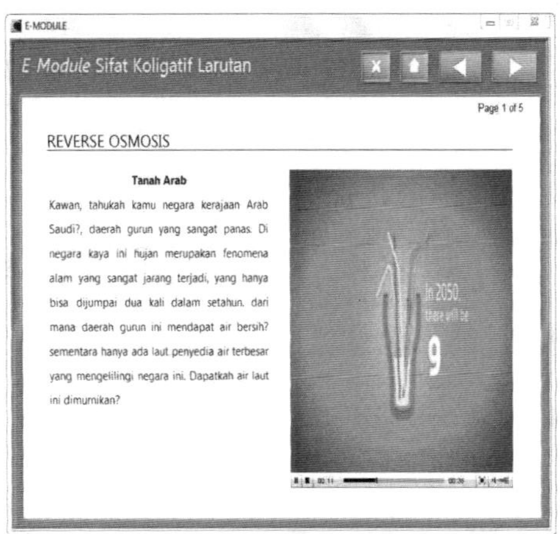

Figure 2. Scientific Context Aspects in Osmotic Reverse Theme

Furthermore, scientific context aspects can be explained by presenting an osmotic reverse process that belongs to scientific process aspects. Meanwhile, scientific content aspects can be explained by presenting sub microscopic animation of osmotic reverse aspects. The visualization of scientific content aspects from osmotic reverse theme can be seen in figure 3.

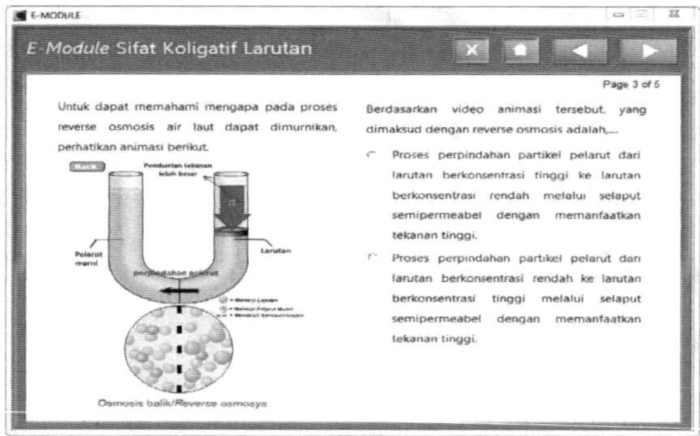

Figure 3. Scientific Content Aspect in Osmotic Reverse Theme

Visualization of the content in *e-module* is generally presented in the form of video or animation with text or pictures with text. The use of this visualization aims to make the user understand the content presented in *e-module*. Feasibility test of this e-*module* is done through two stages namely expert validation and limited trials. There are three validators consists of two chemistry experts and one multi media expert have done the validation. Moreover, a revision over e-module draft has been made based on suggestions given. Then, feasibility test is done in small scale (limited trial) to 20 respondents selected randomly. These trials are done by giving questionnaires and showing e-module completely to respondents in order to obtain an assessment about the ease of using e-module and the ease of understanding the materials presented. The tabulation results of the limited trial data conducted to 20 respondents can be seen in table 2 and table 3. Table 2 gives an overview of respondents' assessment of e-module on the aspect of the material content presentation. The explication of the material description and the suitability of the test questions with the learning objectives appear to be the most prominent criteria.

Table 2. The Percentage of Limited Trial Result on Material Content Aspect

No	Statement	Agreement (%)
1.	The explication of material description	100
2.	The depth of the materials	90
3.	The provision of examples and exercises based on the learning objectives	95
4.	The suitability of the questions with the learning objectives	100
5.	The provision of feedback for students' learning motivation	70
6.	The explication of the targeted learning objectives	85
7.	The ease of learning the materials	70
8.	The suitability of the material presentation order	95
9.	The explication of Al-Quran verses relating to the materials	85
10.	The suitability of the macroscopic presentation	85
11.	The suitability of the submicroscopic presentation	85
12.	The suitability of the symbolic presentation	80
13.	The feedback for the evaluation of attitude and learning performed by students	75
	Average	85,77

Table 3. The Percentage of Limited Trial Result on Display Aspect

No	Statement	Agreement (%)
1.	The ease of using e-module	90
2.	The explication of *E-module* instructions for use	85
3.	The quality of pictorial display presented in relation to the materials	95
4.	The explication of the animation presented in relation to the materials learned	85
5.	The explication of the video in relation to the materials learned	100
6.	The attractive and appropriate color composition	70
7.	The explication of addition value given by pictures to the materials learned	90
8.	The explication of modeling or animation	85
9.	The explication of links available in e-module	85
10.	The suitability of music played	70
11.	The use of the right and correct Bahasa Indonesia	95
12.	The simple yet fascinating e-module design	75
13.	The appropriate layout of font and pictures	100
14.	The use of appropriate font type and size	100
15.	The consistency of button used	85
16.	The explication of navigation buttons	90
17.	The efficiency of text	95
	Average Percentage	87,94

4. Conclusion

The making process of *e-module* consists of analysis, design development, product production, validation, and feasibility test. The visualization and presentation of the concept used in *E-module* refer to learning model oriented to chemical literacy skill with presentation order including scientific context, proses, content, and attitude. Validation result done by chemistry and multi media expert shows that all criteria dividing into two aspects, material content presentation and display, are valid. Feasibility test result states that the material content presentation and display are valid and feasible to use with the value 85,77% and 87,94%. Based on the percentage value, e-module is declared feasible to be used as learning media.

References

[1] S.-Y. Lin, "Chemical Literacy and Learning Sources of Non-Science Major Undergraduates on Understandings of Environmental Issues," *Chemical Education Journal,* vol. 13, no. 1, 2009.

[2] A. Pamoragung, K. Suryadi and M. A. Ramdhani, "Enhancing the Implementation of e-Government in Indonesia Through the High-Quality of Virtual Community and Knowledge Portal," in *6th European Conference on e-Government*, Marburg, 2006.

[3] C. Slamet, A. Rahman, M. A. Ramdhani and W. Darmalaksana, "Clustering the Verses of the Holy Qur'an using K-Means Algorithm," *Asian Journal of Information Technology,* vol. 15, no. 24, pp. 5159-5162, 2016.

[4] M. A. Ramdhani and E. R. Wulan, "The Analysis of Determinant Factors in Software Design for Computer Assisted Instruction," *International Journal of Scientific & Technology Research,* vol. 1, no. 8, pp. 69-73, 2012.

[5] M. A. Ramdhani and H. Muhammadiyah, "The Criteria of Learning Media Selection for Character Education in Higher Education," in *International Conference of Islamic Education in Southeast Asia*, Malang, 2015.

[6] I. Farida, "The Importance of Development of Representational Competence in Chemical Problem Solving Using Interactive Multimedia," in *International Seminar on Science*

Education, Bandung, 2009.

[7] T. Widhiyanti, "Pembelajaran Berbasis Teknologi Informasi untuk Meningkatkan Pemahaman Konsep dan Keterampilan Berpikir Kritis Siswa pada Topik Sifat Koligatif Larutan," Universitas Pendidikan Indonesia, Bandung, 2007.

[8] Mundilarto, "Pendekatan Kontekstual pada Pembelajaran Sains," Sleman Yogyakarta, 2005.

[9] M. D. Gall, J. P. Gall and W. R. Borg, Educational Research: An Introduction, New York: Longman, 2003.

[10] S. Barab and K. Squire, "Design-Based Research: Putting a Stake in the ground," *J. Learn. Sci,* vol. 13, no. 1, p. 1–14, 2004.

[11] T. D.-B. R. Collective, "Design-based research: An emerging paradigm for educational inquiry," *Educational Researcher,* vol. 32, no. 1, p. 5–8, 2003.

[12] R. Marks and I. Eilks, "Promoting scientific literacy using a sociocritical and problem-oriented approach to chemistry teaching: Concept, examples, experiences," *Int. J. Environ. Sci. Educ,* vol. 4, no. 3, p. 231–245, 2009.

[13] M. A. Ramdhani, H. Aulawi, A. Ikhwana and Y. Mauludin, "Model of green technology adaption in small and medium-sized tannery industry," *J. Eng. Aplied Sci,* vol. 12, no. 4, p. 954–962, 2017.

[14] I. Binanto, Multimedia Digital – Dasar Teori dan Pengembangannya, Yogyakarta: ANDI, 2010.

IOP Conf. Series: Journal of Physics: Conf. Series **895** (2017) 012011 doi:10.1088/1742-6596/895/1/012011

Learning Crude Oil by Using Scientific Literacy Comics

R Aisyah*, I A Zakiyah[1], I Farida[1] and M A Ramdhani[2]

[1] Department of Chemistry Education, UIN Sunan Gunung Djati Bandung, Jl. A. H. Nasution No. 105 Bandung 40614, Indonesia.

[2] Department of Informatics, UIN Sunan Gunung Djati Bandung, Jl. A. H. Nasution No. 105 Bandung 40614, Indonesia.

*ririaisyah@uinsgd.ac.id

Abstract. A research has been conducted to create a crude oil learning media in the form of scientific literacy- oriented comic. The research included some phases, namely: concept analysis, material transformation to concept map, indicator identification and science literacy aspect. The product was made based on flowcharts and storyboards that have been validated by expert validators. The product has characteristics namely; 1)Develops indicators and aspects of science literacy, 2) presents the materials in form of story of science fiction genre, 3) has characters adopting levels of scientific literacy, 4) has optional stories, because it depends on questions asked to develop scientific literacy in terms of content, context, process and attitude. Based on feasibility test, the product is feasible to be used as learning media. It is suggested to do an expanded experiment to examine its affectivity in improving scientific literacy and growing students' awareness about the issues of energy crisis and the impacts of fossil fuel use on the environment.

1. Introduction

Crude oil is one of learning material taught in senior high school. In addition to constructing students' knowledge, this learning material can be used to make them care more about the issues of fossil fuel supply and the impacts of using crude oil on the environment [1][2]. Scientific literacy is a skill to use scientific knowledge, identify scientific issues, explaining a phenomenon scientifically, and use scientific evidences in daily life [3]. This learning material includes scientific literacy aspects namely, scientific content, context, process and attitude [4]. However, the learning more emphasizes on understanding the concept than on developing scientific literacy [5]. Moreover, the books that students usually use are less attractive to read [6], so that it becomes the constraints to grow their scientific literacy in relation to crude oil issues.

Students are more likely to be interested in reading pectoral story (comic) rather than academic books [7]. Comic has systematical and orderly story that makes them easier to follow and memorize its content [8] so that they are interested in reading [9]. If designed correctly and appropriately, comic has pedagogic resources in learning [10]. The closing and the story in the comic can be written by carrying pedagogic issues stimulating students' creativity and activeness[11]. The content can be non-fiction and concepts and issues carried in the story can be written as attractive as possible[10][7].

This paper describes the development of design through research to produce a product in the form of a chemistry comic carrying crude oil learning material, oriented to develop students' scientific literacy toward the issues of energy crisis and the impacts of fossil fuel use on the environment.

IOP Conf. Series: Journal of Physics: Conf. Series **895** (2017) 012011 doi:10.1088/1742-6596/895/1/012011

2. Methodology

To create the product, scientific literacy comic on crude oil material, this research used design-based research method [12] [13]. The research included some phases, namely: concept analysis, material transformation to concept map, indicator identification and science literacy aspect. Then, flowcharts and storyboards were arranged as the basis for formulating ideas and stories. The flowchart consisted of the diagram of story plot to develop the plot of every page in the comic. The storyboard consisted of the sketch of pictures arranged systematically based on the planned plot of the comic story. Then, A validation was done by three three experts in the field of this study to review the quality of the comic, the appropriateness between the content and the plot of the story, the aspects of the display, and supporting elements. Based on the result of the validation done by the validators, there should be some improvements in terms of flowchart and storyboard that have been arranged. In addition, the material in the comic was made by using a software, Manga Studio version 5.0.

The product, a comic made in the form of a book, was then tested in relation to its feasibility to obtain an assessment regarding the aspect of learning, the easiness to learn the material, language, and graphics. The respondents in this feasibility test consisted of ten prospective chemistry teachers, two students and one chemistry teacher.

3. Result and Discussion

Based on analysis result toward crude oil concept, there are five sub-concepts, namely: The proses of crude oil formation, proses of crude oil processing, the product of crude oil processing in daily life, fossil fuel, and the impacts of the gases from burning fossil fuel. These five sub-concepts are abstract concepts, concrete example, and the concept based on process[14]. The aspects of content and context of crude oil material are presented in the form of the issues of crisis and energy conversion, crude oil processing, and the impact of fossil fuel on the environment. Furthermore, the aspect of scientific process is presented to develop thinking skill in relation to the process how a phenomenon can occur. As for the dialog in the comic, it is made to provide argumentative thinking that have an impact on the development of scientific attitude. The presentation of these four aspects of scientific literacy are made in a plot of story with pictures based on the character, informative dialog, and argumentative dialog as well.

Figure 1. The Visualization of Comic characters

In general, the idea of the story in this comic is about scientific fiction and adventure. The comic tells about the efforts of a group of teenagers "Earth Ranger" to fight against an attracts from alien named Zirkon. Zirkon as the antagonist came from planet "X" and he wanted to take over earth's energy source, Crude Oil. The teenagers protecting the earth "Earth Ranger" consisted of three characters (picture 1A), namely Baro (rabbit), Una (cat) and Kiki (monkey). Each character represents the levels of scientific literacy, i.e. *conceptual, multidimensional* and *functional scientific literacy* [15]. The use of these characters is for stimulating an attractive dialogue among the characters as we can see on

Figure 1B and 1C. Therefore, this comic is expected to fulfill the characteristics of scientific literacy-oriented comic [4][16].

Figure 2. The example of the display of the content page in the comic

The development of scientific literacy aspects in the comic is made in the form of game capable of challenging readers' thinking capability and growing scientific attitude. In every page, there is half or one page full of pictures with the dialogues of the characters. The flow of the story facilitates the readers to choose their own flow, depending on their capability to answer the questions and challenges. If they are already sure that they already have the right answer, they will be lead to go to certain page or go back to the previous one. This strategy is adopted to encourage readers to think and develop their scientific literacy [10]. On figure 2, we can see the explanation in a page in crude oil comic

Figure 3. The example of the display of content and process aspect game page

Scientific content aspect is described by presenting the explanation of multilevel distillation in the process of crude oil processing. Scientific content consists of scientific concept required to understanding scientific phenomenon [3]. Scientific process is a skill to obtain knowledge according to scientific evidences and facts[3]. The example of the presentation of content and process aspect in the form of game classifies the products of crude oil processing based on their appearance and their length of carbon chain (figure 3). If the readers are already sure about their answer for the questions, they can continue to the next page, but if they are not sure yet, they will be suggested to go back to the previous page.

It is expected that after learning this crude oil material, they will show positive attitude towards knowledge and be aware about the environment and the nature. The visualization of scientific attitude aspect can be seen on figure 4.

Figure 4. The Display of the page of scientific attitude aspect

According to the explanation above, comic learning media is made based on how students learn from picture and the use of graphic during studying [7][11][10]. The pictures and the flow of the story is arranged in order to make students able to interpret and memorize the text content written easier [9]. The flow of the story also deliver some messages that technology should have high adaptability and easy to operate by avoiding the impact on the environment [17].

The validated done to the content by the validators comes up with some advices for the improvement of the product, namely related to the clarity of the content in the dialog, the asking of questions and challenges that should be more focus on examining users' understanding about the content, and the graphics on certain pages. The result of feasibility test shows that 88 % of respondents agree that the questing asked in the game are highly relevant with the indicators of scientific literacy developed, 95 % of respondents agree that the sequence of the material is interesting and challenging, 90 % of respondents agree that the comic is useful for making the learning of the materials about crude oil easier and make them aware about the issues of energy crisis and the impacts of fossil fuel use on the environment. Based on this research results, our product still needs expanded experiment to examine its affectivity in improving scientific literacy, especially in terms of growing students' awareness about the issues of energy crisis and the impacts of fossil fuel use on the environment

4. Conclusion

Through design-based research, we have developed a learning product in the form of scientific literacy-oriented comic. The product has characteristics namely; 1)Develops indicators and aspects of science literacy, 2) presents the materials in form of story of science fiction genre, 3) has characters adopting levels of scientific literacy, 4) has optional stories, because it depends on questions asked to develop scientific literacy in terms of content, context, process and attitude.

Based on validity and feasibility test, the product is feasible to be used as learning media. It is suggested to do a broader experiment to examine its affectivity in improving students' scientific literacy.

References

[1] Y. Karatepe, S. V. Neşe, A. Keçebaş, and M. Yumurtaci, "The levels of awareness about the renewable energy sources of university students in Turkey," *Renew. Energy*, vol. 44, pp. 174–179, 2012.

[2] J. L. Doménech *et al.*, "Teaching of energy issues: A debate proposal for a global reorientation," *Sci. Educ.*, vol. 16, no. 1, pp. 43–64, 2007.

[3] R. Bybee and B. Mccrae, "Scientific Literacy and Student Attitudes: Perspectives from PISA 2006 science," *Int. J. Sci. Educ.*, vol. 331, no. 1, pp. 7–26, 2011.

[4] R. Marks and I. Eilks, "Promoting scientific literacy using a sociocritical and problem-oriented approach to chemistry teaching: Concept, examples, experiences," *Int. J. Environ. Sci. Educ.*, vol. 4, no. 3, pp. 231–245, 2009.

[5] T. Feierabend and I. Eilks, "Teaching the societal dimension of chemistry using a socio-critical and problem-oriented lesson plan based on bioethanol usage," *J. Chem. Educ.*, vol. 88, no. 9, pp. 1250–1256, 2011.

[6] R. Jarman and B. McClune, *Developing Scientific Literacy Using News Media In The Classroom*, 1st ed. Newyork: Mc Graw Hill, 2007.

[7] J. Hosler and K. B. Boomer, "Are comic books an effectiveway to engage nonmajors in learning and appreciating science?," *CBE Life Sci. Educ.*, vol. 10, no. 3, pp. 309–317, 2011.

[8] G. Rota and J. Izquierdo, "Comics as atool for teaching biotechnology in primary schools," *Electron. J. Biotechnol.*, vol. 6, no. 2, pp. 1–6, 2003.

[9] J. S. Park, D. H. Kim, and M. S. Chung, "Anatomy comic strips.," *Anat. Sci. Educ.*, vol. 4, no. 5, pp. 275–9, Sep. 2011.

[10] M. Tatalovic, "Science comics as tools for science education and communication: A brief, exploratory study," *J. Sci. Commun.*, vol. 8, no. 4, 2009.

[11] H. W. Roesky and D. Kennepohl, "Drawing attention with chemistry cartoons," *J. Chem. Educ.*, vol. 85, no. 10, pp. 1355–1360, 1802.

[12] S. Barab and K. Squire, "Design-Based Research: Putting a Stake in the ground," *J. Learn. Sci.*, vol. 13, no. 1, pp. 1–14, 2004.

[13] The Design-based research collective, "Design-based research: An emerging paradigm for educational inquiry," *Educ. Res.*, vol. 32, no. 1, pp. 5–8, 2003.

[14] J. D. Herron, L. L. Cantu, and R. Ward, "Problems Associated with Concept Analysis," *Sci. Educ.*, vol. 61, no. 2, pp. 185–199, 1977.

[15] Y. Shwartz, R. Ben-Zvi, and A. Hofstein, "The use of scientific literacy taxonomy for assessing the development of chemical literacy among high-school students," *Chem. Educ. Res. Pract.*, vol. 7, no. 4, p. 203, 2006.

[16] A. N. Spiegel, J. McQuillan, P. Halpin, C. Matuk, and J. Diamond, "Engaging Teenagers with Science Through Comics," *Res. Sci. Educ.*, vol. 43, no. 6, pp. 2309–2326, 2013.

[17] M. A. Ramdhani, H. Aulawi, A. Ikhwana, and Y. Mauludin, "Model of green technology adaption in small and medium-sized tannery industry," *J. Eng. Aplied Sci.*, vol. 12, no. 4, pp. 954–962, 2017.

Using Android-Based Educational Game for Learning Colloid Material

S Sari[1*], R Anjani[1], I Farida[1] and M A Ramdhani[2]

[1]Prodi Pendidikan Kimia, UIN Sunan Gunung Djati Bandung, Jl. A. H. Nasution No. 105 Bandung 40614, Indonesia.
[2]Jurusan Teknik Informatika, Fakultas Saintek, UIN Sunan Gunung Djati Bandung, Jl. A. H. Nasution No. 105 Bandung 40614, Indonesia.

[*]sari@uinsgd.ac.id

Abstract. This research is based on the importance of the development of student's chemical literacy on Colloid material using Android-based educational game media. Educational game products are developed through research and development design. In the analysis phase, material analysis is performed to generate concept maps, determine chemical literacy indicators, game strategies and set game paths. In the design phase, product packaging is carried out, then validation and feasibility test are performed. Research produces educational game based on Android that has the characteristics that is: Colloid material presented in 12 levels of game in the form of questions and challenges, presents visualization of discourse, images and animation contextually to develop the process of thinking and attitude. Based on the analysis of validation and trial results, the product is considered feasible to use.

1. Introduction

Concept of colloids is part of chemistry closely related to everyday life, for example in industrial chemistry, on the manufacture of various products such as cosmetics, food ingredients, and a number of other products [1]. In colloid material usually students are required to memorize it, without applying it in everyday life [2]. With the difficulties of students applying colloid material in everyday life, it demands teachers to be creative and innovative, one of which makes a media of learning [3]. The growing technology demands teachers to be able to use technology to create and develop the learning media in accordance with technological developments. [4]

Implementation of IT in a field of learning is believed to be able to improve student's learning ability [5]. Learning process by implementing IT will provide the learning materials using various media that will affect the student's learning achievement in a positive way [6]. Learning using multimedia has a positive effect on students' learning outcomes [7]. One example of IT implementation in a field of learning is the making of the educational games.

The making of the game is very important; especially apply it in learning because it will improve the learning outcomes and the students' involvement [8]. One of the technological developments that are currently in demand by many people are smartphones based android. Because smartphones are easier to use, have an open operating system that allows smartphone users can add a variety of applications. [9]. This educational game can be very motivating while still maintaining or even

IOP Conf. Series: Journal of Physics: Conf. Series **895** (2017) 012012 doi:10.1088/1742-6596/895/1/012012

improving the effects of education on the students [10]. Chemical learning on colloid material can be achieved well if the teacher uses the right learning media

Thus, we need efforts to develop the android-based learning media. This paper reports on how to manufacture android-based educational games on a colloidal material to develop students' chemical literacy.

2. Methodology

Methods used in this research were designs based research that will produce the android-based educational game media in the colloidal materials. The subjects of the research were twenty students of Chemistry Education of Faculty of Education and Teacher Training, UIN Sunan Gunung Djati Bandung.

There were six stages in making the android-based educational game media in colloidal materials: (1) analysis of SK and KD of the collodial materials, (2) making process of the android-based educational game media, (3) revision of the android-based educational game media, (4) limited trials, (5) data analysis, and (6) production of the android-based educational game media in the colloidal materials.

3. Result and Discussion

The making process of android-based educational game media in colloidal materials uses research and development. This research method used to produce and examine validity of a particular product [11]. Some phases are conducted in this process. The first stage is analysis of colloidal materials, which aims to plan the sequences of learning process because it plays an important role in learning meaning [12].

In designing stage, the elements needed are integrated in the form of colloidal game by noticing the flowchart and storyboard. Flowchart used to describe the flow from one scene to another which eases the user to operate this colloidal game. The making process of android-based educational game media in colloidal materials is based on the designing stage, such as storyboard and flowchart [13].

The description of scenes, visual and audio design, information, and voice of narration will be made in storyboard stage. Software in this program uses programming language c# c++. After that, the educational game is saved in *.apk format in order to use in android smartphone.

The following is the visualization from android-based educational game media in colloidal materials after finishing some stages.

3.1 Main Scene of Game

In this main scene, there are some buttons of colloidal kinds which displays the menu to enter the game levels according to the colloidal kinds. The characteristic button consists of menu to enter the game levels of colloidal characteristic. Instruction button consists of how the user plays the game. About button consists of application identity. Setting button in the top left consists of sound setting. Exit button used to exit from this application. This can be seen in Figure 2.

Figure 1. Visualization of main scene

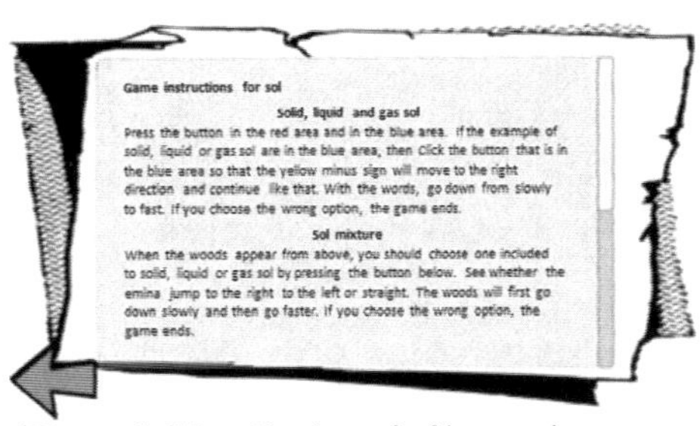

Figure 2. Visualization of of instruction scene

IOP Conf. Series: Journal of Physics: Conf. Series **895** (2017) 012012 doi:10.1088/1742-6596/895/1/012012

3.2 Setting Scene

This scene is a setting for audio and music in the application. This scene used to turn on or off the audio and the music in the application.

3.3 Instruction Scene

This scene is about the instruction how to play the game in some levels in this application. Visualization of instruction scene can be seen in Figure 2.

3.4 About Scene

This scene displays the information and profile of content and design builder of this application.

3.5 Menu Scene

There are two parts of menu scene. The first part of the menu consists of levels of colloidal materials and kinds. Visualization of the first part of menu scene is in Figure 3. The second part of the menu is about the levels of properties and contents of colloid. Visualization of the second part of menu scene is in Figure 4.

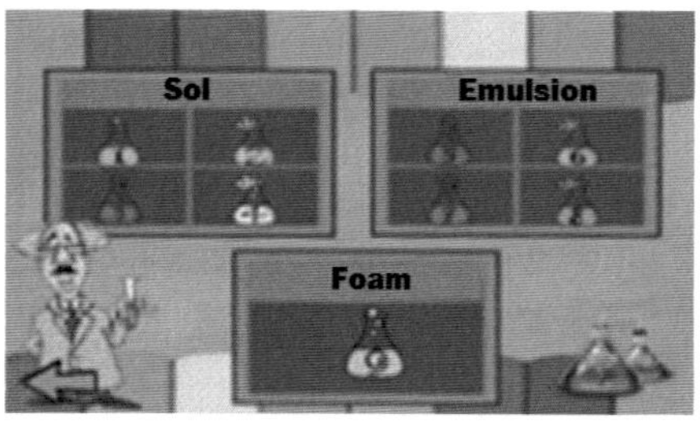

Figure 3. Visualization of colloidal types' menu scene

Figure 4. Visualization of colloidal properties and contents' menu scene

3.6 Levels Scene

This scene displays the game about colloidal kinds, characteristics, and contents. These games provided in text, Figure, and animation which eases the student to learn colloid to develop student's chemical literacy. This game is about the solid soles in which the student can identify the implementation of solid soles in daily life. The game visualization can be seen in Figure 5.

This second level of the game is about liquid soles. Student can identify various things in daily life, thus in the process the student will be able classify colloidal kinds and liquid soles. Visualization of game can be seen in Figure 6.

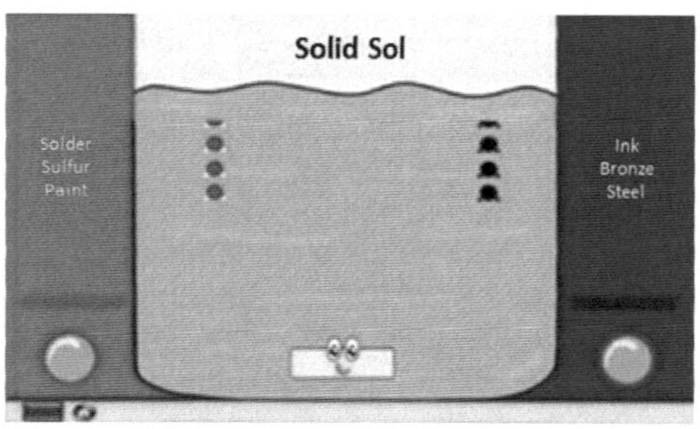

Figure 5. Visualization of level 1 scene

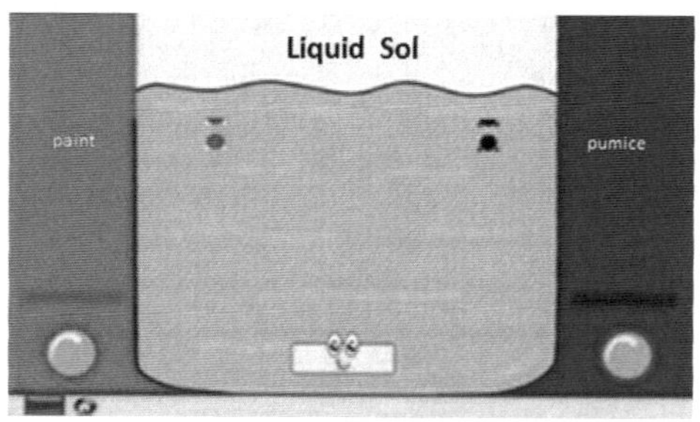

Figure 6. Visualization of Level 2 Scene

This third level of the game is about gas soles. Student can identify various things in daily life, thus in the process the student will be able classify colloidal kinds and gas soles. Visualization of game can be seen in Figure 7.

In the fourth level of the game, student can identify various things in daily life, thus in the process the student will be able classify colloidal kinds, liquid, gas, and solid soles. Visualization of game can be seen in Figure 8.

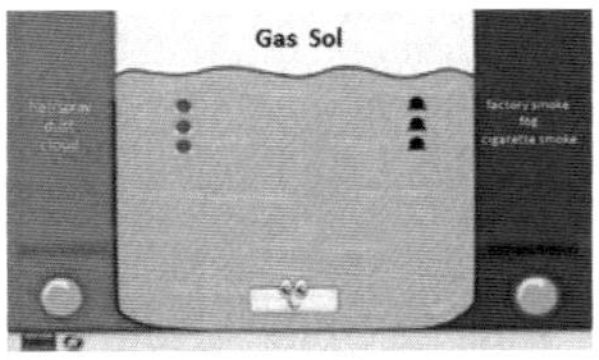

Figure 7. Visualization of Level 3 Scene

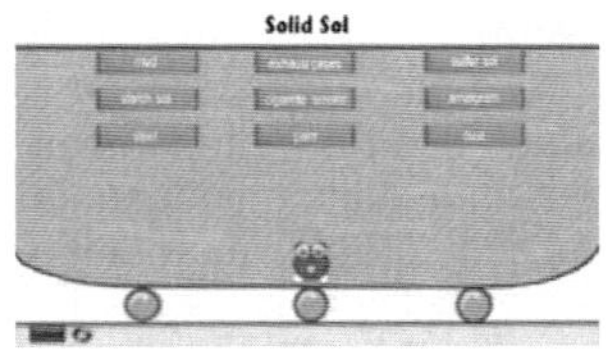

Figure 8. Visualization of Level 4 Scene

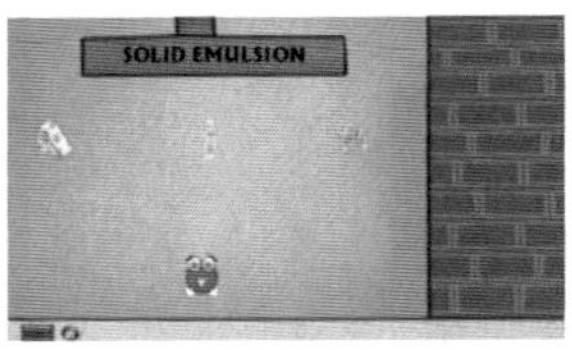

Figure 9. Visualization of Level 5 Scene

Figure 10. Visualization of Level 6 Scene

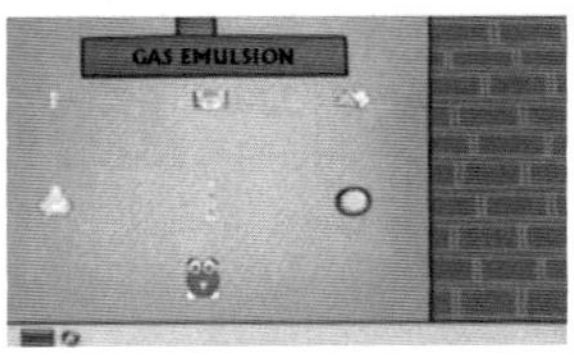

Figure 11. Visualization of Level 7 Scene

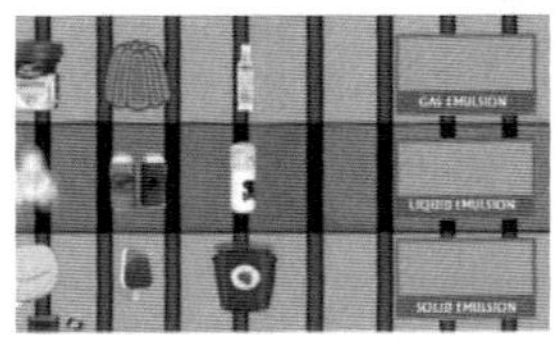

Figure 12. Visualization of Level 8 Scene

In fifth level about solid emulsion, a student can identify many things in daily life to consider an example of solid emulsion. Student can be able to classify colloidal kinds and can explain about the solid emulsion. Visualization of game can be seen in Figure 9.

In sixth level about liquid emulsion, a student can identify many things in daily life to consider an example of liquid emulsion. Student can be able to classify colloidal kinds and can explain about the liquid emulsion. Visualization of game can be seen in Figure 10.

In seventh level about gas emulsion, a student can identify many things in daily life to consider an example of liquid emulsion. Student can be able to classify colloidal kind-sand can explain about the gas emulsion. Visualization of game can be seen in Figure 11.

In eighth level of this game, a student can identify many things in daily life to consider an example of liquid, solid, and gas emulsion. Student can be able to classify colloidal kinds and can explain about the liquid, solid, and gas emulsion. Visualization of game can be seen in Figure 12.

In ninth level of this game is about foams where a student can identify any kinds of things in daily life to be considered as an example of foam. Student also can be able to classify colloidal kinds, can explain materials about foam.

The tenth level of this game is about Tyndall Effect which is visualized the ray path. Then, the student can be able to visualize the ray path, can be able to cope problems with its content. Consequently, student can be able to explain Tyndall Effect.

The eleventh level of this game is about electrophoresis in which displays the animation of electrophoresis cell. Student can be able to conclude the content in a colloid, to explain the materials of electrophoresis.

The twelfth level of this game is about adsorption in which displays the animation of colloidal adsorption. Student can be able to conclude the content in a colloid, to explain the materials of adsorption.

The content presented in the colloidal game is related to the context in everyday life that aims to direct student to understand chemical literacy. This colloidal game presents content in the form of animation, images along with text simultaneously as well as video. Because some colloidal sub material is quite difficult to visualize. Therefore, with this visualization, someone will more easily understand or easily receive information properly if using animations or images with words.

After completing each level, the student is given a video that contains an explanation of the game being played, so that students can realize and appreciate the power of God in creating everything with various benefits.

The context used in this android-based colloidal game lies in the field of food and the environment as targets, learning resources and tools. This is taken because phenomena related to chemistry occur in everyday life, which usually found in the field of medicine, health, and food. Therefore, based on the statement of this android-based colloidal game is made of the colloidal material context contained in everyday life in the field of food and the environment, in order to explore the ability of student's chemical literacy. In addition, this colloidal game has an evaluation level of the previous level to measure the extent to which student master a material.

The results of data processing on the questionnaire by some validators can be obtained that the validator assessment of the android-based colloidal game can be seen on several aspects. It is obtained that the highest r-value is 0,93, meanwhile the lowest r-value is 0,60 with average r-value is 0,77 for whole aspects assessed. Then it can be concluded that the learning media of this android-based colloid game is declared valid.

4. Conclusion

The development of android-based educational game on chemically oriented colloidal materials through the analysis phase and design phase. In the analytical phase, the result is material relevance between the chemical literacy indicator in the material presentation strategy. The design stage generates android-based educational games that have material characteristics presented through questions visualized with games equipped with images and animations so as to increase student's chemical literacy.

References

[1] Hayati, D.K., Sutrisno., dan Lukman, "Development Framework Creative TPACK on Colloids to Enhance Learning Activities for Achieving Student Hots," *Edu-Sains,* vol. 3, p. 29, 2014.

[2] Maulidita Rachmayanti dan Amaria, "Implementation students teams achievement divisions (stad) of cooperative learning in the matter of colloid," *Unesa Journal of Chemistry Education,* vol. 2, no. 2252-9454, pp. 119-12, 2013.

[3] Mulyasa, Pengembangan dan Impelementasi kurikulum 2013, Bandung, Indonesia: PT. Remaja Rosdakarya, 2006.

[4] Muhammad Ali Ramdhani, Hilmi Aulawi Andri Ikhwana, Yusuf Mauludin, "Model of Green Technologi Adaptation in Small and Madium-Sized Tannery Industry," *Journal of Engineering and Applied Sciences,* vol. 12, no. 4, pp. 954-962, 2017.

[5] Slamet, A. Rahman, M. A. Ramdhani and W. Darmalaksana, "Clustering the Verses of the Holy Qur'an using K-Means Algorithm," *Asian Journal of Information Technology,* vol. 15, no. 5159-5162, p. 24, 2006.

[6] M. A. Ramdhani and E. R. Wulan, "The Analysis of Determinant Factors in Software Design for Computer Assisted Instruction," *International Journal of Scientific & Technology Research,* vol. 1, pp. 69-73, 2012.

[7] M. A. Ramdhani and H. Muhammadiyah, "The Criteria of Learning Media Selection for Character Education in Higher Education," in *in International Conference of Islamic Education in Southeast Asia*, Malang, 2015.

[8] Eseryel, D., Law, V., Ifenthaler, D., Ge, X., & Miller, R, "An Investigation of the Interrelationships between Motivation, Engagement, and Complex Problem Solving in Game-based Learning," *Educational Technology & Society,* p. 42–5, 2014.

[9] Y. Prasetyo, "The Development of Android-Based Mobile Learning Media as Chemistry Learning for Senior High School on Acid Base, Buffer, Solution, and Salt Hydrolysis," *Journal Education of Mathematic and Science,* vol. 15, no. Journal Education of Mathematic and Science, p. 18, 2014.

[10] Virvou, M.K.G., dan Manos, K, "Combining Software Games with Education," *Evaluation of its Educational Effectiveness,* pp. 54-65, 2005.

[11] Sugiyono, Metode Penelitian Kuantitatif Kualitatif dan R & D, Bandung: Alfhabeta, 2012.

[12] R. W. Dahar, Teori-Teori Belajar, Bandung: Gelora Aksara Pratama, 2004.

[13] I. Binanto, Multimedia Digital; Dasar Teori dan Pengembangannya, Yogyakarta: Andi, 2010.

IOP Conf. Series: Journal of Physics: Conf. Series **895** (2017) 012013 doi:10.1088/1742-6596/895/1/012013

Corresponding Habits of Mind and Mathematical Ability

G Dwirahayu, D Kustiawati and I Bidari
Mathematics Education Department, Faculty of Tarbiyah and Teachers Training, UIN Syarif Hidayatullah Jakarta, Indonesia

*gelar.dwirahayu@uinjkt.ac.id

Abstract. Objective of learning process regarding system of Indonesia government should be consist of knowledge, attitudes, and skills, or in general we call cognitive, affective and psychomotor. These objective are apply to mathematics education also. Attitude in mathematics known as habits of mind. The teacher should create a learning objective which including all, cognitive, affective and psychomotor. In general, math teachers in Indonesia pay attention to aspects of knowledge, and they ignore other aspects. Habits of mind is term which means the tendency to behave intellectually or intelligently when faced with problems which is immediately known solution. This article examines the math teacher's attention to students' habits of mind. The research done by survey method to 38 students at Islamic School 32 Jakarta academic year 2015/2016 from April to May 2016. Habits of mind are observed in this research restricted to persisting, thinking about thinking, thinking flexible and applying past knowledge to new situation. Based on survey, conclude that teacher, without realizing, they have to improve students' habits of mind, as long as teaching and learning only persisting and thinking about thinking are already well developed, while flexible thinking and applying past knowledge to new situation has not well developed. We hope, math teacher can pay attention not only cognitive aspect but habits of mind also.

1. Introduction

The purpose of education regarding to Indonesian government should create an atmosphere of teaching and learning process, whereas students are actively developing their potential to have spiritual power of religion, self-control, personality, intelligence, character, and skills, and society. This purpose is apply to teaching and learning mathematics. Therefore, math teacher have to think objective of learning process to cover of that. They create learning process to mastery in cognitive, affective and psychomotor [1]. To knowing effectiveness of teaching and learning process, the government holding the final exam. National rate score test, at Banten Provence about 63%. This score is contradiction to math teachers' opinion. They feel difficulty to deliver mathematics because students' understanding on mathematics is still low. There are two fundamental reason why students' success in national exam meanwhile not success in school exam. It is because of instrument test. National exam use multiple choice but school exam use essay test. In other word, students success in cognitive aspect but is not success in other aspect.

As Cuoco said that students have studied mathematics but has very little to do with the way mathematics is created or applied outside of school. One reason for this has been a view of curriculum in which mathematics courses are seen as mechanisms for communicating established results and methods for preparing students for life after school by giving them a bag of facts. [2]. Mason and Watson says that "If students are to make mathematical sense themselves, then they need to be able to assert things for themselves. They need to use technical terms with facility to express their ideas" and if students do not understand theory then when they "come to apply a theorem or technique, they often fail to check that the conditions for applying it are satisfied" [3]. Then, there are 16 indicators habits of mind, such as: persisting, thinking and communicating with clarity and precision, managing impulsivity, gathering data through all senses, listening with understanding and empathy, creating, imagining, innovating, thinking flexibly, responding with wonderment and awe, thinking about thinking (metacognition), taking responsible risks, striving for accuracy, finding humor, questioning and posing problems, thinking interdependently, applying past knowledge to new situations, and remaining open to continuous learning [4], [9].

Based on the explanation before, we conclude that teaching mathematics is not only to make students well in cognitive but they should well in habits of mind. So that, students can apply mathematics concept into daily life problem. Habits of mind is defined as tendency to behave intellectually or intelligently when students faced to problems, especially problems that are not immediately known solution. In other words, habits of mind should be developed in learning process because when students are faced with the problems and occur with anxiety, then it will form a certain intelligent behavior patterns to encourage success in solve the problems. In this research only four aspect of habits of mind, such as: persisting, thinking about thinking, thinking flexibly and applying past knowledge to new situations are used to observe students activity in the math class.

2. Habits of Mind and Mathematical Ability

Habits of mind is defined as a characteristic or a bunch of skills, attitudes, and values is done by intelligent people when they are faced with a problem which is the solution cannot be easy known, with productive action to synergize between intellectual and emotional brain when to face or resolve existing issues [4], [5], [8]. Habits of mind as one of level of learning outcomes, and the position is highest [4]. Look at Figure 1.

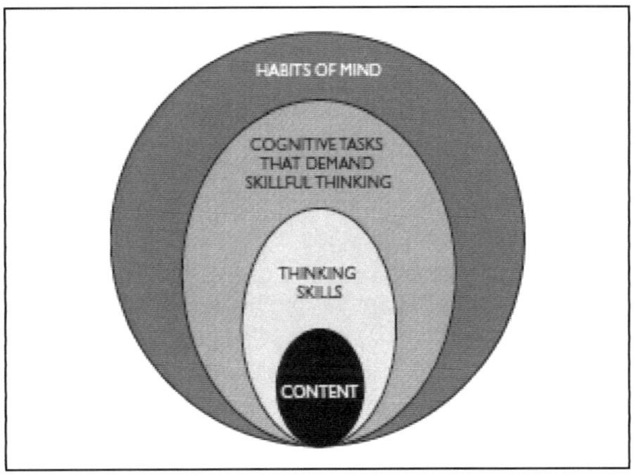

Figure 1. The four levels of learning outcomes

It is clear, Figure 1 explain us that four of learning outcomes should create to improve classroom instruction, curriculum and assessment. In math, teacher must teach content as a basic in math, than teacher train students to use their thinking skills with new problem. After that teacher extend to cognitive task that demand skilful thinking. And in the high level, teacher leads students to habits of mind. Furthermore, habits of minds has five dimensions of learning: (1) Attitude and Perception - learning increases a student's motivation, (2) Acquire and Integrate Knowledge - student connects new learning to what they already know, they learn new skills, practice the skill, learn how to make the skill useful leading to performing the skill easily, (3) Extend and Refine Knowledge – students to get a deeper understanding of their learning, (4) Using Knowledge Meaningful - teachers should provide an opportunity for students to make the knowledge meaningful, and (5) Productive Habits of Mind - Students learn to regulate themselves and think critically and creatively. The fifth dimension of learning are interrelated one another and form a framework that can be used to improve the quality of learning [7]. The four of habits of mind are used in this research: *Persisting*, as thinking diligent and earnest in completing a task through to completion, not easy to give up, able to analyse the issue, making stronger system, structure, and use strategies to solve problems. *Thinking Flexible,* as ability to control of the powerful, able to change the angle of view when receiving new data, search and do a lot of goals and activities simultaneously, and using a set of problem-solving strategies, knowing when to think big or globally and when to think a careful and detailed. *Thinking about Thinking*, as ability to find out what is known and not known, so as to realize the measures/strategies used and to reflect on and evaluate the productivity of their thinking; and *Applying Knowledge to New Situation past*, as ability to extract meaning from experience and being able to apply them to new situations and make analogies [4], [9].

Mathematical ability in this research focus to generalization ability. Generalization ability is defined as a thinking skills that covering distance between previous knowledge and new notions, or make link from previous knowledge to reach new related concepts [11], and also generalization can helps people to blend their experiences with useful knowledge to solve their problems in new conditions and lets them to be more creative [10]. The indicators of generalization are used in this research consist of: perception of generality: students are able to perform the process of identification of the pattern; expression of generality: students can use pattern identification results to determine the structure: symbolic and manipulation of generality: students are able to use generate general rules and use them to solve problems [12].

3. Method

Survey method is used to collect data of habits of mind and generalization ability from April to May 2016. The subject were students 8 grade in mathematics classroom at Islamic School 32 Jakarta for academic year 2015/2016, consist of 38 students. There are three instrument are used in this research: Questionnaire habits of mind, observation sheets and tests of generalization test. Questionnaire habits of mind and observation consist of 16 statement with reliability coefficient r_{11}= 0.88. Generalization test consist of 4 essay with reliability coefficient $r11$= 0.952. Habits of mind are elaborated to be 4 indicator, such as explain in Table 1.

Table 1. Indicator of habits of mind

HOM	Indicator
Persisting	Diligent, using systematic methods to analyse problem, distinguish ideas between good idea or not, look for different ways to solve the task/problem
Thinking about Thinking	Working or doing as a plan, aware to their thinking and actions, devise a strategy to get information to solve the problem, illustrate the steps used to solve problem
Thinking Flexibly	Open-mind, have different ideas to solve a case, change of their idea when you faced to new information, use different ways to solve the problem
Applying Past Knowledge to New Situation	Use own knowledge to understand problem or situation, connect between old and new knowledge, describe of the meaning of an experience to solve new problems

Researchers conducted observations for 5 times, each observation performed about 30 minutes in the early. There are two kind of observation, observation to students' activity and observation to students' worksheet. While questionnaire and generalization test are given to students in the end of teaching and learning.

4. Result and Discussion

As mention by Miliyawati, individual success is determined by the habits, the habitual is done continuously will be stronger and settled on the individual therefore difficult to change [6]. And Marita also find that form covering 16 categories of habits of mind, category of thinking interdependently, and thinking and communicating with clarity and precision shows good performance, while in the other category are less performance [5], in this research show that habits of mind has an impact on mathematical ability for 8 grade students at Islamic School 32 Jakarta. Percentage of the emergence students' habits of mind in five times observation in mathematics class is shown at Table 2.

Table 2. Habits of mind on Worksheet

HOM	Questionnaire	Activity	worksheet
Persisting	75.56	53	52
Thinking Of Thinking	70.60	97	54
Flexible thinking	66.31	29	7
Using past knowledge to new situation	61.15	-	18

From Table 2, researcher observe students activity for three indicator, while questionnaire and students worksheet for four indicator. The result shows that observations on worksheets and activity in the class shows the same score, meanwhile the average values obtained habits of mind in each category did not different significantly. Score of flexible thinking and using past knowledge to new situation is different significantly because all students solve the problem with the same ways similar to the example that's given by teacher. And based on the indicator, we conclude that persisting and thinking about thinking are already well developed, while flexible thinking and applying past knowledge to new situation has not well developed.

***Problem for thinking of thinking**: A block has length 12 cm and a width 9 cm. If the length one of diagonal space is 17 cm, then the area of surface is ...*

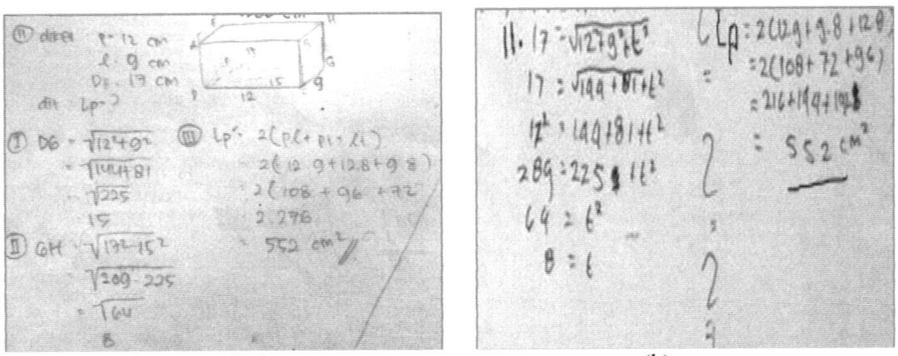

'(a) (b)

Figure 2. Photograph of students' works on thinking of thinking

Figure 4.a shows that students solve problems with systematically, He solve problems using numbering, they think what kind of work that they need done for resolving the problems, and to determine the accuracy of the size of the sides geometrical been obtained aided by drawing a block beforehand, different with other students in the Figure 4.b, he feels confident with the completion strategy because the formula very familiar and he solve the problem with that formula. Habit of writing down the steps of completion will lead to careless students in solving problems.

After teaching and learning process finish, student are given test mathematical ability of generalization. The result shows at Table 3.

Table 3. Description of Mathematical Ability

Statistics	Generalization Test
Mean	59.7363
Std. Deviation	18.26288

Than researcher analyse data of habits of mind and mathematical ability with linear regression using SPSS version 22 [13], then the conclusion is correlation between habits of mind and mathematical ability is described as linear regression Y = (-86.165) +2.116x, with the influence of habits of mind to the mathematical ability was 40.9%.

Constant value of (- 86.165) does not mean that the habits of mind of students is negative, but can be understood that when the ability of the student mathematical value (y) is zero, then the value of habits of mind it:

$$Y = -86{,}156 + 2{,}116x \leftrightarrow -86{,}156 + 2{,}116x = 0$$
$$\leftrightarrow 2{,}116x = 86{,}156$$
$$\leftrightarrow x = \frac{86{,}156}{2{,}116} = 40{,}72$$

In other words, every child has the habits of mind in itself, teachers need to explore and develop, so the potential to develop their knowledge. Cases were found in this study, and mathematics scores of zero, in fact students have habits of mind about 40.71%. This is in line with the results of observations conducted by researchers that the average value HOM that appear in the learning process amounted to 68.96%.

5. Conclusion

It can be concluded that basically everyone has the potential in themselves, whether it is in form of knowledge, attitudes and skills. The results of the analysis showed that the learning process of mathematics of the students' habits of mind are less developed in the learning process. Without any treatment given to the habits of mind. The result shown that impact of habits of mind toward mathematical ability is 40%. If teachers can develop their students' knowledge, attitude and skills holistically, as expectation, then the potential of students to acquire knowledge in mathematics will be higher or better.

Acknowledgments

We are especially grateful to teachers and students at MTs 32 Jakarta, who have given their support during the research, second great thank delivers to our colleagues who have sharing their ideas and validation for the instrument. And we are most grateful to Prof. Dr. U Sumarmo, and Prof. Dr. T Yuniarsih, who have become inspiring and supplies us with some useful material and suggestions.

References

[1] L. W Anderson; D.R Krathwohl. 2001. A Taxonomy for Learning, Teaching and Assessing: A Revision of Bloom's Taxonomy of Educational Objectives. New York: Addison Wesley Longman, Inc.

[2] A Cuoco; E. P. Goldenberg; J. Mark. 1996. *The Journal of Mathematical Behavior*. Habits of mind: An organizing principle for mathematics curricula Volume 15, Issue 4, December 1996, pp. 375-402

[3] A Watson, and J Mason. 2001. *Getting Students to Create Boundary Examples*, MSOR Connections 1(1), pp. 9-11.

[4] A.L Costa. 2000. In A. L. Costa, (Ed.). *Developing minds: A resource book for teaching thinking. Alexandria*: Habits of Mind VA: ASCD, pp. 80-83.

[5] R A S Marita. 2014. *Profile Habits Of Mind XI High School Students in Biology Teaching Method Using Practice and Discussion*. Proceeding of Forum on Mathematics and Sciences. ISBN 978-602-0960-00-5. pp. 441-448

[6] B Miliyawati. 2014. *Urgency Disposition Strategy Mathematical Habits of Mind*. 3(2). pp. 174-188

[7] A Grace. 2015. *Learning Dimension Based Teaching*. National Symposium on Educational Research and Innovation Policy Research, Center of Education. Universitas Pendidikan Indonesia Bandung, Indonesia.

[8] E Susanti. 2015. *Problem of High Order Thinking Skills to Train Mathematical Thinking Habits*. Universitas Sriwijaya, Palembang, Indonesia.

[9] Utari-Sumarmo. 2012. *Education and Development Character and Disposition of Mathematical Thinking in Mathematics Learning*. National Seminar on Mathematics Education, NTT, Indonesia February 25th.

[10] H Nourooz; M S Abu; H Kashefi; K Rahimi. 2013. *Generalization in the Learning of Mathematics*, proceeding of 2nd International Seminar on Quality and Affordable Education (ISQAE), Malaysia.

[11] K Stacey. 2006. *What Is Mathematical Thinking and Why Is It Important?* University of Melbourne, Australia.

[12] H D Putra. 2013. *Teaching Geometry with SAVI based WINGEOM Approach to Improving Students' Mathematical Generalization at Junior High School*. Proceeding National Seminar on Mathematics and Mathematics Education. STKIP Siliwangi Bandung, Indonesia. ISBN 2338-8315.

[13] A Widarjono. 2015. *Application of Multivariate Analysis*. Yogyakarta: UPPSTIM YKPN, Indonesia.

IOP Conf. Series: Journal of Physics: Conf. Series **895** (2017) 012015 doi:10.1088/1742-6596/895/1/012015

Pre-Service Physics Teachers' Perception toward Hands-on Lab Activity and 21st Century Skills

D H Putri*, E Risdianto and S Sutarno

Program Studi Pendidikan Fisika, Jurusan Pendidikan Matematika dan Ilmu Pengetahuan Alam, Fakultas Keguruan dan Ilmu Pendidikan, Universitas Bengkulu, Jl. W.R. Supratman, Kandang Limun, Bengkulu, Indonesia

*dhputri@unib.ac.id

Abstract. This study aimed to describe the hands-on lab activities and 21st century skills of pre-service physics teachers at a university in Bengkulu. The respondents of this study were 113 students who have been finished and were following the laboratory course. The research instrument was questionnaire. The explored aspects of laboratory activities were motivation, the importance of laboratory activities, equipment, laboratory activities process, suitability of curriculum, assessment, laboratory design, and the 21st century skills training. The 21st century skills explored consist of learning and innovation skills, life and careers skills, and media, information and technology skills. The data obtained will be analyzed descriptively. Based on the results of data analysis was obtained that they have a good perception toward the aspect of motivation, the importance of hands-on lab activities, and laboratory activities process; and the perception was fair for other aspects. The lowest perception score was obtained in the aspects of the 21st century skills training. This result was in accordance with the 21st century skills of pre-service physics teachers which were still in moderate category. So it is necessary to develop a model of laboratory activities design that can training and enhancing the 21st century skills for pre-service physics teachers.

1. Introduction

Higher education should do a big effort to anticipate the fast dynamical changing in education. One of the way to do it was by adapting the 21st century learning concept. The concept of 21st century learning was characterized by training and developing the 21st century skills, which were life and carrer skill, learning and innovation skill, and also information, media and technology skill. These skills were united as a scheme called 21st century knowledge-skills [1]. Physics learning as a part of science education has a strategic role in developing 21st century education. In physics learning, the 21st century skills could be developed by learning process in the classroom and hands-on lab activities.

Laboratory activities was an important part in the nature of science. Hands-on lab activity was the main activity in laboratory that could not be separated from physics learning. This activity could build up students motivation and achievement in physics learning, skills developing, based experimenting, practice skill developing, and has a role as a learning mode in science metodh application [2]. The right hands-on lab activity would build and reconstruct knowledge, and strenghten students comprehension of physics [3]. Furthermore, hands-on lab activity was also potential to provide higher order thinking skills [4]. However, there were many studies that found hands-on lab generally has not gave the expected benefit. It is happens because of the many limitations and challenges faced in conducting hands-on lab activities [5].

Several studies have been conducted to explore the pre-service science teacher's responses to laboratory activities. One study result shows that pre-service science teacher's view laboratory activities as indispensable for improving the quality of science learning. Laboratory activities are one of the tools for visualization, application, observation and experimentation related to the concepts of science studied [6]. Meanwhile, other studies show that they feel less comfortable working in laboratories, many challenges to be faced in conducting laboratory activities but nevertheless they realize that laboratory activity can not be separated from science learning. While some of the other pre-service science teacher expressed great pleasure in learning science in the laboratory and it makes the concepts more reasonable [7]. The information is useful for the development of science learning, but not enough to describe the pre-service physics teacher's views on important aspects and challenges faced in conducting laboratory activities

In this study, a more complete investigation was undertaken to explore the teacher's view of laboratory activity on motivational aspects, the importance of hands-on lab activity, laboratory equipments, hands-on lab process, the suitability with curriculum, hands-on lab assessment, and hands-on lab activity design. In addition, this study aims to explore the opinion of prospective teachers related to the provision and development of 21st century skills through laboratory activities. Information regarding pre-service physics teachers' hands-on lab activity and their 21st century skills are important to know. This can be a basic model of a design-oriented to develop 21st century skills. These skills are needed because of the global challenges of the 21st century.

2. Method

This research was a descriptive study to get the description of pre-service physics teachers' perceptions toward hands-on lab activity and their 21st century skills in Physics Education Study Program in one of university in Bengkulu. The respondents that involed in this study were 113 students of even semester in 2016/2017 academic year who has passed the physics hands-on lab course. At first, the questionnaire was distributed to 167 even semester students, but only 133 students (male 52 and female 81, ranging in age from 19 to 22 years) who filled and gave the questionnaire back.

The instrument used for collecting the data was a questionnaire of students perception toward hands-on lab activity and their 21st century skills. The perception scale used 5 level of agreement which were strongly disagree, disagree, less agree, agree, and strongly agree. The questionnaire used in this research has been through the validity and reliability test. The students' perception data toward hands-on lab activity divided into four categories, namely miserably, poor, fair, and good. While the students' perception data toward 21st century skills was divided into very low, low, moderate, and high.

The students' perception questionnaire toward the hands-on lab activity consisted of 8 perception's aspects with total 42 questions. Those aspects were hands-on lab motivation, the importance of hands-on lab activity, laboratory equipments, hands-on lab process, the suitability with curriculum, hands-on lab assessment, hands-on lab activity design, and the 21st century skills' training. The questionnaire of students' perception toward their 21st century skills consisted of two domain skills, which were learning and innovation skills, life and career skills, and media, information and technology skills. Each domain had several aspects of the skills.

Learning and innovation skills consist of critical thinking skills, problem solving skills, communication skills, colaborative skills, learning skills, and creativity and innovation with 33 questions. Life and career skills consists of flexibility and availability aspects, initiative and self-organizing ability, social interaction, productivity and accountability, and leadership and responsibility aspects with 27 as the total number of questions. Media, information and technology skills consist of information literacy, media literacy, and ICT literacy with 16 questions.

The raw data from questionnaire were collected, coded and tabulated to establish the analysis. Further analysis and interpretation of the data was done to obtain the description of students' perceptions of hands-on lab activities and their 21st century skills in accordance with the domain or aspect of skills that were explored.

3. Result and Discussion

3.1. Pre-service Teachers' Perceptions toward Hands-on Lab Activity

Based on the analysis of questionnaires data, it was found that pre-service teachers' perceptions on hands-on lab activities were in good category for motivation aspect (A-1), the importance of hands-on lab activity (A-2), and hands-on lab process (A-4). While on the aspect of laboratory equipments (A-3), suitability with curriculum (A-5), hands-on lab assessment (A-6), hands-on lab design (A-7), and 21st century skills training (A-8), the students' perception was in fair category. The results of data analysis is shown in figure 1.

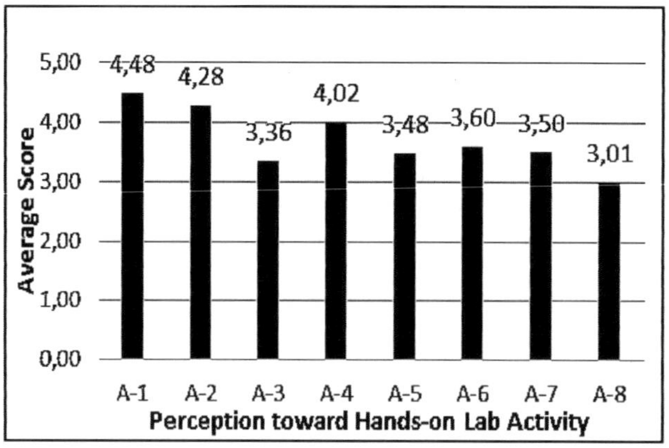

Figure 1. Graph of students' perceptions toward hands-on lab activity.

The highest score of the perception was obtained in the motivation of hands-on lab activity's aspect that was equal to 4.48, and followed by the importance of hands-on lab activity's aspect with 4,28. In general, the students stated that they were happy to do a hands-on lab activity, and hoping that every important concept would be supported by hands-on lab activities. They were challenged in assembling the laboratory equipments, and it was easy to learn physics through hands-on lab activities. In addition, the students believed that hands-on lab activities could improve their concept understanding, train the skills of applying scientific methods, reduce misconceptions, help to understand abstract concepts, and could train a sense of responsibility, honesty and accuracy.

The lowest perception score was obtained in the 21st century skills' training aspect, which was 3.01, and followed by the laboratory equipments' aspect at 3.36. In the aspect of the 21st century skills training, students stated that the hands-on lab activity has not been maximized in trained critical thinking skills, decision-making skills, creative thinking skills, problem-solving strategies, and has not yet maximized in involving computer technology to access information, analyze data, and present the results of the experiment. Furthermore, the students stated that the available laboratory equipments were not yet able to be used to explore the abstract and microscopic concepts, and it was still not fulfilled the standards need for hands-on lab activity.

3.2. Pre-service Teachers' Perceptions toward Their 21st Century Skills

The 21st century skills explored were included the domains of learning and innovation skills (K1), life and career skills (K2), and media, information and technology skills (K3). Based on questionnaires' data analysis, it was obtained that the results of the 21st century skills of the pre-service teachers were in moderate category with an average score was 3.62. The result of data analysis is shown in figure 2. It has seen that 21st century skills domains which were mostly comprehend by the students were life and career skills, and media information and technology skills. While the least-owned domain of students was the learning and innovation skills. However, the final score for those three skill domains were still in moderate category.

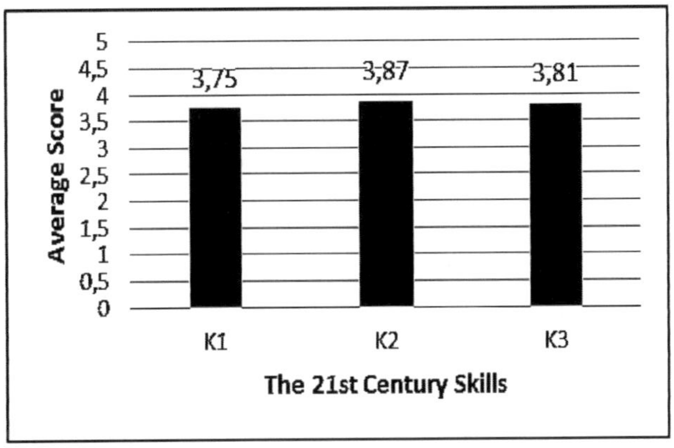

Figure 2. Graph of students' perceptions toward hands-on lab activity.

The domains of learning and innovation skills consisted of critical thinking skills, problem-solving skills, communication skills, collaborative skills, learning skills, and creativity and innovation. Based on the data analysis, the mean score for learning and innovation skills' domain was 3.75 in moderate category. The data of pre-service physics teachers' perception scores on each skills in this domain is shown in table 1. The life and career skills domain consisted of flexibility and adaptability, initiative and the ability of self-management, social interaction, productivity and accountability, and leadership and responsibility aspects. Based on the results of data analysis, it was obtained that the mean score for life and career skills domains was 3.87 in moderate category. The data of pre-service physics teachers' perception score on each skill in this domain is shown in table 2. The domain of media, information and technology skills consisted of media literacy, information literacy, and ICT literacy (information and communication technology literacy). Based on the data analysis, the mean score for media, information and technology skills domain was 3,81 in moderate category. The data of pre-service physics teachers' perception scores on each skill in this domain is shown in table 3.

Table 1. Score and category of learning and innovation skills

Type of Skills	Items	Score	Category
Critical thinking skills	5	3,50	Moderate
Problem-solving skills	6	3,56	Moderate
Communication skills	5	3,98	Moderate
Colaborative skills	5	4,05	High
Learning skills	5	3,85	Moderate
Creativity and innovation	7	3,55	Moderate

Table 2. Score and category of life and career skills

Type of Skills	Items	Score	Category
Flexibility and adaptability	5	3,96	Moderate

Initiative and the ability of self-direction	7	3,89	Moderate
Social interaction	5	3,97	Moderate
Productivity and accountability	5	3,73	Moderate
Leadership and responsibility	5	3,80	Moderate

Table 3. Score and category of media, information and technology skills

Type of Skills	Items	Score	Category
Media literacy	5	3,84	Moderate
Information literacy	5	3,83	Moderate
ICT literacy	6	3,76	Moderate

3.3. Discussion

Hands-on lab is one of the important activities in physics learning. Through this activity, students can study physics through a real experience. Students can replicate how scientists study the universe using the scientific method [3]. For pre-service teacher, hands-on lab activity was not only help them in constructing and reconstructing the concepts of physics, but also as a moderate to understand the nature of physics which consists of product and process domains. These experiences may encourage them to teach physics in a right way to the students. Many benefits of laboratory activity in learning physics encouraged researchers to develop the best way of doing hands-on lab activity [8]. The development of laboratory activity could be based on the main objectives of the learning, the activity design, the students' prior knowledge and experience, the contextual phenomenon, and based on the type of media used. Information related to the perception of pre-service physics teacher toward hands-on lab activities and their early profile of 21st century skills becomes important to know. These information was required as a reference in formulating every aspects of hands-on lab design which would be developed.

This research explored the perception of pre-service physics teacher toward hands-on physics lab activities and their 21st century skills at a university in Bengkulu, Indonesia. The result of data analysis showed that the students' perception toward hands-on lab activity that has been done was in moderate category. Student perceptions of 21st century skills were also still in the same category. If it was reviewed on every aspects of student's perception toward hands-on lab activities, it was only the aspect of hands-on lab motivation, the importance of hands-on lab activities, and the hands-on lab process which were got good category responses. While the student's perception of the other aspects on hands-on lab activities were in moderate category. Based on these results, it was known that generally students have high motivation in following the hands-on lab activities. Students agreed that the hands-on lab is an important activity that is very useful in learning physics. Nevertheless, the results of perceptual data analysis indicated that the laboratory equipments used for hands-on lab activity were still not reached minimal standard of the subject needs, especially for an abstract and microscopic concepts. This was reinforced by the statement that the hands-on lab activites that has been done was in accordance with the objectives of the curriculum but was not able to reach all the main concepts, especially the complex and abstract concepts.

According to the students' perception, hands-on lab activities have been done in a good process, but still in conventional way. Conventional for laboratory activities were characterized by hands-on lab procedures that have been prepared in detail on manual guideline, the unvaried laboratory model, the routine problem's explored, the assessment focused on the lab report and exam, and the lack of ICT technology used in the lab. The results were based on the data analysis of students' perception which showed that as many as 60.2% of the students were less agree on the item which stated that hands-on lab activities has been done in order to train the problem-solving strategy on the real problem than to verify the theory's truth. This means that the laboratory activity is intended to verify and reinforce the material already given in the classroom. Explicitly these results indicate that physics lab activity is generally still done conventionally. This result is consistent with several findings in previous studies [8]. Conventional laboratory activities generally emphasize motor activity rather than metacognition [9] so that it is less able to train high-order thinking [10].

Furthermore as many as 63,7% of students were less agree on the item that stated hands-on lab activity begins by giving contextual problem related to daily life and the solution will be investigated through hands-on lab activity, 67,3% students stated that they were less agree that hands-on lab activities contain the activities of formulating hypothesis and then testing it through observation test, measurement, and data analysis which the procedure was determined by the students themselves, and as 54,9% students were less agree to the item which stated that the hands-on lab activity used a various model. In addition, as many as 82.3% of the students were less agree on the item that said the hands-on lab assessment was mostly prioritized on the science process aspect rather than the hands-on lab report and the test, and as many as 62% of students were less agree on the item which stated that the hands-on lab activity has conducted to encourage the use of computers to organize data, analyze data, and write laboratory reports.

The hands-on lab activity has not been maximal in training and providing 21st century skills. This result was seen from the score of 21st century skills' training aspect through the hands-on lab activity which was in the lowest order (Figure 1). This fact was in line with the analysis results of the 21st century skills' perception questionnaire which showed that students 21st century skills were still in the moderate category (table 1, table 2, and table 3). These results were also in line with the previous research, found that the critical thinking and scientific reasoning of pre-service physics teacher at a university in Bengulu were at low and moderate levels [11].

Based on the explanation above, it can be summarized into some important points such as students have good motivation in the hands-on lab activity, students stated that the hands-on lab activity was very important to do in physics learning, abstract and microscopic concepts could not be explored due to the limitations of laboratory equipments, hands-on lab activity was still done in conventional way, the hands-on lab activity has not maximally provided 21st century skills, and students 21st century skills were still in moderate category. This result could be used as a reference in the development a model of laboratory design which would be used as an alternative solution to various problems obtained.

4. Conclusion

Based on the research results and discussion, it can be conclude that hands-on lab activities that has been done at a university in Bengkulu were in fair category. Students gave the highest perception score on hands-on lab motivation and the importance of hands-on lab activity aspects. While the lowest score was given to the aspects of students 21st century skills' training and laboratory equipments. The 21st century skills of pre-service physics teacher at a university in Bengkulu were in moderate category. The lowest to the highest scores of 21st century skills was gained on the domains of learning and innovation skills; media, information and technology skills; and life and career skills. The results of this study indicated the importance of developing 21st century skills-based laboratory design model.

Acknowledgments

The research was funded by the Ministry of Research, Technology, and Higher Education of the Republic of Indonesia through the Applied Product Research scheme at University of Bengkulu. The research team expressed gratitude for all the facilities provided.

References

[1] Trilling B and Fadel C 2009 *21st Century Skills: Learning for Life in Our Times* (San Francisco: Jossey-Bass A Wiley Imprint)
[2] Hofstein A and Lunneta V N 2004 *Science Education* **88** pp 28-54
[3] Tobin K 1990 School Science and Mathematics **90**(5) pp 403–18
[4] Deacon C and Hajek A 2010 *International Journal of Science Education* pp 1–35

IOP Conf. Series: Journal of Physics: Conf. Series **895** (2017) 012015 doi:10.1088/1742-6596/895/1/012015

[5] Fuccia D, Witteck T, Markic S and Eilks I 2012 Trends in Practical Work in German Science Education

[6] Harman G, Cokelez A, Dal B and Alper U 2016 *Universal Journal of Educational Research* **4**(1) pp 12-25.

[7] Akarsu B 2015 *Journal of European Education* **5**(2) pp 56-61.

[8] Malik A and Setiawan A 2016 The development of higher order thinking laboratory to improve transferable skills of students. *International Conference on Innovation in Engineering and Vocational Education* (Indonesia: Bandung)

[9] Hodson D 1990 *School Science Review* **71**(1) pp 33–40

[10] Heller K and Heller P 2010 *Problem Solving Labs, In Cooperative Group Problem Solving In Physics A: user's Manual* (United State: Departement of Physics University of Minnesota).

[11] Sutarno, Setiawan A and Kaniawati I 2016 Keterampilan berpikir kritis dan penalaran ilmiah mahasiswa calon guru fisika: field study pada salah satu LPTK di Kota Bengkulu *Prosiding Seminar Nasional Pendidikan Dasar dan MIPA* pp 111-22 (Indonesia: Bandung)

IOP Conf. Series: Journal of Physics: Conf. Series **895** (2017) 012016 doi:10.1088/1742-6596/895/1/012016

The Web-based Module of Changes in Objects

R Triayomi
Primary School Teacher Education, Universitas Katolik Musi Charitas, Jl. Bangau No. 60 Palembang 30113, Indonesia.

*riatriayomi@ukmc.ac.id

Abstract. To understand the changes of substances contained in such a kind of substance and substance characteristics then need a deep study of the concept. In this concept is expected to understand the changes of objects such as substance type and substance characteristics. Types of substances and characteristics of substances through physical changes and chemical changes and means of separation consisting of two or more substances. The principle of separation of the mixture is based on differences in physical properties of its constituents, such as substances, particle size, melting point, boiling point, magnetic properties, solubility, and so forth. This study aims to produce a web-based module of changes in objects that are valid, practical, and have effectiveness of student learning outcomes and activities on natural science learning. The experiment was conducted on 30 children in South Sumatera. The case of the development of the learning module of change of the object is influenced by the child's understanding of the concept. Expected to be adapted by world teachers.

1. Introduction

The curriculum has been set by the government to replace the Education Unit Level Curriculum that has been valid for approximately 6 years. The curriculum is set in 2013 so that the curriculum is known as the 2013 curriculum which has three aspects of assessment, namely aspects of knowledge, skills aspects, and aspects of attitude and behavior. The core of the 2013 curriculum is the mindset, strengthening of curriculum governance, deepening and expanding the material, strengthening the learning process. Competence of science for junior high school education level in 2013 curriculum physics, chemistry, and biology aspects have been integrated in science subjects. Science is concerned with how to systematically find out about nature, so science is not only the mastery of a collection of knowledge in the form of facts, concepts or principles but also a process of discovery. The science-learning process emphasizes the provision of hands-on experience to develop competencies to explore and understand the natural world scientifically.

Science course serves to provide knowledge about the natural environment, develop skills, insights, and technology awareness in connection with its utilization for everyday life. Science subjects in elementary schools begin to be taught in the lower classes by giving more knowledge through observation of the various types and temperatures of the natural environment as well as the artificial environment. Science is the building or sequence of concepts and conceptual schemes that are interconnected as a result of experimentation and observation "science has the meaning of referring to the knowledge that is in the system of thought and theoretical concepts in the system, which includes all kinds of knowledge, about anything" [1].

Science is a theoretical approach derived or arranged in a peculiar or special way of observation of experimentation, inference, theorizing, experimentation, observation, and so on one way to another [2]. The same is also expressed science is a knowledge of the universe that relies on data collected through observation and experimentation so that it contains human products, processes, and attitudes [3]. Concept is an idea that unifies the related facts of science then the principle is generalization of the relationship between the concepts of science [1]." Further Referring to the opinion of these experts can be concluded that the science is an organized investigation to find patterns or natural order to provide knowledge. Such knowledge may be facts, concepts, theories, laws, principles of the natural environment in relation to their utilization for everyday life. Material Changes the objects used in the research are material is anything that has mass and occupy space. The mass of a material is not affected by the gravitational force of the Earth. Therefore, the mass of a matter matter will always remain everywhere. The weight of the material will vary depending on the place. Weight is a measure that represents the magnitude of the Earth's gravitational force experienced by a material. Because it is influenced by the gravitational force of the Earth, the weight of matter will vary depending on the position of the material from the center of the Earth. The closer the poles of the Earth, the greater the gravitational force of the Earth, the greater the weight of matter. Conversely, the nearer the equator, the smaller the gravitational force of the Earth, the lower the material weight. So, the same material, if placed in a different place gravitational force then the weight will also be different.

The material in nature can be a single (pure) substance and can also be a mixture. Pure substances are materials that are composed of only one kind of substance. Mixture is a material composed of two or more substances. The change of matter is the change of the nature of a substance or matter into another substance whether it be new or not. Material change is divided into two types, material change physically or physically and chemical material changes. Physical change is the change of a substance that does not produce a new kind of substance. The characteristics of physical changes are not formed a new type of substance, substances that undergo changes can return to their original shape and changes that occur only followed changes in physical properties. Chemical change is the change of a substance that produces a new kind of substance and a change that is eternal. Characteristics of chemical change is the formation of new types of substances, substances that change can not return to their original shape. The changes that occur are followed by changes in chemical properties through chemical reactions, during chemical changes, the mass of substances before equals the mass of substances after the reaction.

The conception of science learning as an increase in knowledge and understanding is more positively correlated with empathy [4]. The study suggests a correlation of empathy with knowledge and understanding. Empathy by creating the desire to help others, experiencing emotions similar to the emotions of others. The relationship between empathy is difficult to achieve with knowledge and understanding. Science learning should be conducted in scientific inquiry to foster thinking ability, work and be scientific and communicate it as an important aspect of Life Skills.

Therefore science lessons in junior high schools emphasize the provision of learning experiences directly through the use of website technology. Learning using technology to support the success of the curriculum 2013. Learners active in gaining knowledge widely. Therefore, based on the above description, the researcher considers the need to conduct research on the development of learning modules of website-based object changes that are tested for validity, tested practicality and have effectiveness. Employing of module in education have been working in a long time. But, in the beginning it is still limited for employing of communication tool among the researchers in form of information exchange with text [5]. Learning by module can be formulated as a complete stand-alone unit and consists of a series of learning activities that are structured to help students achieve a number of clearly defined and specific objectives [6]. The service developing of communication for module based on website develop rapidly in accordance with computer developing become hypermedia [7][8]. The hope after developing a web-based module can improve learning outcomes on natural science subjects.

2. Experimental Method

Design best research by using development of model that oriented to Hannafin and Peck's product. Hannafin and Peck's development model consists of three phases of development: (1) necessary of analysis; (2) design / planning; And (3) development and implementation [9]. Furthermore, for formative evaluation of researcher by using Tessmer. There are five evaluations, namely: (1) self-evaluation; (2) expert's review; (3) one to one evaluation; (4) small group; (5) field test.

For every phase of Hannafin and Peck's development of revision and evaluation were done to reduce errors during development. The phase of development was limited to self-evaluation so that researcher modified it by Tessmer's formative evaluation. The location of the study in SMP Xaverius 1 Palembang. In this study, the subject of the study was the seventh graders of first semester in the academic year 2014/2015. This module based on website would be used by graders to be teaching material alternative for subject of science.

Development of learning modules with several phases of activities were listed in the development phase. Before entering the area of analysis it was determined the subjects to be developed first in the media of this lesson. The lessons that would be developed by researcher was the conversion of object. Phaseof model by Hannafin and Peck namely: 1) phase of analysis necessary, identified the necessary in developing a module including the goal and objective of development. Also, identified the knowledge and skill of graders who would use the computer. The analysis was done by researcher by reviewing journals and articles of teaching material in the form of module that could help graders in the process of learning and interviewing with teachers to find the problems that occurred. Collecting the information to develop goal of module based on website development. Determining the software and hardware that would be developed. Every steps that was done in this phase was evaluated and revised appropriate with literature and expert review; 2) the design / planning, at this phase, the information by the requirements of analysis phase was transferred into the form of document which would be the goal of learning module planning. The goal was identifying and documenting the best rule to rich the goal of the module planning, Hannafin and Peck stated the design phase aimed to make the program. The program had been made after identified the necessary and decision for the goal of learning to conceive the lesson plan.

Basic competence and indicator of the competence achievement was to explain the various of characteristic substances, explaining the definition of conversion of object. Explaining the definition of conversion chemistry and the separation of the mixture. This phase was evaluated and revised subsequently based on reviewed of flowchart and story board was created too during this phase. Flowchart would show the application current while the sample story board was example of learning arrangement and interface layout would be provided to the user. The layout also included multimedia elements that were designed that appropriate to the learning process. The consideration was based on the scientific approach that had been made to ensure the learning objective that was accordance; 3) the phase of development and implementation, the researcher did the income of flowchart, testing as well as formative assessment. The storyboard document would be the base for creating a flowchart that could help the process of creating the module based on website. In this phase, that was prepared an expert validation sheet and questionnaire of the grader's response during the learning process [10].

The results of development to find out the effect for the quality of learning that was included the valid, effectiveness, attractiveness, and effectiveness of learning. To ensure development of application was done correctly, the development phase would be reviewed regularly back to the scientific approach, flowchart and storyboard that had been planned in the phase of design. One important step in this development phase was validation with experts. The process to see if the module based on website was developed by the researcher successful appropriate with expectation or not, the researcher did the evaluation formative.

Formative evaluation phase [11], namely: 1) self-evaluation, in this phase overcome the error of development study at the time, consists of analysis and design were conducted by researcher. In this phase it was done during the phase of necessary analysis and design phase; 2) expert review, the design result that was developed by the researcher was given to the expert to be validated. Validation is a measure that states the validity of an instrument so that it is able to measure what kind of object that will be measured. Validity of instrument test that is used by validity of material test, design of learning and validity of media [12]. Suggestion that was given to the expert review then it was improved by the researcher of the module based on website; 3) one to one test, it was tested by seventh graders consist of three graders in SMP Xaverius 1 Palembang with high ability, average, and low.

One to one test was done to know the practicality. Researcher conducted unstructured interviews and the characteristic was available then it was invited the graders to give some comments/ criticisms, suggestions of the module based on website that they had operated. It was to identify and reduced some errors that exist in in the module based on website so, the product that was resulted could be used by graders easily.

The questionnaires, suggestions and comments would be base to improve the module based on website that was develop by researcher. After the development of product had been improved then conducted the limited test; 4) Small group test, to know if the module available to be used and to know the practical and liveliness of graders in a small scale of module that was developed. The test of was conducted by took eight graders of seventh graders in first semester for the sample. The graders had some characteristic with the others who were subjected to study. At the end of the learning in small groups, the graders were given questionnaire to see students' learning outcomes. Subjects of this phase did not include three students who had participated in one to one evaluations [13]. In addition, researcher was supported by observer who would conduct documentation activities and recorded how graders conducted learning process by using module based on website that was developed by researcher. This study by observation to observe of graders activities that was done by observer and the questionnaire to know the graders response.

After the learning was complete the researcher give the test to see the outcomes of graders learning and were given a questionnaire and asked them to fill it to get information about; (A) whether the learning activity was interesting and systematic; (B) which parts of the material in the multimedia learning that was difficult to understand and what the reason is; (C) whether the questions and instructions were clear enough; (D) which test items were not relevant to the learning material; And (e) If it was relevant between the learning objective and the material contained in the media of learning; (4) organized discussions with the graders to get more detail information when that was possible; (5) analyzed the data that was collected as feedback to revise the media learning product.

The results of the questionnaires, suggestions and comments in this phase were intended to improve and refine the website-based module developed by researcher to be more effective. After the development product had been improved the next phase of conducting into field test; 5) Field test aimed to know the effectiveness of website-based module in the learning outcomes and liveliness of graders. The class was used for the field test by the target class of study. The implementation of the field test was begined by the pre-test and ends by the post-test. Products that had been tested in the field test must fulfill with the quality of criteria. Focus of questions that need to be used as a standard of field test, including (1) ability to be implemented; (2) continuity; (3) effectiveness; (4) compatibility with the environment; (4) acceptance and attractiveness.

According to [12] suggests that there are three quality criteria are: validity, practicality, and effectiveness. The effectiveness would be measured in this field test was the website-based module to the l' learning outcome before using the module and after using the module. If the results was obtained had reached at least 70% then the module declared could be used by graders properly for teaching and learning process science subject. Researcher provide a pretest for measuring the graders' ability first to the learning material that had been developed. The end of the lesson gave the researcher a posttest to know the effectiveness and observations were still conducted during the learning process to see the liveliness of the graders.

Technique of collecting data of this study was done by researcher through the walkthrough data, interview, questionnaire, observation, documentation, and outcome of the learning test. The walkthrough was conducted in the phase of expert review that was used to know the validity of the website based module. Interviews were conducted of necessary analysis and evaluation activities. In the necessary analysis activities were conducted interviews to explore the information was needed in the formulation of the problem. In the evaluation activities of the one to one phase that aimed to know the weaknesses and shortcomings of the website-based module that was developed.

Questionnaire activities in a small group evaluation aimed to get information in the form of comments and suggestions by the graders. Observations were used in the small group evaluation and field test activities. Assessment of graders activities during the learning process to know the practicality. The test of the learning outcome was used during the evaluation of the field test phase.

3. Result and Discussion

Phase analysis needs researchers to collect data by conducting literature study through reference books on php programming to be installed on the computer, besides, researchers also collect journals related to the development of website-based teaching materials through internet media as an additional reference material. At this phase the researcher also formulates the objectives and learning materials of Natural Science class VII especially on the material changes of objects that ultimately produce RPP as a guide in the implementation of learning. The results of RPP attached. At this phase the evaluation and self-revision. After all the needs are identified the next step is to design phase.

The design phase, the information from the needs analysis phase is transferred into a document form. Researchers drafted Flowchart and story board. Flowcharts and story boards will show the flow of apps while the storyboard examples of learning sequences and interface layouts will be provided to users. The layout also includes multimedia elements that are design appropriate to the learning process. The consideration is based on the scientific approach that has been made to ensure the learning process. At this phase the researcher also formulates the objectives and learning materials of Natural Sciences class VII especially on the material changes of objects that ultimately produce RPP as a guide in the implementation of learning. Researchers conducted an evaluation and revision based on literature review.

The development and implementation phase of the researcher will develop the design of website based module design then implemented and evaluated through the pilot phase. One important step in this development phase is validation with experts. The process of obtaining data from experts, researchers use validation sheets that are used to assess whether the design of website-based modules in the system will be more effective than the previous one or not.

The utilization of media that varied had been adapted to the characteristics of the main point of the material. The results of the script on the storyboard found that the teaching materials in the subject of science of the main point in the object conversion. Opening in the open source based website. This phase was the first step in developing web-based modules. Software could be obtained for free by opening the web www.ipasmp.com. Modified Home Page. This phase of web display was designed by researcher while there were home buttons, introductions, materials, and forums. Website design was light green modified with the characteristics of junior high school graders. User registration settings. Every user who joined the module must register first where the username and password were specified by the admin. Importing learning material. Materials that were imported include: text pages, web pages, videos and pictures. Preparation of exercises in order to test graders' skills could be made by format of quiz with multiple choice form. Create and manage tasks. Assessment of graders' answers to the evaluation was done in the website automatically.

The results of the evaluation sent by the graders would be managed by the teacher easily. Every incoming file would be known to the time and who was the sender. The results of expert reviews on the module based website can be seen in table 1.

Table 1. Expert review results on website based module

	Expert	1st Validity	2nd Validity	Average	Category
Media experts	1st Validity	4,76	-	4,76	Very valid
	2nd Validity	3,80	4,50	4,15	Very valid
Material experts	1st Validity	4	4,5	4,25	Very valid
	2nd Validity	3,05	4,7	3,77	Valid
Instructional design experts	1st Validity	4,66	4,83	4,74	Very valid
	2nd Validity	4,42	-	4,42	Very valid
	3rd Validity	3	4,66	3,83	Valid

Based on the table 1. Expert review was conducted with media experts, materials and instructional design. Validation was done in twice. After analyzed, the average of 4.27 with the category was very valid. The advice was given to the expert review then improved by the researcher. Next prototype 1 was tested by involving three graders who had different abilities individually. At the end of the study the researcher conducted an unstructured and open interview.

Next prototype 2 which would be tested of the limited test phase. In this phase consists of eight graders who had similar characteristics by objective study. Furthermore, graders fill out the questionnaire and provide comments and suggestions so, researcher could find out the advantages and disadvantages of the development of website-based module. Graders were given a username and password and how to operate a website-based module. Results in the small group phase could be seen in table 2.

Table 2. The result of the small group phase

Activity	Result	Information
The Graders Observation	88,3%	Very good
The Graders Questionnaire	76%	Effective

Based on the observation result, the average of graders' activity was 88,3% and it was very good, it means the instructional material in the form of module of main point module was quite practical. Graders response data showed the average total Rsmedia 3.8 and in percentage of the average number of graders response equal to 76% or 70% $<$Rsm $\leq$ 85% it means positive response by the graders hence website based module developed by researcher that was effective. Observations made by the observer obtained the average of graders activities by 82.46% as a whole grader, it means the criteria was very good.

After obtaining prototype 3 that was valid and practical, then conducted field test to see effectiveness of the learning result and the graders activities. Website-based modules were tested for graders. Before the learning implemented, the first test to measure the lgraders's ability about conversion of objects by 20 multiple choice questions. The graders conducted trial of physics where job procedures were in website-based modules, the trial result must be uploaded by website-based modules that had been provided. The observation result in this phase could be seen in table 3.

Table 3. The result of observation in the field test phase

Meeting	The Graders Activity	Information
First	85,6 %	Very Good
Second	84,2 %	Very Good
Third	83 %	Very Good
Fourth	88,8 %	Very Good

Based on table 3 the field test phase was conducted during four meetings with an average of 85.4%, it means very good category. At the last meeting conducted the evaluation process by the number of questions as many as 20 multiple choice questions. Researcher conducted the tests to the graders in twice of pretest and posttest. The learning result by using website-based modules could be seen in table 4.

Table 4. The result of pretest and post-test by using website based module

Activity	Result	Information
$\bar{x}$ pre-test score	41,42	Poor
$\bar{x}$ post-test score	85,86	Very good
N-Gain score	0,78	Very good

Based on table 4 the average of value achieved by graders in pretest was 41.42, it means less category, while for posttest result was obtained average of 85,86, it means very good category. If the value of graders posttest eone to one compared with the value of the minimum competency of criteria that must be achieved by the graders was 75, it means that the graders completed and obtained N-gain of 0.78. It showed that the effectiveness of learning by using a website-based module that was developed by the researcher, it means very good in the graders' learning result Website-based module learning of the conversion object was the last product that had some advantages. Website-based modules used a scientific approach. Provide to the graders worksheets. This module contained worksheets and got confirmation directly. Modules could be accessed by smartphone.

4. Conclusion

Based on the results and discussion, the conclusions of the study website-based module of changes in objects development was by researcher declared valid, tested its practicability, and had effectiveness in learning result and liveliness of the graders. Use of a website-based module should use a smartphone that had a high screen resolution. Internet network was very necessary in the using of learning modules. This study could be useful for teachers, graders, schools, and other researchers.

Acknowledgments

Our gratitude goes to Mrs. Siti as the mathematics teacher of SMAN 1 Baregbeg Ciamis who has helped prepare the research object and Dr. Awang kustiawan as the head of LPPM Unigal who has helped provide research funding.

This undergraduate research is entitled the web-based module of changes in objects. In this study, many people have provided motivation, advice, and support for researchers. In this precious opportunity, the researcher intends to express his gratitude and appreciation to all of them. First, the educated appreciation of the researcher is addressed to beloved parents, Johanes Sopiyan (Alm) and Magdalena Saminem for love, prayer, and endless support. Next to my husband Cornelius Hendrik Gunawan, always a good listener for every problem I face, and Thanks to SMPK Xaverius 1 Palembang for giving me the opportunity to research, I would like to thank all the people who are important for this research. This study is far from perfect, but it is expected to be useful not only for researchers, but also for readers. For this reason, constructive thought full suggestion and critics are welcomed.

References

[1] Bundu, P. (2006). Penilaian Keterampilan Proses dan Sikap Ilmiah Dalam PembelajaranSains. Jakarta: Depdiknas.
[2] Aly, A., & Rahma, E. (2008). *Ilmu Alamiah Dasar.* Jakarta: Bumi Aksara.
[3] Carin, A., & Sund, R. (1978). *Developing Questioning Techniques A Self-Concept Approach.* Columbus : Ohio.
[4] Park, J., & Jeon, D. (2015). Correlation of Students' Brain Types to Their Conceptions of Learning Science and Approaches to Learning Science. *EURASIA Journal of Mathematics, Science & Technology Education*, 1141-1149.
[5] Setiawan, D. 2007 *Pengembangan Bahan Ajar.* (Jakarta: PT. Bumi Aksara)
[6] Nasution, S. 2010 *Didaktis Asas-Asas Mengajar.* (Jakarta: Bumi Aksara)

[7] Rohman 2013 Modul Elektronik. *Pengembangan Modul Elektronik Dengan Pendekatan Salingtemas Kompetensi Ekosistem Siswa Kelas X SMA Negeri 8 Malang*, 8.

[8] Sari, R. 2014 Pengembangan Modul. *Pengembangan Modul Berbasis Pendekatan Keterampilan Proses Pada Pokok Bahasan Fluida Statis di SMA Negeri 1 Mlati*, 11.

[9] Rohman, M., & Amri, S. 2013 *Strategi & Desain Pengembangan Sistem Pembelajaran.* (Jakarta: Prestasi Pustaka)

[10] Diah, N. (2012). Journal Education. *A Development of a Computer-Assisted Software (AJaW) That Encourages Jawi Writing for Children*, 130-135.

[11] Tessmer, M 1998. *Planning and Conducting Formative Evaluations: Improving The Quality of Education and Training.* (London: Kogan Page)

[12] Akker, J 1999 Principles and Methods of Development Research. Dalam T. N. Plomp, *Design Approaches and Tools in Education and Training.* (London: Kluwer Academic Publisher).

[13] Indaryati 2008 Pengembangan Modul Pembelajaran Individual dalam Mata Pelajaran Matemati di Kelas VIII Sekolah Menegah Pertama. *Jurnal Pendidikan Matematika*, vol.2 (1) ; 41-56

Discovery Learning with Scientific Approach on Geometry

M R Ramdhani*, B Usodo and S Subanti

Magister Pendidikan Matematika, Universitas Sebelas Maret, Jl. Ir. Sutami No. 36-A, Kentingan Surakarta 57126, Indonesia

*marlisaramdhani2014@gmail.com

Abstract. Geometry as one of the branches of mathematics has an important role in the study of mathematics. This research aims to find out about discovery learning (DL) toward geometry learning achievement. This research was conducted at Junior High School in Karanganyar Regency. Research data are obtained through the test and questionnaire. Furthermore, the data were analyzed by using two-way Anava. The result of this research shows that shows that DL with scientific approach is positive effect towards mathematics learning achievement. DL with scientific approach gives better mathematics learning achievement than direct learning. Students with high self-efficacy category have better mathematics learning achievement than those who have moderate and low self-efficacy category, students with moderate self-efficacy are having the same mathematics learning achievement with those having low self-efficacy category. There is an interaction between discovery learning model with a scientific approach and self-efficacy toward student's mathematics learning achievement. Introduction. Therefore, DL with scientific approach can improve student learning achievement.

1. Introduction

Mathematics is a science that has an important role in education because mathematics equips students to have an ability to think logically, critically, creatively, systematically and innovatively. However, many students find difficulty to understand mathematics because mathematics is abstract, while in their minds are accustomed to thinking about concrete objects. It affects students' low achievement. Based on national examination obtained that the absorption capacity of junior high school students in Karanganyar District in mathematics subjects for geometry and measurement materials has the lowest absorption of 42.01% at the district level, 40.26% at the provincial level, and 47.19% at the national level [1]. Geometry is a branch of mathematics concerned with point, straight line, plane figures, space, spatial figures, and the relations between them [2]. The study of geometry contributes to helping students develop the skills of visualization, critical thinking, intuition, problem-solving, deductive reasoning, logical argument and proof [3]. So, geometry is an important branch of mathematics.

The factors of the teacher becomes one of the decisive factor in the success of student learning. Based on interviews with teachers, it can be seen that teachers are still using direct learners when giving lessons in the classroom. Students are accustomed to only memorizing the formula that is usually given by the teacher so that students do not understand that concept or formula. Mayberry (Zuya and Kwalat) said most students learn geometry based on memorizing [4]. As a result, students can not solve mathematical problems, especially if students face math related to the problems in everyday life. Therefore, teachers must design and implement various models of learning that are appropriate to each subject matter provided. Learning model that can be used as a solution to the problem is Discovery Learning Model (DL). DL is a model for developing an active student learning method by finding out for themselves, self-investigating, the results obtained will be long lasting in their memory, will not be easily forgotten by the students [5]. DL occurs when learners have to discover the knowledge of a target concept in a self-regulated way with only the provided materials [6]. Stages or procedures in DL are stimulation, problem statement, data collection, data processing, verification, and generalization [7]. Joolingen in his research states that discovery learning (DL) is viewed as a promising way of learning for several reasons, the main reason is that students' active involvement with domains will result in better structured bases of knowledge of the learners in opposition to more ways traditional learning, in which knowledge is said to be simply transferred to the students [8]. The study conducted by Alex and Olubusuyi showed that the discovery learning is very useful for students who have a high ability and also can enhance his/her ability for those in the medium category [9]. Other than, Martin and Oyebayi's research shows that discovery learning is more effective than conventional learning and it improves the ability of students [10]. However, Alfieris's research shows that unassisted discovery does not benefit learners [6].

The curriculum is developed at this time was learning based on scientific approach and interactive. Standard Process of Primary and Secondary Education mentions the need for a learning process that is in accordance with the rules of the scientific approach [11]. The six aspects are integrated with five aspects of the scientific approach are observing, asking questions, exploring, associating, and presenting [12]. In the process of a scientific approach, teachers have done a way to enable students in the learning process, so that students can think critically and creatively in solving problems encountered. In'am and Hajar's research show that discovery learning with scientific approach make the students learning result a very good category [12]. Kunsting's research shows that positive effects of metacognitive support on learning outcome and learning emotions [13].

Most students regard mathematics as a difficult and tedious subject. This has an impact on the emergence of students' negative beliefs on math lessons. The difficulty of a mathematical problem depends on how students' confidence in dealing with it. One important part of self-belief is self-efficacy. Self-efficacy is the reflection of an individual's belief in his or her capabilities to complete tasks [14]. Self-efficacy is also the expectation (hope) about how far a person is able to perform one behavior in certain situations [15]. Thus, it can be concluded that self-efficacy is a person's belief in his ability to perform a behavior in certain situations and how to overcome obstacles. Self-efficacy can be grown and studied through four main sources of information, a namely experience of success, vicarious experience, verbal persuasion and physiological state. Self efficacy consists of several dimensions, namely Magnitude, Strength and Generality [16]. In this research, the three aspects will be used as a reference in the self-efficacy questionnaire. The purpose of this research finds out whether discovery learning model provides better mathematics learning achievement than the direct learning model, know students who have mathematics learning achievement, better among students with high self-efficacy, students with moderate self-efficacy or students with low self-efficacy, know an interaction between learning model and self-efficacy toward students' achievement.

2. Experimental Method

This research is a quasi-experimental research. Researchers are unlikely to control all relevant variables. The false experimental goal is to obtain information which is an estimate information that can be obtained with actual experiments in circumstances that are not possible to control and or manipulate all relevant variables [17].

The design of this research using 2x3 factorial with the intention to know the influence of two independent variables to the dependent variable. The independent variable in this research is discovery learning model with the scientific approach for the experimental class and direct learning for the control class, while the other independent variable is the self-efficacy of the students grouped into high, medium, and low self-efficacy. The population of this study is all students of class VIII SMP Negeri 3

Karanganyar in the academic years of 2016/2017, while the sample used is class VIII H which amounted to 32 students as an experimental class and class VIII D which amounted to 32 students as a control class. Mathematics learning achievement data obtained through a test in the form of multiple choice test which contains an indicator of geometry materials, while self-efficacy data are obtained through a questionnaire based on indicator formed from the aspect of self-efficacy.

3. Result and Discussion

Student's Mathematics achievement test is 25 items of multiple choice given to the experimental class students and control class students. Based on the results of the test obtained, the average of students' mathematics learning achievement is 81 for the experimental class and an average of 65,875 for the control class. Prior to the two-way ANOVA test, normality tests were performed on the data. Test of normality of student's mathematics learning achievement data was done using SPSS with significance level 5%. The result of normality test of student's mathematics learning shows that in the experimental class and control class the sig. value. > 0.05 is 0.091> 0.05, and 0.156 > 0.05. Therefore it can be concluded that the sample of achievement tests comes from a normally distributed population. The homogeneity test of student achievement test shows that the value of sig. = 0.387 > 0.05. Therefore it can be concluded that the population has the same or homogeneous variance. In this study data analysis using two-way ANOVA shown in table 1 with the following results

Table 1. The Result of Hypothesis Test

Source	Type III Sum of Squares	df	Mean Square	F	Sig.
Corrected Model	5045.545[a]	5	1009.109	34.064	0.000
Intercept	328585.054	1	328585.054	1.109E4	0.000
Class	3255.493	1	3255.493	109.893	0.000
SE	1009.348	2	504.674	17.036	0.000
Kelas*SE	237.394	2	118.697	4.007	0.023
Error	1718.205	58	29.624		
Total	351920.000	64			
Corrected Total	6763.750	63			

Table 2. The Result of Post Hoc Tests Anava

(I) SE	(J) SE	Mean Difference (I-J)	Std. Error	Sig.	95% Confidence Interval	
					Lower Bound	Upper Bound
High	moderate	8.63*	1.705	0.000	4.53	12.73
	low	11.13*	1.623	0.000	7.23	15.03
Moderate	high	-8.63*	1.705	0.000	-12.73	-4.53
	low	2.50	1.687	0.307	-1.56	6.56
Low	high	-11.13*	1.623	0.000	-15.03	-7.23
	moderate	-2.50	1.687	0.307	-6.56	1.56

Table 3. Homogeneous Subsets Result

SE	N	Subset	
		1	2
Low	23	68.87	
Moderate	19	71.37	
High	22		80.00
Sig.		0.301	1.000

Based on the calculation above it can be seen that the effect of discovery learning model with a scientific approach to mathematics learning achievement has a value of F obtained = 109.893 with sig. = 0.000 < 0.05, so H0 is rejected and Ha is accepted. So it can be concluded that there is a difference between students' learning achievement of mathematics in experiment class using learning discovery learning model-scientific approach with student's mathematics learning achievement in control class using direct learning. Whereas based on the average of students' mathematics achievement test, group of students with discovery learning model-scientific approach is greater than the student with direct learning model. The effect of self-efficacy on mathematics learning achievement has a value of F obtained of 17.036 and sig. = 0.000 < 0.05, so it can be concluded that there is a difference between categories in self efficacy. Based on Table 2. It can be seen that there is a difference between high self-efficacy against moderate and low category, but there is no difference between moderate and low category self-efficacy, whereas based on Table 3. It shows that high category self efficacy is better than moderate self-efficacy and low, but the self-efficacy of the moderate category is the same as the low self-efficacy category. Based on table 1. In column Class * SE obtained sig. = 0.023 <0.05, so it can be concluded that there is an interaction between discovery learning model-scientific approach and self efficacy toward student's mathematics learning achievement.

The use of learning model aims to overcome the problems in the learning process so that the subject matter can be delivered optimally. Most students learn geometry based on memorizing. Students just memorize a concept, so when faced with problems that use some concept, students have not understood. By Ozerem opinion that the student may hold the visualization and the verbal definition, but prefer the visual prototype when classifying and identifying geometric figures [18]. The reasons students use memorize is the inability to understand the concept of geometry. Therefore, teachers implement discovery learning (DL) to teach material about geometry. The principles learned of DL that is material to be delivered is not in final form, but students are encouraged to identify what it is like to note continued looking for information on its own then organized the already known and understood in a final form. The use of DL can change the passive learning conditions become active and creative. Based on the results of the research indicate that discovery learning model and self-efficacy influence mathematics learning achievement. This is in line with Kamel who states that the discovery learning strategy helps recruit activities where students learn to themselves and apply what they know in a new situation, which in turn led to achieving effective learning [19]. The results of Gunay's research shows that there is a significant difference in favor of the experimental group over the control group regarding the average of academic achievement, the scores of retention of learning, and the perception of discovery learning skills scores, Both on cognitive and affective levels [20]. Zedan and Bitar's research states that the findings pointed to strong positive correlations between the dimensions of class climate: satisfaction and enjoyment, the teacher's support, rules and instructions, competitiveness and between mathematical self-efficacy and achievement in mathematics[21]. Liu and Koirala's research shows that mathematics self-efficacy and mathematics achievement were positively related. Students with high mathematics self-efficacy were associated with high mathematics achievement [22].

One of the benefits that can be gained from DL is the emergence of student scholarship attitudes, such as objective attitude, curiosity to solve problems well, and critical thinking. Students' curiosity causes students to be more eager to learn and know his knowledge. Student curiosity provides motivation for students to seek answers to the questions or problems that arise and face it. By Jones opinion that presenting geometry in a way that stimulates curiosity and encourages exploration can enhance student's learning and their attitudes towards mathematics [23]. If the students can find their own concepts in mathematics, then the results obtained will last long in the memory and students' understanding ability will increase so that the students can solve math problems well. Learning with a scientific approach requires students to be active and critical in facing a problem, so that when the learning model combined with a scientific approach will form a belief in students' self-understanding of the capabilities so students can find appropriate solutions to overcome these shortcomings. Student self-confidence will make students feel confident to express their opinion and not ashamed to ask when something is not understood.

4. Conclusion

Based on the result of the research, the conclusion that can be taken are there is the difference of students' mathematics learning achievement between the group of students with discovery learning model and direct learning, discovery learning model gives better learning achievement of mathematics than direct learning. There is the difference of students' learning achievement between students having a high, medium and low category, but students with the same self-efficacy with the students with a low category. There is an interaction between discovery learning model and self-efficacy to student's mathematics learning achievement. Based on the discussion of research results indicate that the influence of discovery learning model on learning mathematics can improve student's mathematics learning achievement higher than direct learning.

Acknowledgments

The authors thank to the Department of Magister Mathematics Education, Faculty of Teacher Training and Education, Sebelas Maret University.

References

[1] BSNP 2016 *Laporan Hasil Ujian Nasional Tahun Pelajaran* 2015/2016

[2] Biber, C., Tuna, A. & Korkmaz, S. 2013 *The mistakes and the misconceptions of the eighth grade students on the subject of angles* European Journal of science and mathematics education, 1 (2), 50-59.

[3] Jones, K 2002 *Issues In The Teaching and Learning of Geometry* in Linda Haggarty (Ed), *Aspects of Teaching Secondary Mathematics: perspectives on practice*. London: RoutledgeFalmer. Chapter 8, pp 121-139. ISBN: 0-415-26641-6

[4] Zuya H E and Kwalat S K 2015 *Teacher's Knowledge of Students about Geometry* International Journal of Learning, Teaching and Educational Research Vol. 13, No. 3, pp. 100-114

[5] Prince, J.M. 2006 *Inductive Teaching and Learning Methods* Definitions, Comparisons and Research Bases. *J. Engr. Education*. Vol. 95. Nomor. 02, p 123-138

[6] Alfieri, Brooks, Aldrich, and Tenenbaum 2011 *Does Discovery-Based Instruction Enhance Learning?* Journal of Educational Psychology American Psychological Association 2011, Vol. 103, pp. 1, 1–18

[7] Hosnan, M. 2014 *Pendekatan Saintifik dan Kontekstual dalam Pembelajaran Abad 21* (Bogor: PT.Ghalia Indonesia)

[8] Joolingen, V.W. 1999 *Cognitive tools for discovery learning* International Journal of Artificial Intelligence in Education (1999), 10, 385-397

[9] Alex A M and Olubusuyi M F 2013 *Guided-discovery Learning Strategy and Senior School Students Performance in Mathematics in Ejigbo, Nigeria* Journal of Education and Practice Vol.4, pp.12, 2013

[10] Martins O and Oyebanji R K 2000 *The effects of inquiry and lecture teaching approaches on the cognitive achievement of integrated science students* Journal of Science Teachers' Association of Nigeria. 35 (1&2) 25-30

[11] Kemendikbud 2013 *Permendikbud No.65 tentang Standar Proses Pendidikan Dasar dan Menengah* (Jakarta: Kementerian Pendidikan dan Kebudayaan)

[12] In'am A dan Hajar S 207 *Learning Geometry through Discovery Learning Using a Scientific Approach* International Journal of Instruction Vol.10, No.1 p-ISSN: 1694-609X pp. 55-70

[13] Kunsting J, Kempf J, and Wirth J 2013 *Enhancing scientific discovery learning through metacognitive support* Contemporary Educational Psychology 38 (2013) 349–360

[14] Bandura A 1997 *Self-efficacy: The exercise of control* (New York, NY: W. H. Freeman)

[15] Friedman, Howards S. & Miriam Schustack 2006 *Kepribadian (Teori Klasik dan Riset Modern* (Jakarta: Erlangga)

[16] Aimee D, Steven R H, Richard D J 2014 *Exploring the Dimensions of Self-Efficacy in Virtual* World *Learning: Environment, Task, and Content* MERLOT Journal of Online Learning and Teaching Vol. 10, No. **2**

[17] Budiyono 2003 *Metodologi Penelitian Pendidikan* (Surakarta: UNS Press)

[18] Ozerem A 2012 *Misconceptions in geometry and suggested solutions for seventh grade students.* International Journal of New Trends in Arts, Sports & Science Education, 1(4), 23-35

[19] Kamel. Abdelrahman 2014 *The Effect of Using Discovery Learning Strategy in Teaching Grammatical Rules to first year General Secondary Student on Developing Their Achievement and Metacognitive Skills* International Journal of Innovation and Scientific Research ISSN 2351-8014 Vol. 5 No. 2 Jul. 2014, pp. 146-153

[20] Gunay, Ali 2009 *The Effects of Discovery Learning on Students'Success and Inquiry Learning Skills* Eurasian Journal of Educational Research, Issue 35, Spring 2009, 1-20

[21] Zedan, Raed dan Bitar, Jarmas 2014 *Environment Learning as a Predictor of Mathematics Self - Efficacy and Math Achievement* American International Journal of Social Science. Vol. 3, No. 6; November 2014.hal 85-97

[22] Liu, Xing and Koirala, Hari 2009 *The Effect of Mathematics Self-Efficacy on Mathematics Achievement of High School Students* Northeastern Educational Research Association(NERA) Annual Conference Proceedings 2009, pp. 10-22

[23] Jones K 2002 *Issues in the Teaching and Learning of Geometry.* In: Linda Haggarty (Ed), Aspects of Teaching Secondary Mathematics: perspectives on practice. London: RoutledgeFalmer. Chapter 8, pp 121-139. ISBN: 0-415-26641-6

IOP Conf. Series: Journal of Physics: Conf. Series **895** (2017) 012040 doi:10.1088/1742-6596/895/1/012040

Students' Mathematical Communication Ability using Learning Cycle 7E on Junior High School

N H D Jati[1*], Budiyono[2] and I Slamet[2]

[1]Magister Pendidikan Matematika, Universitas Sebelas Maret, Jl. Ir. Sutami No. 36A, Kentingan, Surakarta 57126, Indonesia
[2]Magister Pendidikan Matematika, Universitas Sebelas Maret, Jl. Ir. Sutami No. 36A, Kentingan, Surakarta 57126, Indonesia

*nurhidayatdamarjati06@gmail.com

Abstract. This study aims to determine whether there are differences in the effect of applying learning cycle 7E and direct learning to students' mathematical communication ability in junior high school. This research was conducted at junior high school 16 Surakarta. This study was conducted in the even semester of 2017. This research uses the experimental method. The study sample consisted of two classes taken randomly from the eighth grade. The experimental class is treated 7E learning cycle and the control class is given direct learning. Methods of data collection in this study using the method of documentation and test methods. The data analysis in this research uses t-test. Data obtained from experiment class and control class. The results of this study conclude that there are differences in the effect of applying the learning cycle 7E and direct learning to students' mathematical communication ability. This proves that students 'mathematical communication ability using 7E learning cycle is better than students' mathematical communication ability using direct learning. This happens because the Learning cycle 7E there are steps that can be used to improve the ability of mathematical communication such as for elicit, engage, explore, explain, elaborate, evaluate and extend. Learning cycle 7E can be used to improve students' mathematical communication ability of junior high school.

1. Introduction

One of the subjects on geometry and measurement materials is solid geometry. Problems associated with solid geometry are often found in everyday life. The problem is generally in the form of a story. Solving the story problem required mathematical communication ability. Lomibao states that mathematical communication ability is the ability of students to express ideas, describe, and discuss mathematical concepts coherently and clearly [1]. Nartani states that students' mathematical communication ability is the main process that must be possessed by students to improve thinking ability in math lesson [2]. Some research on mathematical communication has also been implemented. The result of lomibao research is that students with learning mathematics communication have high learning achievement and understanding and students' anxiety can decrease [1]. One of the learning that can be used to improve the mathematical communication is constructivist learning. Constructivist learning has an important contribution to student achievement [3]. In constructivist learning, teachers can design teaching in accordance with learning materials so that students' understanding of the material will last long [4]. Qarareh states that constructivism learning is important for students.

Constructivism learning makes students' learning process active where students build new knowledge and prior knowledge [13].

One of the constructivism learning is learning cycle 7E. Some studies have shown that learning cycle 7E is better than other learning. Khaskan's research is comparing the learning cycle 7E and the traditional method of learning mathematics in preparatory year students at King Saud University. The results of the study showed that the learning cycle 7E is better than the traditional method on student's mathematics learning achievement [5]. In Balta study shows the 7E learning cycle is useful for the science curriculum [6]. Shaheen's research results show student achievement in biology lessons taught with learning cycle 7E is better than traditional instructional [7]. The results of Siribunnam's research that analytical thinking, learning achievement and attitudes in chemistry on students who were taught by learning cycle 7E were higher than KWL methods and conventional approaches [8]. However, there are studies that suggest that learning cycle 7E is no better than other learning. Polyiem's research results are learning achievement, science process skill, and moral reasoning on socioscientific issue-based learning better than learning cycle 7E [9].

In this study used learning cycle 7E and direct learning to mathematical communication ability in junior high school on solid geometry materials. This research uses the experimental method. The study sample consisted of two classes taken at random from six classes. The experimental class is given learning cycle 7E and control class is given direct learning. Methods of data collection in this study using documentation method and test method. Documentation method used as initial data. The test method is used as the final test. The final test used is a test of students' mathematical communication ability of the experimental and control classes. An indicator of mathematical communication ability in Nartani research is (1) student can express ideas with verbal mathematical sentences, (2) students can have active discussion about math, (3) Students can formulate definitions and generalizations about mathematics, (4) students can define mathematical definitions using their own sentences [2]. The aspects of mathematical communication in Widjajanti's research are (1) students 'ability in writing statements, reasons, or explanations, and (2) students' ability to use terms, notations, tables, diagrams, graphics, drawings, illustrations, mathematical models, or mathematical formulas [10]. The indicators of mathematical communication ability in this study are: (1) students can express the problem given in the form of drawing, (2) the student can change and interpret the mathematical information of an image in the mathematical representation and (3) the student can express the idea or mathematical idea In the form of mathematical writing. Data analysis in this research uses t-test. The results of this study conclude that there are differences in the effect of applying the learning cycle 7E and direct learning to the mathematical communication ability on solid geometry material. Students' mathematical communication ability using the learning cycle 7E is better than students' mathematical communication ability using direct learning. Learning cycle 7E can be used in learning mathematical communication in junior high school students.

2. Experimental Method

Type of research used in this study is the experimental method. This research was conducted at junior high school 16 Surakarta in 2017. Population in this study was all eight graders of junior high school 16 Surakarta consisting of 6 classes. An average number of students per class is 30 students. In this study, two classes were selected as experimental class and control class. Experimental class was treated with learning cycle 7E and control class was given direct learning treatment. Methods of data collection in this study using documentation method and test method. Documentation method used as initial data. The test method is used as the final test. The final test in this study is a test of written mathematical communication ability. Final test consists of 6 item essay test on solid geometry material. An instrument has been validated by a lecturer of Mathematics Education Sebelas Maret University. An instrument has been tested and analyzed. Instrument analysis consists of reliability test with Alpha Cronbach formula. Result of instrument analysis that is r_{11} = 0,733. Reliability value used to describe the usefulness of an item in measurement. It means that the result of measurement having a reliability index 0.70 or more then instrument can be used to measurement [12]. Preliminary analyzes were performed to determine whether the two samples were from the same initial condition. In the initial analysis using normality test, homogeneity test, and t test. In the final analysis used normality test, homogeneity test and t test.

3. Result and Discussion

3.1. Preliminary data analysis

In this section, we describe the initial data analysis. In this study, preliminary data is taken from final exam score of the previous semester. In table 1, the results of normality test on experimental and control class test were tested with SPSS 16.0.

Table 1. Normality test in the experimental and control class preliminary data

Class	Statistic	df	Sig.
Experimental	0.158	30	0.055
Control	0.166	24	0.087

Based on table 1 shows normality test in the experimental class that is sig value = 0.05 and the control class sig value = 0.087. On both sig values obtained both of more than 0.05 then data of two classes are said to be normally distributed. Test homogeneity of variants of both classes using Levene's test with SPSS 16.0. The result of homogeneity variance test is sig value = 0.893. Sig value more than 0.05 then data of both classes have the same variance.

Table 2 describes the t test results with SPSS 16.0 to find out whether there are differences in the experimental and control class.

Table 2. T test on initial ability of experimental and control class

Class	N	Mean	SD	t- value	df	Sig
Experimental	30	71.50	10.351	0.163		0.871
Control	24	71.04	10.212			

The results of hypothesis test in table 2 got sig value = 0.871. Sig value more than 0.05 then it can be said there is no difference in both classes. Both classes have the same initial capability. Furthermore, both classes were treated for experimental classes given learning cycle 7E and for control classes were given direct learning

3.2. Analysis of final data

Table 3 presents normality test of experimental and control class final tests. The final test is obtained from mathematical communication ability test in experimental and control class.

Table 3. Normality test in the experimental and control class final tests

Class	Statistic	df	Sig.
Experimental	0.146	30	0.102
Control	0.122	24	0.2

Result in table 3 shows sig value in experimental class is 0.102 and control class is 0.2. Both sig values are more than 0.05 then it is said the final test data is normally distributed. Test homogeneity of variants of both classes using Levene's test with SPSS 16.0. The result of homogeneity variance test is sig value = 0.111. Sig value more than 0.05 then data of both classes have the same variance.

Table 4 shows the results of t test on the final test of experimental and control class with SPSS 16.0. Final test data is a test of mathematical communication ability in experiment and control class.

Table 4. T test on initial ability of experimental and control class

Class	N	Mean	SD	t- value	df	Sig
Experimental	30	72.67	8.33	2.613	52	0.012
Control	24	65.33	11.169			

The result in table 4 gets sig value 0.012. Sig value is less than 0.05 then it is said there is a difference in the final test of experiment and control class. Based on an average value obtained, it is said that students' mathematical communication ability in experimental class is better than control class.

The result of the research shows that use of learning cycle 7E is more effective than direct learning of students' mathematical communication ability on solid geometry material. This is related to learning cycle 7E in experiment class become more active students. Students become the center of teaching and learning process where students learn by exploring solid geometry material. In elicit step, the teacher gives a question that stimulates the student's initial knowledge. Students answer questions from teachers according to their understanding. Students answer questions from teachers according to solid geometry material in real life. In engaging step, the teacher motivates students by explaining solid geometry material in real life Students see and listen to the teacher's explanation with enthusiasm. Of the two steps, early knowledge of students began to form. In next step the students more easily in investigating the concept of solid geometry. In the learning cycle, 7E students get the initial knowledge that will help him to build further knowledge [5]. Initial knowledge of students makes it easier for students in mathematical communication. Mathematical communication is required in solid geometry. Students can describe the problem of geometry into images. Students can write the problem of geometry drawing on mathematical sentence. Students can express mathematical ideas in writing. In explore step, the teacher provides an opportunity for students to observe and analyze solid geometry samples to determine the parts and their measurements. Students observe solid geometry samples given by the teacher to know the part and size.

In explain step, the teacher assigns a task to students to convey their findings in explore phase. Some students actively convey their idea of geometry size to all students and teachers in the classroom, while some students expressed different opinions. Furthermore, teachers together with students summarize the results submitted by students about a size of solid geometry. In elaborate step, the teacher gives students the opportunity to apply parts and sizes of solid geometry material to some story questions. Students answer some questions given by teachers on solid geometry material. Some students discussed to answer the problem. In evaluating step, the teacher checks the students' understanding by assigning tasks to the students as a form of learning assessment. Each student does the task of the teacher about learning on solid geometry material. In the extending step, the teacher gives some examples of application solid geometry material to more complex issues. Students pay attention to teachers about examples of solid geometry material applications. Furthermore, students try to solve complex solid geometry material problems. Some of these steps make students have good mathematical communication ability. Students are actively discussing mathematical images and writings on solid geometry material. Polyiem states that learning cycle 7E is beneficial for students. Student learning ability can increase. Students gain new knowledge through various learning sources. Students become able to apply learning methods in real life [9]. In control, class is given direct learning to students.

In the first step, the teacher prepares students to learn and convey the purpose of learning solid geometry material. Students prepare to carry out learning and listen to explanations from teachers. In the second step, the teacher explains to the student on solid geometry material. Students see and listen to the teacher's explanation on solid geometry material. In the third step, the teacher gives an example and guides students to work on solid geometry material. Students follow example described by the teacher. In the last step, the teacher assigns the task to students as a learning assessment on solid geometry material. Students do the tasks assigned by the teacher. These learning steps make students passive. Students only record, listen and rarely ask questions when students cannot do the teacher's questions. Students have less discussion with other friends. Teachers are very dominant and students are not asked to be active in a learning process. Some students have difficulty in direct learning. Students have difficulty in changing the problem into mathematical sentences. Students have difficulty in visualizing the problem. In control class of students' mathematical communication ability are not as good as in experimental class. Based on table 4, average students' mathematical ability in experimental class is better than control class. Several other studies got results that were consistent with this study. Muyono research is the use of effective learning cycle 7E on self-regulation and problem-solving ability in mathematics learning [12]. Khaskan research is the use of learning cycle 7E is more effective than a traditional method on mathematics learning [5].

4. Conclusion

Learning cycle 7E is effective in learning students' mathematical communication abilities and on learning geometry material. Learning cycle 7E has effective learning steps. In elicit and engage steps, students' initial knowledge on the learning materials will be well established. In explore and explain steps, each student will be creative to find the concept of material based on student's initial knowledge. In elaborate and evaluate steps used by students to practice applying learning materials. In the extend step, students can learn more widely and complex. Based on the research findings, the researcher suggested that learning cycle 7E can be used in the learning of solid geometry material. Teachers can use the learning cycle 7E to improve students' mathematical communication ability. Suggestion for further research is the researcher doing learning cycle 7E on other material and use learning cycle 7E to improve other mathematical ability.

Acknowledgments

The authors thank Sebelas Maret University and junior high school 16 Surakarta due to give permission for this study.

References

[1] Lomibao L S, Luna C A and Namoco R A 2016 *American Journal of Educational Research* **4**(5) 378-382
[2] Nartani C I, Hidayat R A and Sumiyati Y 2015 *International Journal of Innovation and Research in Educational Sciences* **2**(4) 284 - 287
[3] Ayaz M F 2015 *The Turkish Online Journal of Educational Technology* **14**(4) 143-156
[4] Jones M G and Araje L B 2002 *American Communication Journal* **5**(3)
[5] Khaskan K 2016 *Journal of Education and Practice* **7**(36) 40-52
[6] Balta N & Sarac H 2016 *European Journal of Educational Research* **5**(2) 61-72
[7] Shaheen M N U K and Kayani M M Kayani 2015 *Mediteranean Journal of Social Science* **6** (4) 471-481
[8] Siribunnam R & Tayraukham S 2009 *Journal of Social Sciences* **5**(4) 279-282
[9] Polyiem T Nuangchalerm P and Wongchantra P 2011 *Australian Journal of Basic and Applied Sciences* **5**(10) 257-262
[10] Widjajanti D B 2013 *Jurnal Teknologi (Social Sciences)* **63**(2) 39-43
[11] Budiyono 2015 *Pengantar Penilaian Prestasi belajar* (Surakarta: UNS Press) p 55
[12] Mulyono and Noor N L 2017 *IOP Conf. Series: Journal of Physics: Conf. Series* **824** 1-6
[13] Qarareh A 0 2012 *International Journal of Educational Sciences* **4**(2), 123-132

IOP Conf. Series: Journal of Physics: Conf. Series **895** (2017) 012045 doi:10.1088/1742-6596/895/1/012045

Problem Solving Reasoning and Problem Based Instruction in Geometry Learning

F Sulistyowati [1*], B Budiyono [1], and I Slamet [2]

[1]Departemen Pendidikan Matematika, Program Pascasarjana Universitas Sebelas Maret, Jl. Ir. Sutami No. 36A Jebres, Surakarta 57126, Indonesia
[2]Departemen Matematika, Universitas Sebelas Maret, Jl. Ir. Sutami No. 36A Jebres, Surakarta 57126, Indonesia

*sulistyowati.fitria9@student.uns.ac.id

Abstract. This research aims to analyze the comparison Problem Solving Reasoning (PSR) and Problem Based Instruction (PBI) on problem solving and mathematical communication abilities viewed from Self-Regulated Learning (SRL). Learning was given to grade 8th junior high school students. This research uses quasi experimental method, and then with descriptive analysis. Data were analyzed using two-ways multivariate analysis of variance (MANOVA) and one-way analysis of variance (ANOVA) with different cells. The result of data analysis were learning model gives different effect, level of SRL gives the same effect, and there is no interaction between the learning model with the SRL on the problem solving and mathematical communication abilities. The *t*-test statistic was used to find out more effective learning model. Based on the test, regardless of the level of SRL, PSR is more effective than PBI for problem-solving ability. The result of descriptive analysis was PSR had the advantage in creating learning that optimizing the ability of learners in reasoning to solve a mathematical problem. Consequently, the PSR is the right learning model to be applied in the classroom to improve problem solving ability of learners.

1. Introduction

Geometry is sometimes thought of an investigation or discovery of pattern and relationship in shape, size and place (position) [1]. In other words, in geometry will be studied the relationship between point, line, plane and space to conduct the investigation. There are many activities related to geometry that are easy to find in everyday life. For example, when someone wants to go somewhere by looking at the map, he will measure and estimate the shortest distance that can be taken to save time. The example shows that geometry is a part of mathematics that is often applied in human life.

Polyhedron is one part of a geometry consisting of prisms and limas. In Indonesia, Polyhedron has been introduced since elementary school to high school level. The complex problem is in categorizing Polyhedron. During interviews with students and teachers, the researchers gave some pictures of Polyhedron. When students are given Figure 1 and 2, they answer with confidence that the shapes in Figure 1 are cubes while Figure 2 is a prism. When the researcher asked "are prisms a cube?" they answer "no". When researchers ask "is the cube a prism?" they also answer "no". There are even teachers who answer the same. This shows that in the learning of geometry there is still a mistake in understanding the concept. Therefore, the need for an alternative learning that is able to overcome the problem is Problem Based Learning (PBL). PBL is an instructional learner-centered approach that empowers learners to conduct a research, integrate theory and practice, and apply knowledge and skill to develop a viable solution to defined problem [2]. Based on the definition, it can be seen that the PBL focuses on the students to conduct independent activities and compile their knowledge of these activities.

1.1 PSR and PBI

Reasoning encompasses the cognitive procedures we use to make inferences from knowledge and draw conclusions [3]. Reasoning is a part of thinking that is above the level of retention or recall, reasoning includes: basic thinking, critical thinking, and creative thinking [4]. The problem-solving learning model is a learning model that focuses on teaching and problem-solving skills followed by skill enhancement [5]. Problem solving is one of the problem-based teaching strategies where teachers help learners to learn to solve problems through hands-on learning experiences [6]. Hands-on learning is a learning that requires active learners to solve problems.

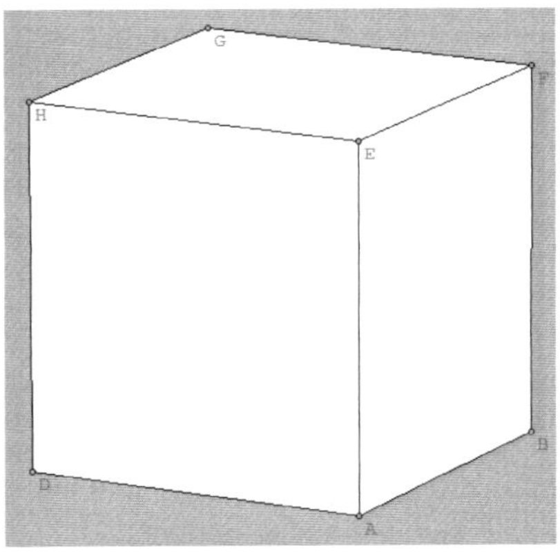

Figure 1. Cube

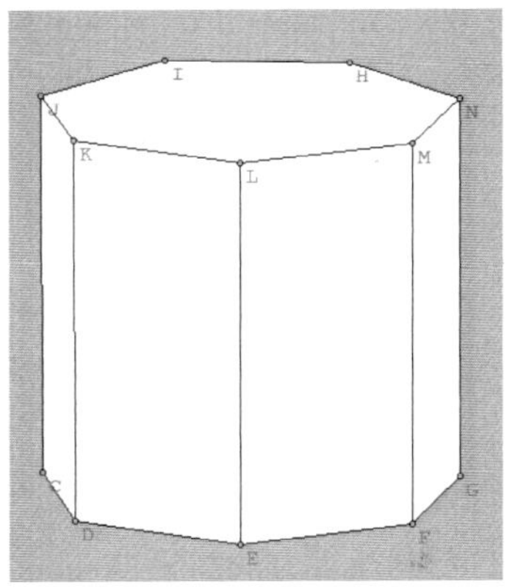

Figure 2. Prisms

Based on the description above, PSR is a problem solving learning that is learning that focuses on teaching and problem-solving skills followed by strengthening of skills through hands-on learning experiences equipped with steps to support reasoning ability include: (1) basic thinking; (2) critical thinking; and (3) creative thinking.

The problem-based teaching (PBI) is a learning approach whereby learners work on authentic issues with the intent to develop their own knowledge, develop inquiry and higher-order thinking, develop self-reliance and self-confidence. The goals in PBI learning are: (1) helping learners develop thinking skills and problem-solving skills; (2) learning the role of an authentic adult; (3) being an independent learner [7]. Characteristics of the PBI's learning model are: (1) asking questions or problems; (2) focusing on interdisciplinary linkages; (3) authentic investigation; (4) produce the product / work and show it off; (5) cooperation [8]. The benefits of PBI are: (1) realistic with real life; (2) concepts according to the needs of learners; (3) fostering the nature of the inquiry of learners; (4) strong concept retention; (5) cultivate problem-solving abilities. The weaknesses of the PBI are: (1) complex learning preparation (tools, problems, concepts); (2) difficulty in finding relevant problems; (3) miss-conception often occurs; (4) time consumption, where this model takes a long time in the process of investigation, so much time is consumed for the process [7].

2. Methods

This research used a quasi-experimental methods, and then with descriptive analysis. The data were analyzed using two-way MANOVA and one-way ANOVA with different cells to determine the effect of two independent variables on two dependent variables. The independent variables in this research are learning model and Self-Regulated Learning (SRL), while the dependent variables are problem solving ability and mathematical communication ability on Polyhedron discussion.

Learning model applied to grade 8th of junior high school students year 2016/2017 in Magelang City, Central Java, Indonesia. Sampling technique in this research is Cluster Random Sampling and got experimental group 1 given PSR model, and experiment group 2 given PBI model. Data collection in this research using test method and questionnaire. The test method is used to get the data of problem solving and mathematical communication abilities of the students done twice. The questionnaire method is used to get student's SRL level data. Descriptive analysis was conducted to determine and describe the advantages of more effective learning models are applied in Polyhedron learning.

3. Result and Discussion

The prerequisite test results that all samples were from normally distributed populations. The results of the equilibrium test show that the two experimental groups have the same problem-solving and mathematical communication abilities. Then, data were analyzed using two-way MANOVA with different cells.

Table 1 is a overview of MANOVA that indicating that H_{0A} is rejected because $F_{obs} > F_{tab}$, meaning that there is a difference in the effect of the learning model on problem-solving and mathematical communication abilities. Therefore, it is necessary to do further test using ANOVA to know the effect difference of learning model on each dependent variable. In the same table shows that H_{0B} and H_{0AB} are not rejected because $F_{obs} < F_{tab}$, meaning there is no difference effect in level of SRL, and there is no interaction between learning model with SRL on problem-solving and mathematical communication abilities.

There are several reasons why there is no difference effects in level of SRL on problem-solving and mathematical communication abilities, one of which is the characteristics of students in the samples classified as an independent learner. The statement was according with the opinion of the guidance and counseling teachers who explained that most of the students in the school where the research always take the initiative in learning, both in daily and private lessons outside school hours. In other words, most students of the samples have a high level of learning independence.

The main factor causing no interaction between learning model and SRL is the less of good cooperation between researchers with students. This is because the study only takes no more than two months. The absence of long-term attachment makes the researcher difficulty in understanding the characteristics of each learner which will certainly affect the selection of appropriate attitude toward the learners.

Table 1. Overview of MANOVA

	F_{obs}	F_{tab}	Decision
Factor A (learning model)	7,198	3,070	H_{0A} rejected
Factor B (SRL)	1,334	2,450	H_{0B} not rejected
AB (interaction of learning model with SRL)	2,225	2,450	H_{0AB} not rejected

Table 2 is overview of ANOVA shows that there is a difference in the effect of learning model on problem solving ability and there is no difference effect of learning model on mathematical communication ability. Because the learning model gives different effects on problem solving ability, it is necessary to do further test to find out more effective learning model using *t*-test statistic. The test gives results that regardless of the level of the SRL, the PSR learning model is more effective than the PBI learning model on problem-solving ability.

Table 2. Overview of ANOVA

Dependent variable	Source	F_{obs}	F_{tab}	Decision
Problem solving ability	Factor A (learning model)	8,950	4,080	H_{0A} rejected
Mathematical communication ability	Factor A (learning model)	1,872	4,080	H_{0A} not rejected

3.1. The reasons why PSR is more effective than PBI

From the previous data analysis, it was found that the PSR learning model is more effective than PBI in problem solving ability. At this section, we will explain the advantages of PSR and PBI learning models by analyzing the results of the Student Activity Sheet (LAS) when the learning model is applied.

Photograph 1 is an answer of student activity on learning geometry using PSR, while photograph 2 is the answer of student activity on learning geometry using PBI. Photograph is one step in finding the formula to determine the volume of the cube. In photograph 1 it can be seen that the student's ability to describe the reason for finding the formula is better than photograph 2. This shows that PSR is able to optimize the students to hone their brain ability to solve a given question or problem.

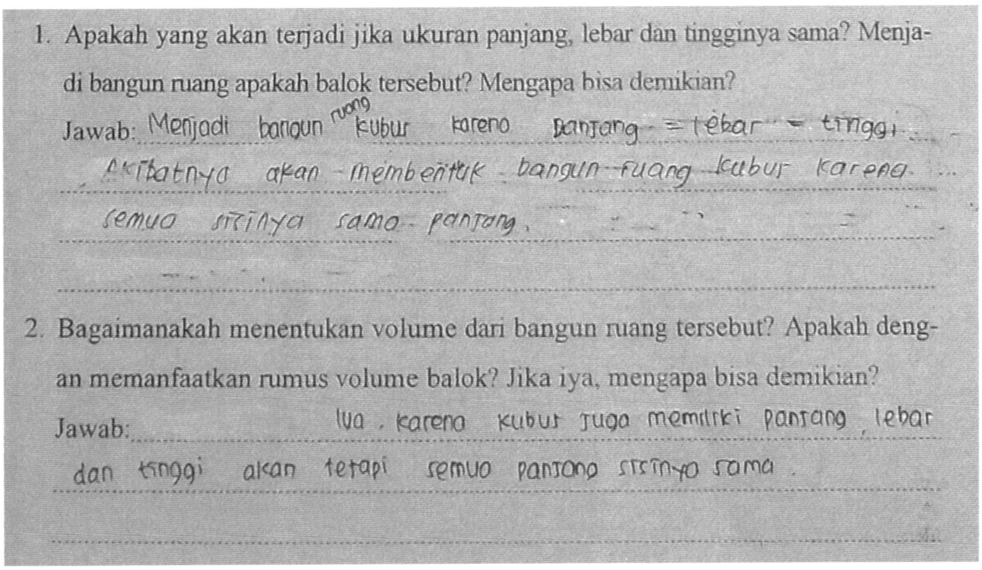

1. Apakah yang akan terjadi jika ukuran panjang, lebar dan tingginya sama? Menjadi bangun ruang apakah balok tersebut? Mengapa bisa demikian?
Jawab: Menjadi bangun ruang kubus karena panjang = lebar = tinggi. Akibatnya akan membentuk bangun ruang kubus karena semua sisinya sama panjang.

2. Bagaimanakah menentukan volume dari bangun ruang tersebut? Apakah dengan memanfaatkan rumus volume balok? Jika iya, mengapa bisa demikian?
Jawab: Iya, karena kubus juga memiliki panjang, lebar dan tinggi akan tetapi semua panjang sisinya sama.

Figure 3. Photograph 1

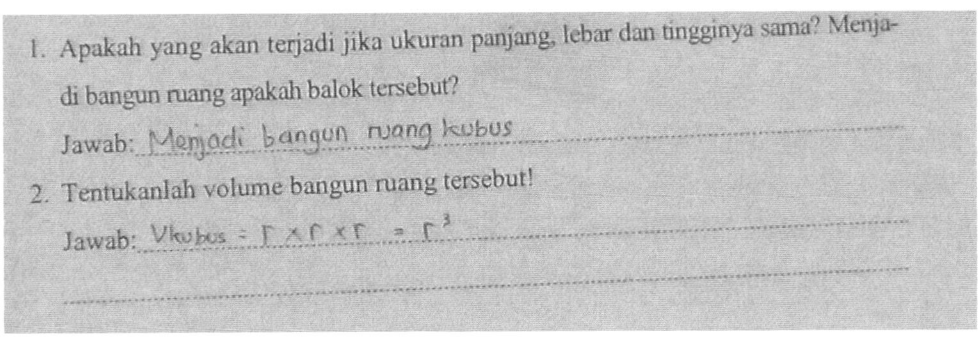

1. Apakah yang akan terjadi jika ukuran panjang, lebar dan tingginya sama? Menjadi bangun ruang apakah balok tersebut?
Jawab: Menjadi bangun ruang kubus

2. Tentukanlah volume bangun ruang tersebut!
Jawab: $V_{kubus} = r \times r \times r = r^3$

Figure 4. Photograph 2

4. Conclusion

Based on the result of the research, can be concluded that PSR is the right learning model to be applied in the classroom to improve problem-solving ability of learners. Because, PSR better optimize students in reasoning than PBI, so PSR is very influential for increasing problem solving ability.

Acknowledgments
The successful implementation of this research can not be separated from the help of various parties. Therefore, on this occasion the authors would like to thank. (1) parents and families who always provide support; (2) mentors who always provide criticism and suggestions; (3) comrades who always give advice; (4) ICMScE which provides an opportunity for publication, Hopefully the good deeds of all those parties get a reply from God Almighty.

References
[1] Tom Cooper, 1986 *Geomet: Space and Shape in The Primary School* (Brisbane: Brisbane C.A.E)
[2] Cindy E H, and Peggy A E 2015 *Essential Reading in Problem-Based Learning* (West Lafayette: Purdue University Press)
[3] Dunbar K, and Fugelsang J 2006 *Problem Solving and Reasoning Appear in E.E. Smith & S. Kosslyn an introduction to cognitive psychology* http://www.utsc.utoronto.ca/
[4] Sanjaya, W 2006 *Strategi Pembelajaran Berorientasi Standar Proses Pendidikan* (Jakarta: Kencana)
[5] Shoimin A 2014 *68 Model Pembelajaran Inovatif dalam Kurikulum 2013* (Yogyakarta: Ar-ruzz Media)
[6] Jacobsen D A, Eggen P, and Kauchak D 2009 *Methods for Teaching* (Terj A Fawaid & C. Anam. Yogyakarta: Pustaka Pelajar)
[7] Trianto 2014 *Mendesain Model Pembelajaran Inovatif, Progresif, dan Kontekstual: Konsep, Landasan, dan Implementasinya pada Kurikulum 2013 (Kurikulum Tematik Integratif/TKI)* (Jakarta: Prenadamedia group)
[8] Ibrahim M, and Nur M 2004 *Pengajaran Berdasarkan Masalah* (Surabaya: Surabaya University Press)

Problem Posing with Realistic Mathematics Education Approach in Geometry Learning

R Mahendra*, I Slamet and Budiyono

Magister Pendidikan Matematika, Universitas Sebelas Maret, Jl. Ir. Sutami No. 36A, Surakarta 57126, Indonesia

*renggamahendra7@gmail.com

Abstract. One of the difficulties of students in the learning of geometry is on the subject of plane that requires students to understand the abstract matter. The aim of this research is to determine the effect of Problem Posing learning model with Realistic Mathematics Education Approach in geometry learning. This quasi experimental research was conducted in one of the junior high schools in Karanganyar, Indonesia. The sample was taken using stratified cluster random sampling technique. The results of this research indicate that the model of Problem Posing learning with Realistic Mathematics Education Approach can improve students' conceptual understanding significantly in geometry learning especially on plane topics. It is because students on the application of Problem Posing with Realistic Mathematics Education Approach are become to be active in constructing their knowledge, proposing, and problem solving in realistic, so it easier for students to understand concepts and solve the problems. Therefore, the model of Problem Posing learning with Realistic Mathematics Education Approach is appropriately applied in mathematics learning especially on geometry material. Furthermore, the impact can improve student achievement.

1. Introduction

Geometry is one of the branches of mathematics that students must study at the junior high school level. In addition, geometry is also a branches of mathematics that touches most aspects of human life [1]. Studying geometry is very important for students because students' geometry learning will be faster in developing the ability of logical thinking and problem solving [2, 3]. Achieved or not the purpose of learning geometry one of them can be seen from the success of students in applying concepts and formulas geometry in solving the problem. As an effort to realize the purpose of learning geometry is required a good conceptual understanding in studying geometry. This is because conceptual understanding is the basic capital for students in learning the next mathematical concept or material [4]. However, in fact during this time students still have difficulty in learning geometry [5]. One of the difficulties of students in the learning of geometry is on the subject of plane that students must to understand abstract matters. The material discussed in the plane includes the properties, circumference, breadth, and application of the plane. Furthermore, the plane material in this study will be focused on the properties, circumference, area of triangle and quadrilateral at the junior high school level. The difficult of students in understanding the material plane may be influenced by many factors, one of which lessons is the way of teacher delivery the material in the class isn't right. During this time the learning process still tends to focus on the teacher or in other words there are still many teachers who apply direct learning [6, 7]. This causes the students tend to be passive and less motivated in the learning process. Therefore, it is necessary to make changes in the learning of geometry, especially in the model of learning applied by the teacher to improve students' conceptual understanding, especially on the plane material.

The learning model need an innovative learning model that can make students active in learning and train students to think creatively in proposing and problem solving [6, 8, 9]. As an effort to realize changes in the learning of geometry, the possibility of learning model that can be used is Problem Posing. Problem Posing in principle is an innovative learning model that make students to be able to reformulate the problems that have been given with the aim of improving the understanding and facilitate students in solving the problem [10, 11]. Students are invited to be active in learning so that information is not only obtained from the teacher, but the students are also required to construct their own new knowledge with capitalize knowledge that has been previously owned. However, in fact the application of Problem Posing still has many detriments among students still have difficulty to understand the problem because the problems given are sometimes difficult to imagine by the students. Based on that, the measure that can be done is to modify Problem Posing with Realistic Mathematics Education Approach. The Approach of Realistic Mathematics Education is one of the mathematical approaches that always connect mathematics materials to the world of students so that students feel more motivated and easier to imagine the material or problems given [12, 13].

Based on the explanation, the application of Problem Posing with Realistic Mathematics Education Approach that invites students to actively propose and solve realistic problems ascending to be studied. The purpose of this study is to determine the effect of Problem Posing learning model with Realistic Mathematics Education Approach in geometry learning. Furthermore, the impact can improve student achievement.

2. Experimental Method

This research is a quasi experimental research. In this research, the independent variable is the learning model which consists of Problem Posing learning model with Realistic Mathematics Education Approach as the experimental class and the direct learning model as the control class. The dependent variable in this research is the students' conceptual understanding on the plane material. The population of this research is all seventh graders of State Junior High School 1 Gondangrejo, Karanganyar, Indonesia academic year 2016/2017. The sample was taken by stratified cluster random sampling technique and consist of 63 students. The first class ($n = 31$) as the experimental class and the other class ($n = 32$) as the control class. This research took place from November 2016 until June 2017 which is divided into 3 stages. The first stage is preparation of research which took place in November 2016 until February 2017. The second stage is the implementation of research which took place in March 2017 until April 2017. The third stage is data analysis and reporting that took place in May 2017 until June 2017. The instrument for collecting data in this study is a conceptual comprehension test on a plane material. Implementation of the test in this study was conducted twice that is pretest and posttest. Pretest were performed to obtain conceptual understanding of the student data prior to treatment. This data is then used as a basis for normality and homogeneity testing. Normality test in this research using Lilliefors method with 5% significance level and homogeneity test using Bartlett test with 5% significance level. Posttest were performed to obtain conceptual understanding data after being subjected to treatment. Furthermore, pretest and posttest results were used as the basis for hypotheses tested by using t-test with significance level of 5%.

3. Result and Discussion

Based on normality test the result of using Lilliefors method toward student pretest data on plane material showed that the sample of research came from normally distributed population. Then, based on the homogeneity test the results of using Bartlett test of the students' pretest data on the plane material showed that the population variance is homogeneous. Furthermore, the research hypothesis test is processed using t test on the pre-test and posttest data of students on the plane material can be seen in Table 1 below.

Table 1. Pretest data and posttest data

Groups	N	Mean		t_{obs}		t_{tabel}
		Pretest	Posttest	Pretest	Posttest	
Experimental	31	75,96	79,19	-3,5568	-4,8168	-1,99962 or 1,99962
Control	32	62,5	64,06			

Table 1 shows the average pretest of the experimental class is 75,96 and the average of pretest of control is 62,5. In addition, Table 1 also shows the average posttest of the experimental class 79,19 and the average posttest of the control class is 64,06. Furthermore, Table 1 also shows that $t_{obs\ pretest}$ = (-3,5568), $t_{obs\ posttest}$ = (-4,8168), and t_{table} = (-1,99962 or 1,99962) or in other words $DK = \{t \mid t < -1.99962 \text{ or } t > 1.99962\}$. This means $t_{obs\ pretest} \in DK$ so H_0 rejected, it means that there is a difference in conceptual understanding between experimental class students and control class prior to treatment, especially on the subject of a plane. It is same with the result of the post-test shows that $t_{obs\ posttest} \in DK$ so H_0 rejected, it means that there is a difference in conceptual understanding between the experimental class and control class after being subjected to treatment, especially on the subject of a plane.

Based on the results of the study, it can be see that the experimental class students after being subjected to the treatment experienced increased conceptual understanding is more significant than the control class students. Students of experimental class are students who are subject to Problem Posing learning model with modification of Realistic Mathematics Education Approach. This modification between the model and the learning approach has a positive impact in improving students' conceptual understanding especially on the plane matter. There are some factors increases of conceptual understanding of students who are subject to Problem Posing learning model with Realistic Mathematics Education Approach are as follows. (1) Students tend to be more motivated and challenged in the learning process. It is cause, in the material lesson is not delivered directly but the material is delivered by giving problems to the students, so that from the problem students can construct their knowledge and find the concept to be taught (characteristic Approach of Realistic Mathematics Education: utilization of student construction result). (2) Students tend to be easier in understanding concepts and learning becomes more meaningful. It is cause students in learning are always given realistic problems (characteristic Approach of Realistic Mathematics Education: the use of contextual problems) so as to make students become easier to imagine the problems. (3) Students tend to be easier in problem solving. It is cause students tend to be easier in performing mathematical modeling (characteristic of Realistic Mathematics Education Approach: using model for progressive mathematization) so as to facilitate students in problem solving. (4) Students become more active in learning. It is cause students in learning will be divided into several small groups that can be used for discussions related to the given problem, so that learning becomes more alive (characteristic Approach of Realistic Mathematics Education: interactivity).

Different with students in control classes who are subject to direct learning models. Students in the control class tend to be passive in their learning because learning is only focused on the teacher. In addition, students in the control class also tend to be not challenged in their learning because the learning materials are explained directly by the teacher. Furthermore, students in the control class tend to be less motivated because learning only focus one way so that students only writing and listening, they aren't construction their own new knowledge by capitalizing on the previous knowledge.

Based on the description, it can be concluded that the students who are subjected to the Problem Posing learning model with Realistic Mathematics Education Approach have conceptual understanding better than the students who are directly model on the subject of plane. The results of this study are appropriate with the research conducted [14] can concluded, Problem Posing learning has a positive effect to improving learning achievement, problem solving skills, problem asking, and demeanor toward mathematics, and [15] which concluded, students who were taught Problem Posing learning has better learning achievement than students who are taught by direct learning.

4. Conclusion

The conclusion of this research is Problem Posing with Realistic Mathematics Education Approach can improve students' conceptual understanding significantly in geometry learning especially on plane topics. It is because students on the application of Problem Posing with Realistic Mathematics Education Approach are become to be active in constructing their knowledge, proposing, and problem solving in realistic, so it easier for students to understand concepts and solve the problems. Therefore, the model of Problem Posing learning with Realistic Mathematics Education Approach is appropriately applied in mathematics learning especially on geometry material. Furthermore, the impact can improve student achievement.

References

[1] Schmitt J M 2006 Developing Geometric Reasoning Washington DC: GED Mathematics Training Institute
[2] Clements D H and Battista J M 1992 Geometry and Spatial Reasoning Hand Book of Research on Mathematics Teaching and Learning, New York: Macmillan Publishing Company
[3] Van de Walle and J A 1994 Elementary School Mathematics New York: LONGMAN
[4] Kilpatrik J, Swafford J, and Findell B 2001 Adding It Up: Helping Children Learn Mathematics Washington: National Academy Press
[5] Adolphus T 2011 Problem of Teaching and Learning of Geometry in Secondary Schools in Rivers State, Nigeria *International Journal of Emerging Sciences* **1**(2) p 143-152
[6] Effendi 2012 Pembelajaran Matematika dengan Metode Penemuan Terbimbing untuk Meningkatkan Kemampuan Representasi dan Pemecahan Masalah Matematis Siswa SMP *Jurnal Penelitian Pendidikan* **13** p 1-10
[7] Komalasari K 2012 The Effect of Contextual Learning in Civic Education on Students' Civic Skills *International Journal for Educational Studies* **4**(2) p 179-190
[8] Godfrey 2013 Effect of Active Learning Teaching Methodology on Learner Participation *Journal of Education and Practice* **4**(4) p 157-168
[9] Ahmad S and Zanzali N 2006 Problem Posing Abilities in Mathematics of Malaysian Primary year 5 Children: An Exploratory Study *Jurnal Pendidikan Universitas Teknologi Malaysia* p 1-9
[10] Kar T 2016 Prospective Middle School Mathematics Teachers' Knowledge of Linier Graphs in Context of Problem Posing *International electronic Journal of Elementary Education* **8**(4) p 643-658
[11] Arikan E E and Unal H 2015 Investigation of Problem–Solving and Problem-Posing Abilities of Seventh Grade-Students *Educational Sciences: Theory & Practice* **15**(5) p 1403-1416
[12] Köğce D, Yıldız C, Aydın M, and Altındağ R 2009 Examining Elementary School Students' Attitudes towards Mathematics in Terms of Some Variables *Procedia Social and Behavioral Sciences* **1**(1) p 291-295
[13] Gravemeijer K P E 1994 Developing Realistic Mathematics Education Utrecth: Freudenthal Institute
[14] Rosli R, Capraro M M, and Capraro R M 2014 The Effects of Problem Posing on Student Mathematical Learning: A Meta-Analysis *International Education Studies* **7**(13) p 227-241
[15] Guvercin S and Verbovskiy V 2014 The Effect of Problem Posing Task Used in Mathematics Instructions to Mathematics Academic Achivement and Attitudes Toward Mathematic *International Online Journal of Primary Education* **3**(2) p 59-65

Developing Statistical Physics Course Handout on Distribution Function Materials Based on Science, Technology, Engineering, and Mathematics

M A Riandry, I Ismet*, and H Akhsan

Department of Physics Education, Universitas Sriwijaya

*ismet@fkip.unsri.ac.id

Abstract. This study aims to produce a valid and practical statistical physics course handout on distribution function materials based on STEM. Rowntree development model is used to produce this handout. The model consists of three stages: planning, development and evaluation stages. In this study, the evaluation stage used Tessmer formative evaluation. It consists of 5 stages: self-evaluation, expert review, one-to-one evaluation, small group evaluation and field test stages. However, the handout is limited to be tested on validity and practicality aspects, so the field test stage is not implemented. The data collection technique used walkthroughs and questionnaires. Subjects of this study are students of 6th and 8th semester of academic year 2016/2017 Physics Education Study Program of Sriwijaya University. The average result of expert review is 87.31% (very valid category). One-to-one evaluation obtained the average result is 89.42%. The result of small group evaluation is 85.92%. From one-to-one and small group evaluation stages, averagestudent response to this handout is 87,67% (very practical category). Based on the results of the study, it can be concluded that the handout is valid and practical.

1. Introduction

According to Physics Education of Sriwijaya University curriculum, statistical physics course has three semester credit units and is taught in semester V [1]. Statistical physics is part of science which describes micro states collection of system of gas, atoms or subatomic particles [2]. It can be viewed as statistics with physical boundary condition, so it needs high abstraction to be understood and many student do not have the abstraction [2]. From informal interview with students of physics education of 6th semester of academic year 2016/2017 at Sriwijaya University who have taken statistical physics course, they had difficulities to understand statistical physics course, especially on distribution function materials. Based on final examination of statistical physics course academic year 2015/2016 which tested out the understanding of distribution functions materials, the achievement of students of physics education were still low with mean score that is 65.07. These difficulities because of statistical physics books are too difficult to be comprehended and not comprehensive enough on the science and technology application materials. If the materials are uninteresting and complicated, then teaching and learning becomes a dull and monotonous activity [3].

STEM is an integrated learning approach of science, technology, engineering, and mathematics. Study shows that this integrated approach can increase learning interest in science and technology lessons. Many benefits have been associated with STEM education, such as providing opportunities for more student-centered, meaningful, engaging, and less fragmented learning experiences involving higher-level thinking and problem-solving skills [4]. Experience in STEM learning can prepare students for the global economy of the 21st century [5]. Students need a solid STEM knowledge to become ready for employment [6]. STEM country of origin, United States, the application of STEM in the field of education is being held intensively. As the President's Council of Advisors on Scientific and Technology (PCAST) recommended to form the "STEM Master Teachers Corps" that assess and reward 5% of the best STEM teachers in America [7]. PCAST is also planning in the next decade to recruit 100.000 teachers to teach by using the STEM approach [7]. Distribution function materials require deep mathematical and physical skill to be understood and it have many applications in sciences and technologies, such as solar cell and superconductor material. Thus, STEM approach is very suitable to be implemented in distribution function materials.

A lack of commercial materials forces teachers to fall back on their own resources and designing their own teaching materials can enable them to make best use of the resources available in their teaching context [8]. Developing teaching materials by educators can be a solution to the lack of teaching materials. Studies that provide teaching materials have been done, one of them is done by Kurniasari et al [9]. Based on this, it is important to produce teaching materials on distribution function materials which contain also explanation about physics and technology application materials, so the students can get meaningful learning and the objectives of learning can be achieved. Similar with that study, researcher tried to develop teaching materials in the form of handout on distribution function materials based on STEM for Physics Education of Sriwijaya University.

2. Research Method

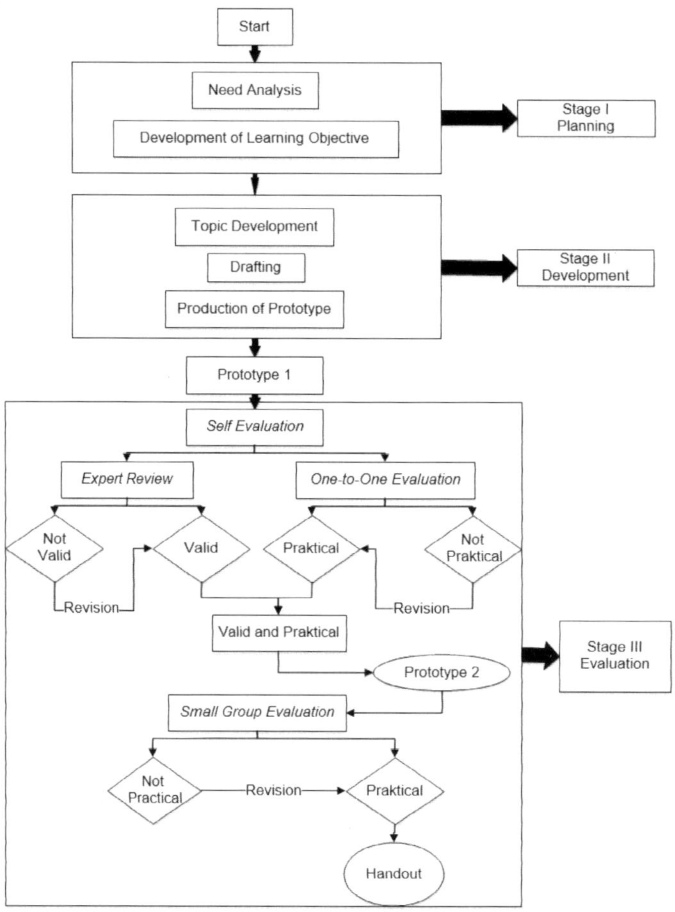

Figure 1. Study Design Flow

This study is a development research by using the model of Rowntree development. Rowntree development model consists of three stages. They are planning, development, and evaluation stages [10]. Materials that are developed are statistical physics materials with their applications materials. The materials are based on STEM and they are tested on validity and practicality aspects. At the evaluation stage of the study uses Tessmer formative evaluation. The evaluation stages are self-evaluation, expert review, one-to-one evaluation and small group evaluation [11]. Self-evaluation and expert review stages test the validity aspect. One-to-one and small group evaluation stages test the practical aspect. The flow of this study can be seen in Figure 1.

For the data availability in this study, researcher used data collection techniques. They are validations and questionnaires sheets. Subjects of this study are physics eduaction students of 6th and 8th who has taken statistical physics course. There are three criterias which are tested in this study, they are content, design and language criterias by each expert. Each criteria has several indicators. Experts answer the indicators in the form Likert Scale which is shown in the Table 1.

Table 1. Validation category value [12].

Answer category	Score of question
Strongly agree	5
Agree	4
Uncertain	3
Disagree	2
Strongly disagree	1

The indicators are also made in percentage. Furthermore, an average score percentage from expert validation results is obtained by using equation (1) and (2) [13].
1. Determine Criteria of Total Percentage (CTP).

$$CTP(\%) = \frac{Total\ score\ of\ each\ indicator}{Maximum\ total\ score\ of\ each\ indocator}\ x\ 100\% \quad (1)$$

2. Percentage of Final Score from Expert Validations (FSEV).

$$FSEV(\%) = \frac{Total\ score\ of\ each\ criteria}{Maximum\ total\ score\ of\ each\ criteria}\ x\ 100\% \quad (2)$$

Final Score from Expert Validations (FSEV) is also grouped according to Table 2. Suggestions comments are also written on sheets of expert validation and then they were used as reference to n revisions of the handout.

Table 2. Final Score from Expert Validations (FSEV) category [13].

Percentage of Validations (%)	Category
$86 \leq FSEV \leq 100$	Very valid
$70 \leq FSEV < 86$	Valid
$56 \leq FSEV < 70$	Less valid
$0 < FSEV < 56$	Invalid

In One-to-one and small group evaluation stages, students have a chance to fill the questionnaire and give comments and suggestion about this handout. The data obtained through the questionnaire was obtained with Likert Scale which is shown in the Table 3 to get the opinion of students on the handout. Each indicator is made in percentage. Furthermore, an average score percentage from student response is obtained by using equation (3) and (4) [13].

1. Determine Score Percentage of Student Evaluation (SPSE).

$$SPSE(\%) = \frac{Total\ score\ of\ each\ indicator}{Maximum\ total\ score\ of\ each\ indicator}\ x\ 100\% \quad (3)$$

2. Percentage of Final Score from Student Responses (FSSR).

$$FSSR(\%) = \frac{Total\ score\ of\ student\ evaluations}{Maximum\ total\ score\ of\ student\ evaluations}\ x\ 100\% \quad (4)$$

Table 3. Student response category value [12].

Answer category	Score of question
Strongly agree	5
Agree	4
Uncertain	3
Disagree	2
Strongly disagree	1

The Result of Final Score of Students Response (FSSR) form one-to-one and small group evaluation stages was made in percentages then grouped according to Table 4. Comments and suggestions from students were filled in the questionnaires, they are purposed to revise the handout to become better.

Table 4. Final Score of Students Response (FSSR) category [13].

Percentage of Questionnaire (%)	Category
$86 \leq FSSR \leq 100$	Very practical
$70 \leq FSSR < 86$	Practical
$56 \leq FSSR < 70$	Less practical
$0 < FSSR < 56$	Not practical

3. Result and Discussion

3.1 Planning Stage

At the planning stage, the result of need analysis is students who have taken statistical physics course have difficulities to understand the physics, mathematics and application concepts which are taught through lecture or discussion. From the students' opinion, the most difficult materials in statistical physics course are distribution functions materials, because they are more complicated to be learned than the others materials. STEM (science, technology, engineering and mathematics) integrated approach is very much suitable for distribution function material characteristics which require deep understanding of science and mathematics to be studied and has many applications in the fields of science and technology. Handout was selected to be developed because it is flexible in form, content and using. Furthermore, at this stage, the syllabus analysis is done by developing indicators of the Standards Competencies (SC) and Basic Competencies (BC), as well as the formulation of learning objectives.

3.2 Development Stage

The result of development stage of this study consists of topic development, drafting and prototype production stages. At the topic development stage, researcher prepared Materials Enrichment (ME) and Content Outline Handout (COH) on the course of statistical physics in distribution function materials. They are used as a reference to drafting handout. ME contains distribution function and application materials which contain STEM (mathematics, science, engineering and technology), they are: Maxwell-Boltzmann distribution function (science and mathematics elements), Maxwell-Boltzmann distribution function on speed variable (science and mathematics elements), entrophy (mathematics and science elements), Doppler effect (mathematics and science elements), laser (mathematics, science, engineering and technology elements), Bose-Einstein distribution functions (mathematics and science elements), black body radiation (mathematics and science elements), spesific heat capacity (mathematics and science elements), Fermi-Dirac distribution function (mathematics and science elements), electron gas (mathematics and science elements) and metal heat capacity materials (mathematics and science elements). Not all of the materials consist STEM elements such as engineering and technology. Many materials of statistical physics are too statistical matematics. However, not all of STEM elements are needed in the certain materials, but STEM elements can be found in this handout. Beside of that in the class, approaching all STEM elements through engineering design is not always possible [14].

ME focuses on six basics competencies that must be owned by students. While COH contains indicators, learning objectives, materials, assessment and reference. The drafting stage developed handout cover, study guides, table of contents, introduction, main map, content standards, material descriptions, sample questions, exercise questions, exercise key and reference. The handout is written by using 12 pt Times New Roman and it is written coherently based on curriculum of statistical physics with STEM elements combination. The result of production stage is draft which is made in the form of printed material (handout). The result of this development stage is called prototype 1, it contains three chapters about Maxwell-Boltzmann, Bose-Einstein and Fermi-Dirac distribution function and their application materials in each chapter.

3.3 Evaluation Stage

Table 5. Expert review stage result.

No	Indicator	Percentage of result (%)
1	Use science element	90,00
	Use technology element	90,00
	Use engineering element	85,00
	Use mathematics element	90,00
	Compliance with basic competencies and learning objectives	90,00
	Compliance with student needs	90,00
	Compliance with handout needs	95,00
	Truth of content of learning materials	90,00
	Benefits to add knowledge	100,00
	Average percentage of content criteria (CTP)	91,11
2	Order of materials	80,00
	Completeness of information	100,00
	Use font: type and size	90,00
	Lay *out*	90,00
	Illustrations, pictures and photos	80,00
	Design view	80,00
	Average percentage of design aspect (CTP)	86,67
3	Legibility	86,67
	Clarity of information	80,00
	Conformity with Indonesian language rules	80,00
	Use language effectively and efficiently	90,00
	Average percentage of language criteria (CTP)	84,16
	Average percentage of total validation (FSEV)	87,31

The prototype 1 that has been developed is then evaluated based on Tessmer formative evaluation. At the self-evaluation stage, the researcher conducted self-checking on the prototype 1. The results are finding typing error in handout, handout cover was still less interesting and images on the sub material were still less clear. After the handout was revised, expert review stage was conducted. At this stage, prototype 1 was validated by experts consisting of content, design and language aspect. The result of the it can be seen at Table 5. From this stage, the highest score indicator is given benefits to add knowledge that is 100%, because it has many application materials in this handout can extent knowledge about distribution function materials of students. There are four indicators which are given lowest score 80%, they are order of materials, illustrations, pictures and photos, design view and clarity of information indicators, because index of handout, some illustrations, design and material informations were needed to be repaired to become clearlier. The results of expert review stage is 87,31%, so it can be concluded that this handout is very valid based on Table 2. Experts also gave suggestions and comments to make this handout became better. Superconductor (science, engineering and technology elements) and spectrograph materials (science, engineering and technology elements) have been added and some illustrations are repaired to read clearly.

Table 6. Result of one-to-one and small group evaluation stage.

No	Indicator	Percentage (%) of one-to-one evaluation	Percentage (%) of small group evaluation
1	Benefits to add knowledge	93,33	95,56
2	Clarity of information	84,44	85,92
3	Use STEM elements	93,24	91,67
4	Giving motivation	89,83	86,67
5	Use language effectively and efficiently (clear and concise)	89,83	83,33
6	Clear clue of handout usage	100,00	86,67
7	Use of font: type and size	96,66	92,22
8	Lay out	89,83	84,94
9	Illustrations and pictures	80,00	73,33
10	Design view	79,83	74,44
11	Exercise using	86,33	82,22
	Average percentage of score (SPSE)	89,42	85,92
	Average total percentage of student responses (FSSR)	87,67	

The one-to-one evaluation stage was implemented to test the practicality of handouts (prototype 1). This stage is tested on three students of 8th semester in Physics Education Study Program of Sriwijaya University who have taken statistical physics course with different abilities. Those are low, medium and high ablility. Students also gave comments and suggestions about this handout. The result of this stage can be seen on Table 6. Students appreciated this handout because it gave clear clue to use the handout, this indicator got the highest score that is 100%. The lowest indicator score is gotten in design view that is 79,83%, because some fonts needed to be repair and too few pictures in the cover of handout. After prototype 1 was revised based on expert review and one-to-one evaluation stages, prototype 2 was generated and small group evaluation was started.

Furthermore, prototype 2 was tested on students at small group evaluation stage which aimed to test the practicality of handout (prototype 2). In this stage, handout was tested on nine students of 6th semester in Physics Education Study Program of Sriwijaya University who have taken the course of statistical physics. They were divided into three groups based on different levels of ability, those are low, medium and high abilities. The result of this stage can be seen on Table 6. From this stage, the highest indicator score is given benefits to add knowledge that is 95,56%, because many applications in this handout could extent knowledge about distribution function materials. Some illustrations and drawings still needed to be repaired to read clearly, so this indicator got 73,33%. Students also gave comments and suggestions to this handout. Glosarium has been added to help students understand difficult terms in the handout. After prototype 2 was revised based on small group evaluation, prototype 3 was generated and it is the final product of this study. Based on one-to-one and small group evaluation stage, student response of this handout is 87,67% which is based on Table 4, this handout is very practical.

IOP Conf. Series: Journal of Physics: Conf. Series **895** (2017) 012047 doi:10.1088/1742-6596/895/1/012047

4. Conclusion

Based on the results of development research using Rowntree development model and Tessmer formative evaluation that has been done, it can be concluded that it has successfully been developed a valid and practical handout of statistical physics on distribution function materials based on STEM. Because of that, the handout can be used in statistical physics lesson at Physics Education Study Program of Sriwijaya University and it can be implemented in field test stage to test the effectiveness of this handout.

References

[1] FKIP 2016 *Buku Pedoman FKIP Universitas Sriwijaya* Inderalaya Universitas Sriwijaya p 63

[2] Abdullah M 2007 *Pengantar Fisika Statistik untuk Mahasiswa (Dilengkapi Contoh Soal)* Bandung Penerbit ITB p 1

[3] Dar F 2012 Textbook Materials and Their Successful Application In The Classroom: Implications for Language Development *Journal of Educational and Instructional Studies In The World* **2 (4)** 109-114

[4] Stohlmann M, Moore T J, and Roehrig G H 2012 Considerations for teaching integrated STEM education *Journal of Pre-College Engineering Education Research* **2(1)** 28–34

[5] Cachaper C, Spielman L J, Soendergaard B D, Dietrich C B, Rosenzweig M, Tabor L, Fortune J C and Edmister W 2008 *Universities as Catalysts for Community Building among Informal STEM educators: The Story of POISED* Paper Presented at the American Educational Research Association Conference

[6] Becker K and Park K 2011 Effect of integrative approaches among science, technology, engineering, and mathematics (STEM) subjects on students' learning: A preliminary meta analysis *Journal of STEM Education* **12** 23-37

[7] President's Council of Advisors on Science and Technology 2010 *Prepare and Inspire: K-12 Education in Science, Technology, Engineering, and Math (STEM) for America's Future* Retrieved from http://www.whitehouse.gov/sites/default/files/microsites/ostp/pcast-stemed-report.pdf pp 11-12

[8] Howard J and Major J 2004 Guidelines for designing effective English language teaching materials. *The TESOLANZ Journal* **12** 50–58

[9] Kurniasari D A D, Rusilowati A, and Niken S 2014 Pengembangan Buku Suplemen IPA Terpadu dengan Tema Pendengaran Kelas VIII *Unnes Science Education Journal* **3(2)** 462-467

[10] Prawiradilaga D S 2009 Prinsip Desain Pembelajaran Instructional Design Principles Jakarta Kencana Prenada Media Group p 46

[11] Tessmer M 1993 *Planning and Conducting Formative Evaluation* London Routledge p 62

[12] Sugiyono 2012 *Metode Penelitian Kuantitatif, Kuantitatif, dan R&D* Bandung Alfabeta p 135

[13] Wiyono K 2015 Pengembangan Model Pembelajaran Fisika Berbasis *ICT* Pada Implementasi Kurikulum 2013 *Jurnal Inovasi Dan Pembelajaran Fisika* **2 (2)** 123-131

[14] Kelley T R and Knowles J G 2016 A conceptual framework for integrated STEM education *International Journal of STEM Education* **3 (1)** 1–11

IOP Conf. Series: Journal of Physics: Conf. Series **895** (2017) 012048 doi:10.1088/1742-6596/895/1/012048

Adult Gesture in Collaborative Mathematics Reasoning in Different Ages

M S Noto[1,6*], Y Harisman[2,6], L Harun[3,6], A Amam[4,6], and S Maarif[5,6]

[1]Departement of Mathematics Education, Universitas Swadaya Gunung Djati, Jl. Perjuangan No.1 Cirebon 45134, Indonesia
[2]Departement of Mathematics Education, STKIP PGRI Sumatera Barat, Jl. Gunung Pangilun Padang Utara Kota Padang Sumatera Barat 25000, Indonesia
[3]Departement of Mathematics Education, Universitas PGRI Semarang, Jl Sidodadi Timur 24-Dr.Cipto, Karang Tempel, Semarang Timur, Kota Semarang, Jawa Tengah 50232, Indonesia
[4]Departement of Mathematics Education, Universitas Galuh, Jl. R.E Martadinata No. 150 Ciamis 46274 Jawa Barat, Indonesia
[5]Departement of Mathematics Education, Universitas Muhamadiyah Prof. Dr. Hamka. Jl. Limau 2 Kabayoran Baru Jakarta 12130, Indonesia
[6]Departement of Mathematics Education, Universitas Pendidikan Indonesia, Jl. Dr. Setiabudi No. 229, Bandung 40154, Indonesia

* msubalinoto@fkip-unswagati.ac.id

Abstract. This article describes the case study on postgraduate students by using descriptive method. A problem is designed to facilitate the reasoning in the topic of Chi-Square test. The problem was given to two male students with different ages to investigate the gesture pattern and it will be related to their reasoning process. The indicators in reasoning problem can obtain the conclusion of analogy and generalization, and arrange the conjectures. This study refers to some questions—whether unique gesture is for every individual or to identify the pattern of the gesture used by the students with different ages. Reasoning problem was employed to collect the data. Two students were asked to collaborate to reason the problem. The discussion process recorded in using video tape to observe the gestures. The video recorded are explained clearly in this writing. Prosodic cues such as time, conversation text, gesture that appears, might help in understanding the gesture. The purpose of this study is to investigate whether different ages influences the maturity in collaboration observed from gesture perspective. The finding of this study shows that age is not a primary factor that influences the gesture in that reasoning process. In this case, adult gesture or gesture performed by order student does not show that he achieves, maintains, and focuses on the problem earlier on. Adult gesture also does not strengthen and expand the meaning if the student's words or the language used in reasoning is not familiar for younger student. Adult gesture also does not affect cognitive uncertainty in mathematics reasoning. The future research is suggested to take more samples to find the consistency from that statement.

1. Introduction

The gesture is a movement towards the body that person does in understanding problems. According to Alibali and DiRusso [1], the functions of gesture can help children to an understanding of numbers to others. In other words, the gesture may help children to pass on their knowledge to the respondent. According to Reynolds and Robert [2], the students make a gesture of collaborating is a source of diagnosis or as information for teachers. This research makes a study of two graduate students gestures collaborate in conducting mathematical reasoning with the statistical material. The three functions of gesture that is: first, the gesture is used to achieve, support, and return the focus to the problems. Second, gesture strengthens and expands the meaning of the word if the student or the language used in problem-solving is not familiar. Third, it is possible in some circumstances, the gesture is an index of an uncertainty level cognitive that replaces the understanding (the changes cognitive). Reynold and Robert [2] see three of these functions on students who collaborate in solving problems of graph speed time and graphical representation of a bus trip. This research addresses the different perspectives, by observing the changes of understanding about what and how an understanding of the changes that occurred when two students collaborate in generalities (reasoning) of a problem on the topic of chi-square test in statistics courses.

The research of Reynold and Robert [2] gives the conclusion that the gesture of the students made an impact in the case of collaborative learning mainly limited to formal language in understanding problems. Students must seek to dynamically together understanding the issue through gesture to solve it. This study remains a question, whether the gesture is unique to each, or whether it is possible to find patterns of the gesture used students, from gesture between different age, gender, etc. Harisman, Noto, Bakar and Amam [3] have studied the patterns of inter-gender gesture. The conclusions of the study, some gesture showed by female students are: often lower his head, less confident, shy, often nodded when the transfer of cognitive understanding, advancing the body movements directly towards the interviewer and ask questions if there are things that are not understood. Some gestures shown by male students are: play the hand, still see the problem, think long when it cannot answer the question, stay quiet when changing cognitive understanding. The researchers giving two students seven questions (problem-solving) on the topic of geometry. In this study, the students do not collaborate in solving the problems is given, but working on the problem separately. Furthermore, each student interviewed with semi-open questions, and the conclusion is about gesture patterns of genders. The gesture almost a consistent pattern to the seven questions provided.

In this study, we see how the gesture of two graduate students (with different ages) collaborates in the reasoning problem. Things that will be observed adapted to gesture function expressed Reynold and Robert [2], namely: (1) how the gesture used to achieve, support, and return the focus on the problem that sees from gradations of age-associated mathematical reasoning processes? (2) How gesture to strengthen and expand the meaning if the words of the student or the language used in solving unfamiliar problems as seen from the age gradation attributed to the process of mathematical reasoning? (3) How gesture affects cognitive uncertainties that may gesture can be used as a vehicle to replace the understanding (cognitive change) as seen from the age gradation attributed to the process of mathematical reasoning?

The reference to this study is research by Graham [4], which examines how the spontaneous gesture of children of all ages on the counting topic. The findings of this research are the role of gesture might serve to see the children's understanding of counting they do. Furthermore, Alibali, Flevares, and Meadow [5] conducted a study in which 20 teachers and 20 students see the sketch video footage of the 12 children that explains the solution to the problem of the equation. Based on this gesture the children can show important information about their problem-solving strategies. Six children revealed the same strategy in speech and gesture and the other six states of different strategies. The study investigated whether children from disclosing information may include a gesture that is accessible and understandable to others. The results of the study revealed that adults are not trained to express their understanding of the gesture.

This research will study whether to stimulate the understanding of others in learning collaborative need the gesture shown by adults (graduate students). Graduate students that have been subjected to this research are two male students with a different age. One of them is an adult that has had experience in teaching for 19 years, and the other has three years of teaching experience. A graduate student that is the subject of research is a lecturer at different universities.

This study is important to a university professor, whether adults need to express their understanding by using gestures or body movements when the student is difficult to understand the spoken language of their lecturers. To assess that, starting with assessing whether adults are going to show their gesture when faced with the problem or if they are not accustomed to trained in using the gesture as disclosed Alibali. Caused by these reasons, this study is very important to do as instructional decision-making in teaching. Why viewing from the perspective of collaborative inter-age is to answer whether age adults affects them in demonstrating gesture when confronted with problems. Various kinds of questions come to mind is whether older people will always keep their movements or even more expressive than children. When adults are expressive in understanding the problem, whether it will affect change or stabilization of their understanding of the opponent's discussion, in the context of school, of course, other people it is their students. Not many researchers that speak about the gesture of the level of adults.

Research Alibali and Meadow [6] conducted entitled Gesture-speech mismatch and mechanisms of learning: what the hand reveals about a child's state of mind are still talking at the level of children that study about the relationship between their saying with a gesture they have shown when explaining their understanding of the material in the series. Previous research by the Church and Meadow [7] also conducted a study evaluating the discrepancy between gesture and spoken language of children as the index changes in the knowledge that they have. The study examined two implications mismatch between gestures and speech of children in explaining a concept, namely: (1) Is child gesture and speech, show an underlying inconsistency in understanding a child's understanding of the concept? (2) Do mismatch gesture and speech is a result reflection of this inconsistency, reflecting the acceptance or rejection of the concept that was their understanding?

Meadow and McNeill [8] also still researching about the gesture's function in children. Beside the gesture is a sign that children's understanding of the concept, gesture also serves as an identifier of cognitive changes in children. That gesture has privileged access, meaning children may know about a concept or information, but can't say. Thus, it can be said that gesture serves as an extra window to make a child's mind develops. The gesture probably is more than just to show an understanding of the child, but involved in the process of cognitive change itself.

The study further revealed that this question would guide research on gesture will enter the new millennium. It has also been disclosed by Reynolds and Robert on the research before, probably because such research is doing in 2002 which has entered the new millennium on research gesture. Why study the gesture only at the level of the children? Goodwyn and Acredolo [9] confirmed they show that children learn sign language to develop signs to understand something at the early age of development, rather than to understand something in words. It is also revealed by Iverson and Meadow [10] who studied children born blind, with the aim of exploring the use of gesture to people who are blind. Four children born blind tested for three situations discoursed (narrative, reasoning, and spatial) and compared with a group of children that see but closed their eyes. Blind children do the gesture, although not all children make the gesture, and the gesture resembles the children that see but blindfolded. The results of this study show that the gesture can replace the function of communication.

There is some research that examines the gesture teenager like Roth [11] that conduct research on students. As a result, developmental psychology indicates shift function and frequency of use of the cue from an early age to early adolescence. Kelly and Church [12] also conduct research gesture towards adults, where the study was to compare the ability of children and adults to detect the information conveyed through hand movements. According to the Meadow and McNeill [8], the gesture is the language that is easier for the baby to master cognitive skills using the gesture conveyed by others. If the gesture is part of communicating then this research will review for the perspective, how two adults exposing their gesture when faced with problems such as reasoning. This is done because if a teacher or lecturer communicates processed must show whether the gesture so children easily understand the material if there is a verbal language that is less familiar to them or teacher or lecturer is quite silent and only read out the learning material in front of the class.

2. Method

This study was a descriptive study to see gesture two graduate students in solving the problems of mathematical reasoning with the topic of chi-square. How the two post-graduate students collaborate to understand something is by design the reasoning problems. This study saw a different perspective because most research always looks at how children are three aspects of it in solving problems. It is forgotten is that if children can understand the problem with the gesture, the teacher should teach to develop and train their gesture to make students or their students better understand the material presented. In the first study, we will see how the gesture of adults with different ages in reasoning problems. It is important to give an idea to train adults (teachers) in issuing a gesture to solve problems. The study involved two students of different ages because researchers wanted to see whether age affects adults gesture that they show in understanding the problem. Based on that reason, there is a big question on us how adults (teachers) to communicate to their students, whether requiring gesture or just sit quietly in front of the class.

Participants in the study were two post-graduate students of different ages. One of them was 41 years old and the other 28 years old. The age difference is to choose to find how age affects the gesture made when collaborating in reasoning problems. Do mature a lecturer/teacher will increasingly leave their gesture of solving problems. Judging from teaching experience, participants aged 41 years have had experience teaching for 19 years and the other for three years teaching experience. Both participants have a GPA of almost the same order do not to affect their academic ability in understanding and reasoning problems. Both students also have gained an advanced statistics course in the second semester.

The task given is a matter of reasoning, containing the indicators proposed by Sumarmo [13], namely: First, draw conclusions analogy, generalisation, and arrange conjecture. Second, draw a logical conclusion based on inference, check the argument's validity, and make a valid argument. Third, compile evidence directs, indirect, and mathematical induction. In a matter of reasoning, just load indicator (1) and (2). However, before the process of reasoning takes place, the two students should understand the problem first. Researchers chose questions about indicators reasoning because it is the ability to help high-level participants to think deeply. The task given is to generalise how to find the expected frequency (f_e) of observed frequencies (f_o) of the topic's square test.

The formula of square test: $X^2 = \sum_{i=1}^{k} \frac{(f_O - f_e)^2}{f_e}$

The form of the first and second question will be given the shape of its expected frequencies while in the third form, participants will find its frequency expectations. Once they find the answers to expected frequencies, we ask participants to generalize counting or how a formula to find the frequency of a given expectation. Here the one examples of three given problems is displayed as follows.

Table 1. We collect the following observations to decide whether gender and happiness of married have correlation and will feed all the observed frequencies and the expected frequencies

(f_O)	Gender		
The Satisfaction of Married	Men	Women	Total
Very Happy	30	45	75
Happy Enough	50	45	95
Not Happy	20	10	30
Total	100	100	200

Table 1. Cont.

(f_e)	Gender		
The Satisfaction of Married	Men	Women	Total
Very Happy	37,5	37,5	75
Happy Enough	47,5	47,5	95
Not Happy	15	15	30
Total	100	100	200

The source of the problem: Wahyudin [14]

After understanding the case of one, two and three, determine the formulation of how to determine the expected frequency of observation.

Both students must understand the problem with ten minutes. After that, the two students collaborate in solving the problem. They are free to debate, influencing the understanding, or change the understanding. We tell the students to only do discussions without notifying them to use gesture, so the gesture that occurs are natural.

We collect the data by providing reasoning test of both students. Furthermore, the two students will collaborate in solving problems of reasoning. The students' collaboration will be recorded from the beginning to the end of the discussion. We analyse the video footage to assess the students' gesture shown to answer the three problems outlined in the introduction. The result of collaboration between the two students will be presented in prosodic cues such as time and conversation. The form of gesture that appears will helpful in understanding the gesture given by two students.

3. Results and Discussion

After giving the reasoning problem to the BM (41 years) and AMH (28 years old), we obtain the gesture form from his understanding of the reasoning for 10 minutes until obtaining generalisation about the current form of collaboration is as follows.

Table 2. Protocol from Conversation and Gesture

Time	Conversation	Gesture BM (41 years old)	Gesture AMH (28 years old)
00:01sec-00:34 Sec	There has been no conversation	Silent, calm, but occasionally looked nervous and less confident 	Playing a pen and end at 34 seconds into nodded while playing lip 
00: 35sec-00:41 Sec	AMH : Do you understand problem number one? BM : I was even scared about the problem number three; could you please to read it. AMH : Oh yes, the first paragraph, it is dangerous BM : Hahahaha.. There is a typo, how can I read it, I failed to focus. AMH : Let's just say it's true BM : Oh yes, let's just say it's true	The thrusting paper shows the writing error. 	Stay play a pen and quiet. 

Table 2. Cont.

00: 42 Sec-01: 28 Sec	BM AMH BM	: : :	How can I solve problem number one? Is there a formula? Not a paper to write off? Just write out here, try number one, please teach me	Switch position sat approached AMH 	Put the paper on the table and get ready to write 
01: 28 Sec-01:48 Sec	AMH BM AMH	: : :	It's the data, from here to here, we find the empty, well how do we solve it? Means we have to find from where it can be 39.4? Yeah, well, we suppose these data a, b, c, d for its observation frequency, then $a, b, c,$ and d also for frequencies up to his expectations, the expected frequency of a is already there, just search for the others?	More subduing the body of the paper and to consider seriously 	Cross out the f_o and f_e on the matter of naming the columns of the same location with the letter a, b, c, and d
01:48 Sec-02:30 Sec	AMH BM AMH BM AMH	: : : : :	Well now how do I find the value b, c, and d here by looking at the data before What should I use? Can I use a comparison? I already know how to solve this. How do I solve it? Want to know the formula?, Please try first, because we want to shape the understanding. Please try to solve it by yourself.	Move his legs to the left while straightening 	Forming quotes by hand 
02:30 Sec-02:31 Sec	BM AMH	: :	You mean this 39.4? Yeah, where is the result come from? Why if the observed frequencies have a value 54, the frequency of its expectations is 39.4?	Continue to pay attention to problems seriously and listen to explanations AMH 	Stay calm and pointing to the problem on paper 
02:32sec-03:16sec	AMH BM AMH BM	: : : :	Is there a calculator? There should be. There should be a formula, if there is no formula, we cannot solve? Yeah, but the instruction of the problem is to find the formula. Oh, to find the formula ...	Opened the jacket and seemed enthusiastic, and participate put out smartphone 	Turning on the calculator contained in the smartphone from his pants pocket and keep calm

Table 2. Cont.

03: 16 Sec – 04:00 Sec	AMH	:	We try first; I test my guess, I already know the formula, it seems true, that's right, I got the formula	Glancing an eye toward AMH to ask for an explanation	Stay calm 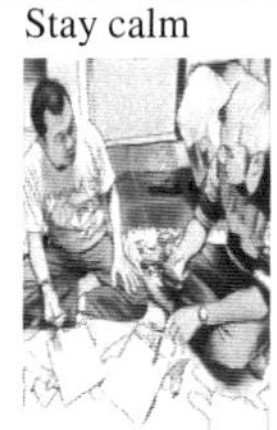
	BM	:	I'm lazy to think right now, how do I solve it?		
	AMH	:	Please don't cheat on me, later, the reasoning problem can not be executed		
	BM	:	No problem, we're being collaborate		
04:01 Sec – 06:00 Sec	BM	:	I will make corrections, how your logic?	Enthusiastic and play hands like to request an explanation 	Describe the pattern of answering on paper 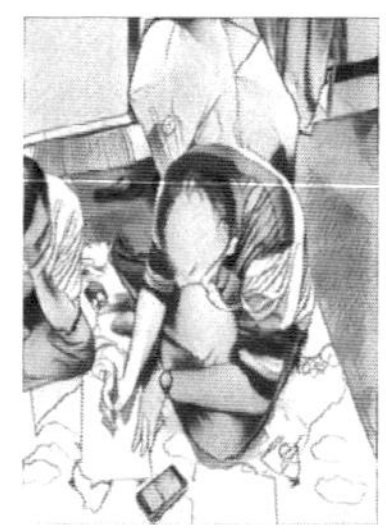
	AMH	:	There is value 54, and there was 39.4 right?. Relating to 54 it was 85, 83, and 179. So the chances of 34.9 are formed by the numbers right?, I created the matrix, the matrix is answered 39.4, right? From where? Using existing value, it turns out there is 83 divided by 179 times 85.		
	BM	:	83 divided by 179 times 85. So that, now how did that happen? Was the term already directly into the formula right? How the logic of thinking?		
06:00 Sec- 08:17 Sec	BM	:	Why would it be 83 divided by 179 times 85?	Keep writing, and occasionally nodding and smiling 	Stay calm, pointing to some of the numbers in problems
	AMH	:	Well, that was the expected frequency, some rows, divided by the total, times the number of columns, well, how about this b problem? 83 divided by?		
	BM	:	179, Yes 179 times?		
	AMH	:	31?		
	BM	:	No		
	AMH	:	It becomes this, be this, it means times?		
	BM	:	times 31?		
	AMH	:	No		
	BM	:	Oh, times 94 right?	BM : 31?	
	AMH	:	Yeah, like that, for the c, how?	AMH : No	
	BM	:	Means, 96 divided by 179, then times?	BM : 85? AMH : Yeah, okay the last one, the d?	
	AMH	:	Come on, times with?	BM : Okay, the last one right, 96 divided by 179, times 94? AMH : Correct, finished	

Gesture analysis from the Table 2 is to answer the three question and associating it with the reasoning process that draws conclusions analogy, generalisation, arrange conjecture and draw logical conclusions based on inference, check the argument validity, and make the valid argument. Before the process of reasoning, the two students must understand the problem. In this process, we see how the gesture uses to achieve, maintain, and re-focus on the problem to solve. It can be seen at minute 00: 01 - 00: 34, BM appeared calm and occasionally looked nervous, but AMH seems to play pen and at the end of 00:34 minutes showed a nod. If viewed from a gesture that indicated at the beginning of understanding,

IOP Conf. Series: Journal of Physics: Conf. Series **895** (2017) 012048 doi:10.1088/1742-6596/895/1/012048

suspicions seemed BM has not yet understood the problems and AMH seem to have understood the issues indicated. If seen how the gesture shows the focus on the problem is we can see in minute 00: 35 Sec - 00: 41 Sec, BM shows a gesture of paper handed to AMH and spoke context typing errors, while AMH does not realise it and stay calm and focus on the problem.

BM saw not focus on the question because he was talking about things outside the context of the question. AMH looks very focus, he showed a quiet gesture, focus on the question and does not even realise it before typing errors notified by BM. When viewed from the perspective of age, in the context of this study shown that age does not affect one's understanding. Furthermore, the process to generalize the results related to the process of mathematical reasoning and associate it with how the gesture (movement) to strengthen and expand the meaning if the words of AMH & BM or the language they use in problem-solving is not familiar, can we see from the gesture in the 01: 28 Sec - 01:48 Sec, 2:32 Sec – 3:16 Sec, 4:01 Sec – 6:00 Sec. At this time-intervals AMH try to expand the meaning of intent about the BM that seems less familiar with the language and numbers on the question, where the gesture showed AMH is crossing out on (f_e) and (f_0) on the question about naming the columns in the same place with the letter a, b, c, and d, then pull out a calculator in his smartphones from his pocket and keep quiet, and a portrait of answer patterns on paper. In Figure 1, we can see AMH shape pattern made on the process to generalize and find the pattern.

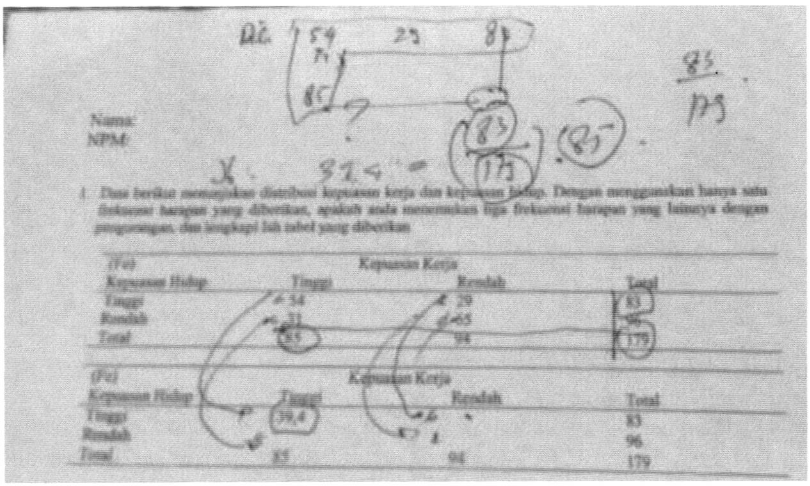

Figure 1. The pattern created by AMH to expand (clarify) the meaning of the BM

Furthermore, BM shows a gesture of curiosity to the explanation or expansion meaning that AMH described. Gesture shown by BM is, move closer to a sitting position on the AMH 00 min: 42 Sec-01: 28 Sec. BM opens the jacket, looked enthusiastic, come pick up and turn on HP Sec, 02:32 Sec-03: 16 Sec, and enthusiastic as well as playing a hand like asking for an explanation at minute 04:01 Sec - 6:00 Sec. All the gesture is made by both students in the process of obtaining a general formula for determining the expected frequency of the Chi-Square test. The terms of the age are seen that the age factor does not affect the gesture of the two students when they want to generalise the right formula.

Furthermore, when seen from how the gesture (movement) affects cognitive uncertainty, perhaps the gesture can be used as a vehicle to replace the understanding (cognitive change), seen from the gesture given by AMH affect cognitive certainty BM. When looking at AMH quiet gesture, BM immediately moves a seated position for an explanation. Furthermore, BM also shows the gesture of not understanding and not being able to do reasoning with the context of generalisation. It can be seen in Sec- 06:00 08:17 Sec where BM continued to write and occasionally nodded and smiled, indicating that it may generalise the problem well. Furthermore, AMH continues to bear the quiet gesture indicating that AMH has been able to generalise the problem from the beginning. It can be demonstrated by a second gesture on Sec 06:00 - 08:17 remained calm, pointing to some numbers in problems. It indicates AMH unchanged or unaffected by BM understanding. When viewed from the age, it is concluded that gesture on students being more mature age did not affect changes in the understanding of the younger students, even younger students who are influencing changes in the understanding of the students in a more mature age.

After a process of discussion, some questions posed to the students, why it happened, based on interviews revealed that AMH is a lecturer of statistics and BM is teaching calculus course. At the beginning of the study, it is ignored because our view is academic ability, as well as two students alike, have obtained an advanced statistics course in the second semester.

4. Conclusions and Implications

From the results of the study, we conclude that age is not the main factor affecting gesture, which appeared at the time of the reasoning problems given. In this case, the gesture shown by students who are more mature age did not show the first achieve, maintain, and focus on solving the problem. The gesture also did not strengthen and expand the meaning, if the words of the student or the language used in problem-solving are not familiar to younger students. The gesture more mature age student also does not affect cognitive uncertainty for younger students at the time of mathematical reasoning. Particularly, in this case, it turns teach experience that became the dominant factor of the gesture shown as more aware of the problem, capable of clarifying and expanding its meaning, and cognitive changes may affect other students. For further research, it is advisable to take more samples to see the consistency of the statement.

Acknowledgments

Thanks to Mr. M. Win Afgani (UPI graduate student, doctoral program) already provide feedback on this article and assisting during the research process. Thanks to Mrs. Marwia Tamrin Bakar (UPI graduate student, the program as a doctor) who have been willing to accompany the researchers recorded the collaborative process from the beginning to the end.

References

[1] Alibali M W and DiRusso A A. 1999. *The Function Of Gesture In Learning To Count: More Than Keeping Track.* Cognitive Development **14** 37-56.

[2] Reynolds F J & Robert A. 2002. *Journal of Mathematical Behavior* **20** 447–460.

[3] Harisman Y., Noto M S., Bakar, M T., and Amam A. 2017. *Journal Of Physics: Conference Series. IOP Conf. Series* **812** (012039).

[4] Graham A. 1999. *Journal of Experimental Child Psychology* **74** 333–355.

[5] Alibali M W., Flevares L M., and Meadow S G. 1997. *Journal of Educational Psychology*. **89** 183-193.

[6] Alibali M W and Meadow S G. 1993. *Gesture-Speech Mismatch and Mechanisms of Learning: What the Hand Reveal About a Child's State of Mind.* Cognitive Psychology *25*. p. 468-523.

[7] Church R and Meadow S G. 1986. *The Mismatch Between Gesture and Speech as an Index of Transitional Knowledge.* Cognition **23** 43-71.

[8] Meadow S G and McNeill D. 1999. *The Role Of Gesture And Mimetic Representation In Making Language The Province Of Speech. In Michael C. Corballis and Stephen Lea (eds.) Evolution of the Hominid Mind*, pp. 155-172. Oxford University Press.

[9] Goodwyn S W and Acredolo L P. 1993. *Symbolic Gesture versus Word: Is There a Modality Advantage for Onset of Symbol Use?. Child Development* **64** 3 pp. 688-701.

[10] Iverson J M and Meadow S G. 1997. *What's Communication Got to Do With It? Gesture in Children Blind From Birth.* Developmental Psychology **33** 3 pp. 453-467.

[11] Roth W M. 2000. *Journal of Pragmatics* **32** 1683-1714.

[12] Kelly S D and Church R B.1998. *A Comparison between Children's and Adults' Ability to Detect Conceptual Information Conveyed through Representational Gestures*. Child Development **69** 1 pp 85-93.
[13] Sumarmo U. 2006. *Pembelajaran untuk Mengembangkan Kemampuan Berpikir Matematik*. Presented in pada Seminar Nasional Pendidikan Matematika dan Ilmu Pengetahuan Alam, FPMIPA Universitas Pendidikan Indonesia.
[14] Wahyudin. 2016. *Statistika Terapan Lanjut*. Mandiri: Bandung.

International Conference on Mathematics and Science Education (ICMScE) IOP Publishing
IOP Conf. Series: Journal of Physics: Conf. Series **895** (2017) 012050 doi:10.1088/1742-6596/895/1/012050

Validating the Assessment for Measuring Indonesian Secondary School Students Performance in Ecology

A Rachmatullah[1], F Roshayanti[2] and M Ha[1*]

[1]Division of Science Education, College of Education, Kangwon National University, Chuncheon-si, Republic of Korea
[2]Department of Biology Education, Faculty of Mathematics, Science and Information Technology Education, Universitas PGRI Semarang, Semarang, Indonesia

*arifraach@gmail.com

Abstract. The aims of this current study are validating the American Association for the Advancement of Science (AAAS) Ecology assessment and examining the performance of Indonesian secondary school students on the assessment. A total of 611 Indonesian secondary school students (218 middle school students and 393 high school students) participated in the study. Forty-five items of AAAS assessment in the topic of Interdependence in Ecosystems were divided into two versions which every version has 21 similar items. Linking item method was used as the method to combine those two versions of assessment and further Rasch analyses were utilized to validate the instrument. Independent sample t-test was also run to compare the performance of Indonesian students and American students based on the mean of item difficulty. We found that from the total of 45 items, three items were identified as misfitting items. Later on, we also found that both Indonesian middle and high school students were significantly lower performance with very large and medium effect size compared to American students. We will discuss our findings in the regard of validation issue and the connection to Indonesian student's science literacy.

1. Introduction

Ecology is one branch in Biological science that addresses the inter-correlation between all the living and non-living beings exist in the Earth or frankly could be called as the study about ecosystems. Besides that the idea of ecosystems is one out of four core life sciences ideas that is necessary to be delivered to children in the school levels, from elementary to high school [1]. In addition Barman & Mayer [2] in their study found out that most of Biology teachers think that ecosystems concept and other related-concepts are important topics being learned by students because it provides the inter-connection of science discipline. Thus, students could learn the integration of theories while learning about ecosystems that can establish the so called meaningful learning. There are several sub-concepts addressed when learning about ecology and ecosystems, but the most ardently-discussed topic is the topic about interdependence of ecosystems. Interdependence of ecosystem is the topic in ecology whose discussions are related to how every component in ecosystems interact and influence one another. This discussion is needed to be delivered to student because it is related to the problems that are being faced by the degree of biodiversity in this world, particularly in regard to the loss of biodiversity due to habitat destruction [3]. The loss of biodiversity may be impacted by the loss of components of ecosystems that could be assumed as the results of human activities, such as exploiting and even hunting animal in the exaggerated ways in order to fulfill their economy or even just for their enjoyment [4].The loss of biodiversity is strongly related to the equilibrium of life in the Earth which directly or indirectly may influence human in the later time. By having more knowledge about this issue of interdependence, students are expected to have wiser and better attitude towards the ecosystems and environment. Given that Indonesia is known as the second most biodiverse country in the world after Brazil, thus it is important to keep the richness of flora and fauna in Indonesia by educating youth to prevent the loss of biodiversity.

In addition, having more knowledge about interdependence in ecosystems that influence environmental attitude and behavior is one of the ways becoming the scientifically literate society. Because as what mentioned by Bybee, McCrae & Laurie about becoming scientifically literate person, student needs to be able to implement what they know about particular knowledges in their daily life [5]. The issue on producing the scientifically literate society is emphasized in the newly published Indonesian Curriculum and become one of the goals of the curriculum. It responses to the internationally published results about Indonesian student performance in Trends in International Mathematics and Science Study (TIMSS) and The Programme for International Student Assessment (PISA), in the science literacy sections that Indonesian students are always placed in the bottom rank [6]. Thus, the newly implemented Curriculum 2013 attempts to increase the achievement of Indonesian students in those international assessment programs [7].

Assessment is one of the important aspects on tackling the issue of science literacy and on finding out how far Indonesian student's science literacy is increased. Heretofore, there are few studies which developed the assessment to measure student's performance on the concept of interdependence on ecosystems and one of them is American Association for the Advancement of Science (AAAS). AAAS has developed one assessment in the topic of interdependence of ecosystems for American students. It has been known that the assessment is valid and reliable for American students. But, there is still no study uncovering whether the instrument is also valid for students in other countries, especially for Indonesian students, given that there is still no study examining how well Indonesian students on doing the other internationally published assessment. Therefore in this current study we are focused to examine the fitness of AAAS Ecology assessment for Indonesian secondary school students, the interaction effect between gender and grade on Indonesian secondary students' ecology performance and also the differences between Indonesian and American students on the AAAS Ecology assessment.

2. Method

2.1. Participants

The participants joining this study consisted of 611 Indonesian secondary students. The secondary school level in Indonesia is divided into two levels which are middle and high school students. Among the total sample of participants, 218 students were middle school students and 393 students were high school students. They were in their second grade, 8th and 11th grade. In terms of gender, the participants were made up by 36% male students and 64% female students. They were recruited from two private and two public schools in West Java province.

2.2. Research Instrument

The instrument administered to students was the Ecology assessment that developed by American Association for the Advancement of Science (AAAS) and publicly published in their website (http://assessment.aaas.org/topics/IE#/). The Ecology topic that addressed in the assessment was about Interdependence on Ecosystems and was consisted of 45 items that divided into three key ideas of interdependence of ecosystems which are (1) food web, (2) competition in environment and (3) contributing factors on growth rate and mortality in ecosystem. The assessment was in the multiple choice form with four options in for every item.

The total of forty-five items, we believe that it would take more time and give more cognitive pressures to students if all of the items were given to them in one time. Thus, we decided applying one of the Rasch methods called *linking item methods* to reduce the cognitive pressures obtained by students when doing on the assessment. The linking item method is the method that partitions the instrument into several packages whose some of the items are called as common items because those items could be found in other packages too. We divided the total of forty-five items into two packages (package A and package B). Each package was consisted of 33 items, and 21 items among those 33 items were common items/linking items. Frankly, the linking item method that utilized in our current study is depicted in Figure 1.

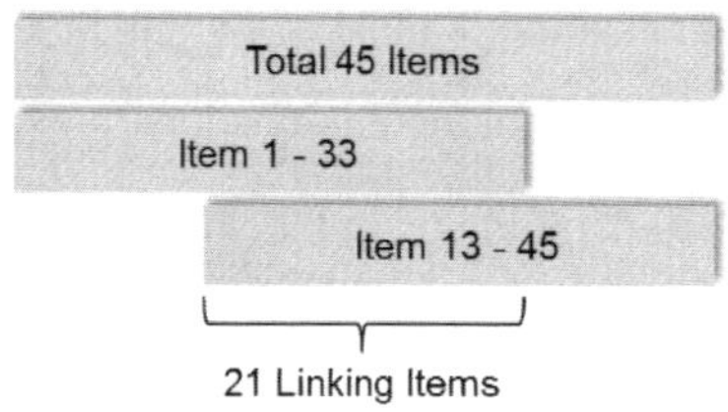

Figure 1. Linking Items

2.3. Data Analyses

In terms of examining the validation of AAAS Ecology assessment, first Rasch analyses through Winstep V.3.92.1 software were performed. Finding out whether the linking item methods was successfully done or not is one of the important issues to be addressed and it can be explored by performing the correlation test to item difficulties of common items in the package A and package B. The following analysis was determining the misfitting items by investigating infit and outfit mean-square (MNSQ) values. The cut-off that suggested by Linacre [8] was used to address this misfitting items. The cut-off for infit and outfit MNSQ that could produce the productive measurement is between 0.5 – 1.5. Therefore, item with MNSQ value beyond that range is called as misfitting item. Another consideration for addressing validity issue of the instrument is generalizability of the item or whether bias exists in the item or not or even whether some particular items could be well-answered by some particular groups, we used Differential Item Functioning (DIF) to find out this issue by testing the gender bias. The DIF contrast cut-off suggested by Boone *et al.* [9] was used in our current study. The cut-off is 0.64, which means that the value higher than 0.64 indicates the existence of item bias with medium magnitude. The validation of the assessment is not limited to construct validity, the issue of internal consistency is also important. Thus, we also reported item and person reliability with using the standard of interpretation that suggested by DeVellis [10].

The raw data obtained from the student answer is in the form of categorical data which are right and wrong answer. Data in the form of categorical data cannot be directly analysed by using statistical analyses, especially parametric test, because in order to perform parametric test, the data should be in the form of interval data. Therefore, using the raw score is not a right way to run statistical test. This issue can be tackled by performing Rasch analyses. A set of value, which one for each student, can be obtained after performing Rasch analysis. The values are called as person measure that corresponds to every student raw score but the values have been converted to interval form. We used these values to perform further statistical analysis. Examining the interaction between gender and grade was done by performing analysis of variance (ANOVA) test. Answering the difference in performance between Indonesian students and American students was done by running the Independent sample t-test. The comparison was done by comparing the mean of every item achieved by Indonesian students and for American students we used the percentage of students responding correctly reported in the AAAS website. All statistical analyses were done through SPSS V.22.

3. Results and Discussions

The thematic method will be used to report the findings of our study. The findings of our study will be directly followed by the discussions. The first finding is related to the validation of AAAS Ecology assessment and the issue of misfitting items that for further analysis will be deleted. Following to the validation findings, the finding of interaction effect between gender and grade will be reported. Later, in the last section the comparison between Indonesian students and American students will also be reported and discussed.

3.1. Instrument Validation

The first findings that responses issue addressed in the first research question are related to the validation of the instrument. Aforementioned, linking item was utilized in our current study. To find out whether the linking item is working we did correlation test for the difficulty of common items in the both packages. We found that the correlation of the common items in significantly correlated ($p < 0.001$) with the coefficient correlation 0.971. This highly correlated indicates that the difficulty level of common items are similar in both packages, thus we can combine the data by putting missing value on the data of students who just did one particular package and so for students who just did another package.

After combining the data, Rasch analysis was run to find out the whether there are some misfitting items or not and whether some items are biased or not. Based on the findings, in terms of infit MNSQ which is related to the pattern of responses, all the values are in the range of 0.5 – 1.5, which is indicating most of student pattern responses are fit with the Rasch model. In contrast, we found some misfitting items based on outfit MNSQ. Outfit MNSQ is sensitive to outliers or unpredicted responses or in short it is indicating that high achievers did not answer correctly in the items that they are expected to correctly answer those items or vice versa for the low achievers. Based on results shown in Table 3, there are three items outside the range of 0.5 – 1.5 which are item coded A07 with outfit MNSQ 2.68, item coded B07 with outfit MNSQ 1.79 and item coded B09 with outfit MNSQ 1.94.

The second issue of validating the instrument is generalizability of the instrument. As noted above, this generalizability issue is related to the existence of biases in items. This generalizability is identified by using DIF. We attempted to identify the bias issue in the group of gender, male and female. There were three items whose DIF contrast exceeded the benchmark 0.64. Those three items are C11 (0.74), B07 (0.87) and B11 (0.99). We considered the outfit value first then DIF values as additional consideration when deciding and refining the package of the instrument. The issue of outfit value is crucial because it shows how well the item can differentiate the high achiever and low achiever [9]. The DIF issue actually could be handled by analyzing the wording of the item, when the wording of the item is still not indicating inequality of particular groups, in our analysis is gender, the item could be still included in the measurement. In addition the items that are indicated the occurrence of bias mentioned above, the DIF contrast values are still near the cut-off, thus we still considered to include the items. Therefore, we just deleted three misfitting items based on outfit MNSQ values for further analyses. Another thing related to validating instrument, that is not less important, is reliability or internal consistency. There are two values of reliability computed through Rasch analysis, item and person reliability. Those reliabilities can also be interpreted by using common reliability interpretation benchmark [11]. The item reliability before and after deleted the three misfitting items are similar, which is 0.98. This value of item reliability is categorized as the 'excellent' reliability [12], which indicates the difficulty level's probability of every item would remain same if the instrument used in the different population [9]. Therefore, the instrument shows not being dependent on respondents. Similar to the finding from item reliability, the person reliability values are also similar between overall and refined version. The person reliability values are 0.83. Based on DeVellis [10], the coefficient reliability above 0.8 indicates a "very good" instrument. In addition, based on interpretation from Fisher, the value between 0.81 – 0.90 indicates as "good instrument" that can differentiate participants into three categories based on their ability [12].

3.2. Indonesian Secondary Student's Performance

In response to the second research question addressing the effect of interaction of gender and grade on the performance, ANOVA test was performed. Based on ANOVA test we found that in the pool of middle school students, male sample (M = –0.533, SD = 0.887) has lower performance than female one (M = –0.285, SD = 0.964). Contrarily, in the pool of high school student, male sample (M = 0.628, SD = 1.323) has higher mean than female students (M = 0.467, SD = 1.082). In addition, we did not find the significant effect of gender on the performance ($F[1, 561] = 0.20$, $p > 0.05$, $\eta_p^2 = 0.000$), while we found significant effect with large effect size ($F[1, 561] = 94.94$, $p < 0.001$, $\eta_p^2 = 0.145$).

Ultimately, we also found that the interaction effect of gender and grade was statistically significant on the performance ($F[1, 561] = 4.34$, $p < 0.05$, $\eta_p^2 = 0.008$).

Further interpretation of the findings, we assume that the performance on ecology assessment does not depend on gender, but it depends on educational level. It is shown in the Figure 2 that most of middle school student's ability is below 0 (zero) point, which indicates that most of them just could correctly answer less than 50% of total items. In contrast, there are less high school students who are below zero point zone and most of them are above the zero point zone. This indicates that many high school students could correctly answer the questions more that 50% of total items. It aligns with the theory of learning progression whereby the higher level of education has higher level of performance. Another interesting finding is the interaction effect between gender and grade, as shown in Figure 3a, the mean of male students in middle school level was lower than female students while in the high school level the mean of male students was higher than female student. The reason of this finding is still unclear, given that we do not have any interview. Thus, further study is needed to scrutinize this finding.

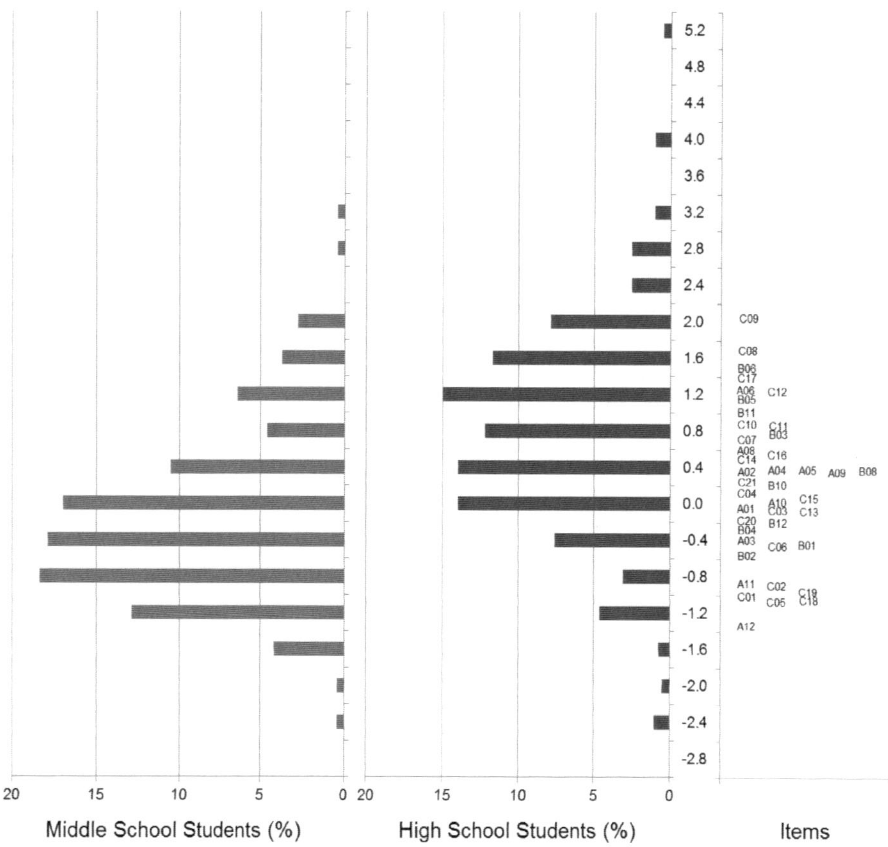

Figure 2. Wright Map Indonesian Middle and High School Students

3.3. Comparison with American Students

The instrument was originally used for American students, thus here we attempted to compare the performance of Indonesian secondary students to American secondary students. The comparison is based on the mean of traditional difficulty level. Higher mean indicates that students have higher performance and lower mean indicates having lower performance. Based on our analysis, we found that in the pool of middle school student, American sample has higher mean ($M = 0.585$, SD = 0.128) than Indonesian sample (M = 0.424, SD = 0.127) and the difference was statistically significant with very large effect size (t = 5.784, p < 0.001, d = 1.28). Similar to the finding from middle school level, we found American high school students (M = 0.674, SD = 0.128) also have higher mean than Indonesian high school student (M = 0.595, SD =0.174). The difference was also statistically significant with medium effect size (t = 2.355, p < 0.05, d = 0.52). The results are visualized in the Figure 3b.

The AAAS Ecology assessment addresses the questions in the type of high order thinking (HOT) where the answer is not beyond the article of the question instead the answer is in the text itself [13]. The significant lower performance of Indonesian students compared to American students indicates that Indonesian students are still lack of using their high order thinking ability or it could be caused by the lack of emphasizing high order thinking in biology classes. It is particularly for middle school students, given that the effect size shows 'very large' differences. Besides high order thinking ability, the questions of the assessment also address system thinking, where students are asked to investigate the effect of one component if the particular component is added or removed from the system of ecosystems. The low performance of Indonesian students indicated that they are lack of thinking a system which later impacted their ability to precisely predict. These high order thinking and system thinking are the necessary ability to be scientifically literate students [14]. Therefore, in response to the low performance of Indonesian students in science literacy assessment that reported by PISA and TIMSS, in order to increase the performance we believe that the teaching activities with emphasizing the higher order thinking and system thinking ability is suggested to be used in Indonesian science classroom, especially biology classrooms. Several teaching methods and approaches that cover those kinds of ability are problem based learning and project based learning.

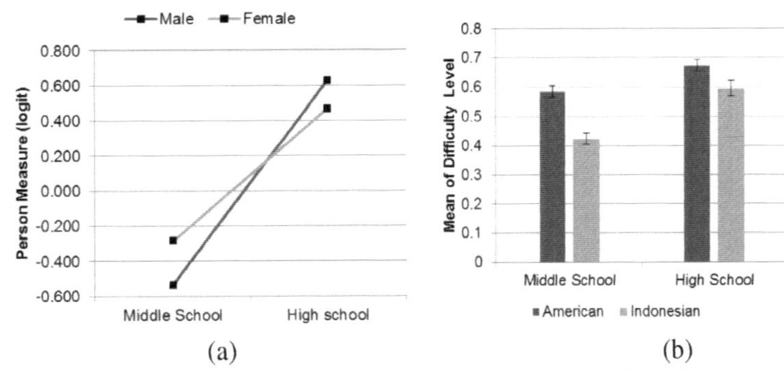

Figure 3. (a)The interaction effect between grade and gender on AAAS Ecology assessment, (b) the comparison between Indonesian and American secondary school students

4. Conclusion

Based on our current study, we found that most of questions in AAAS Ecology assessment in the topic of interdependence in ecosystems are psychometrically valid to be administered to Indonesian middle and high school students. Besides that we also found that educational level give more effect on students' performance on this assessment than gender. We found that Indonesian secondary students were significantly lower performance than American students. These findings could be used to response the longstanding issue of Indonesian student's science literacy, that always place in the bottom rank. We believe that more assessment and teaching and learning activities that employ and address higher order thinking and system thinking are needed to be implemented in Indonesian science and or biology instructions.

References

[1] N National Research Council 2012 *A Framework for K-12 Science Education: Practices, Crosscutting Concepts, and Core Ideas (*Washington, DC: The National Academies Press)

[2] Barman CR and Mayer DA 1994 *The American Biology Teacher* **56** 160-3.

[3] Coleman FC and Williams SL 2002 *Trends in Ecology & Evolution* **17** 40-4.

[4] Allesina S and Bodini 2004 *Journal of Theoretical Biology* **230** 351-8.
[5] Bybee R, McCrae B and Laurie R 2009 *Journal of Research in Science Teaching* **46** 865-83.
[6] Rachmatullah A, Octavianda RP, Ha M, Rustaman NY and Diana S 2017 *IOP Journal of Physics: Conf. Series* **812** 012009
[7] Mendikbud [Indonesian Ministry of Education and Culture] 2016 *Salinan Lampiran Permendikbud No.21 th 2016 tentang Standar Isi Pendidikan Dasar dan Menengah* [A copy of the additional documents of Indonesian ministry of education and culture regulation number 21 year 2016 on Content Standards for Primary and Secondary Education].
[8] Linacre JM 2002 *Rasch Measurement Transactions* **16** 878
[9] Boone WJ, Staver JR and Yale MS 2014 *Rasch Analysis in The Human Sciences* (Dordrecht: Springer)
[10]DeVellis RF 2003 *Scale Development: Theory and Applications 2nd edn* (Thousand Oaks,CA: Sage)
[11] Fisher WP Jr 2008 *Rasch Measurement Transactions* **22**1160-3
[12] Fisher WP Jr 2007 *Rasch Measurement Transactions* **21** 1095
[13] Zohar A and Dori YJ 2003*The Journal of the Learning Sciences* **12** 145-181.
[14] Norris SP and Phillips LM 2003 *Science Education* **87** 224-40.

IOP Conf. Series: Journal of Physics: Conf. Series **895** (2017) 012052 doi:10.1088/1742-6596/895/1/012052

Developing Technological Pedagogical Content Knowledge (TPACK) in Animal Physiology

F Pusparini[1*], R Riandi[1] and S Sriyati[1]

[1]Departemen Pendidikan Biologi, Universitas Pendidikan Indonesia, Jl. Dr. Setiabudi No. 229, Bandung 40154, Indonesia

[*]fitriapusparini@student.upi.edu

Abstract. The purpose of this study is to describe pre-service teacher's learning during lecturing Animal Physiology and investigate it's impact on pre-service teacher's technological pedagogical content knowledge (TPACK). How was the lecturing process can improve TPACK of preservice teacher on Biology education espescially in Animal Physiology. There are four experiment classes using Solomon four group design, there are pedagogic treatment, content treatment and technological treatment, the last class without any treatment. Both quantitative and qualitative data were collected. Quantitative data were collected through a questionaire of TPACK. Qualitative data were collected through a lesson plan and teaching simulation. Findings has revealed that participants experienced significant gains in all TPACK constructs. Both of pedagogic and technology treatment is better than others, but pedagogical treatment didn't also increase PCK most of participants. Findings has implications for teacher education programs to be a professional teachers and for researchers interested.

1. Introduction

The 21st century demands qualified human resources, generated by professionally managed institutions to produce superior results. Education as part of efforts to improve the quality of human resources also changes and follows the times. Therefore, professional teachers are required. To prepare a professional teacher must be prepared from the beginning, that is when they are still students (preservice teachers). During the course of lectures, preservice teachers of the University of Galuh Ciamis Biology are equipped with knowledge about Biology as content, pedagogy, and technology. These three types of knowledge interact and form new knowledge known as Technological Pedagogical Content Knowledge (TPACK) (Mishra and Koehler, 2006). TPACK which is owned by biology preservice teacher has an important role, because TPACK will influence the student how preservice teachers in teaching (Srisawasdi, 2012). Effective teachers not only know their subject matter but they also able to create a stimulating learning environment and apply pedagogical strategies, including technology-rich strategies that engage students while helping them improve their achievement (NCATE, 2010).

The TPACK framework was developed by Mishra and Koehler based on Lee Shulman's concept of PCK by adding technology. One form of TPACK application in learning is the use of technology in teaching certain materials. Technology is considered important to be integrated in learning to respond to the challenges of the 21st century. Teachers are not only required have the ability of PCK but also apply technology in learning, so technology, pedagogy and content has become a part of teacher education programs to prepare preservice teachers. Applying technology in its teaching process. This research was conducted in order to analyze TPACK development process in Biology preservices teacher and to analyze how the training received during lecturing can support TPACK of Biology preservice teacher in animal physiology material (digestive system and circulatory system).

2. Experimental Method

This research is examining how preservice teachers develop and apply technological pedagogical content knowledge (TPACK) troughout their teacher preparation program. As part of the research plan, we constructed the Survey of Preservice Teacher's Knowledge of Teaching and Technology to collect data on preservice teacher's self-assessment of the seven knowledge domains within the TPACK framework. These knowledge domains include: technology knowledge (TK), content knowledge (CK), pedagogical knowledge (PK), pedagogical content knowledge (PCK), technological content knowledge (TCK), technological pedagogical knowledge (TPK), and finally, technological, pedagogical content knowledge (TPACK). This instrumen adopted from TPACK instrument developed by Widodo (2017).

Respondents who participated in this study are biology preservice teacher enrolled in a fourth semester in Biology Departemcnt of Universitas Galuh Ciamis. The participants were students of the researcher's classes, the first class was designed to introduce participants the technology that can use to teach animal physiology, the second class was designed to introduce participants focus on pedagogical tasks, the third class is combine both technological and pedagogical treatment, and the other class doen't get any treatment. At the beginning and end of course participant complete the TPACK instruments consist of questionaire and open ended question. The instrument constructed contained 35 items for measuring preservice teachers's self-assessments of the seven TPACK domains: 6 TK items, 5 CK items, 5 PK items, 6 PCK items, 4 TCK items, 4 TPK items, and 5 TPACK items. For these 35 items, participants answered each question using the following five-level Likert scale:

1. Strongly disagree
2. Disagree
3. Neither agree nor disagree
4. Agree
5. Strongly agree.

3. Result and Discussion

3.1 Technological Knowledge (TK)

Technology knowledge refers to the knowledge about various technologies, ranging from low-tech technologies such as pencil and paper to digital technologies such as the Internet, digital video, Interactive whiteboards, and software programs. Based on analysis, the profile of TK preservice teachers of Biology is presented in table 1.

Table 1. Technological Knowledge Score Biology Preservice Teachers

No	Statement	M	SD	Criteria
1	I follow the latest technological developments	4,00	0,68	Good
2	I am proficient at using word processing programs, number processors, and presentations like Ms. Office	4.14	0,53	Good
3	I can store data on digital media (upload or download)	4,07	0,73	Good
4	I can use internet as a tool for communication such as having email and social networking	4,57	0,51	Good
5	I can use printers, scanners, projectors and cameras	4,00	0,88	Good
6	I can create a discussion forum on email or social networking	3,71	0,83	Good
	Mean	4,08		Good

The average score TK of preservice teachers is 4.08, it is in the good criteria this indicates the ability of technological preservice teacher has been good average that the highest on the item statement is "I can use internet as a tool for communication such as having email and social networking"

3.2 Pedagogical Knowledge (PK)

Pedagogical knowledge refers to the methods and processes of teaching and includes knowledge in classroom management, assessment, lesson plan development, and student learning. Based on analysis, the profile of PK preservice teachers of Biology is presented in table 2.

Table 2. Pedagogical Knowledge Score Biology Preservice Teachers

No	Statement	M	SD	Criteria
1	I can use a variety of learning strategies	3,50	0,76	Adequate
2	I can manage and master the class well	3,50	0,76	Adequate
3	I can use various assessment methods and techniques	3,43	0,76	Adequate
4	I am aware of the possibility of misconceptions and learning difficulties experienced by students	3,57	0,85	Good
5	I do reflective actions to improve the quality of learning	3,57	0,76	Good
6	I can use a variety of learning strategies	3.51	0,76	Good
	Mean	3,50		Adequate

The average score PK of preservice teachers is 3.50, it is in the good criteria this indicates the ability of pedagogical preservice teacher has been good average that the highest on the both item statement is "I am aware of the possibility of misconceptions and learning difficulties experienced by students" and "I do reflective actions to improve the quality of learning"

3.3 Content Knowledge (CK)

Content knowledge is the "knowledge about actual subject matter that is to be learned or taught" (Mishra & Koehler, 2006, p. 1026). Teachers must know about the content they are going to teach and how the nature of knowledge is different for various content areas. Based on analysis, the profile of CK preservice teachers of Biology is presented in table 3.

Table 3. Content Knowledge Score Biology Preservice Teachers

No	Statement	M	SD	Criteria
1	I understand the concept of digestive system and transportation system and its application	3,57	0,65	Good
2	I follow the development of science, especially related to the digestive system and the circulatory system	3,57	0,76	Good
3	I use the latest reference sources (such as books and journals) to augment the knowledge that I have	3,71	0,83	Good
4	I attended a seminar related to the field of Biology	3,43	0,65	Adequate
5	I can design and implement Biology experiments for learning purposes	3,57	0,85	Good
	Mean	3,57		Good

The average score CK of preservice teachers is 3.57, it is in the good criteria this indicates the ability of content knowledge preservice teacher has been good average that the highest on the item statement "I use the latest reference sources (such as books and journals) to augment the knowledge that I have".

3.4 Pedagogical Content Knowledge (PCK)

Pedagogical content knowledge refers to the content knowledge that deals with the teaching process (Shulman, 1986). Pedagogical content knowledge is different for various content areas, as it blends both content and pedagogy with the goal being to develop better teaching practices in the content areas. Based on analysis, the profile of PCK preservice teachers of Biology is presented in table 4.

Table 4. Pedagogical Content Knowledge Score Biology Preservice Teachers

No	Statement	M	SD	Criteria
1	I can help students keep understanding the material of the digestive system using various strategies without using technology	3,79	0,70	Good
2	I can engage students actively in meaningful discussions on the material of the digestive system even without the use of technology	3,86	0,77	Good
3	I can make students actively solve real problems on the topics of the digestive system without using technology	3,64	0,74	Good
4	I can help change the thinking process of students in order to master a difficult topic on the material of the digestive system without using technology	3,29	0,61	Good
5	I can evoke meaningful reflection on the students on the material of the digestive system without using technology	3,50	0,52	Good
	Mean	3,61		Good

The average score PCK of preservice teachers is 3.61, it is in the good criteria this indicates the ability of pedagogical content knowledge preservice teacher has been good average that the highest on the item statement "I can engage students actively in meaningful discussions on the material of the digestive system even without the use of technology".

3.5 Technological Pedagogical Knowledge (TPK)

Technological pedagogical knowledge refers to the knowledge of how various technologies can be used in teaching, and to understanding that using technology may change the way teachers teach. Based on analysis, the profile of TPK preservice teachers of Biology is presented in table 5.

Table 5. Technological Pedagogical Knowledge Score Biology Preservice Teachers

No	Statement	M	SD	Criteria
1	I use computer apps in every lesson	3,79	0,70	Good
2	I choose technology that suits the approach and strategy of learning in the classroom	3,36	0,50	Adequate
3	I use internet facilities (like social networks, emails and blogs) to communicate with students	3,86	0,66	Good
4	I use online quiz (such as Quipper School, Edmodo, Socrative) to do the tests	3,43	0,76	Adequate
	Mean	3,61		Good

The average score TPK of preservice teachers is 3,61, it is in the good criteria this indicates the ability of technological pedagogical knowledge preservice teacher has been good average that the highest on the item statement "I use internet facilities (like social networks, emails and blogs) to communicate with students".

3.6 Technological Content Knowledge (TCK)

Technological content knowledge refers to the knowledge of how technology can create new representations for specific content. It suggests that teachers understand that, *by* using a specific technology, they can change the way learners practice and understand concepts in a specific content area. Based on analysis, the profile of TCK preservice teachers of Biology is presented in table 6.

Table 6. Technological Content Knowledge Score Biology Preservice Teachers

No	Statement	M	SD	Criteria
1	I use technology to help understand the concept of the Digestive System and Circulatory System	3,29	0,61	Adequate
2	I can download the application of computer and animation related to Digestive System and Circulatory System	3,57	0,51	Good
3	I can develop student activities and tasks that involve the use of technology	3,79	0,70	Good
4	I can create self-learning activities for students using ICT devices (eg webquest or flip)	3,43	0,65	Adequate
	Rata-rata	3,52		Good

The average score TCK of preservice teachers is 3.52, it is in the good criteria this indicates the ability of technological content knowledge preservice teacher has been good average that the highest on the item statement "I can develop student activities and tasks that involve the use of technology".

3.7 Technological Pedagogical Content Konwledge (TPACK)

Technological pedagogical content knowledge refers to the knowledge required by teachers for integrating technology into their teaching in any content area. Teachers have an intuitive understanding of the complex interplay between the three basic components of knowledge (CK, PK, TK) by teaching content using appropriate pedagogical methods and technologies. Based on analysis, the profile of TPACK preservice teachers of Biology is presented in table 7.

Table 7. Technological Pedagogical Content Knowledge Score Biology Preservice Teachers

No	Statement	M	SD	Criteria
1	I understand how to integrate Biological knowledge (especially the Digestion System and Circulatory System), pedagogical knowledge and technological knowledge	3,50	0,85	Adequate
2	I can choose a learning strategy and technology that match the material of the Digestive System and Circulatory System that will be delivered in class	3,50	0,65	Adequate
3	I can integrate knowledge of the content, pedagogical knowledge and knowledge of technology to realize meaningful learning	3,71	0,61	Good
4	I can apply different learning strategies and varying technologies in the implementation of Biology learning	3,50	0,52	Adequate
5	I can enable students to construct deep and meaningful insights into the material of the Digestive System and Transportation System using various technologies (such as Google site or concept maps)	3,71	0,73	Good
	Mean	3,58		Good

The average score of TPACK Biology teacher candidates is 3.58 are in good criteria this indicates TPACK candidate ability Biology teacher is good. Although the means to integrate the three content knowledge, pedagogy and technology are still sufficient criteria, the selection and application of technology in learning strategies is also sufficient, but prospective teachers are good enough in constructing meaningful knowledge in learning the transport system and the digestive system.

The preservice teachers has different capability in each component of TPACK. Preservice teacher has very good knowledge about technology, it shown by the technological knowledge mean was the highest. But alhtough the TK is high, the TCK is lower than TK it means the preservice teachers still doubt to apply technology in content of animal physiology. Both of pedagogic and technology treatment is better than others, but pedagogical treatment didn't also increase PCK most of participants. The lowest is pedagogical knowledge, it is cause most of the participants doesn't has an experience in teaching. The other component of TPACK (CK, TPK and PCK) almost the same. The capabilities of preservice teacher in every components of TPACK is describe in Figure 1.

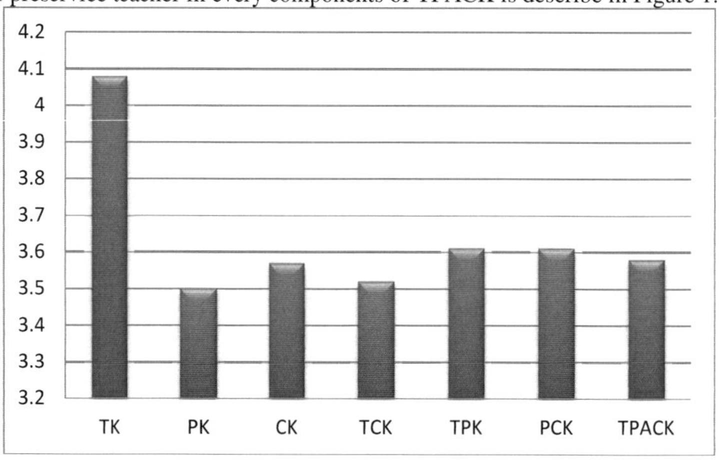

Figure 1. The capability of preservice teacher for each component TPACK

In general, the ability TPACK prospective teachers Biology University Galuh has good criteria with an average of 3.58. Prospective Biology teachers who are analyzed do not have a long teaching experience (less than 2 years old), and some do not have teaching experience. Animal Physiology Lectures that provide knowledge of content related to material that is mastered by the teacher form Content Knowledge preservice teacher. Technological treatment is given so that preservice teachers can improve their technology knowledge and pedagogical treatment is given so that preservice teachers have pedagogical knowledge. Thus it is hoped that preservice teachers can integrate all three when he will teach animal physiology.

4. Conclussion

The aim of this study is preservice teachers can certainly develop their technological, pedagogical, and content knowledge separately, but integrating these types of knowledge through the development of their TPK, TCK and TPACK gives them a more holistic view of their teaching and helps them transition from learners of Biology to teachers of Biology. Our data show that close attention must be paid to the relationship between the university classroom and the field placement; ideally, every preservice teacher would see that what they learn in the university classroom has an impact on their work in the field. Field placements are where preservice teachers face the reality of a classroom and experience first-hand that how they design tasks affects student learning. Using advanced methods classes puts preservice teachers in the position of being learners. This allows them to pay explicit attention to developing their TCK, which in turn encourages them to reflect on their PCK and CK. Thinking about, and engaging with, advanced technologies gives preservice teachers advantage point to examine their beliefs about, and attitudes towards, what it means for their students to be successful.

References

[1] Chai, C.S., Koh, J. H. L. & Tsai, C.C. Fascilitating Preservice Teacher's Development of Technological Pedagogical and Content Knowledge (TPCK). 2010. Journal Educational Technology and Society 13 (4), 63 – 73

[2] Harris, J., Mishra, P., & Koehler, M. Teacher's technological pedagogical content knowledge and learning activity types: Curriculum-based technology integration reframed. 2009. *Journal of Educational Computing Research*, 32 (4), 329-340.

[3] Lestari, Suci. Analisis Kemampuan Technological Pedagogical Content Knowledge (TPACK) pada Guru Biologi SMA dalam materi Sistem Saraf. 2015. (Tesis). Sekolah Pascasarjana, Universitas Pendidikan Indonesia, Bandung.

[4] Leys, J.M. & Marx, R.W. Teacher Knowledge of Educational Technology: A Case study of Student/Mentor Teacher Pairs. 2002. Journal Educational Computing Research. Vol. 26 (4), hlm. 427-462

[5] Mishra, P. & Koehler, M.J. What happens when teachers design educational technology? The development of Technological Pedagogical Content Knowledge. 2005. Journal Educational Computing Research. 32 (2), hlm. 131-152

[6] Mishra, P. and Koehler, M.J. Technological Pedagogical Content Knowledge: A Framework for Teacher Knowledge. 2006. Teachers College Record. 108 (6), 1017–1054.

[7] Mishra, P. and Koehler, M.J. What is Technological Pedagogical Content Knowledge? Contemporary Issues in Technology and Teacher Education. 2009. 9 (1), 60 – 70.

[8] NCATE. What make a Teacher Effective? National Council for the Accreditation of Teacher Education. Retrieved 29.04.17 from http://www.ncate.org/LINKCLICK.aspx?fileticket=JFRrmWqa1jU%3d&tabid=361

[9] Pujiastuti, E., Raharjo, T.J., Widodo, A.T. Kompetensi Profesional, Pedagogik Guru IPA, Persepsi Siswa tentang Proses Pembelajaran dan Kontribusinya terhadap Hasil Belajar IPA di SMP/MTS Kota Banjarbaru. 2012. Innovative Journal of Curriculum and Educational Technology Vol. 1 (1), hlm. 22-29.

[10] Srisaswadi. N. The Role of TPACK in Physics Classroom: Case Study of Preservice Physics Teachers. 2012. Procedia Social and Behavioral Sciences. [Online: 46 (3235 – 3234). Doi: 10.1016/j.sbspro.2012.06.043.

[11] Suryawati, E., Hernandez, Y. Firdaus L.N. Analisis Keterampilan Technological Pedagogical Content Knowledge (TPCK) Guru Biologi SMA Negeri Kota Pekanbaru. Jurnal Biogenesis. 2014. 11 (1), hlm. 67-72

[12] Widodo, A. 21QL Integration (TPACK) Survey. 2017.

Implementing Google Apps for Education as Learning Management System in Math Education

S Widodo*

Universitas Pendidikan Indonesia, Jl. Dr. Setiabudhi No. 229, Bandung 40154, Indonesia

*supri@upi.edu

Abstract. This study aims to find the effectiveness of math education using Google Apps for Education (GAFE) as learning management system to improve mathematical communication skill primary school preservice teacher. This research used quasi-experimental approach, utilizing the control group pre-test – post-test design of two group of primary school preservice teachers at UPI Kampus Purwakarta. The result of this study showed that mathematical communication skill of primary school preservice teacher in the experiment group is better than the control group. This is because the primary school preservice teacher in the experiment group used GAFE as a tool to communicate their idea. The students can communicate their idea because they have read the learning material on the learning management system using GAFE. All in all, it can be concluded that the communication tool is very important, beside the learning material, and also the options to choose the learning model to achieve the better result.

1. Introduction

Out of the many mathematical skills that must be mastered by students as stated in the objectives of the 2006 Indonesian Curriculum (KTSP), mathematical communication skill is one of the skills that is expected to be obtained after students learn mathematics [1]. This also corresponds to Wichlet that communication is a vital skill in mathematics [2], and NCTM which state that mathematical communication is a way of sharing ideas and clarifying understanding. Through communication, ideas become objects of reflection, refinement, discussion, and amendment [3]. Mathematical communications skill include: 1) understand math concepts, explains the relationship between concepts, and apply the concepts or algorithms in a flexible manner, accurate, efficient, and precise in problem solving; 2) Using the pattern and nature of the reasoning, mathematical manipulation in making generalizations, compile evidence or explain ideas and mathematical statement; 3) Solve problems that include the skill to understand the problem, devised a mathematical model, solve the model and interpret the obtained solution; 4) Communicate ideas with symbols, tables, diagrams, or other media to explain the situation or problem; 5) Have respect for mathematics in life, which has a curiosity, attention and interest in studying mathematics, as well as a tenacious attitude and confidence in problem solving [1].

Several studies and efforts have been conducted to develop students' mathematical communication skills related to the use of models and media such as [4], [5], [6], [7], only few with regard to the use of technology. However, University has a very high dependency on the use of ICT in many aspects, both in the terms of the learning process as well as in management education. Utilization of ICT in learning has become a primary need for students and teachers, so that the availability and the selection of appropriate *tools* to be used will also facilitate the process of education in the University. The availability of online learning for students should be a main concern. In addition, ICT will provide the online learning process with learning resources that can be accessed anytime and anywhere. ICT also develop a university quality in providing infrastructure by the use of technology in learning. *Google Apps for Education* (GAFE) is an information technology solution that can be used by schools/colleges so that the entire academic community can communicate and collaborate. Students and lecturers can share ideas more quickly and effectively in obtaining information, creating documents, spreadsheets, presentations, and collaborating with each other in real time.

This research uses Google Site as one of the features provided by GAFE. Google site on Google Apps and other support tools, had a possibility to develop a learning management system, because GAFE could provide learning materials, assignments, discussion forums, quizzes and others. This study analyze and design a *learning management system* for math education as a case to implement online learning that provide by Google Apps for Educations

1.1 Learning Management System

Learning Management System (LMS) is needed to support online learning. The LMS should meet some requirement such as: possibility for administrators, teachers and students to register on curriculum; provide a link between curriculum and instructional administration on learning content; supporting virtual learning; provide for various analyzes and measurements, as well as group or sole progress tracking; administering tests, assessments and reports; administrative system; management of capabilities so that an organization can measure learning needs and identify development needs[8],

There are few things that must be considered when developing LMS [9] that can be reviewed by these following factors: 1) User and availability: This factor can be seen from the speed of internet connection required; Special system requirements on client; The use of web browser; Other functional requirements on LMS; Documentation and help for users; 2) Usability: Some of the important things on this factor include the user interface, and the most important thing in the interface is the navigation that should be easy to learn and use, has a consistent structure and tailored to the general habits of the users; Presence of site-map; the availability of search function; help functions and language options; 3) Learning environment: The learning environment factors has several functions for teachers and students and divided into a communication environment where teachers and students are not simultaneously learning (asynchronously) and in real time (synchronous); Distribution environment, ie in what form/learning format distributed to the user; test environment, ie how many types of tests/ tests that can be done; the interaction environment, which is the possibility for participants to influence the environment.

A focus on LMS development to support student learning in higher education institution should paid attention to the theoretical, pedagogical and technological problems because most of problems reported concerning LMS, center on underutilization of interactive features and the low level of student engagement of the system [10]

This study implemented GAFE as LMS using flipped classroom model because GAFE offers embedded features that promote collaboration and exchange of information. The integration of the various tools comprised in the Google Apps portfolio offers a powerful framework that ranges from basic communication to video streaming tools. Flipping a classroom becomes much simpler when these features are present. [11]

2. Experimental Method

This study was implemented from July - December 2014 at UPI Kampus Purwakarta. This study used a quasi-experimental approach and the Control Group Pre-test – Post-test Design [12]. The first group of subjects as the experimental group use the LMS that developed by using GAFE and the second group as a control group use conventional learning as the control group. In order to obtain the necessary data, this study used research instruments such as achievement test of mathematical communication skill, the format of observations during the learning process, and student journal.

The research procedure started by the identification of the problem to several classes of primary school pre service teacher level 2 UPI Kampus Purwakarta, then the activities carried out with details as follows: literature study, develop research instruments and instructional design; system development; testing system with black box testing; Implementation; data analysis and hypothesis testing of data that has been collected; building conclusions to obtain the expected goals. To obtain the necessary data, this research using the research instruments in the form of test and observation format during the learning process.

This study determine how GAFE implemented in the mathematic classroom to support the regular class and the effect of its implementation to pre service primary school teachers mathematical communication skill improvement.

3. Result and Discussion

The LMS design is made specifically for learning that is able to display and manage learning by two role of users, that is tutors and students. In general, the LMS is designed in accordance with the standard features of a distance learning. The LMS designed adjustable with a W3C consortium standard for a site and regard to the Best Practice Statement and the Default Delivery Context. The Implementation of LMS Mathematics Education develop by Google Site which was very easy to use, it is equipped with others features of GAFE such as Google Drive, Google Calendar, Google docs, etc. In general, there are no significant obstacles during the system *deployment* process with sufficient connection to the implementation environment.

The first step of this research is develop the instructional design, test instruments and observation guide and then develop a system process that deployed at http://www.supri.upi.edu, and the learning content entirely created using tools provided by GAFE such as Google Docs, Google Videos, and Google Slide.

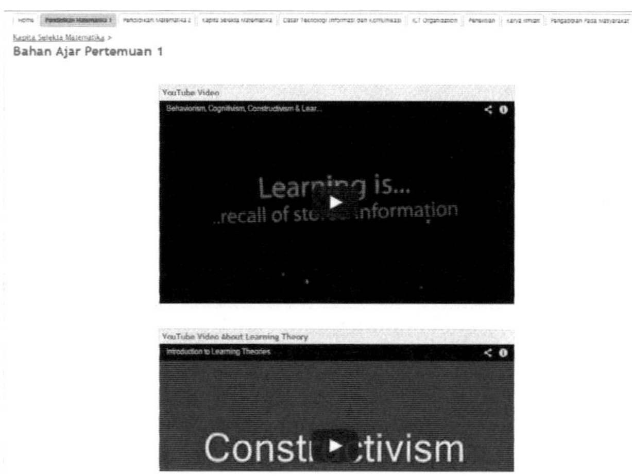

Figure 1. The LMS with a learning content sample development using google site and youtube videos

After the completion of the learning content required deployed, the next step is to test this LMS with the black box testing scenarios. Black box testing is a term of testing a system by simply observing the results of execution through the data test and functional check of the system. Black box testing identical with the evaluation process from the point of view of the outside (interface) and its functionality without know what actually happened in the full process [12].

The Implementation of the learning process using GAFE as LMS conducted in accordance with a schedule. The learning steps in the experimental group using *flipped classrooms* as follow: 1) The students are required to learn independently at home about the learning materials, watch instructional videos or presentations provided by the LMS and communicate through google talk or email; 2) In the classroom, the students were divided into several heterogeneous groups; 3) The role of the lecturers during the learning activities is to facilitate the ongoing discussions with *cooperative learning* methods. In addition, teachers will also prepare some questions (questions) of the material; 4) Lecturer give quizzes / tests, as a facilitator in helping students in learning and problem solve problems related to the material using Google Form, Google Docs and Google Sheet.

In the learning process, students were given opportunity to communicate orally and in writing related course materials. Student activity is more dominant, debriefing and very active discussion conducted in this group because they already know and some have understood the material that will be discussed.

The learning process in the control group is using conventional design. The learning process is generally performed by the discussion method and debate. But in this class, although the opportunity has been given the widest possible to do a question and answer, discussion and reflection, student activity is still not as good as the experimental class

3.1 Learning Effect

The pretest data obtained by providing mathematics communication skill test to students in the experimental and control group. It aims to determine the initial student mathematics communication skill.

Table 1. The score of student mathematics communication skill pre test

	N	Minimum	Maximum	Sum	Mean	Std. Deviation	Variance
Experiment group	40	2	12	210	5.25	2.968	8.808
Control Group	38	2	12	200	5.26	3.142	9.875

From the data shown in table I the initial skill of student mathematical communication in learning mathematics has similarities, that was very lack. The mean different test of the pre-test data by U-test showed that there is no different mean between the experiment and control group pre-test score.

The post-test data obtained by providing mathematics communication skill test to students in the experimental and control group after using the GAFE as LMS in the experiment group and direct instruction in the control group, and the result showed in the table 2.

Table 2. The score of student mathematics communication skill post test

	N	Minimum	Maximum	Sum	Mean	Std. Deviation	Variance
Experiment group	40	11	15	544	13.60	1.194	1.426
Control Group	38	5	15	429	11.29	2.470	6.103

Table 2 shows that there is a different mean of post test score of student mathematical communication skill. The t-test used to test to the data show that there is a significant mean difference between the experiment and control group on the LoC 95%. So that the instructional using GAFE as LMS with flipped classroom model has a good effect on student mathematical communication skill. The student mastery learning in the experimental group show that 100% primary school preservice teacher pass the exam with minimum C score, while the control group only 60.53%.

The pre-test, post-test data showed that there was a significant improvement of student mathematics communication skill in the experiment group. This was caused that the student in the experiment group has been given a new method of learning math using learning management system developed from GAFE. Students learn math individually from the LMS to learn the learning material repeatedly, and they can communicate, collaborate and sharing their documents needed using google mail and google talk to discuss with their friends. The students also communicate their idea, making a sentences or paragraphs to solve a problem together with graphs, symbols, tables using google docs and google sheet. The students were provided by the exercise so they can try to understand the concept, reasoning, and learn the pattern as the curriculum goals with google form. [1].

4. Conclusion

The availability of online learning for students in higher education (university) should be a priority because it can be accessed anytime, anywhere. In addition, LMS was the most frequently used tool in higher educational learning [10]. Beside the learning model and method, the result of this study shows that GAFE is one LMS that proves to be used in learning math and improve communication skills of primary school preservice teacher. Further research can be focused on other mathematical abilities with a wide range of development and materials

Acknowledgments

This work is supported by the University of Indonesia Education for young lecturers and development of mathematics learning at UPI Purwakarta Campus. The appropriate author is Suprih Widodo, supri@upi.edu

References

[1] Depdiknas. 2006. *Kurikulum Tingkat Satuan Pelajaran*. (Jakarta: Depdiknas)
[2] Wichlet, L. 2009. Communications: A Vital Skill in Mathematics (University of Nebraska – Action Research Project 18)
[3] NCTM. Principles and Standards for School Mathematics. NCTM: Reston VA
[4] Cai J, Jakabcsin S, Lane S. 1996. *J. School Science and Mathematics*. 96 5
[5] Bell, M. & Carr, P. 2014. J. *Innovation in Science and Mathematics Education*. 22 (4), 65-78
[6] Clark, K.K. et all. 2005. J. *Current Issues in Middle Level Education*. 11(2), 1-12
[7] Rahmawati, F. 2013. *Proc. Nat. Conf. on Math and Informatics*. (Lampung: Indonesia/UNILA)
[8] Dye, A dan Fagerberg, T. 2004. *Exploring online service in a mobile environment*. (Technical Working Paper: NKI Project)
[9] Hultin, J. 2006. LMS, A Review. Retrieved from http://se2.isn.ch/serviceengine/FileContent?serviceID= 18&fileid=09FD0FDC-723F-5E63-CCDB-6AFCDA98BCC8&lng =en.
[10] Alhazmi A. K. & Rahman, A.A. 2012. IEEE *Symp. On E-Learning, E-Management and E-Services* (Kuala Lumpur: Malaysia/ IEEE Xplore)
[11] Ferreira, J M M. 2014. *Proc. Int. Conf. on Remote Engineering and Virtual Instrumentation*. (Porto: Portugal/IEEE Xplore) p 204
[12] Borg, W.R. & Gall, M.D. 1989. *Educational Research. (*New York: Longman)
[13] Patton, R. 2000. *Software Testing*. (Indianapolis: Sams Publishing)

Contribution of Auditory Learning Style to Students' Mathematical Connection Ability

Karlimah and F Risfiani
Department of Elementary School Teacher Education, Universitas Pendidikan Indonesia, Tasikmalaya Campus
Jl. Dadaha N0. 18 Tasikmalaya 46115 Indonesia

*karlimah@upi.edu

Abstrak. This paper presents the results of the research on the relation of mathematical concept with mathematics, other subjects, and with everyday life. This research reveals study result of the students who had auditory learning style and correlates it with their ability of mathematical connection. In this research, the researchers used a combination model or sequential exploratory design method, which is the use of qualitative and quantitative research methods in sequence. The result proves that giving learning facilities which are not suitable for the class whose students have the auditory learning style results in the barely sufficient math connection ability. The average mathematical connection ability of the auditory students was initially in the medium level of qualification. Then, the improvement in the form of the varied learning that suited the auditory learning style still showed the average ability of mathematical connection in medium level of qualification. Nevertheless, there was increase in the frequency of students in the medium level of qualification and decrease in the very low and low level of qualification. This suggests that the learning facilities, which are appropriate for the student's auditory learning style, contribute well enough to the students' mathematical connection ability. Therefore, the mathematics learning for students who have an auditory learning style should consist of particular activity that is understanding the concepts of mathematics and their relations.

1. Introduction

Mathematics is one of the structured subjects, having a nature that is continuous from simple concepts to more complex concepts. This nature indicates that interdisciplinary concepts in math subjects are related to each other or having mathematical connections. The ability to make connections in mathematics is a standard of the process of mathematical power required by everyone in solving math problems and daily life problems [5]. This mathematical connection can be developed in students in effective mathematics learning. One of the requirements to implement effective learning, according to Slameto [8], is "teachers need to consider individual differences". Learning style is one of the individual differences that must be considered by the teacher. In line with that statement, May [4] states, "we all use a range of learning styles at different times, and the most effective mathematical thinkers are flexible". This statement shows that by utilizing the learning style of students, we will have more effective math learning. Another opinion is given by May [4], "pupils may adopt different styles as they first explore and understand, and then rehearse and apply, each new concept". These two opinions indicate that learning styles can determine how students understand something initially, as an ingredient to be applied in other new concepts. This application of a new concept is known as students' mathematical connection ability. Thus, there is a link between learning styles and the students' mathematical connection ability.

The mathematical connection ability is very important for students. The linkage between learning styles and mathematical connection ability can be utilized by teachers to develop mathematical connection ability. The main thing that teachers need to know is the students' learning style. The statement about the need for teachers to understand the learning styles of students is based on relevant research results obtained previously. In the studies conducted by Halim [3] and Sagitasari [7], they found that learning styles can influence students' learning outcomes and achievement. According to Purwanto [6], learning outcomes are often used as a measure to find out how far a person has mastered the material already taught. Thus, based on the learning results, the students' mathematical connection ability can be determined.

Based on these statements, teachers must recognize the learning styles of students and develop what kind of learning can be carried out in accordance with the learning style of students to develop students' mathematical connection ability. Since there are no studies which have revealed the contribution of learning styles to students' mathematical connection ability and also considering the importance of teachers' knowledge about the learning style of students to create effective learning, it was necessary to conduct a research to find out how much contribution of auditory learning style to students' mathematical connection ability. This was done as an effort to create effective learning to build students' mathematical connection ability in accordance with students' learning style.

2. Experimental Method

In this research, the researchers used a combination model or sequential exploratory design method which is the combination of qualitative and quantitative research methods in sequence. Qualitative research methods were used to set the setting of auditory learning styles and the mathematical connection ability. The instruments used were a learning style questionnaire and math test for students who had an auditory learning style. The findings of hypotheses in the first stage of the research were then tested by the subsequent research by using quasi-experiment nonequivalent control group design. The population was the sixth-grade students of an elementary school in Kecamatan Tamansari Tasikmalaya City. The sample was determined by purposive that is the sixth-grade students of SD Setiamulya. Then, the data collection was conducted by giving the test, then continued with data analysis by using SPSS software. Here are the main steps in this type of research [9].

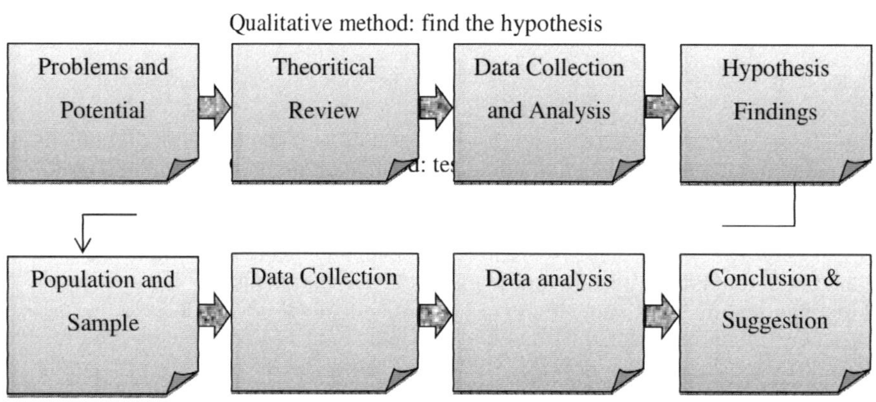

Figure 1. Combination Method, Sequential Exploratory Design

3. Result and Discussion

Based on the results of data collection, some findings were found related to the auditory learning style in elementary school in Kecamatan Tamansari Tasikmalaya city. The home room teacher at the school, had an auditory teaching style that was recognized from some characteristics appeared during the lesson. When teaching, the teacher did not tend to use the media, the use of books and student worksheet did not dominate the learning process, the teacher also created self-made questions spontaneously and asked them orally, the learning activities were mostly lectures and question and answer sessions, and discussing problems in a discussion was more preferable. These characteristics made it possible to create a learning environment which was suitable for students who had an auditory learning style. The learning conditions were formed based on the principle of bringing the world of students to the world of teachers and delivering the world of teachers to the world of [2]. Each interaction with the students, each curriculum design, and each interactional method was made by considering the student's character and the demands of the curriculum as well as the teaching facilities implemented by the teacher in accordance with the student's learning style. In this case, the teacher had learned the auditory learning styles possessed by the students which were seen during the learning, so the teacher provided auditory learning facilities. Thus, the provision of learning facilities in the classroom was implemented in accordance with students' learning style. Adjustment of learning to the students' learning style is a balancing effort between internal and external stimuli [1]. Therefore, a learning, which is implemented in accordance with the students' learning styles, can enhance motivation, increase the value of learning, increase self-confidence, improve self-respect, maintain a positive attitude and continue to utilize skills [1]. Therefore, learning in the auditory classroom with auditory teaching style can influence well the development of students' mathematical connection ability.

Furthermore, to determine the students' mathematical connection ability with learning facilities that suited the students' learning style, a test was conducted. The results of the tests, which performed before (pretest) and after the learning (posttest), showed the existence of students' mathematical connection that was at medium qualification with the frequency was 12 students (44,4%). These results indicated that student's mathematical connection ability was almost sufficient.

In developing the mathematical connection ability, varied learning activities were undertaken in accordance with students' learning style, so the results were better than before (Table 1 of the Posttest column). Based on the pretest and posttest data, it was found that students' mathematical connection ability after the learning demonstrated good progress and there were also a change that is an increase in high-level qualification as much as 22.1% and a decrease in very low and low level of qualification. So that the provision of learning facilities in accordance with auditory learning style indicated the development of mathematical connection ability in students who possessed very low, low and high level of qualification of mathematical connection ability. It can be seen in Table 1 as follows:

Table 1. Recapitulation of Frequency of Students' Mathematics Connection Ability in Auditory Class

No	Qualification	*Pretest*		*Posttest*	
		Student Frequency	Percentage	Student Frequency	Percentage
1	Very High	0	0 %	0	0 %
2	High	2	7.5 %	8	29.6 %
3	Medium	12	44.4 %	12	44.4 %
4	Low	7	25.9 %	5	18.5 %
5	Very Low	6	22.2 %	2	7.5 %
	Total	27	100 %	27	100 %

To obtain valid results, a covariance analysis was performed with the aim of controlling the effects of other uncontrolled variables. For example, the view that each student has another learning style (visual or kinesthetic), in addition to the auditory learning style. Thus, in order to overcome the effects of other learning styles that might have an effect on the students' mathematical connection ability during pretest, the control of possible influences was made, then a covariant analysis (ANKOVA) was performed. The covariance analysis was performed by using SPSS 16.0 application and the following regression result was obtained:

Regression equation : $Y = (\beta\_o + \beta\_i) + (\delta\_o + \delta\_i) * GB$

Description : β_o = Intercept of Learning Style

B_i = Learning Style Coefficient

IOP Conf. Series: Journal of Physics: Conf. Series **895** (2017) 012058 doi:10.1088/1742-6596/895/1/012058

Δ_o = Covariates
Δ_i = Correlation coefficient
$Y = (23.901 + 0) + (0 + 0.638) * GB2.00 = 23.901 + 0.638GB2.00$

Based on the regression equation, it can be seen that the coefficient of students' auditory learning style was categorized as high. Although there were other things that affected (23,901). Furthermore, other SPSS output results showed R Square value of 0.507. This value indicates that the learning style contributes moderately to the students' mathematical connection ability as much as 50.7%. (Table 2)

Table 2. Tests of Between-Subjects Effects

Dependent Variable: Mathematical Connection

Source	Type III Sum of Squares	df	Mean Square	F	Sig.
Corrected Model	6521.164[a]	2	3260.582	26.779	.000
Intercept	3237.302	1	3237.302	26.588	.000
GB	1387.436	1	1387.436	11.395	.001
X	4406.535	1	4406.535	36.191	.000
Error	6331.350	52	121.757		
Total	181789.300	55			
Corrected Total	12852.514	54			

a. R Squared = .507 (Adjusted R Squared = .488)

Furthermore, the significance level for the mathematical connection ability of auditory learning style of 0.001 was higher than other learning styles of 0,000 (Table 3). It means that there was an influence on the mathematical connection ability of the learning given by the teacher in accordance with the learning style of the students. This indicated that the learning, which was in accordance with the students' learning style, affected the development of students' mathematical connection ability. The learning that was not in accordance with the learning style of students did not result in the better mathematical connection ability compared to the learning which was in accordance with the students' learning style.

Table 3. Parameter Estimation for Advanced Test

Dependent Variable: Final Mathematical Connection Ability

Parameter	B	Std. Error	T	Sig.	95% Confidence Interval	
					Lower Bound	Upper Bound
Intercept	21.083	5.120	4.118	.000	10.810	31.356
[GB=1.00]	10.128	3.000	3.376	.001	4.107	16.148
[GB=2.00]	0[a]	.	.	.	.	.
X	.710	.118	6.016	.000	.473	.946

a. This parameter is set to zero because it is redundant.

4. Conclusion

The mathematics learning with auditory learning style facilities for students who had an auditory learning style contributed well to students' mathematical connection ability. The good mathematical connection ability of students who have an auditory learning style was determined based on the initial mathematical connection ability which was in sufficient level. Based on the pretest and posttest results of students with auditory learning style, there was an increase in the mathematical connection ability.

Based on the initial findings, students had medium level of qualification of mathematical connection ability. However, after the learning with auditory learning style, the average mathematical connection ability of students with auditory learning style increased in high, very low and low level of qualification.

Based on the results of data analysis, it was found that the learning style contributed 50.7% to the students' mathematical connection ability. The results showed that learning, which was ajusted to the learning style of students, made good contribution to the ability of students' mathematical connection. Therefore, to improve the mathematical connection ability of students who have an auditory learning style, it is necessary to provide learning that match the learning style of the students as an effort to realize effective learning and optimal results. The main conclusion obtained from this research is that the learning activities carried out by the teacher, which are in accordance with students' learning style, affect the development of students' mathematical connection ability.

Acknowledgments
Hopefully, this article can encourage school teachers to implement mathematics learning which takes into account learning conditions in which can provide suitable learning facilities and learning targets to develop mathematical components (mathematical connections) of students. Thus, the effectiveness of learning can be achieved, and the results of mathematics learning will be not only about how to solve math problems but also able to achieve students' math power as well. Therefore, I would like to thank you for the trust in me to submit this article.

References

[1] DePorter, B. & Hernacki, M. (2011). *Quantum Learning*. Bandung: KAIFA
[2] DePorter, B., Reardon, M., & Nourie, S.S. (2010).*Quantum Teaching*. Bandung: KAIFA
[3] Halim, A. (2012). Pengaruh Strategi pembelajaran dan Gaya Belajar terhadap hasil Belajar Belajar Fisika Siswa SMPN 2 Secanggang kabupaten Langkat. *Jurnal Tabularasa PPS UNIMED*, 9(2), hlm.142-158.
[4] May, T.C. (2005).*Teaching Maths to Pupils with Different Learing Styles.*London: Paul Chapman Publishing.
[5] National Council of Teachers of Mathematics (NCTM). (2000). *Principles and Standards for School Mathematics*. Reston, Va.styles
[6] Purwanto. (2014). *Evaluasi Hasil Belajar*. PUSTAKA PELAJAR: Yogyakarta
[7] Sagitasari, D.A. (2010). *Hubungan antara Kreativitas dan gaya Belajar dengan Prestasi Belajar Matematika Siswa SMP*. (Skripsi). Universitas Negeri Yogyakarta.
[8] Slameto. (2010). *Belajar dan Faktor-faktor yang Mempengaruhinya*. Jakarta: Rineka Cipta.
[9] Sugiyono. (2015). *Metode Penelitian Kombinasi (Mixed Methods)*. Bandung: Alfabeta

A Cognitive Analysis of Students` Mathematical Problem Solving Ability on Geometry

N A Rusyda*, K Kusnandi and S Suhendra

Departemen Pendidikan Matematika, Sekolah Pascasarjana, Universitas Pendidikan Indonesia, Jl. Dr. Setia Budhi No. 229, Bandung 40154, Indonesia

*nurul140892@gmail.com

Abstract. The purpose of this research is to analyze of mathematical problem solving ability of students in one of secondary school on geometry. This research was conducted by using quantitative approach with descriptive method. Population in this research was all students of that school and the sample was twenty five students that was chosen by purposive sampling technique. Data of mathematical problem solving were collected through essay test. The results showed the percentage of achievement of mathematical problem solving indicators of students were: 1) solve closed mathematical problems with context in math was 50%; 2) solve the closed mathematical problems with the context beyond mathematics was 24%; 3) solving open mathematical problems with contexts in mathematics was 35%; And 4) solving open mathematical problems with contexts outside mathematics was 44%. Based on the percentage, it can be concluded that the level of achievement of mathematical problem solving ability in geometry still low. This is because students are not used to solving problems that measure mathematical problem solving ability, weaknesses remember previous knowledge, and lack of problem solving framework. So the students` ability of mathematical problems solving need to be improved with implement appropriate learning strategy.

1. Introduction

Mathematical problem solving ability is one of process standard in learning mathematics [1]. The ability of mathematical problem solving is the heart of mathematics [2]. Given the importance of problem solving ability in learning mathematics, it is necessary to do an effort that aims to determine the extent to which the problem solving ability based on students` answer on geometry. These efforts can help educators to find ways to solve the problems in mathematical problem solving of students. In fact, students` mathematical problem solving ability is still low. The students have difficulties to solve a problem that measures problem solving ability. Based on the research, the students` problem solving ability has low score on aspect of make model that is 23,53% and interpret result obtained that is 32,35% which including very bad qualification [3]. Other research said that, students' problem solving ability is low because students are not familiar to solve problem solving problems (non-routine problem) [4]. If the mathematical problem solving ability compared to other mathematical abilities such as mathematical communication ability, students' problem-solving ability also has lower average score than the average score of mathematical communication skills. This is seen in the results of research, the average score of pretest and postes in the control class and experiment to measure the problem solving ability is lower than the ability of mathematical communication. So that, we should to analyze the ability of mathematical problem solving if we want to improve the student`s ability[5].

Another thing that is important in view of the extent of the problem solving ability of students is see what kind of the students` errors that student made in solving mathematical problems. Error analysis has proven to be an effective method for identifying patterns of students' mathematical errors. Based on that opinion research related error analysis is very important and should be developed so that the teachers know what mistakes made by the students so that teachers can identify and review mistakes that are often made by students who in the end of the teaching process teachers can choose the right strategy for students no more mistakes [6]. The type of error is categorized into 4; facts error, concepts error, operations error and principles error [7]. Other opinion said that student errors are classified into several types of errors: type I, type II, and type III errors. Type I errors include the concept of understanding that students know what is known and know what is being asked. Type II errors include stages of problem solving that is students make a mathematical model, have accuracy in solving the problem, and true or not in operate. Type III errors are true or not students in conclusion [8]. This research is focused on the analysis of students errors in solving mathematics problems with the type of errors are concept error, principle error, operation error, and mistake of conclusion. The purpose of this research is to analyze of mathematical problem solving ability of students in one of secondary school on geometry.

2. Experimental Method

This research was conducted by using quantitative approach with descriptive method. Population in this research was all students of one secondary school and the sample was twenty five students that was chosen by purposive sampling technique. Data of mathematical problem solving were collected through essay test. The topic of the test tested was geometry "quadrilateral".

3. Result and Discussion

The results of this research showed the percentage of achievement of indicator mathematical problem solving ability of students were: 1) solve closed mathematical problem with context in math was 50%; 2) solve closed mathematical problems with contexts outside mathematics was 24%; 3) solving open mathematical problems with contexts in mathematics was 35%; And 4) solving open mathematical problems with contexts outside mathematics was 44%. Based on the percentage of achievement of each indicator, it can be concluded that the level of achievement of mathematical problem solving ability in solving the problem in segiempat topic by using conventional learning still low with the average indicator was 38.25%, so the students` ability of mathematical problem solving on quadrilateral topic need to be improved. This can be seen from the analysis of the following student answers.

The problem number 1 with indicators solving closed mathematical problems with contexts in mathematics is : " Length DC = OE = 5cm, FO = OD = 6cm, dan OG = GA = 8cm. If the area of the shape below is 238cm^2, calculate the area of shaded shape !

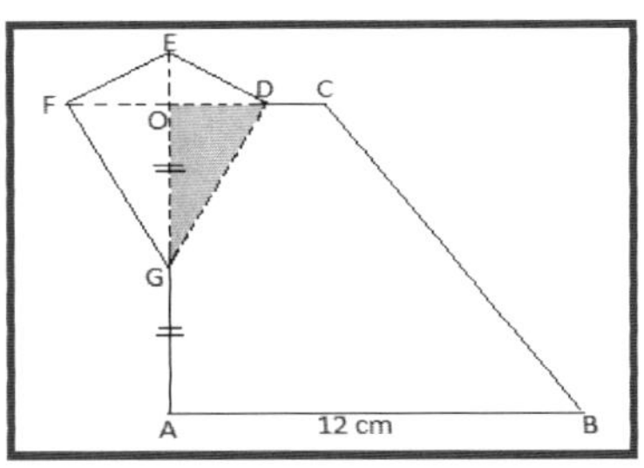

Figure 1. The first problem of mathematical problem solving based on first indicator

Most of students have difficulties because the average score of this indicator was 50%. This can be seen from the students' answers. Based on these answers, students can be grouped as follows. Figure 2 shows that, the students simply rewrite the question without answering questions. This caused by the students were not able to write down ideas or planning in solving the problem. So, we can conclude that the students` mathematical problem solving ability are still low.

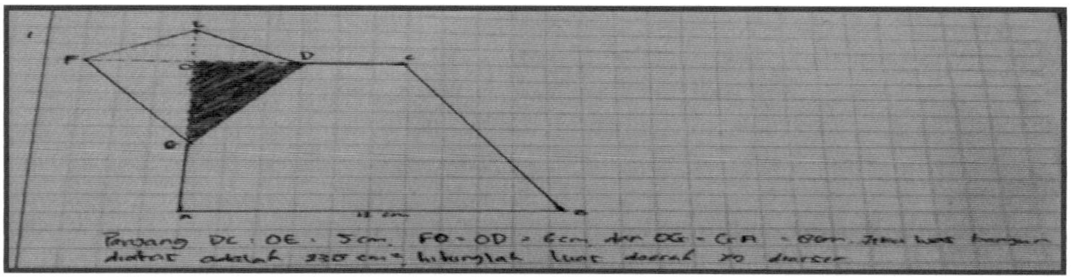

Figure 2. Student`s answer

Figure 3 shows that, the student only determine the area of trapzoid and paralellogram without determine the shaded area. Students are still having difficulties to understanding the questions. But, the students already understand the procedure to get the correct answer. It is seen from the student's answer that wrote the right formula to solve that problem.

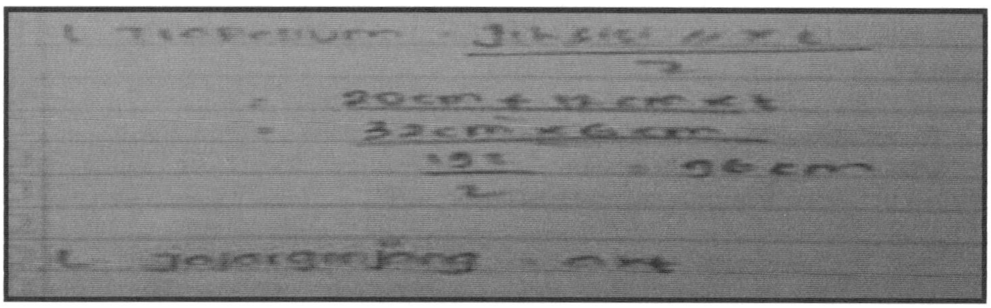

Figure 3. Student`s answer

Figure 4 shows that, students have difficulties to understand the picture in that problem so that the students wrong in choosing the correct formula to determine the area. The students have difficulties to remember the formula on segiempat because there are so many formula for each plane.

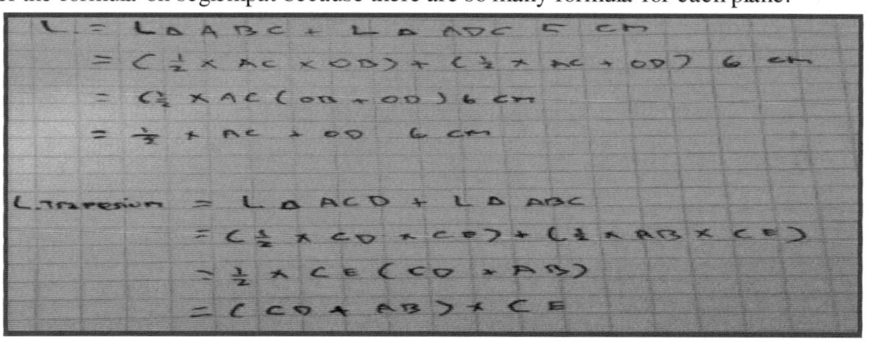

Figure 4. Student`s answer

Figure 5 shows that, it appears that most of students make miscalculations in operating numbers. Based on all the examples of students` answer, we can conclude that there are some students can answer with the correct answer and some students answer with the wrong answer. The most students` error in this problem is operation error.

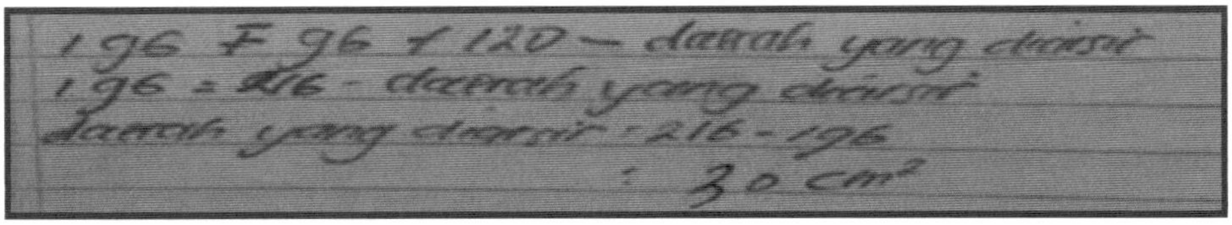

Figure 5. Student`s answer

The problem number 2 with indicators solving closed mathematical problems with contexts outside mathematics is: "The floor of living room shaped square with a length of side is 5 m. The living room will be fitted with tiles (sized 50 cm x 50 cm). Calculate how the many tiles needed to cover the floor! If 1 box of tiles cost Rp 70.000,00 and contains 4 pieces of tile. Calculate the total price paid for the purchase of tiles!"

Most of students have difficulties to answer the problem number 2. It can be seen by the average score of this indicator was 24%. The average score in this indicator is the lowest than other average score This can be seen from the students' answers. Figure 6 shows that, it appears that students have difficulties in calculating the many tiles required. It seen from some students only calculate the area of floor and the area of tile, but not proceed to the next procedure. So that, the student does not know how many tiles are required and the total price is paid.

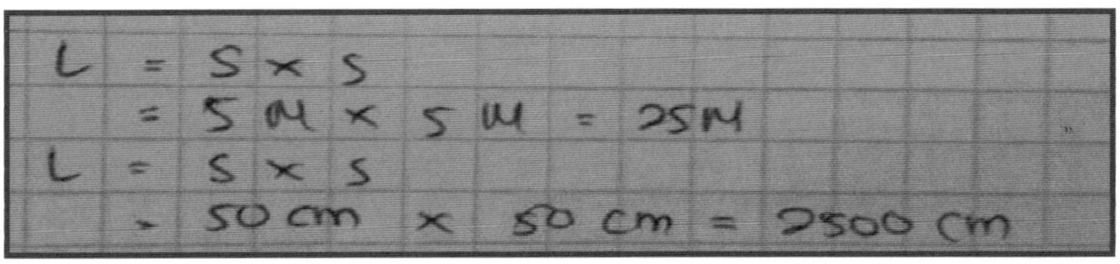

Figure 6. Student`s answer

Figure 7 shows that, the students are only guessing the problem solving. This was evident from the student's answer that does not lead to the correct answer. The student does not determine the area of floor and the area of tile.

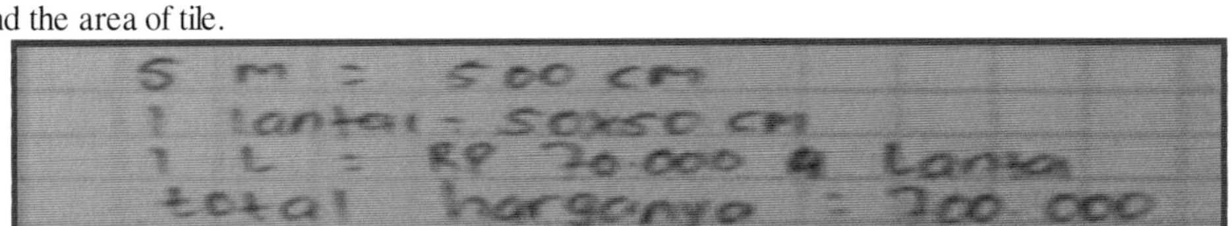

Figure 7. Student`s answer

Figure 8 shows that, it appears that most students make miscalculations in operating numbers. The student already answer by the correct procedure. The student already calculate the area of the floor and change the unit of area from *m* to *cm* to determine how many tiles are required by the correct procedure, eventhought the student still wrong in write the unit of area (do not write cm^2 or m^2). But, the student make miscalculation in operating numbers. The student wrong in determine the total price is paid. The correct answer is Rp 1.750.000,00.

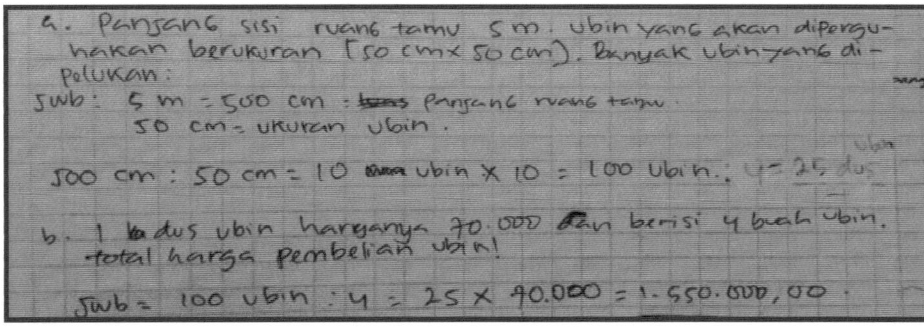

Figure 8. Student`s answer

The problem number 3 with indicators solving open mathematical problems with contexts in mathematics is: "The area of rhombus ABCD is = 300cm^2. Calculate the length of diagonals!". Most students have difficulties because the average score of this indicator was 35%. This can be seen from the students' answers. Figure 9 shows that, it appears that students simply write down the formula without proceeding to the proper procedure to find the correct answer. Students do not understand that the given problem was a matter of open ended, so the students were confused about what the question asked. Most of the other students were able to answer correctly by guessing although not using the proper procedure.

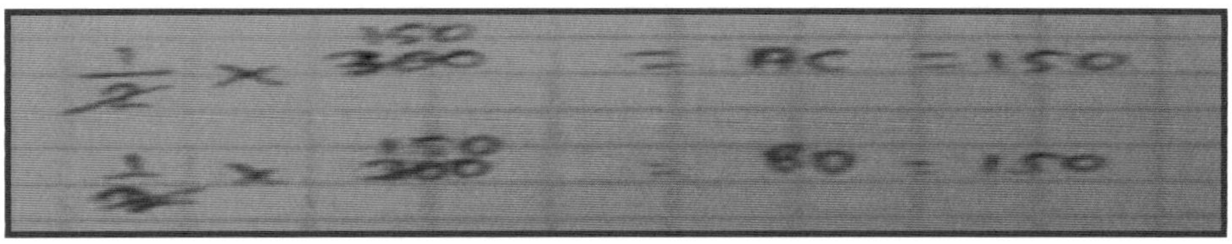

Figure 9. Student`s answer

The problem number 4 with indicators solving open mathematical problems with contexts outside mathematics is: " A yard shaped rectangular with the area is 100 m^2. If around the yard will be placed flower pots with a distance between pots 1 m. How many flower pots are needed?". Most of students have difficulties because the average score of this indicator was 44%. This can be seen from the students' answers. Figure 10 shows that, it appears that the student just wrote the formula without continuing to the proper procedure to find the correct answer. Students do not understand that the given problem was a matter of open ended, so the students are confused about what the question asked. Most of the other students were able to answer correctly by guessing despite not using proper procedures.

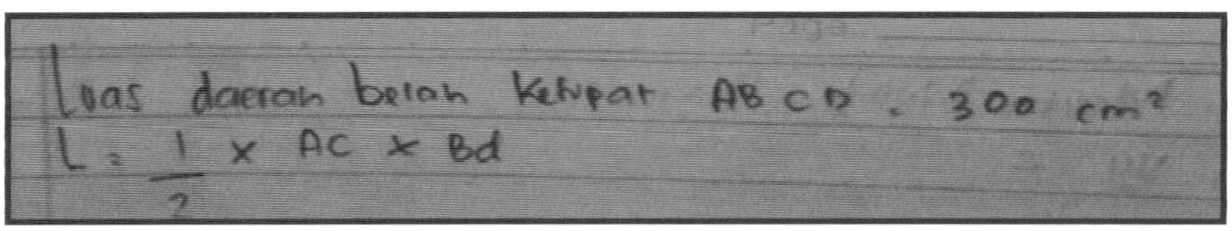

Figure 10. Student`s answer

Based on the analysis of student answers as a whole above, can be seen some mistakes made by students. The most common mistakes were concept errors and operation errors. The concept error that students know what was known and know what was being asked. Operation errors were about students who have accuracy in solving the problem, and true or not in operate. Based on the previous description shows that students' mathematical problem solving ability have not been good, therefore it takes an effort to help the students in developing problem solving ability. Several efforts to improving mathematical problem solving ability are the teachers should not only pay attention to students who have high but also low ability, and should be able to design innovative learning that suits all the characteristics of students. Teachers need to guide students not to be easily discouraged in improving problem-solving ability by providing training and teaching routinely in solving problems that measure problem solving ability. Teachers should also ask to get back to check what has been done to minimize errors in the solution of the problem.

4. Conclusion

Based on the percentage achievement of each indicator, it can be concluded that the level of achievement of students' mathematical problem solving ability on geometry was still low with an average of the indicator was 38.25%. This is because students are not used to solving problems that measure mathematical problem solving ability, weaknesses remember previous knowledge, and lack of problem solving framework. So the students` ability of mathematical problems solving need to be improved with implement appropriate learning strategy. So the ability of mathematical problem solving need to be improved. Alternative ways that can be done to develop the ability of mathematical problem solving was train the students by getting used to solving problems that measure the problem-solving's ability and familiarize students to re-check the results of student's task. Based on this research, the teacher can choose the correct strategy and plan the learning activity for learning mathematics that can improve ability of mathematical problem solving.

Acknowledgments

The authors would like to express their appreciate to students and also mathematic`s teacher that help on get the data this research. The highest appreciation is present to two lecturers, Mr. Dr. Kusnandi, M.Si., and Mr. Suhendra, M. Ed, Ph.D., that have guided researcher from beginning research until the completion of this article.

References

[1] NCTM 2000 *Principle and Standars for School Mathematics* United States NCTM

[2] Branca, N A 1980 *Problem Solving as A Goal, Process and Basic Skill, Problem Solving in School Mathematics Reston* VA NCTM

[3] Noor A J and Norlaila 2014 Kemampuan Pemecahan Masalah Matematika Siswa Dalam Pembelajaran Matematika Menggunakan Model *Cooperative Script Jurnal Pendidikan Matematika* vol 2 no 3 pp 250-259

[4] Windari F, Dwina F, and Suherman 2014 Meningkatkan Kemampuan Pemecahan Masalah Matematika Siswa Kelas VIII SMPN 8 Padang Tahun Pelajaran 2013/2014 Dengan Menggunakan Strategi Pembelajaran Inkuiri *Jurnal Pendidikan Matematika* vol 3 no 2 pp 25-28

[5] Anisa W N 2014 Peningkatan Kemampuan Pemecahan Masalah Dan Komunikasi Matematik Melalui Pembelajaran Pendidikan Matematika Realistik Untuk Siswa SMP Negeri Di Kabupaten Garut *Jurnal Pendidikan dan Keguruan* vol 1 no 1 artikel 8
[6] Brown J and Kim S 2016 *Mathematics: Identifying and Addressing Student Errors* bisa diakses di http://iris.peabody.vanderbilt.edu
[7] Fitria T N 2013 Analisis Kesalahan Siswa dalam Menyelesaikan Soal Cerita Berbahasa Inggris Pada Materi Persamaan dan Pertidaksamaan Linear Satu Variabel *MATHEdunesa Jurnal Ilmiah Pendidikan Matematika* vol 2 no 1
[8] Utami N W 2014 Analisis Kesalahan Penyelesaian Masalah Matematika Siswa. *Jurnal Media Pendidikan Matematika* vol 3 no 2 pp 457-460

IOP Conf. Series: Journal of Physics: Conf. Series **895** (2017) 012090 doi:10.1088/1742-6596/895/1/012090

Effects of Cooperative Learning STAD on Mathematical Communication Ability of Elementary School Student

E Maelasari [1*] and Wahyudin [2]

[1] Departemen Pendidikan Dasar, Sekolah Pascasarjana,Universitas Pendidikan Indonesia, Jl. Dr. Setiabudi No. 229, Bandung 40154, Indonesia

[2] Departemen Pendidikan Matematika , Universitas Pendidikan Indonesia, Jl. Dr. Setiabudi No. 229, Bandung 40154, Indonesia

*maygunawan@student.upi.edu

Abstract. The aim of this study is to describe the increasing of mathematical communications capability, and difference between students who get STAD cooperative learning and students who receive Direct Instruction. This study is a quasi-experimental study with pretest posttest study design. Subjects in this study in one of the fifth grade elementary school located in Cibeureum District, Kuningan. The research instrument used was a written test mathematical communication skills. The results showed that the improvement of mathematical communication capabilities Direct Instruction students who scored significantly better than students who learned with STAD cooperative learning. By grouping according to the ability of students will show a positive impact on student achievement in the classroom.

1. Introduction

Learning mathematics in primary schools has significance for the students, because in this level of mathematics learning was first laid out multidisciplinary basic abilities about mathematics in the environment. Mathematics in primary school aims to master the competency standards have been set and the children are able to solve simple problems systematically. Because mathematics is the basic science and the development has an important role in a variety of disciplines and promote the power of human thought. As mentioned in *The Ontario Curriculum* 1] that *"The of mathematics equips students with the knowledge, skills, and habits of mind that are essential for successful and rewarding participation in such a society"*. Learning mathematics equips students with the knowledge, skills and habits of thought which is essential for a successful and useful in society.

Mathematics is a language. Mathematics is a language that symbolizes the series of meanings of the statement to be conveyed [2]. The symbols are artificial and will have a meaning after a meaning given to it, without the math merely a collection of axioms, definitions, theorems and formulas that are less meaningful. Math is a language, a way of expressing or explaining a certain way [3]. The mathematics languages are the terminology, notation and mathematical symbols. The capability of comunication is one of important aspect for students to have a mathematical problem solving abilities [4]. Without communication in mathematics, we will only have a bit of information, data, and facts about the students' understanding of the process and the application of mathematics [5]. This argument implies meaning that the communication of mathematics, teachers are helped to understand students' abilities when interpreting and expressing his understanding of mathematical ideas that are being or they have learned during the learning process. As for the creation of more learning situations provide a conducive atmosphere to optimizing the mathematics communication ability of students, students can be organized in small groups. Model learning in small groups enables the emergence of communication and interaction between students in a higher quality.

Recognizing the importance of active students to develop communication skills and mathematical understanding of students, then teachers need to choose a model or approach which are suitable to engage students in learning groups to create a discussion. Activities of students in small groups provide an opportunity for students to perform mathematical communication through a number of metacognitive questions that focused on the problem, build prior knowledge with new knowledge, and the use of appropriate strategies to solve problems [6]. The learning which led the students in the learning setting in a small group called cooperative or collaborative learning.

Cooperative learning has various types, one of which is *Student Team Achievement Divisions* (STAD). *Student Team Achievement Divisions* (STAD) is one of the most cooperative learning method is simple and good for beginer teachers who use a cooperative approach [7]. STAD cooperative learning is considered suitable for the class which is unfamiliar or rare to the discussion or groups learning. Thus, it makes easier to explain the rules of learning in the new learning conditions.

By looking at the background above, the researcher is interested to assess the extent of "The Effect of Cooperative Learning STAD on Mathematical Communication Skills for Elementary School Students"

The purpose of this study is to determine whether there are the distinction in improvement of mathematical communication capabilities between students who get STAD cooperative learning with Learning Directly. Communication is one of the factors that are important in the mathematics learning process inside or outside the classroom. Communication plays an important role in mathematics. Communication is considered as a tool that can help teachers and students to share learning processes, understanding and learning mathematics. Everyone with an interest in mathematics will require communication in finding more information.

Related to the role of communication, one of the aim to be achieved in mathematics is to provide greater opportunities for learners to develop and integrate communication skills [8]. There are many ways to develop the communication skills is through speech and writing, modeling, drawing, and presented what they have learned.

Communication ability of mathematics is the ability to include and contain a variety of opportunities to communicate in the form of: (1) reflect the real objects, pictures or ideas of mathematics, (2) create a model of a situation or problem using the method of oral / oral, written, concrete, graphics, and algebra, (3) using the skills of reading, writing, and studying to interpret and evaluate ideas, symbols, terms and mathematical information, (4) respond to a statement / problem in the form of a convincing argument.

The indicators used in this study are: 1) The ability to express mathematical ideas through speech, writing, and demonstrate and describe it visually; (2) The ability to understand, interpret, and evaluate mathematical ideas, either orally, in writing, or in other visual forms. Both indicators adapted to the material that will be taken on the geometry. [9]

Cooperative learning STAD *(Student Teams Achievement* Division) is a method developed by Slavin, this learning method involves competence between groups. Students are grouped in various ways based on ability, gender, race, and ethnicity. Students learn the material together with friends in the group. In the process of discussion, groups of students who already understand the material presented, shall teach the material to members of the group who do not understand the material presented, then they are tested individually through the quiz. STAD advantages of this method are: all students be ready in learning; in the process of discussion going on cooperation and mutual communication between students; and can motivate students to learn. In addition STAD also has the disadvantage that students are familiar with reward and less conducive classroom conditions at the time of division of the group, thereby reducing the effectiveness of the learning.

STAD has the learning steps as follows: Submission of Interest and Motivation, Distribution Group, Presentation of Teachers, Learning Activities Team (Teamwork), Quiz (Evaluation) and Award Presentation Team. [7].

Based on the explanation above, the STAD cooperative learning is intimately connected with the students' mathematical communication skills.

2. Experimental Method

This study uses a quantitative approach with quasi experimental method. Design of experiments in this study refer to the *Nonequivalent Control Group Design* with two variables: learning discourse with the strategy *of mathematical bet line* (X) as the independent variable *(Independent Variable)* and the ability of mathematical communication of students (Y) as the dependent variable *(Dependent variable)*.Here is a picture of quasi-experimental research design that will be done.

O	X	O
O		O

Figure 1. *Nonequivalent Control Group Design* [10]

Description:
X : The treatment use cooperative learning STAD type
O : Pre-posttests communication ability mathematical

The subjects were fifth grade students in one of the public elementary schools in the Cibeureum district. In this study, one class were being an experimental class will be given STAD cooperative learning, while the other class will be used as a control class given direct instruction.

The test instrument used in this study is a mathematical communication test. Problem mathematical communication ability tests are given to the experimental class and control class in the beginning of treatment as a pretest and a posttest treatment in the end. The tests are given to students in order to determine upon ability mathematical communication. Therefore, it is based on indicators of mathematical communication skills.

3. Results and Discussion

The study is about the application of learning cooperative courses STAD on mathematical learning about the concept of geometry in 5th grade of elementary school. The research was carried out during eight meetings. Before doing the study, the researchers first do a pre-test of the two classes. Results of preliminary data analysis indicate that the data are not normally distributed in experimental class, while the control class data is normally distributed. Then *Mann Whitney* was tested to determine the difference average of mathematical communication ability to control class and experimental class before the action is taken. The results showed the Sig. *(2-tailed)* are 0,038 lower than $\alpha = 0.05$ significance level, so that H_0 is rejected. This means that there are no differences in average mathematical communication ability between the experimental class and control class before the class action or both originate from the same condition.

In this study, the data were analyzed overwhelm the score of pre-test and post-test communication capabilities. Based on the score of pre-test and post-test gain value calculated normalized (N-gain) the ability of mathematical understanding well the experimental class and the control class. The following descriptive statistics score pretest, posttest, and N-gain the experimental class and control class.

Table 1. Descriptive statistics Communication Ability Mathematical

	Experiment class				**Control class**			
	Total students	**Min. score**	**Max. score**	**Average**	**Total students**	**Min. Score**	**Max. score**	**Average**
Pre Test	20	7	18	11,2	20	2	17	10,55
Post Test	20	10	20	15,4	20	10	20	15,25
N-gain	20	0,11	1	0,50	20	0,2	1	0,264

After getting a different treatment, it appears that the data *post-test* mathematical communication capabilities of the two classes are different. This is also evidenced by the statistical test with the result that the two classes are not normally distributed. So, the next test was to test the difference average n-gain with *the Mann-Whitney U* test. The test of the difference in the average *N-gain* mathematical communication skills was done to prove the research hypothesis, that there is a difference mathematical communication ability of students who received STAD cooperative learning and improvement of mathematical communication skills students who got the direct instruction. The test results showed the Sig. *(2-tailed)* that is 0,038 less than the significance level $\alpha = 0.05$ so that H_0 is rejected. This means that there are differences in the average increase in the ability of mathematical understanding between the experimental class and control class.

Based on the results of data processing and analysis, there was found an increasing in mathematical communication skills of students who take STAD cooperative learning and the students who take the *Direct* Instruction. The improvement of communication of experimental class mathematically proved with *N-gain* is 0.49 is in the middle criteria. Meanwhile, the control class *with N-gain* of 0.67 is in the middle criteria. The test results of statistical testing showed that the ability of understanding mathematical communication of the fifth grade students in a public elementary school in cibeurerum district who follow the direct instruction better than the students who take STAD cooperative learning geometry.

Based on the analysis, the steps of learning affect the improvement of communication capabilities. Overviewing the steps of learning, the *Direct Instruction* provides a freedom for students to communicate with each other. Students who have the low proficiancy can learn how to communicate mathematically in high proficiency with the other students. Communication primarily looks dominant in step *Direct Instruction* that is currently structured drills and exercises under the guidance of teachers. The step of *Direct Instruction* requires students to be more communicative with their peers, then the planning, the implementation, and finding report assigned indicated that these students can finish well. While at STAD, the freedom to communicate between students just showed only in a group discussion. Because the material given to the SATD group was already given classically by the teacher. The freedom of communication on *the Direct Instruction,* this is according to the researchers, it raise the difference quality and improvement on the students' mathematical communication skills.

The results of the study suggests that the effect of STAD cooperative learning and *Direct Instruction* for high-ability students. In a various group of high-ability students often as a "teacher" and explain the concept shown to other students strengthen their understanding of the role that [11]. In a joint study group, students teach the knowledge shown to the group, and when students teach something is shown to other students, they tend to learn more in depth [12].

In addition to the above effect is seen on the *n-gain* and pretest posttest scores of students, the effect of both learning to high-ability students are also visible from the motivation factor. A highly capable student other than internal motivation to learn, it is because of the influence of cooperative learning, especially in this study is STAD and *Direct* Instruction will increase the external motivation of students. This is due to that the cooperative groups will provide an assessment of the group in addition to an assessment of the individual. This will give a sense of responsibility over to the students, especially high-proficiancy students to get the maximum score. High proficiancy students will usually do something to get a good score, avoid punishment, to get *reward* or any other reason that he and other members will strive to get good grades. Learning which there is co-operation will generate extrinsic motivation of students, because there the students will perform a challenge not only for personal use but also for the benefit of the group [11].

From the explanation above, it is clear that the measures and the characteristics of cooperative learning effect for students to improve achievement or ability class, both for their cooperative groups or because of *rewarding* of the two learning makes students are challenged to get the achievement of a better capability. In this research is directed to students' communication abilities.

4. Conclusion

Based on the research that has been conducted on improvement of communication capabilities mathematical students between groups of students who received STAD cooperative learning and student groups who received direct learning can be concluded from the results that, there are the differences in improvement of mathematical communication capabilities of the students who treated with STAD cooperative learning and improvement of communication ability of students who treated with direct instruction. By grouping according to the ability of students will show a positive impact on student achievement in the classroom. Direct learning can be used to develop the skills of mathematics. In direct instruction the teacher can better control the students and the ability to be achieved.

Acknowledgments

We are very thanked the students and teacher who participate actively in this research.

References

[1] Tn 2005 *The Ontario Curriculum Grades 1-8.* p 3

[2] Suriasumantri J S 2007 *Filsafat Ilmu Sebuah Pengantar Populer* (Jakarta: Pustaka Sinar Harapan) pp 13-20

[3] Alisah dan Dharmawan 2007 *Filsafat Dunia Matematika Pengantar untuk Memahami Konsep Konsep Matematika* (Jakarta: Prestasi Pustaka) pp 54-63

[4] Baroody A J 1993 *Problem Solving, Reasoning, and Communicating* (New York: Macmillan Publishing) pp 87-95

[5] Peressini and Bassett 1996 *Mathematical Communication in Student's Responses to a Performance-Assesment Task* Communication in Mathematics K-12 and Beyond (Virginia: NCTM) p 34

[6] Kramarski and Mizrachi 2004 *Enchancing Mathematical Literacy With The Use Of Metcognitive Guidance Forum Discuccion*. Proceedings of the 28th Conference of the International Group for the Psychology of Mathematics Education, Vol 3 pp 169–176

[7] Slavin 1995 *Cooperative Learning. Theory, Research, and Practice: Second Edition* (Boston: Allyn and Bacon) pp 26-45

[8] Huinker and Laughin 1996 "*Talk Your Way Into Writing*" Communication in Mathematics K-12 and Beyond. (USA: NCTM) p 102

[9] NCTM 2000 *Principles and Standards for Schools Mathematics* (USA: NCTM Inc) p 3

[10] Creswell 2010 *Research Design Pendidikan Kualitatif, Kuantitatif, dan Mixed* (Yogyakarta : Pustaka Pelajar) pp 18-21

[11] Woolfolk 2004 *Educational Pschology 9th Edition* (USA:Pearson) pp 257-260

[12] Santrock 2007 *Child Development, eleventh edition* (New York :The McGraw-Hill Companies) pp 115-117

IOP Conf. Series: Journal of Physics: Conf. Series **895** (2017) 012099 doi:10.1088/1742-6596/895/1/012099

Applying Authentic Data Analysis in Learning Earth Atmosphere

H Johan[1,2,*], A Suhandi[3], A Samsudin[3] and A R Wulan[4]

[1] Sekolah Pascasarjana, Universitas Pendidikan Indonesia, Jalan. Dr. Setiabudi No. 229, Bandung 40154, Indonesia
[2] Progran Studi Pendidikan Fisika, Jurusan Pendidikan MIPA, FKIP, Universitas Bengkulu, Jalan. Wr Supratman, Kandang Limun, Kota Bengkulu, Indonesia
[3] Departemen Pendidikan Fisika, Fakultas Pendidikan MIPA, Universitas Pendidikan Indonesia, Jalan. Dr. Setiabudi No. 229, Bandung 40154, Indonesia
[4] Departemen Pendidikan Biologi, Fakultas Pendidikan MIPA, Universitas Pendidikan Indonesia, Jalan. Dr. Setiabudi No. 229, Bandung 40154, Indonesia

*hennyjohan@student.upi.edu

Abstract. The aim of this research was to develop earth science learning material especially earth atmosphere supported by science research with authentic data analysis to enhance reasoning through. Various earth and space science phenomenon require reasoning. This research used experimental research with one group pre test-post test design. 23 pre-service physics teacher participated in this research. Essay test was conducted to get data about reason ability. Essay test was analyzed quantitatively. Observation sheet was used to capture phenomena during learning process. The results showed that student's reasoning ability improved from unidentified and no reasoning to evidence based reasoning and inductive/deductive rule-based reasoning. Authentic data was considered using Grid Analysis Display System (GrADS). Visualization from GrADS facilitated students to correlate the concepts and bring out real condition of nature in classroom activity. It also helped student to reason the phenomena related to earth and space science concept. It can be concluded that applying authentic data analysis in learning process can help to enhance students reasoning. This study is expected to help lecture to bring out result of geoscience research in learning process and facilitate student understand concepts.

1. Introduction

One of the important things in science education is learning process that can stimulate students to use the ability of thinking and reasoning in constructing a concept that has a different character. Concept of Earth is an imaginary in the sense that most of it could not be observed directly [1]. The earth science included earth atmosphere concepts have character which cannot be observed directly and related closely to natural phenomena. For example in Figure 1, solar radiation and earth radiation play a role at a temperature in the upper atmosphere. Ozone layer also gives impact to earth temperature. How they give impact to earth temperature could not be observed directly. This concept also related phenomena of rill condition of earth atmosphere. It affected the student's conception thus an alternatives conceptions besides the scientific conception aroused [2, 3]. Some research had shown

IOP Conf. Series: Journal of Physics: Conf. Series **895** (2017) 012099 doi:10.1088/1742-6596/895/1/012099

that most of earth and space science concepts were still considered to be difficult and there were many misconceptions happened, such as moon phase concept [2, 4, 5, 6].

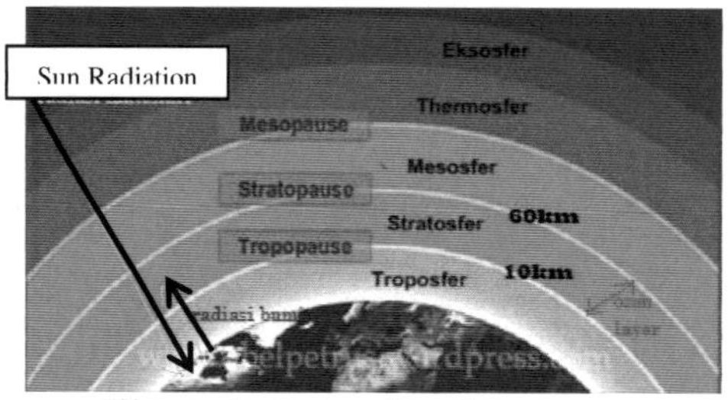

Figure 1. Atmosphere layer

The reasoning ability could be developed by analogizing the concept with a scientific model for better understanding a concept [7]. In addition, some research suggests that analogizing the reasoning concepts could be developed during the representations exploration process of concepts [8, 9]. Some implementations to improve the reasoning ability in the geological concept had been done. Analogizing, scientific models, and the graphical representation had been used in geological learning to improve student reasoning ability [7, 10, 11]. Although media visualization had been used widely in learning however it still needed encouraged for using. This research provided media visualization from authentic data to support other media visualization that used to help student increase their reasoning ability. It expected can give different way to help student understand and reason the concept clearly.

Several studies which concern in reasoning had been done. Scientific model had been used as visualization to help student reach their scientific conception [6]. Students' sketches were explored to investigate the student's alternative conceptions about plate tectonics using visualization from in learning activity [11]. In addition, collaboration science research and learning research just had been trend among education study. Result of science research used to facilitate student thinking process. Werts S and L Hinnov presented visualization in the form of three dimension graph from authentic data that was analyzed by MATLAB [12]. Ellwein A L, Hartley L M, Donovan S, & Billick concerned to develop module based data to facilitate students whom never worked with authentic scientific data [13]. This study provides earth and space science learning program using multi modus visualization such animation, picture, and graph to enhance reasoning. Visualization in this study also came from considering authentic data about Indonesian atmosphere using Grid Analysis Display System (GrADS). Novelty in this study is compilation research science from analyzing authentic atmosphere data using GrADS in learning earth and space science. Advantage of GrADS is easier to operate than using MATLAB. Display result from GrADS closely similar to display result from MATLAB.

2. Experimental Method

2.1. Research Design

Quasi experimental with one group pre-test and post-test design was implemented in this research. Before treatment, authentic data was examined using Grid Analysis and Display System (GrADS). Data of atmosphere's temperature and atmosphere's pressure on 1990 and 2010 were explored using GrADS. Pre-test was conducted before treatment process. Student engaged to construct the concept interactively with help of authentic data. Students were also engaged to find correlation among concept using visualization from GrADS about real condition of Indonesian atmosphere. Questions and feedback were done along learning activity. Classroom interactive by using questions and feedback was presented to support student improve their reasoning skill. Students were given question to explore the value in atmosphere phenomena. Students were engaged to interpret visualization about real condition of Indonesian atmosphere from GrADS. After treatment process, pos-test was conducted. Last step were analysis and conclusion.

2.2. Research Subject
23 pre-service physics teacher were involved to implement the learning model. Research subject never got earth and space science course before. Convenience sampling was conducted in this research.

2.3. Instrument Collecting Data
For collecting data, some instruments were constructed. Three types of instrument collecting data were developed. It includes essay test, open ended question, and observation sheet. An essay test was conducted to collect data reasoning. Open ended question was castoff to collect data about students' responses. Observation sheet was used to encounter all phenomena during learning process. Before used, these instruments were validated. 4 experts participated for validating the instruments. These instruments were tested to 34 pre-service physics teacher students who had passed earth and space science course in Bengkulu University, Indonesia. The result showed that instruments were valid and reliable. Detail of developing instrument doesn't be a focus discussion in this article.

2.4. Learning Process
Research based material in learning process came from science research which analyzed authentic data using GrADS (Grid Analysis Display System). This research tried to harness science research which analyzed authentic data using GrADS and learning activity to help students developed their reasoning skill. Students were encouraged to interpret visualization from GrADS analysis, correlated it to earth concept that learned. In learning activity, various media visualization included media visualization from investigation authentic data using GrADS were charity. Interactive discussions are done during learning activity.

2.5. Analysis Data
Students' responses from open ended were coded to find the pattern of the responses about learning program. Students' reasoning skill was examined quantitatively, category for reasoning skill followed criteria from Furtak (2010). Category, score, and description of reasoning skill can be seen in Table 1.

Table 1. Category of reasoning skill [14]

Category	Score	Description of Indicators
Inductive/ deductive rule-based reasoning	4	The rationale consists of a comprehensive data analysis supported by a principle, theory, law, or definition that are relevant to the data and problem being solved. The scientific correctness of the theory, law, etc., used in this reason is not important.
Evidence based reasoning	3	The rationale has considered an amount of data (including implicit data) and applied a relevant data analysis, but not enough to solve the problem correctly.
Data-based reasoning	2	The rationale relies on limited data or the surface feature of the problem.
No reasoning	1	The rationale, if any, is merely a restatement of the claim (response) or not clearly related to the problem nor clear in meaning.
Unidentified	0	Student's answer sheet is blank

3. Result and Discussion
Distribution of students in each level of reasoning criteria before and after implementation of learning model is shown in Table 2.

Table 2. Distribution of students' reasoning before and after implementation

Level of Reasoning Criteria	Score	Before implementation	After implementation
Inductive/ deductive rule-based reasoning	4	0	21.70%
Evidence based reasoning	3	0	60.90%
Data-based reasoning	2	0	17.40%
No reasoning	1	21.70%	0
Unidentified	0	78.30%	0

The shifted of students' reasoning were obtained from difference in scores pre-test and post-test. Most of students' reasoning levels before implementation were unidentified (state in level 0). It means students' answer sheet was blank. Just little of them had no reasoning (state in level 1). Their answer was just a restatement of the claim (response) or not clearly related to the problem. According Table 2, before implementation almost students reasoning were unidentified. They leave their answer sheet blank. Data in Table 3 couldn't show students' prior reasoning clearly. It indicates that learning model designed gives positive impact on students' reasoning development. Students' reasoning improved form no reasoning to inductive/ deductive rule-based reasoning, evidence based reasoning, data-based reasoning. Student response indicates that student knows prominence principle such solar radiation is absorbed by ozone at stratosphere layer. Temperatures of troposphere are predominantly affected by earth radiation. The air density decreases by elevation. Earth radiation of heat is transferred by air convection. Besides that, student can explain about correlation of temperature and elevation, correlation of stratosphere temperature and ozone layer, and correlation among air density (density is closely related to temperature), temperature, and heat radiation that absorbed. Student's response indicates that is correlation between air temperature and air pressure.

Media visualizations such animation, graph, table, or picture in learning process were aimed to visual how concepts related one each other to display an atmosphere phenomena. Visualization from GrADS supported visualizations in concept focus. Allegedly, visualizations used in this study helped student to develop their thinking skill and to reason concepts in atmosphere phenomena. Media visualization is technological utilization. Santos, Valéria Campos and Agnaldo Arroio argued that technological resources helped student to get a complete understanding in learning process [15]. Analysis data using GrADS is sample of using technology in learning process. The findings indicated that impact of learning process was as expected. Result of this study relevant on dual coding system theory, human cognition included two cognition systems, verbal system and visual system. Visual system included animation, picture, graph, table, formula mathematic. Based on findings, visualization included visualization from analysis authentic data using GrADS in this study supported human visual system to improve their cognition. Visualization from analyzing authentic data of Indonesian atmosphere's temperature and pressure also help student understand about temperature and pressure of atmosphere.

Based on Table 2, more than half students (60.9% of participant) were at evidence based reasoning level. Allegedly, their thinking skill improved as impact of learning process in learning model that designed in pre implementation step. Visualizations from authentic data analysis were used to help students understood clearly how an atmosphere phenomena was formed. Visualizations used in this learning process also could show correlation among concepts. This situation facilitated student to reason based on correlation among concept. Werts and Hinnov also reported that correlation among concepts could be explained by animation used in learning activities [12]. This study supported thinking skill that stated by Presseisen, he stated that thinking was cognition process as mental activity to construct knowledge [16]. Cognition process included perception, reason, and intuition. One of the thinking skills were correlation and causality. Interactive process especially in concept focus using visualizations facilitated student to engage their mental process in thinking. So that, this learning model in Table 2 allegedly facilitated students to construct their knowledge. Learning model designed was harnessed be a factor that support to improvement of students' reasoning. This is seen from N-gain value that was obtained in category high. Synergy with result from Carolan, Prain & Waldrip that stated learning process focused on thinking skill and reason ability also supported student understand concept well [17]. Johan [a,b,c] argued that media visualizations were success to help student improved their reasoning on earth and space science [18, 19, 20].

Finding in this study are supported by data from observation during learning process. Base on phenomena encountered during learning activities, shifted of students' reasoning was also caused by using visualization from investigating authentic data of earth atmosphere. Visualization atmosphere data from GrADS were harnessed to show how temperature and pressure were related each other in real condition of Indonesian atmosphere. Sample of using visualization from authentic data can be seen in Figure 2.

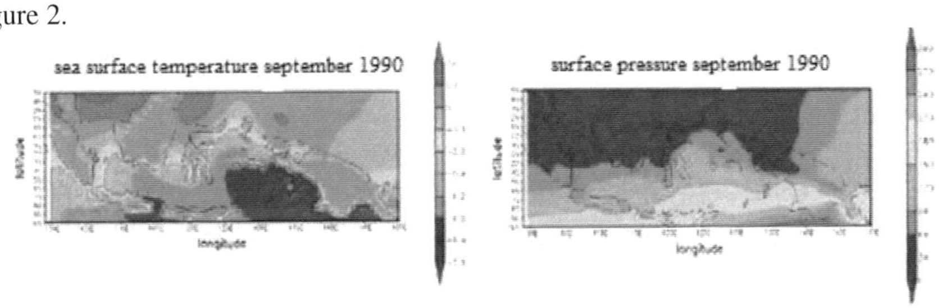

(a) (b)

Figure 2. Display for: (a) temperature in September 1990; (b) pressure in September 1990

According Figure 2, real condition of atmosphere in September 1990 showed that temperature and pressure had inverse correlation. Based on color of visualization, in September northern Indonesia was warmer (represented in red color) than southern one but the pressure was lower in northern than southern one. Werts and Hinnov also reported that visualization from MATLAB could show correlation between slope of incidence angle and incoming solar radiation [12]. Allegedly, visualization from authentic data of atmosphere supported student to understand and reason concept. This allegation supported by data from students' responses. They testified that visualization from GrADS used in learning activities helped them to understand and reason well. Students were also encouraged to interpret different between earth surface temperature at 1990 and 2010, the factor of it, and the cause of it. Visualization between earth surface temperature at 1990 and 2010 can be seen in Figure 3.

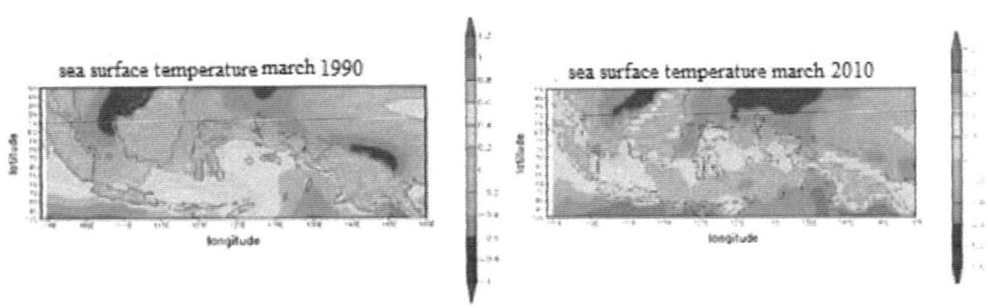

(a) (b)

Figure 3. Different between earth surface's temperatures: (a) March1990; (b) March 2010

Based on findings and phenomena encountered by observation during learning activities, we argue that applying authentic data analysis in learning process gives some benefits, include: helping students to reason and to understand concepts, giving them different experience learning about real condition of Indonesian atmosphere, showing correlation among concepts, and providing enjoyable learning process. Kortz and Kraft argued that supporting science research in learning activities gave benefit such as improved understanding concepts, skill, thinking, and motivation [21]. Combination between science research and learning research had been trend. Relevance results were also shown in study by

[13]. Kastens revealed that authentic scientific data from scientific research in learning activity could provide an engaging different learning experience. Enjoyable learning could bring out their motivation to learn more active [22]. Allegedly, this motivation contributed to improvement students' reasoning. This result supported Pratama & Corebima who found that self-motivation also had a contribution in learning to improve cognitive [23]. Data from open ended questions supported these findings where student gave evidence that learning activity was more enjoyable by using animations. It also brings out curiosity, interactive, and interesting. Similarly, Pratama & Corebima stated that geoscience learning using visualization could bring out motivation and encourage like geoscience learning [23]. Motivation was important factors in learning process to be success [24].

4. Conclusion

Based on the analysis of findings, it can be concluded that learning model supported by authentic data analysis using GrADS is potentially enhance students' reasoning. After implementation, students' reasoning improved from no reasoning to inductive/ deductive rule-based reasoning, evidence based reasoning, and data-based reasoning. No one of student was in level 1 (no reasoning). Visualization from GrADS supported student to understand and reason concept. Visualization in concept focus step and in research based material were supported each other to help student understand and enhance student's reasoning. Visualization from GrADS clearly shows the real correlation between air temperature and air pressure in Indonesian atmosphere. Visualization from GrADS clearly shows the real correlation between air temperature and air pressure in Indonesian atmosphere.

Acknowledgments

The authors thank to the Ministry of Research and Technology and Higher Education (Research-Higher Education) for the Doctoral Dissertation Grant Funds. The authors also thank to Septi Johan, M.Si who help to analysis authentic data using GrADS.

References

[1] Sunderlin D 2009 *Journal of Geoscience Education* 57(1) 73-81.
[2] Miller, Brian W and W F Brewer 2010 *International Journal of Science Education* 32(12) 1549-1560.
[3] Jolley A, Lane E and Kennedy B, Seneclauze, and Tom-Piere Frappe 2012 *Journal of Geoscience Education* 60(1) 83-91.
[4] Gross N A and Lopes R E 2009 *Astronomy Education Review* (8).
[5] Libarkin J C, Asghar A, Crockett C and Sadler P 2011 *Astronomy Education Review* 10(1) 1-25.
[6] Park Su-Kyeong 2013 *Eurasia Journal of Mathematics Science & Technology Education* 9(3) 285-299.
[7] Sibley D F 2009 *Journal of Geoscience Education* 57(4) 255-263.
[8] Prain V, Russel Tytler and Suzanne Peterson *2009 International Journal of Science Education* 31(6) 787-808.
[9] Smith G A and Bermea S B 2012 *Journal of Geoscience Education* 60(4) 350-359.
[10] Smosna R and Bruner K R 2007 *Journal of Geoscience Education* 55(1) 17-21.
[11] Jee B D, Uttal D H, Dedre G, Cathy M., Thomas F S, Basil T, Carol JO and Bradley S 2010 *Journal of Geoscience Education* 58(1) 2-13.
[12] Werts S and L Hinnov 2011 *Journal of Geoscience Education* 59 219–228.
[13] Ellwein A L, Hartley L M, Donovan S and Billick I 2014 *Journal of Geoscience Education* 62 578–586.
[14] Furtak E M, Hardy I, Beinbrech C, Shavelson R J and Shemwell J T 2010 *Educational Assessment* 15(3) 175–196.
[15] Santos, Valéria Campos and Agnaldo Arroio 2016 *Journal of Turkish Science Education* 13(1) 3-18.
[16] Costa A L 1988 *Developing Mind: A Resource Book for Teaching Thinking* (Alexandria: ACCD).

[17] Carolan J, Prain V & Waldrip B 2008 *Teaching Science* 54(1) 18–23.
[18] Johan H, A Suhandi and A R Wulan 2016 *Proceeding National Seminar of Elementary Education Mathematic and Natural Science Bandung Indonesia* 123-131.
[19] Johan H, A Suhandi and A R Wulan 2016 *Proceeding National Seminar of SNIPS Bandung Indonesia* 385-389.
[20] Johan H, A Suhandi and A R Wulan 2017 *Proceedings International Seminar on Mathematics, Science, and Computer Science Education* 175-181.
[21] Kortz K M and KJ van der Hoeven Kraft 2016 *Journal of Geoscience Education* 64(1) 24-36.
[22] Kastens Kim 2010 *Journal of Geoscience Education* 58(2) 52-57.
[23] Pratama A T and A D Corebima 2016 *International Journal of Environmental & Science Education* 11(18) 11007-11017.
[24] Mc Connell D and Khatrien, J V H K 2011 *Journal of Geoscience Education* 59(3) 106-110.

Analysis of Critical Thinking Skills on The Topic of Static Fluid

I. Puspita*, I. Kaniawati and I. R. Suwarma

Sekolah Pascasarjana Universitas Pendidikan Indonesia, Jl. Dr. Setiabudi No. 229, Bandung 40154, Indonesia

*ikeu_puspita@ymail.com

Abstract. This study aimed to know the critical thinking skills profil of senior high school students. This research using a descriptive study to analysis student test results of critical thinking skill of 40 students XI grade in one of the senior high school in Bogor District. The method used is survey research with sample determined by purposive sampling technique. The instrument used is test of critical thinking skill by 5 indicators on static fluid topics. Questions consist of 11 set. It is has been developed by researcher and validated by experts. The results showed students critical thinking skills are still low. Is almost every indicator of critical thinking skills only reaches less than 30%. 28% for elementary clarification, 10% for the basic for decisions/basic support, 6% for inference, 6% for advanced clarification, 4% for strategies and tactics.

1. Introduction

Static fluid is a physics concept always related to daily life. Therefore, it takes Critical thinking (CT) to be able to understand the concept well. The pressure exerced by a static fluid depends only upon the depth of the fluid, the density of the fluid and acceleration of gravity. The pressure in a static fluid from the weight of the fluid and is given by the expression:

$$P_{\text{static fluid}} = \rho\ g\ h\ = mgh/V \text{ becasue } V= Ah, \text{ so } P_{\text{static fluid}} = m\ g/A$$

CT is a process of challenging the individual to reflect reflective, reasonable, rational thinking to collect, interpret and evaluate information in order to obtain a decision [2]. Reflective here means to actively consider carefully and cautiously against all alternatives before making a decision. CT skill has five indicators [7] and become a key component of recent education [4]. So CT skill is very important to be controlled by all people to be able to protect themselves and make decisions wisely in everyday life [5] And able to compete in the global world [1]. Therefore the most important final goal in the curriculum of 2013 is to print graduates who have critical, logical and creative thinking [6].

The economy of the 21st century, has shifted into a service economy. The shift is due to the development of information, knowledge, and innovation. The economic shift caused all countries to compete globally. This completion spawned challenges that took place in the workplace and working communities. Because it is, demanding and encouraging everyone as citizens to have ingenuity, alertness and skill. To compete as a nation to advance the economy and its industry [1]. It is same as what Greenspan said in the global competition requires everyone to have 21st century skills to be able to hold new technology in order to survive in that era.

2. Experimental Method

Researcher conducted a descriptive study to analysis CT skill of 40 students XI grade in one of the senior high school in Bogor District. The subjects taken are XI class students who have not received static fluid lessons in senior high school. This study was conducted once. The method used is survey research with sample determined by purposive sampling technique. The instrument used is a CT skill test by 5 indicators on static fluid topics. Questions consist 10 essay sets. it has been developed and validates by team of experts.

Meanwhile, to investigate score of CT skill using score the rubric was developed based on the answer each item and validated by experts (table 1).

After scoring, the next step is do the percentage of each CT skill indicator, the percentage of the process id done by using the following formula:

$$\% = \frac{score\ each\ indicator}{total\ score\ each\ indicator} \times 100\%$$

There after do the performed categorization **[11]** for each indicator, namely grouping scores obtained by students in the category of very high, high, fair, low and very low. Guidelines for the categorization of students' CT skill such as in Table 1 below.

Table 1. Categorization of students' CT skills

Percentage score (%)	Category
81 – 100	Very high
61 – 80	High
41 – 60	Fair
21 – 40	Low
0 – 20	Very low

Data student score and the percentage will be used by the researcher to conduct descriptive analysis. Descriptive analysis was conducted to provide an overview and obtain complete information related to profile of CT skill.

3. Result and Discussion

3.1 *Results*

Figure 1 below shows the average percentage of students' critical thinking skills on the concept of hydrostatic pressure, pascal principles, and Archimedes principles. At the time of learning related concepts are taught in high school but at least have been taught in the previous ladder. The average percentage of all indicators is still low. It is seen that the elementary clarification indicator is the highest indicator compared to other indicators. Followed by the basic for decisions / basic support, advanced clarification and strategy and tactics.

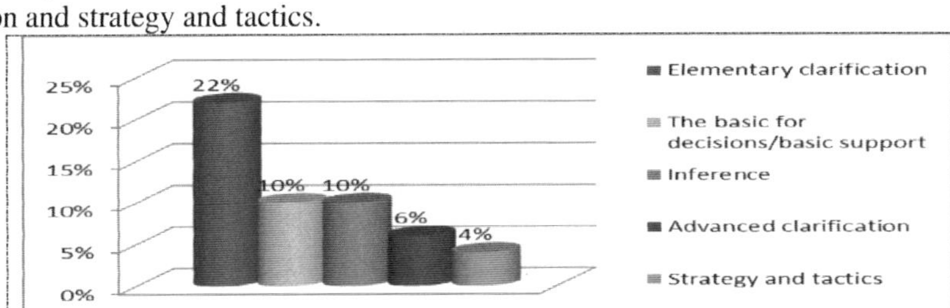

Figure 1: Percentage Of Average Students' Critical Thinking Skills For Each Indicator

Figure 2 shows the average percentage of students in answering each item. The average student can answer high enough for item 2 item compared to the result of student answer on other item. However, when compared to the total ideal value that should be in the value, this value is still relatively low. Problem 2 is a matter related to the concept of hydrostatic pressure. Meanwhile, very few items answered by the students are no 7 and no 11, which items that contain hydrostatic concepts with the basic indicators for the decisions and strategies and logic. These are questions that relate to the process of discovery.

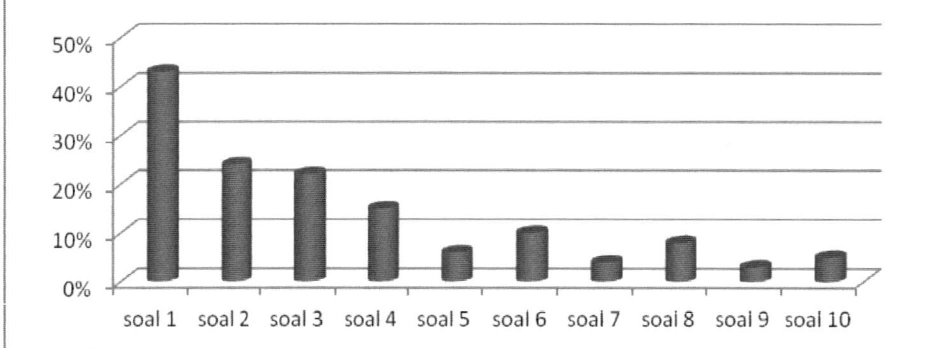

Figure 2: *Average percentage of students in answering each item*

3.2 Discussion

Critical thinking skills are part of high-level thinking skills and 21st century skills that the best way to tackle 21st century skills is to apply discovery-based learning processes that focus on real global problems [8]. Similarly, physics as a process, which views that physics is a science that contains natural phenomena are processed in the physical and empirical assessment which became known as the concepts, principles, law and principles which then used its application in technology, so that physics can be said there Because of the process of the invention [1]. Concerning the concept of hdrostatic pressure, the principles of Archimedes, and the pascal principle have at least been studied while at lower levels of junior high school, but that does not necessarily make them able to criticize the related issues of those concepts.

Is a challenge learning model (CBL) which is one of the learning models that can facilitate students actively involved in the process of discovery of real global problems (related to daily life). Because if students are active in learning, it can train students' scientific reasoning in developing their critical thinking skills [8]. This model is the first model developed by the education team at Apple Inc. Where CBL is built on experimental learning that upholds freedom, this model is effective in learning to solve real-world challenges that have efficient and effective frameworks that are organized into 3 phases: Engage, investigate, and act [9]. The phases consist of collaborative and hands-on, asking students to identify big ideas, asking good questions, identifying and solving challenges, deepening the realm of knowledge, and developing 21st century skills, and sharing experiences gained to audiences. CBL is an innovative learning model that combines the three learning models that are problem based learning, project based learning, and contextual teaching and learning modes. In this challenge based learning problem related to real world problem is said as challenge. Meanwhile, [10] concluded that the project undertaken in learning with the integration of CBL approaches in it can improve the retention of long-term student information.

4. Conclusion

Critical thinking skills need to be trained on students in one of the high schools in the Bogor district. And the best way to practice them is by applying active learning that facilitates students in the process of discovery of global problems, namely CBL which is integrated with the project.

Acknowledgments
The author would like to say thank to the students and teacher in our research classroom at Senior High School Bina Mulia in Bogor District, who have help the author, and also author want to say thank to Bu Ida and Bu Irma whose generous advices guided the research and development of the manuscript.

References

[1] Partnership 21st century skill 2008 *21st Century Skills Education & Competitiveness.* Diakses dari: http://files.eric.ed.gov/fulltext/ED519337.pdf
[2] Ennis R 1993 Critical thinking assessment. *Theory into practice* **32** 3 Ohio state university: college of education. Diakses dari: http://www3.qcc.cuny.edu/WikiFiles/file/Ennis%20Critical%20Thinking%20Assessment.pdf
[3] Mitrevski B and O Zajkov 2012 Physics lab, critical thinking and gender differences *Macedonian Physics Teacher* **48**
[4] Liliasari, Setiawan A and Widodo A 2007 *Pembelajaran Berbasis TI untuk Mengembangkan Keterampilan Generik Sains dan Berpikir Tingkat Tinggi Pebelajar.* Laporan Penelitian Hibah Pascasarjana, Jakarta DIKTI.
[5] kemendikbud
[6] Ennis H 1985 *A logical besis for measuring critical thinking skills.* The Association for supervicion curriculum development.
[7] Alice R and Maurícia O 2003 *The role of critical thinking In physics learning.* Diakses dari: http://lsg.ucy.ac.cy/girep2008/papers/THE%20ROLE%20OF%20CRITICAL%20THINKING.pdf
[8] Nichols M, Cator K and Torres M 2016 *Challenge Based Learner User Guide.* Redwood City, CA: Digital Promise. Diakses dari: http://cbl.digitalpromise.org/wpcontent/uploads/sites/7/2016/10/CBL_Guide2016.pdf
[9] Kukreti A 2015 Integrating the Challenge Based Learning Approach in a Freshman Engineering Foundations Course: Project Team Perspective. *Proceedings of the 2015 ASEE*
[10] Firdaus 2015 Developing Critical Thinking Skills of Students in Mathematics Learning. *Journal of Education and Learning* **9** 3 226-236

IOP Conf. Series: Journal of Physics: Conf. Series **895** (2017) 012101 doi:10.1088/1742-6596/895/1/012101

Profile of Students' Mental Model Change on Law Concepts Archimedes as Impact of Multi-Representation Approach

M Taher[1,2*], I Hamidah[1], I R Suwarma[1]

[1]Departemen Fisika Sekolah Pasca Sarjana Universitas Pendidikan Indonesia Jl. Setiabudi no 229 Bandung 4015.Jawa Barat Indonesia
[2]Sekolah Menengah Pertama Negeri 1 Keritang. Kabupaten Indragiri Hilir Provinsi Riau

*taher79@student.upi.edu

Abstract. This paper outlined the results of an experimental study on the effects of multi-representation approach in learning Archimedes Law on students' mental model improvement. The multi-representation techniques implemented in the study were verbal, pictorial, mathematical, and graphical representations. Students' mental model was classified into three levels, i.e. scientific, synthetic, and initial levels, based on the students' level of understanding. The present study employed the pre-experimental methodology, using one group pretest-posttest design. The subject of the study was 32 eleventh grade students in a Public Senior High School in Riau Province. The research instrument included model mental test on hydrostatic pressure concept, in the form of essay test judged by experts. The findings showed that there was positive change in students' mental model, indicating that multi-representation approach was effective to improve students' mental model.

1. Introduction

Until today, Physics is still considered a difficult, tedious, and daunting subject in schools. Students believe that physics can only be understood by special people with natural gift of extraordinary intelligence. That is why most students dislike physics subject so much. It is widely known that this phenomenon occurred in almost all educational institution in Indonesia, from primary schools to universities. Therefore, the government and all elements of education process have a huge homework to solve this physics education issue [1].

In this regard, teachers has important role and responsibility; that is to develop physics learning as well as to improve its negative image in students' eyes. This is in line with Minister of Education and Culture's Regulation No. 65/2013 on Process Standards. The regulation stipulates that every educator in schools is responsible for formulating systematic and comprehensive lesson plans so that the learning process could be interactive, inspirational, fun, challenging, motivating students' active participation, and providing enough opportunities for students to develop their skills, creativity, and independency according to their own talent, interest, and physical and psychological development. To do so, teachers need to implement appropriate learning approach, among others, which may facilitate students to understand Physics concepts in a more fun and easier way [2].

The multi-representation approach could be implemented as an approach in Physics learning. Presenting various representations to instill a concept will facilitate students to develop scientific mental model of the concept. Heuvelen and Zou argued that when students learned using several types of representation, their comprehension increased [3].

Stages teach the concept of archimedes using multi-repetition approach can be done by conveying the legal understanding of archimedes verbally *"If an object is immersed in a liquid, it will receive a force called buoyant force (upward force) of the weight of the liquid it displaces"* then presents the physical phenomenon of a plastic ball dipped into water, then students are required to squeeze the ball into the water slowly while feeling the reaction of the ball as in figure 1.

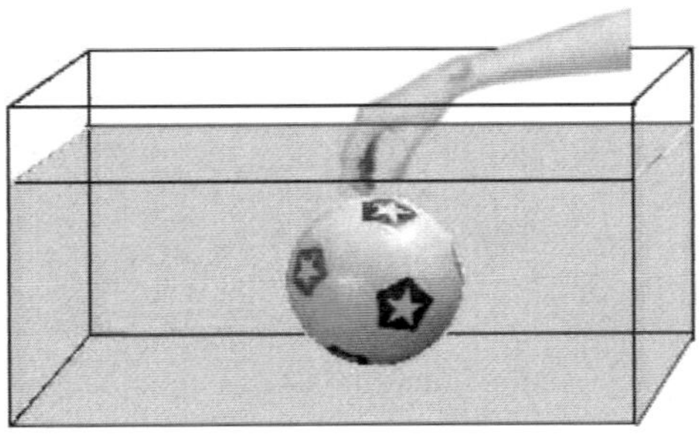

Figure 1. The plastic ball dipped in water

The upward force experienced by objects when in water is called the Archimedes style. The great style of Archimedes is formulated in terms of mathematical equations as follows.

$$\mathbf{F_a = \rho . g . V_{bc}}$$

Information:

ρ : Mass of a liquid (kg/m^3)

g : Acceleration of gravity (10 m/s^2)

V_{bc} : Volume of dyed objects (m^3)

In addition, the material can also be presented in the form of pictures and graphs as in the figure 2 and 3 and its application in everyday life as in figure 4 so that students' understanding will be built intact to the concept delivered.

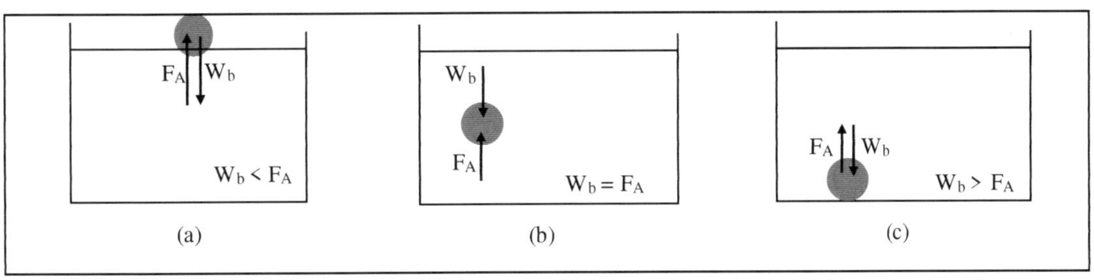

Figure 2. Position of the object (a) Floating, (b) Floating, (c) Sinking

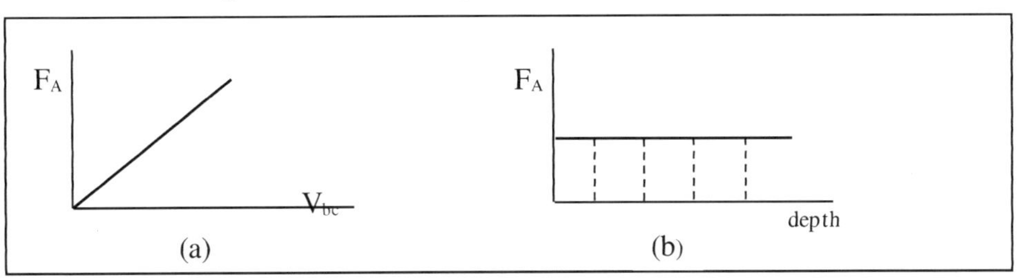

Figure 3. Graph (a) relationship Between buoyant force (FA) to the volume of dyed objects (b) relationship Between buoyant force (FA) to depth

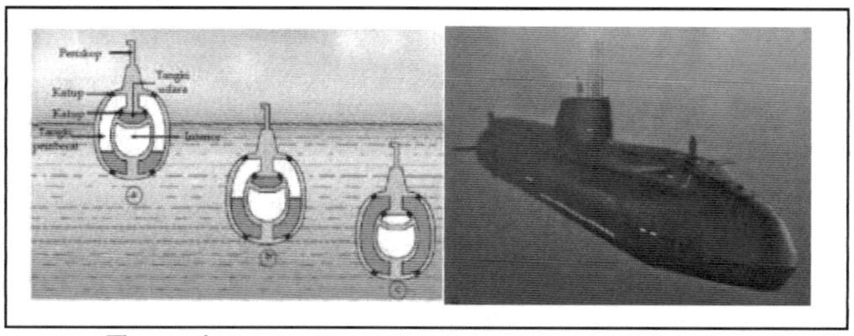

Figure 4. Application of Archimedes' Law on submarines

This was because each student had unique and specific skill that was more prominent than the other skills. Some students had more prominent verbal skill than spatial or quantitative skills, and vice versa [4]. Students should have scientific mental model of physics concepts. Developing mental model is the core of meaningful learning, in which student develop a mental model in his mind to un derstand and reason various micro and macro components of a system.

Mental model is interesting to study for two reasons. First, mental model affects cognitive function. Second, mental model may provide useful information for teachers and researchers of science education concerning students' conceptual structure [5].

This is in line with the findings of Supriyatman which stated that students' mental model profile could be used as an alternative of appropriate approach in teaching electricity and magnetic concepts [6]. Kurnaz found that students' mental model on solid friction concept was on scientific level (17.21%), synthetic level (56.28%) and initial level (26.51%). More than half of the students had synthetic level of mental model, nearly a quarter of them had initial mental model, and a fifth of the students had scientific mental model on solid friction concepts [7].

These previous studies provided information that many students still did not have scientific mental model, indicating that there was a problem in physics lessons they had received.

Among various researchers studying multi-representation approach in learning, and its effects on various learning aspects, was Sunyono they studied the effects of multi-representation approach implemented in teaching atomic structure concept on students' mental model [8]. Their statistical analysis result indicated that multi-representation approach in learning was more effective in developing students' mental model to comprehend atomic structure concepts, compared to conventional learning. Chittleborough and Treagust reported that students' mental model could be developed through interpretation, comprehension, and explanation of phenomenon for sub-micro representation [9]. However, most students preferred to use their mental model in simple representation phenomenon, for instance through suitable visualization for certain topic.

The present study was conducted to examine the implementation of multi-representation approach in Physics lessons in Senior High School. The primary aim was to see the effectiveness of this implementation in improving students' mental model. The problem was formulated in the following research question: "How is the effectiveness of multi-representation approach implementation in improving students' mental model on Archimedes Law concepts?"

This paper outlines the process and result of the study on multi-representation approach implementation in Physics learning and its effects on students' mental model improvement concerning Archimedes Law concepts.

2. Methodology

The methodology employed in the present study was quasi-experimental methodology with one-group pretest-posttest design. In this design, pre-test was first administered to the subjects, followed by a treatment of cooperative learning using multi-representation approach. Then, post-test was administered to measure the effect of multi-representation approach on students' mental model improvement on Archimedes Law. Pre-test and post-test results were then analyzed to find the answer to the research question. Figure 5 displays the research design implemented in this study.

	Pre-test	**Treatment**	**Post-test**
Subject	Test	Treatment of cooperative learning using multi-representation approach	Test

FIGURE 5. Research Design

The multi-representations implemented in discussing Archimedes Law concepts included verbal, pictorial, mathematical, and graphical representations. The subjects of this study were eleventh grade students in a Senior High School in Riau Province. Total population was 32 students. For data collecting purpose, a research instrument was constructed, i.e. a mental model test on Archimedes Law in the form of essay test. This instrument had been validated by experts and field-tested.

The effectiveness of multi-representation approach implementation in improving students' mental model on Archimedes Law was determined based on the percentage of students achieving ≥ 80 test score in a scale of 100, with the criteria shown in Table 1.

TABLE 1. Criteria of learning effectiveness (Source: A. Suhandi and F. C. Wibowo. 2012)

Number of students achieving ≥ 80 score	Learning effectiveness criteria
>75 %	High learning effectiveness
50% - 75%	Medium learning effectiveness
<50%	Low learning effectiveness

3. Result and Discussion

The diagram in Figure 6 displayed the percentage of students achieving each score range in mental model tests on Archimedes Law.

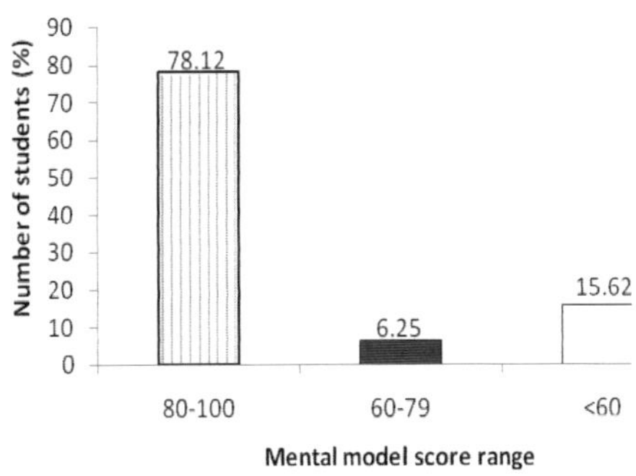

FIGURE 6. Bar diagram of students' percentage in achieving each score range in Archimedes Law mental model test.

Figure 6 showed that there were 78.12 % students (or 25 out of 32 students) achieving the 80-100 score range. This indicated that most students had good comprehension of the concepts in Archimedes Law. Figure 7 displayed students' percentage in each mental model level, based on pre-test and post-test result.

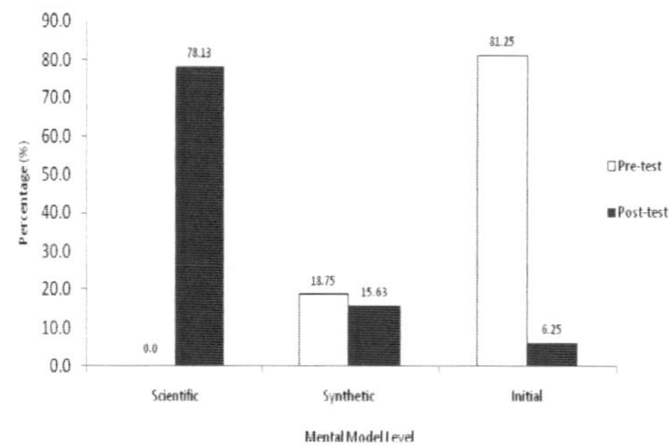

FIGURE 7. Bar diagram of students' percentage in each mental model level

Figure 7 showed that in the pretest, no student achieved scientific mental model level. However, post-test result indicated that 78.13% students managed to achieve scientific mental model. On synthetic level, there were 18.75% students in pre-test. This number decreased to 15.63% in the post-test result. Meanwhile, the 81.25% students achieving initial level in pre-test decreased to 6.25% in the post-test. This number indicated that students' mental model on Archimedes Law improved greatly. The high percentage of students achieving 80-100 score range on mental model test (Figure 6), and the high percentage of mental model level improvement (Figure 7), indicated that the multi-representation approach implemented in cooperative learning was highly effective in improving students' mental model on Archimedes Law.

This was understandable, considering that using various representations in explaining a concept would facilitate students' comprehension. When a student had not fully comprehend a concept explained using a representation, the use of another representation would reinforce the concept and help the student to understand. Therefore, students would have deep understanding [4]. This finding was in line with Mayer who stated that multi-representation could facilitate deep conceptual understanding construction [10]. Savinainen and Viiri argued that the use of several representation in teaching provided great benefits for students' deep conceptual understanding [11]. In regard to multi-representation function in learning, Ainsworth (Nieminen) stated that the implementation of multi-representation could complement other representations, limit other representations, and build comprehensive understanding [12].

4. Conclusion
Based on the findings and discussion, it could be concluded that multi-representation approach implemented in cooperative learning had quite high level of effectiveness in improving students' mental model on Archimedes Law. Therefore, the researcher recommended that multi-representation approach be implemented in Physics and other lessons.

Acknowledgement
The researcher would like to offer his deep gratitude to Headmaster of Public Senior High School 1 Keritang, Riau Province, and all the teachers for providing him with the opportunity to conduct this study. The researcher would also like to send his gratitude to Riau Province government for funding this study.

References
[1] M. Faizi 2012 *Ragam Metode Mengajarkan Eksakta Pada Murid*, 1st ed. Yogyakarta: DIVA Press.
[2] "Peraturan Menteri Pendidikan dan Kebudayaan Nomor 65 tahun 2013 tentan Standar Proses," 2013.
[3] A. Van Heuvelen and X. Zou 2001 "Multiple representations of work – energy processes Multiple representations of work – energy processes," *Am. J. Phys.*, vol. 69, no. 2, pp. 184–194.
[4] A. Suhandi and F. C. Wibowo 2012 "Pendekatan Multirepresentasi dalam Pembelajaran Usaha-Energi dan Dampak Terhadap Pemahaman Konsep," *J. Pendidik. Fis. Indones.*, vol. 8, pp. 1–7.
[5] L. A. R. Laliyo 2011 "Model Mental Siswa Dalam Memahami Perubahan Wujud Zat," *J. Penelit. dan Pendidik.*, vol. 8, no. 1, pp. 1–12.
[6] Supriyatman, A. Suhandi, A. Rusli and D. Rusdiana 2012 "The Profile of Student Physics Education Mental

Model in Electricity and Magnetism Concepts Using Problem Solving Test," *Int. Jounal Sci. Res.*, vol. 3, no. 8, pp. 2093–2097.

[7] M. A. Kurnaz 2015 "An Analysis of High School Students ' Mental Models of Solid Friction in Physics," *Educ. Sci. Theory Pract.*, vol. 15, no. 3, pp. 787–795.

[8] Sunyono, L. Yuanita and M. Ibrahim 2015 "Supporting Students in Learning with Multiple Representation to Improve Student Mental Models on Atomic Structure Concepts," *Sci. Educ. Int.*, vol. 26, no. 2, pp. 104–125.

[9] G. Chittleborough and D. F. Treagust 2007 "The modelling ability of non-major chemistry students and their understanding of the sub-microscopic level," *Educ. Res.*, vol. 8, no. 3, pp. 274–292.

[10] R. E. Mayer 2003 "The promise of multimedia learning : using the same instructional design methods across different media," *Learn. Instr.*, vol. 13, pp. 125–139.

[11] A. Savinainen and J. Viiri 2004 "A Case Study Evaluating Students Representational Coherence of Newtons First and Second Laws," *Proc. Phys. Educ. Res. Conf. 2003, AIP Conf. Proc. No. 720, Ed. by J. Marx. S. Franklin, K. Cummings (AIP, New York, 2004),* no. 720, p. 77.

[12] P. Nieminen, A. Savinainen and J. Viiri 2010 "Force Concept Inventory-based multiple-choice test for investigating students ' representational consistency," *Phys. Rev. Spec. Top. - Phys. Educ. Reseach*, vol. 6, pp. 1–12.

IOP Conf. Series: Journal of Physics: Conf. Series **895** (2017) 012102 doi:10.1088/1742-6596/895/1/012102

Understanding of Prospective Physics Teachers Students Toward Pedagogical Content Knowledge on Optical Geometry Materials

E Erwin[1,2,*] and NY Rustaman [1]

[1]Program Studi Pendidikan Ilmu Pendidikan Alam, Sekolah Pascasarjana Universitas Pendidikan Indonesia, Jl. Dr.Setiabudi No. 229, Bandung 40154, Indonesia.
[2] Sekolah Tinggi Keguruan dan Ilmu Pendidikan Nurul Huda, Kabupaten Ogan Komering Ulu Timur, Provinsi Sumatera Selatan

*erwinpohan74@student.upi.edu

Abstract. This article discusses about Pedagogical content knowledge (PCK) profile of prospective physics teachers on optical geometry materials. Data collected using interview and questionnaire, and the data were analyzed descriptively. The results showed that PCK is an unfamiliar term to students. The extreme findings in this study is the lack of understanding of PCK by prospective physics teachers relating to the importance of recognizing the characteristics of students and how to manage questions from students, which teacher has to directly answer questions from students, and how to respond to the students' incorrect answer, mostly prospective physics teachers assume that in case of the students answer incorrectly, the students should be directly blamed. Prospective physics teachers have not yet integrated the pedagogical knowledge with the content knowledge in their possess learning it he optical geometry material.

1. Introduction

PCK is a combination of two competencies, namely pedagogical knowledge (pedagogic competence) and content knowledge (content competence). PCK for teacher is very important due to create meaningful learning. content knowledge includes knowledge of concepts, theories, ideas, framework of thinking, method of verification and evidence, while pedagogical knowledge relates with the ways and processes of teaching which include knowledge of classroom management, tasks, effective lesson planning, and student learning and assessment [1]. Initially, education of teacher emphasized more of teacher's knowledge of subject matter [1,2] However, over the past decade, education of teacher has begun to emphasize on the effectiveness of general pedagogical methods which include the use of questions, and the design of an assignment and the curriculum, also the assessment of student's independent performance on each subject matter [2, 3]. The experts now recognize that both knowledge of subject matter and pedagogical knowledge are crucial in a good teaching and in enhancing students' understanding [2]. A good teacher should be able to be an expert of the content (subject matter) and master the science of teaching (pedagogy).

In accordance with the National Education Regulation of the Republic of Indonesia number 16 of 2007 about Academic Qualification Standards and Teacher Competencies, it is stated that the teacher's pedagogy competence includes seven competencies and in this study, PCK components are: Mastering the characteristics of learners, Mastering the learning theory and principle of educational learning, Curriculum development, Educational learning activities, Development of learners' potentials, Communication with learners, and Assessment and evaluation. The pedagogical competence in which is applied in teaching special subject matter will actualize the ability of PCK [4]. Inculcating PCK to teachers is most likely to be done while they are taking the education for prospective teacher. The teacher education institution (Teacher Training Institute), which is an institution that produces educators (teachers), should provide the PCK acquaintance to prospective teacher before they are graduated from Teacher Training Institute. To facilitate the development of prospective teachers' PCK, educational models prospective teachers need to be gradual and structured early gives experience to prospective teachers to interact and gain experience in schools [5]. As a form of Teacher Training Institute responsibility to the prospective teacher related to the ability of their PCK, and the institute ought to assess the ability of prospective teachers' PCK until being declared as a competent prospective teacher.

Several studies on how to assess PCK have been done using a variety of instruments, including: Essay open ended test [6], Content Representation (CoRe) [7], Observation [8, 9, 10], Interview [8, 9, 10, 11, 12]. But there have not been research that reveals about the understanding of PCK prospective physics teacher on optical geometry material, then based on background described, this article will discuss about how the PCK profile of prospective physics teacher on optical of geometry materials..

2. Method

In The research was conducted at one of the High School of Teacher Training and Education in South Sumatera Province, involved 30 respondents, specifically prospective physics teachers who had followed the pedagogy and Field Experiment Program I and II, so it was assumed to have a sufficient understanding of PCK. Respondents were askedto provide information on how they understood the importance of PCK. The instruments used for data collection were questionnaires and interview guidelines.

Questionnaire consisted of 45 items of statements on optical geometry materials, with a choice of answers strongly agree, agree, disagree, and strongly disagree. Scoring method used Likert scale of 4 points. A score of 4 was given on the option strongly agree for a positive statement and strongly disagree for a negative statement. Score 3 was given on the option of agreeing to a positive statement and not agreeing to a negative statement. Score 2 was given on the option of disagreeing for a positive statement and agreeing to a negative statement. As well as score 1 was given on the option strongly disagree for a positive statement and strongly agree for a negative statement.

Data analysis was done descriptively by presenting the findings from the research. Data derived from the questionnaire were verified and calculated on average PCK score of prospective teacher and presented in percentage form.

3. Result and Discussion

3.1. College profile of research location

Students of Physics Education study program at High School of Teacher Training and Education who became research object mostly came from farmer family, so many students used their time to work part time as farm labours or helped their parents farming.

The location of the college was located at in a village about 30 km from the district capital, and was ± 200 km from the capital city of South Sumatera Province, Palembang. The regency was an ex-transmigration area whose inhabitants were almost 80% were transmigration community from ethnic Javanese. People's livelihood was generally rice farmer, paddy field spread in this regency was technical irrigation rice field.

The college students generally were students who did not pass the entrance examination public universities, although there were some who did choose that college as the main goal for some reason.

For example, because of that college was under control of a boarding school, so the Islamic nuances more felt compared with other colleges nearby. Another reason was because that college was located near the location of student residence and for students from outside the region, there was dorm boarding school available for students' residence.

Input of students at that college was not as good as the input of students at college which hold very strict selection and tight competition, but the college made its special standard, with respect to the ability to read Al-Quran for graduate candidates, in which student who was Muslim wouldn't be allowed to take final examination or thesis court, if the student had not passed the Al-Quran literacy test, and his graduation was proved by the certificate of pass reading Al-Quran issued by high school based on the recommendation of lecturers who were appointed as the testers and mentors of reading Al-Quran

3.2. Profile of Pedagogical Content Knowledge (PCK) prospective physics teachers

3.2.1. Comprehension the concept of Pedagogical Content Knowledge (PCK)

Based on the result of interviews, prospective physics teachers had their own perceptions about the concept of pedagogical knowledge (PK). most of the prospective physics teachers who were interviewed (7 out of 10) did not understand the pedagogical understanding well, even though they had heard the term from the lecturers who taught the subjects of education, they commonly thought pedagogy was teaching. Even when all students had not fully understood the notion of pedagogy, the understanding of prospective physics teachers was only limited to how to manage the class during learning activities, not comprehensively comprehending about pedagogy which included 7 competencies. According to the regulation of education and culture minister number 16 of 2007 [4], that is: Mastering the characteristics of students, mastering learning theories and principles of educational learning, development of curriculum, educational learning activities, development of learners' potential, communication with learners, and assessment and evaluation.

Student understanding about content knowledge (CK) which in Minister of National Education number 16 of 2007 was named by the term of professional competence also still very minimal [4]. All of the prospective physics teachers stated that CK was only about understanding the content of subject that would be taught to the students.

Students claimed that the CK was the ability of the teacher to understand the material regardless of the depth, breadth, and how to package the subject material that would be taught to the students, so the learning could be effective and meaningful. Subject material was presented to students depended only on the teacher's curriculum or handbooks.

PCK was an unfamiliar term for students who were interviewed. It was revealed from their statement as a respond to the question; ''I just heard the term –pedagogical content knowledge-, which usually I had heard only pedagogical term.''. All the interviewed students gave a statement with a same meaning, based on the statement, neither understand the definition or concept of PCK, nor even did they hear the term before.

The term of PCK during lectures had never been introduced by lecturers, and after making confirmation with lecturers who taught the subjects on physics education, they absolutely never had introduced the term to the students. The reason why lecturers had not introduced the term was because they themselves also did not understand the term of PCK, even it was the first time of hearing the term.

3.2.2. Prospective physics teachers feedback on the importance of pedagogical content knowledge (PCK)

At the first time of the interview as stated preceding, it was revealed that prospective physics teachers were still unfamiliar with the term of PCK. Based on the interview guidelines, if the they were not familiar with the term, then the PCK definition: "Pedagogical content knowledge was a special combination between pedagogy and content or a mixture of content and pedagogy to form a knowledge of how a topic, problem, or issue was organized, represented according to the learner's ability" plus a simpler language explanation indicating that the PCK was used specifically on a particular material does not apply equally to all material. The interviewer's explanation did not impact the respondent, but only provided a theoretical understanding of PCK, so that the respondent understood what the term meant and could give any knowledge based feedback about PCK. After the respondent understood the meaning or definition of PCK, then respondents were asked for their responses about whether or not the PCK was provided for the prospective physics teachers.

Based on interviews with prospective physics teachers indicating that an understanding of PCK was considered to be very important. PCK provided provision for prospective teachers to understand how to package a subject matter combined with pedagogical knowledge to teach a subject to produce a meaningful learning for students. This was as revealed by one the students who stated that:

''I just heard the term PCK. I once read that a learning model could not be used on all materials, because a learning model was used on materials that matched with the characteristics of the material. If PCK was used only on certain materials, it meant what I had read and that was important for students.''

The average score and percentage of scores based on the potential physicist PCK component were presented in the following Figure 3.1 and 3.2.

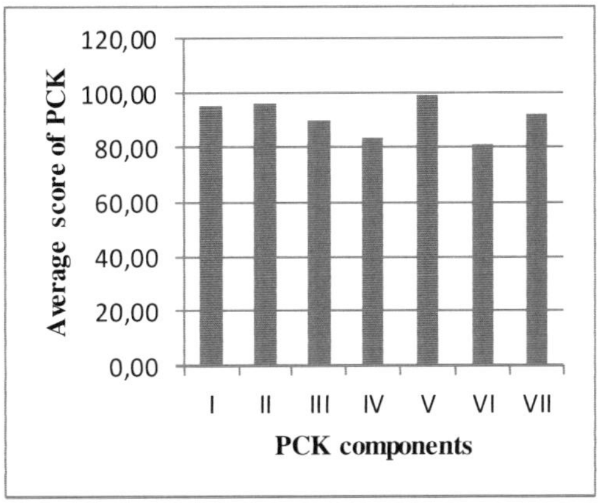

Figure 3.1. The average score based on PCK components

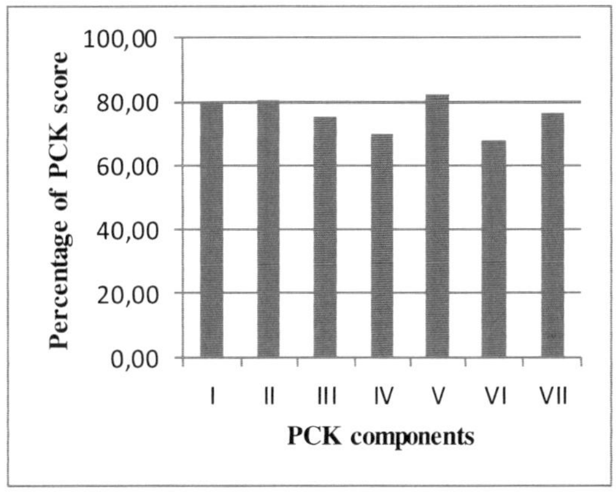

Figure 3.2. Percentage score based on PCK components

PCK components as shown in the figure 3.1 and 3.2 above were: I is mastering the characteristic of learners, II is mastering the learning theories and principles of educational learning, III is curriculum development, IV is educational learning activities, V is development of learners' potential, VI is communication with learners, and VII is asssessment and evaluation.

Based on the picture 3.1 and 3.2 above, it turns out PCK component most understood by physics teacher candidate is on the component of potential development of learners, while the lowest understanding is the component of establishing communication with learners. The response of respondents who were interviewed as a while showed similarity of opinions, PCK was an important knowledge to be given to prospective physics teachers, as the ability to be able to carry out better learning. During lectures, students were provided with pedagogical knowledge which includes: Introduction to educational sciences, study and learning, educational psychology, fundamentals of teaching process I & II, curriculum review, learning evaluation, and field experiment program I & II.

Besides pedagogical knowledge, the students were also equipped with knowledge of physical content, but both components were provided separately, according the concept of Pedagogical Content Knowledge, meanwhile, according to Shulman learning of a particular material is done by amalgamation between pedagogical knowledge and content content, so that learning becomes more meaningful.Knowledge of Pedagogical content knowledge (PCK) prospective physics teachers [1].

The result showed that the average understanding of students about PCK on optical geometry material was generally quite good, which obtained the average questionnaire score 89.76 of a maximum score 120 or the percentage of 74.80%. PCK of prospective physics teachers which revealed in the study was divided into 7 competencies, according to the regulation of the national education minister number 16 of 2007 [4], the average score of understanding of PCK by prospective physics teachers on the competence; Mastering the characteristics of the learners is 95.50 or 79.58%, mastering learning theories and principles of educational learning is 80.17 or 96.20%, curriculum development is 90.00 or 75.00%, educational learning activities is 83.69 or 69.74%, development of potential learners is 99.00 or 82.50%, Communication with learners is 81.00 or 67.50%, assessment and evaluation is 92.00 or 76.67%.

The surprising findings were maintained based on a questionnaire on six items; *first*, item number 4, which reads; 'in the course of learning materials of optical geometry, the teacher must understand the students' habits in his environment' only got the percentage score of 56.67%, it indicated that the prospective teacher had not understood the introduction of the characteristic of learners as a whole, whereas the introduction of the characteristics of learners on aspects of learners' habits or culture was important to provide appropriate treatment to student related to optical geometry material. *Second*, Item number 13, which reads; 'in the course of implementing the learning of optical geometry materials, the student's questions should be answered by teacher immediately' only got the percentage score of 46.67%, this showed the understanding of PCK prospective physics teachers for strategies to answer students' questions on the material of optical geometry materials was not sufficient, students considered a good way to manage learner questions was directly answered the questions. *Third*, Item number 15, which reads; 'the reason for using experimental method in optical geometry learning was because the curriculum demand' got percentage 58.33%. *Fourth*, Item number 16, which reads; 'the reason for using experimental method in optical geometry learning was that the method used by teachers to be varied' got percentage 48.83%. *Fifth*, Item number 17, which reads; 'the reason to use experimental method on optical geometry learning was order to student practice' earned a percentage of 47.50%. The three items of questions number 15,16, and 17 related to the reason of using the method of learning used was the experimental method on optical geometry materials. Students had not understood well the reason for the use of experimental methods that were closely related to inquiry. Instead, prospective physics teachers understood the use of experimental methods only because the curriculum demands, so the methods used in learning vary and so students practice. that was not a fundamental reason for the use of experimental methods in the learning of optical geometry. The reasons for using experimental methods in optical geometry learning which should be understood by the prospective teacher, based on pedagogy or psychological function were; experiments developed causal and functional thinking and creativity, experiments could develop the ability to work in teams, experiments motivated, increased variation, and aroused interest, experiment gave experience learning, and etudents were actively involved. Meanwhile, based on functions relating to the subject or epistemology the use of experimental methods on optical geometry materials were; the experiments supported the study of scientific research methods, experiments were established methods of acquiring knowledge in physics (generating hypotheses and working with them), experiments made a concrete visual physical facts, experiments made physical facts or relationships reasonable or explained it, experiments supported the formation of concepts, and experiments could cause cognitive conflicts. Whereas, based on built practical functions, the use of experimental methods on optical geometry materials were; students practiced how to handle data and done data analysis, students practiced to dealing with differences orrelationships with them, and haptic or developing psychomotor aspects [6], and *Sixth*, Item number 32, which reads; "if a student gave incorrect answers when learning optical geometry, teacher should immediately blamed him/her, so the student would know that he/she was wrong" only got the percentage score of 55.83%, it indicated the lack of understanding of prospective physics teachers to do educational learning process related in building student knowledge. The student's incorrect answer should not be directly adjudged, but the teacher should guide the student to find the correct answer, so that the student's understanding was built by constructing his/her own knowledge and through his/her own process of thinking and understanding.

4. Conclusion

The understanding of prospective physics teachers about PCK on optical geometry material is quite good, but needs to be improved so that the understanding of PCK more deeply. Nevertheless, the terms in PCK are still not well understood by student prospective teachers. The understanding of Prospective physics teachers is very low on some component of PCK that is: *first*, the lack of understanding of PCK about the importance of knowing the characteristics of students relating to the students' habits, or cultural environment that the students consider unimportant in carrying out optical geometry. *Second* how to manage student questions that teachers directly answer. *Third,* the reason to use of experimental methods on the learning of optical geometry, students assume because of the demands of the curriculum only, in order for the method to vary, and to have students practice. *The last*, respond to the students' incorrect answers, where most students understand that if the learners are incorrect to answer the question, then the student must be blamed immediately.

Acknowledgments

Infinite thanks to colleagues at the College who have assisted in the collection and processing of this research data. Without the help provided by friends this research could not possibly be done well.

References

[1] Shulman, L.S. 1986 *Those who understand: knowledge growth in teaching*. Educational Researcher, 15, 4–14

[2] Cochran, et al., 1993. Pedagogical Content Knowing: An Integrative Model for Teacher Preparation. *Journal of Teacher Education*, 44, 263–272

[3] Ball and McDiarmid, G.W. 1990. *The Subject-matter Preparation of Teachers. In W.R. Houston (Ed.), Handbook of research on teacher education* (pp. 437-449). New York: MacMillan

[4] *Regulation of the Minister of National Education of the Republic of Indonesia number 16 of 2007 on standard of academic qualification and teacher competence*. Ministry of National Education Republic of Indonesia.

[5] Anwar, Yenny, Nuryani Y Rustaman and Ari Widodo 2014 Hypothetical Model to Developing PedagogicalContent Knowledge (PCK) Prospective Biology Teachers in Consecutive Approach *International Journal of Science and Research* Volume 3. Pg. 138-143

[6] Kirschner, et al., 2016 Developing and Evaluating A Paper-and-Pencil Test to Assess Components of Physics Teachers' Pedagogical Content Knowledge *International Journal of Science Education* Vol. 38. (1-30)

[7] Niño, et al., 2015 *Initial Characterization of Colombian High School Physics Teachers' Pedagogical Content Knowledge on Electric Fields*. Research in Science Education

[8] Yuenyong and Thathong 2015 Physics Teachers' Constructing Knowledge Base for Physics Teaching Regarding Constructivism in Thai Contexts. *Mediterranean Journal of Social Sciences*. Vol. 6(2)

[9] Qhobela and Moru 2014 *Examining Secondary School Physics Teachers' Beliefs About Teaching and Classroom Practices in Lesotho as A Foundation for Professional Development*. Vol. 12

[10] Seung, Eulsun, Lynn A. Bryan and Mark P. Haugan 2012 Examining Physics Graduate Teaching Assistants' Pedagogical Content Knowledge for Teaching A New Physics Curriculum *Journal Science Teacher Education*. Vol. 22

[11] Alonzo and Jiwon Kim 2015 Declarative and Dynamic Pedagogical Content Knowledge as Elicited Through Two Video -Based Interview Methods *Journal of Research in Science Teaching*. doi:10.1002/tea.21271

[12] Nilsson & Vikström 2015 Making PCK Explicit-Capturing Science Teachers' Pedagogical Content Knowledge (PCK) in the Science Classroom *International Journal of Science Education* Vol. 37

Trained Inquiry Skills on Heat and Temperature Concepts

U Hasanah[1*], I Hamidah[2] and S Utari [1]

[1] Departemen Pendidikan Fisika, Universitas Pendidikan Indonesia, Jl. Dr. Setiabudi No. 229, Bandung 40154, Indonesia
[2] Departemen Pendidikan Teknik Mesin, Universitas Pendidikan Indonesia, Jl. Dr. Setiabudi No. 229, Bandung 40154, Indonesia

*uswatun198944@gmail.com

Abstract. Inquiry skills are skills that aperson needs in developing concepts, but the results of the study suggest that these skills haven't yet been trained along with the development of concepts in science feeding, found the difficulties of students in building the concept scientifically. Therefore, this study aims to find ways that are effective in training inquiry skills trough Levels of Inquiry (LoI) learning. Experimental research with one group pretest-postest design, using non-random sampling samples in one of vocational high school in Cimahi obtained purposively 33 students of X class. The research using the inquiry skills test instrument in the form of 15questions multiple choice with reliability in very high category. The result of data processing by using the normalized gain value obtained an illustration that the ways developed in the LoI are considered effective trained inquiry skills in the middle category. Some of the ways LoI learning are considered effective in communicating aspects through discovery learning, predicting trough interactive demonstration, hypotheses through inquiry lesson, and interpreting data through inquiry lab, but the implementation of LoI learning in this study hasn't found a way that is seen as effective for trespassing aspects of designing an experiment.

1. Introduction

To build knowledge scientifically and use it in decision making or solving problems then in need of inquiry skills that include asking questions, planning and conducting investigations, using appropriate tools and techniques to gather data, thinking critically and logically about relationships between evidence and explanations, constructing and analyzing alternative explanations, and communicating scientific arguments [1]. Inquiry skills are the skills needed to conduct a scientific inquiry process. These inquiry skills are developed and experienced through working collaboratively with others and so communication, teamwork, and peer support are vital components of inquiry classrooms [2].

Based on the results of observation analysis of physical learning in vocational high school conducted by researchers to physics learning activities in the classroom obtained that the inquiry aspects of student's skills are not trained on the process of physical learning in class because based on the observation of physics learning process is still centered on the teacher. The role of more students as recipients of what information is conveyed by teachers, student activities in this learning is listening to information, record teacher explanations, reading books and exercise questions. Based on the result o the score of inquiry skills test, it was found that the inquiry skills test results predicting, hypotheses, design an experiment, interpreting data, and communicating the students incluiding very low category. This is because in the learning process students do not trained aspect of inquiry skills, teachers rarely do practical activities that can develop aspects of student inquiry skills, one cause is due to inadequate laboratory facilities. Learning strategies implemented by teachers have not trained aspects of student inquiry skills.

Helping students engage in scientific inquiry and develop science inquiry skills in the context of learning science is one of the most important goals of science education [3]. One characteristic of inquiry learning is that students are fully involved in the active learning process. Students who are making observations, collecting data, analysing data, synthesizing information, and drawing conclusions are developing problem-solving skills. These skills fully incorporate the basic and integrated science process skills necessary in scientific inquiry [2]. Scientific inquiry is often presented as a set of unorganized but interconnected procedures [4]. Teachers and prospective teachers are regulary encouraged to use inquiry processes in demonstrations, lessons, and laboratories, but few organized patterns are provided to link inquiry into this approach. This often raises questions for teachers and prospective teachers about the differences in demonstrations, lessons, and labs, and what role each inquiry possesses for each of these activities. The science teacher in the world who uses an inquiry-oriented teaching approach should have a comprehensive understanding of relationships among them [5]. Therefore, a construction is needed that can link these procedures in an organized manner so that it is easier for teachers to apply an inquiry activities into science learning. Teachers should have a comperehensive understanding of the hierarchical nature and the relationship between the various pedagogical practices and the inquiry process if they will teach science effectively by using inquiry [4]. Levels of Inquiry learning can train inquiry skills, students have the opportunity to observe, formulate predictions, collect and analyse data develop scientific principles, synthesize laws, and creat and test hypotheses to produce explanations [6,7]. Associated with material contens characteristics temperature and heat inquire needs of skills related to vocational TPTU (Teknik Pendingin Tata Udara), then in this research aspects of inquiry skills revealed are aspects of predicting, hypotheses, designing an experiment, interpreting data, and communicating. The five aspects of inquiry skills in this study trained by applying the Levels of Inquiry (LoI) learning which consists of Discovery Learning, Interactive Demonstration, Inquiry Lesson, and Inquiry Lab.

2. Learning Material

The learning material at the first encounter relates to the material of expansion on a substance, as for the material review is as follows: if the object is in heat, then the object will experience temperature rise and expansion, or change its form. The graph $\Delta V = f(\Delta T)$ is a linear function, it can be obtained through the graph curve equation. The expansion coefficient is a characteristic of the physical properties of an object, the object can be seen expanding the length (α), expansion area (β), and expansion of space (γ), with the ratio $\alpha : \beta : \gamma = 1:2:3$. Characteristic properties of materials This can be used on the manufacture of bimetal switch, fuse. The material review at the second meeting: If the object is in heat, then the object will experience a temperature rise or a change in form. The graph $Q = f(\Delta T)$ is a linear function, hello expansion can be obtained through the equation of the graph curve line. The effect of heat on an object can be $Q = m\ c\ \Delta T$ (not the important factors of heat). Material review at the third meeting: Heat moves from a high-temperature object to a low-temperature object. Conduction is a heat transfer process without particle transfer. The advanced factors of heat conduction rate: the temperature difference between the two surfaces, the length of the cross section, the cross-sectional area, and the thermal conductivity of the substance, the heat conduction rate formula is $\frac{Q}{t} = \frac{k.A.\Delta T}{l}$. Convection is the process of transfer of heat from one part fluid gets another fluid by the movement of the fluid itself. The advanced factors of heat convection rate: the temperature difference between the two surfaces, the cross-sectional area, and the following convection fluid, the heat convection rate formula is $\frac{Q}{t} = h.A.\Delta T$. Radiation is the process of heat transfer in the form of electromagnetic waves. The advanced factors of the heat conduction rate: surface area, temperature, and emissivity, the radiation heat rate formula is $\frac{Q}{t} = e.\sigma.A.T^4$.

3. Experimental Method

Given this research aims to get an overview of ways to improve inquiry skills through the LoI approach on temperature and heat material for TPTU vocational students, then this research was conducted by quasi experimental method with one group pretest-posttest design [8], and the research will produce conclusions on the limited scope.

The population used in this study were the students of class X TPTU of vocational high school in Cimahi city with the number of 70 students, while the sample used was based on the purposive sampling based on the vocational curriculum which has the material of temperature and heat with 33 samples [8].

Based on population percentage and sampling and purposive sampling techniques according to research needs, this sample has drawn population [8].

The inquiry skills instruments developed based on the inquiry aspects of skills [9] totaling 15 multiple choice questions are reasoned. Instrument testing is done through two stages, the first step is to test the validity of the construct and this through expert judgment by using triangulation technique [8]. The second test is carried out by field testing using method of re-test so that obtained value of reliability equal to 0.91 with very high category. Based on this information then the instrument is worthy of use [1]. The sampling technique using non random sampling samples in one of vocational high school in Cimahi obtained purposively by department so that there is one class of 33 students. The study used the inquiry skills test instrument in the form of 15 multiple choice questions reasoned. Question of inquiry skills have been tested the validity of construct by three lecturers who are competent and have been tested by using the re-test method to obtain reliability of 0.91 which is included into the category very high.

In trained inquiry skills students applied Levels of Inquiry (LoI) learning for three meetings. The study of Levels of Inquiry (LoI) in this study is limited from the stages of Discovery Learning, interactive demonstration, Inquiry Lesson, and Inquiry Lab, this is due to the characteristics of the ability of vocational high school student in Indonesia. Physical material taught at the first meeting on expansion, at the second meeting on heat, and the third meeting on heat transfer. In addition to the inquiry skills test, the supporting instruments used are student worksheets which contain aspects of inquiry skills trained on the students and the implementation of Levels of Inquiry (LoI) learning as well as the questionnaire of student responses to learning.

4. Result and Discussion

4.1 Recapitulation score average normalized gain inquiry skills test

Associated with the results of the pretest and posttest skeptical inquiry skills of the skills in the samples tested were obtained by the unrealized gain values in the following Table 1.:

Tabel 1. Recapitulation score average normalized gain inquiry skills test

Test	X_{Ideal}	X_{min}	X_{max}	$\overline{X}$	Normalized average gain score category <g>	Category
Pretest	45	2	15	7,18	0,63	Medium
Posttest	45	17	43	31,18		

Based on Table 1., then it can be said that the ways that apply by using the LoI approach is considered effective to increase the inquiry skills students of vocational high school trhough the topic of temperature and heat in the medium category. Some of the ways LoI learning are considered effective in communicating aspects through discovery learning, predicting trough interactive demonstration, hypotheses through inquiry lesson, and interpreting data through inquiry lab, but the implementation og LoI learning in this study hasn't found a way that is seen as effective for trespassing aspects of designing an experiment.

4.2 Recapitulation the acquisition of localized gains for every aspect of inquiry skills

If we distinguish for each aspect of inquiry skills, we get the value of the uniform gain $<\hat{g}>$ as in the following Figure 1:

IOP Conf. Series: Journal of Physics: Conf. Series **895** (2017) 012103 doi:10.1088/1742-6596/895/1/012103

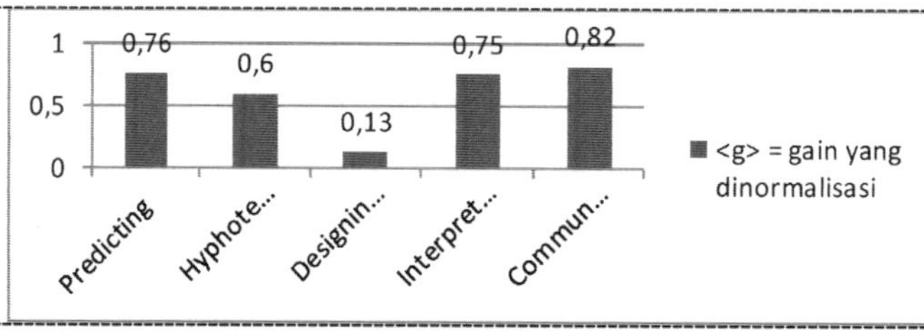

Figure 1. Recapitulation of normalized gain scores on each aspect of inquiry skills

Based on the picture above, the discussion that can be expressed in this research are:
Ways that are built with the LoI approach have a low category in improving design an experiment skills. The related instruments designing experiment are as follows:

8. To investigate the heat capacity of a liquid by using the set of tools and materials already provided, then the experiments that Anton does must be repeated using an electric calorimeter, then how to control the variables associated with the investigation....

A. The voltage source, the current flowing, and the time must be kept constant
B. The voltage source, the current that flows must be kept fixed while the time must change
C. The voltage source, the current that flows, and the time must change
D. The voltage source must change, while the current that flows and the time is kept constant
E. The voltage source is kept constant while the current that flows and the time must change.

Give the reason why you choose that answer!

Figure 2. Sample of designing an experiment aspect

And the example of the answer expressed by the party is as follows:

Examples of student answers 1:

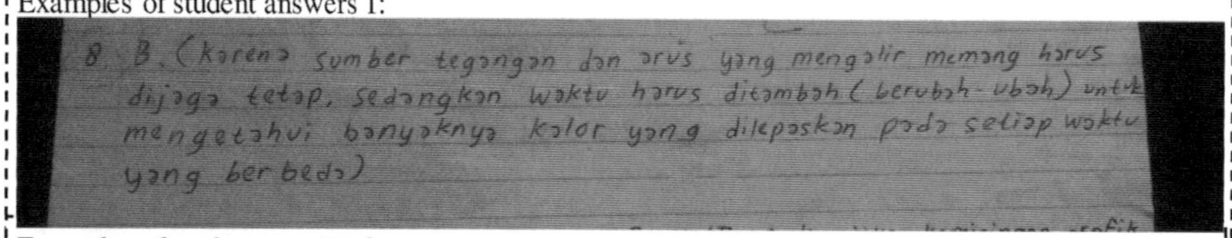

Examples of student answers 2:

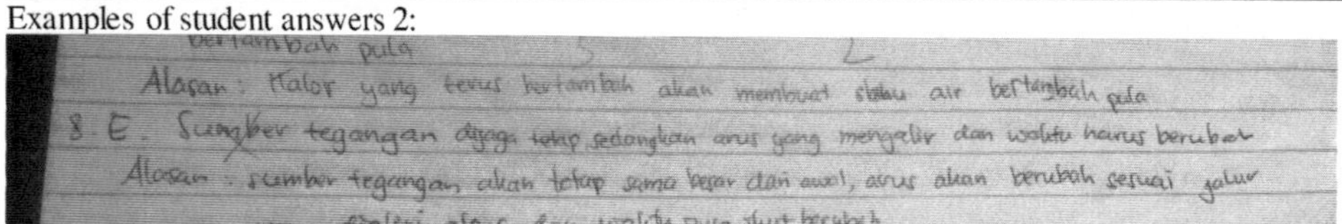

Figure 3. Example of the answer designing an experiment aspect

If we observe the result of learning activity, this skill planning training is at inquiry stage, the observation result of the implementation shows students are still mistaken in answering questions on aspects of designing experiments, this gives an indication of the ways in which the LoI approach is perceived to be ineffective. If we observe the results of side performance through worksheet that used, it can be disclosed the students do not know the variables investigation, where variables cause and effect dan mana variables that must controlled. The role o interactive demonstration does not provide for inquiry lesson, for novice students this may have to be trained in a real model that students can follow, an interactive demonstration must provide a clear example of independent, bonded and controlled variables, and then repeated in the inquiry lesson. Teaching modelling learning by Harlen makes it easy for students to receive ways to designing an experiment [10]. If children are to develop the ability to plan there must be opportunities for them to start from a question and work out how to answer it. To take these steps by themselves is asking a great deal of young children and of older ones unused to devising inquiries. A set of questions related to the decisions they have to take provides a starting point. For instance, planning a fair test can scaffolded using a planning board on which children pin labels showing what they are going to change, what variables must stay the same and what has to be measured or observed to find the result [10].

Figure 1. shows that ways that are effective in Levels of Inquiry (LoI) learning can tap into the aspects of communicating through discovery learning, predicting through interactive demonstrations, hypotheses through inquiry lesson, and interpreting data through inquiry lab. It can be seen from the normalized average gain score on the communicating aspect that is 0.82 and included into normalized gain of predicting aspect is 0.76 included into the high category, the normalized average gain score aspects of the data interpret namely equal to 0.75 incluided into the high category, the average score of normalized gain on hyphothesized aspect that is equal to 0.60 included into the medium category. Aspects of inquiry skills communicating is to place information or data obtained from our observations into several forms that we can understand later. Students learn to communicate in many ways. They learn to draw diagrams, and accurate maps, make the right graph; create an accurate experiment models; and use clear language when describing objects or events. While the predicting aspect is a forecast of future observations based on the conclusions of available data. The more data available, the more confident we are in predicting, but it can also apply otherwise. Without some data we can only guess about the future; To predict is not possible when students put their data in form of graphs, the usually have many opportunities to predict. Students often need help when predicting, a simple diagram can help them reason through data. If they can not calculate the correct predictions, ask them to predict the direction of change that is useful to them [10]. The hypothesized aspect is a prediction of the outcome of an experiment [9]. In the Levels of Inquiry (LoI) study the skills of interpreting data are trained at the inquiry lab stage. In fact, tracing the hypothesized and interpreting aspect of the data requires an example or guidance from the teacher on how to hypothesize an experiment, and how to interpret the data correctly, since the characteristics of high school students in Indonesia still can not hypothesize and interprate the data themselves before given by example teacher. However, after several examples they are accustomed to hypothesize and interprat the data properly and correctly. In this study for three meetings each student is required to complete the student worksheet which contains aspects of inquiry skills trained including aspects of hypotheses and interpreting data, because they are accustomed to complement the student worksheet in making hypotheses and interpret the data although it must be guided by the teacher, so that students obtain average gain scores in the medium categoryon aspects of the hypotheses.

While the implementation of Levels of Inquiry (LoI) in this study has not found a way that is considered effective to trespass aspects of trial experiment. Normalized average gain scores on the design aspect of 0.13 are included in the low category. The inquiry skills aspect of designing experiments belongs to an integrated Levels of Inquiry skills aspect [9]. Aspects of designing experiments to design experiments required several skills including: student's skills in establishing investigative variables, inquiry procedures, preparation of necessary tools and materials. Variables are objects that are observed and investigated. The types of variables in the investigation are independent variables, dependent variables, and control variables. The independent variables is the treatment variable or the manipulation variable, the dependent variable as the result variable, while the control variable is the controlled variable so it does not participate effect on the dependent variable. Students skills of working procedures, can guide students to discover their own concepts of nature and to test hypotheses and predict expected outcomes. The inquiry skills aspect of designing an experiment is a very complex aspect, so it takes a long time to get it done so that the student has the inquiry aspect of designing an experiment well. In fact the students still can not control the variables in the experiment.

Judging from the overall activity of teachers and students when learning can be seen that the learning of physics by using Levels of Inquiry (LoI) can be done well, although there are still shortcomings.

5. Conclusion

The result of data processing by using the normalized gain value obtained an illustration that the ways developed in the LoI are considered effective trained inquiry skills in the middle category. Some of the ways LoI learning are considered effective in communicating aspects through discovery learning, predicting trough interactive demonstration, hypotheses through inquiry lesson, and interpreting data through inquiry lab, but the implementation of LoI learning in this study hasn't found a way that is seen as effective for trespassing aspects of designing an experiment.

Acknowledgments

The author wishes to acknowledge the helpful contributions of Mrs. Ida Hamidah and Mrs. Setiya Utari who reviewed the original manuscript for accuracy and made numerous suggestions for improvement. The author hereby expresses his thanks for their contributions..

References

[1] National Science Education Standards 2012 *Science Content Standards*. Washington, DC: The National Academies Press

[2] Harrison C 2014 Assessment of Inquiry Skills in the SAILS Project. *Science Education International* Vol. 25, Issue 1

[3] National Research Council (NRC) 2007 *Taking science to school: Learning and teaching science in grades K–8.* Washington, DC: National Academies Press

[4] Wenning, C.J. 2005 *The Levels of Inquiry Model of Science Teaching.* Illinois State University

[5] Wenning, C.J. 2011 *The Levels of Inquiry Model of Science Teaching.* Illinois State University

[6] Wenning, Carl J. 2007 *Assessing inquiry skills as a component of scientific literacy*. Illinois State University

[7] Wenning, C. J. 2011 *Levels of Inquiry Model of Science Teaching: Learning sequences to lesson plans.* Illinois State University

[8] Creswell, J.W. 2009 Research Design: Qualitative, Quantitative and Mixed Method Approach. Third Edition. California: SAGE Publication

[9] Prentice Hall 2006 *Science Explorer Inquiry Skills Activity Book*. Boston: Pearson Education

[10] Harlen, W 2014 Helping children's development of inquiry skills. *Inquiry in primary science education (IPSE)* Pri-Sci-Net

[11] George, E 2005 *Scientific skills development* Miami, Floridina International University

Analysis of Student's Skills on the Concept Dynamic Electricity

I Safrina[1,2,*], J Maknun[1] and L Hasanah[1]
[1] Departement of Physics Education, Universitas Pendidikan Indonesia, Jl. Dr. Setiabudi No. 229, Bandung 40154, Indonesia
[2] Madrasah Tsanawiyah Islamiyah Ciputat

*i2n_mars@yahoo.com

Abstract. Physics becomes one of the science lessons in schools that guide student to apply materials in everyday life and commnicating the results natural phenomena. This study intended to figure out understanding skill and argumentation skill on the concept dynamic electricity. This study was descriptive research of senior high school students in South Tangerang. Technique of collecting data was done by test method and interview. The understanding skill refers to indicators of conceptual understanding in the cognitive process of Bloom's Taxonomy Revision, which contains the indicators were explaining, interpreting, summarizing, comparing, classifying, and inferring. While the argumentation skill refers to Toulmin's Argumentation Pattern (TAP) which contains the components of claim, evidence, warrant, and backing. The result showed that ability of explaining was higher and ability of comparing was lower. The component of claim was higher and backing was lower. Based on the result, it can be conclude that understanding skill and argumentation skill still less. They have been teached to student with inovation learning method.

1. Introduction

Physics becomes one of the science lessons in schools that guide students to be able to apply materials that have been studied in everyday life, so that the quality of human resources increases. In addition, physics learning examines students in concluding and communicating their research results from natural phenomena that have been studied. The initial understanding that has not been compatible can make the students have difficulty in doing the learning process [1]. Learning science to build arguments and active discussion can improve students' conceptual understanding of the subject matter [2]. Therefore, it needs information about students' understanding ability and argumentation skills.

Based on interviews of teachers and students, in one public school in South Tangerang obtained information that physics learning more emphasis on the completion of physics problems, especially the calculation, the lack of delivery of physics concepts, the lack of examine the process of thinking students in discussions and argue scientific, learning activities that rarely construct Students through experimentation, and the lack of explanation of conceptual interrelation with applications as well as natural phenomena in everyday life. In the science classroom, teachers often lack the resources to challenge student statements, because of the argument becomes a monolog - a one-way conversation where the pupil cannot be involved in the original question about it. As a result, everything is judged right and wrong not because of scientific claims, warant, and precise data [3]. Students with high-argument ability have few misconceptions on a concept compared with students lacking in argument, conceptual understanding is very important to be possessed by students [4].

The ability to understand the incompatible beginning can make the students have difficulty in learning process [5]. Thus, understanding students' concepts becomes the basis for developing other abilities, including the ability of argumentation. In accordance with the results of Sadler and Zeidler (2005) studies that students who have a high understanding of a particular topic, they will have better analytical and decision-making skills to argue [6].

Understanding the concept makes it easier for students to submit their arguments appropriately. Improved students' understanding and argumentation skills are known when teachers know how much they get it. So, it takes a figure of understanding ability and argumentation skills of students, especially the concept of physics.

Research on understanding electricity is focused on current, potential difference and especially brightness of bulbs in parallel or series DC circuits [7,8]. Beside that, to understand the concept of alternating current (AC) electric and direct current (DC) electric in everyday life. The ability to understand the relationship between electrical voltage, electrical resistance, and electricity, direct current electricity law, calculate power and direct current electric energy, is needed by students. Understanding of the process of electrical occurrence allows students to explain the phenomenon of electricity in everyday life. Therefore, it takes a figur of understanding ability and argumentation skill of student of dynamic electrical concept.

2. Experimental Method

This research was conducted on 80 students class XI in one of the public high school in South Tangerang academic year 2016/2017. This research uses descriptive research to find out the understanding ability and argumentation skill of student on dynamic electrical concept. In this research, the sampling technique is purposive sampling technique. The flow of research conducted in the study of literature relates to the focus of research, interviews with some physics teachers and students on physics learning in schools, and provides research instruments. The research instrument is a multiple choice question to measure the understanding ability and essays to measure argumentation skills. The instrument of unde rstanding ability refers to the cognitive process of Bloom's Taxonomy Revision which is to interpreting, explaining, summarizing, comparing, classifying, and inferring [9]. Assessment of argumentation skills refers to Toulmin's Argumentation Pattern (TAP) which contains the components of claim, evidence, warrant, and backing [10].

Before this instrument was used in the research, first it tested the validation of experts to correct the problem that has not been in accordance with the measured indicator. Then test the reliability by test retest to see the level of the tested matter being tested with the same group and different time. The results of reliability test obtained at 0.9 with very high category. This means that the instruments of understanding ability and argumentation skills are worthy of use in this study.

3. Result and Discussion

3.1 Figure the Profile Understanding ability of Student

The ability to understand concepts becomes the basis of students' ability to apply science better and achieve learning goals. Students easily inform, interpret, integrate, and construct existing knowledge. It allows students to understand the concept. Learning science requires students to understand both theoretical and application concepts in solving the problems encountered. The problems in everyday life require students to understand the concept of science more deeply. It allows students to understand the concept. The result of the students' ability to understand assesment on the cognitive process of taxonomy Bloom revision on dynamic electric concept is presented as follows:

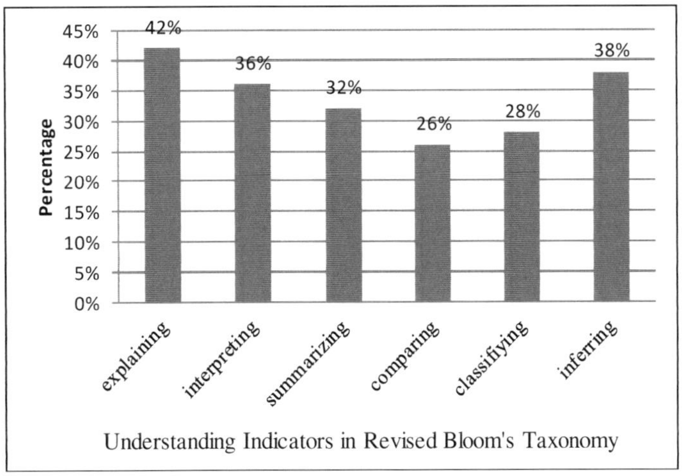

Figure 1. Profile Ability Understanding students to class XI of Dynamic Electric Concepts

Figure 1 shows the ability to understand students who are still low. This ability is less examined by teachers, so the achievement of physical values is still low. It is known that the understanding indicator on the aspect explaining is 42%, interpreting is 36%, summarizing is 32%, comparing is 26%, classifiying is 28%, and inferring is 38%.

Based on the above percentage, it can be seen that students' ability in explaining aspects is still lacking, meaning that students have not been able to build and use a concept on causal model. The interpretive aspect of students is still lacking, meaning that students have not been able to change the information from one form to another, for example from graph to verbal, table to graph, and others. This is because the lack of examined by teachers to translate information, students only focus on the matter of calculation only. Aspects summarize is still lacking, meaning that students have not been able to put forward a sentence representing the infromation of an existing problem. Students only see general information only, without considering the implicit information contained in the problem. The comparing aspect is still lacking, meaning students have not been able to see the similarities and differences between two or more objects, events, and problems. The classification aspect is still lacking, meaning the students have not been able to group the characteristics or patterns that fit in accordance with certain concepts. Aspects of drawing conclusions are still lacking, meaning students have not been able to summarize a concept from a series of events and then drawn relationships according to the characteristics of the series of events. This is in line with previous research that Learning and understanding helps students to process information, interpret, integrate with old knowledge, and manage the memory location of the information [11] .

The ability to understand the lowest is to classify. Students have not been able to classify and characterize cases in dynamic electricity. For example, grouping the magnitude of electric current flowing in the carrier with different types of barriers.

Based on the explanation, there should be a solution that can improve and increase students' understanding ability. Therefore, the ability to understand this profile is expected to improve the quality of better learning in schools.

3.2 Figure the Profile argumentation skills of student

The argumentation skills measured in this study refer to Toulmin's Argumentation Pattern (TAP). The result of the research is the percentage of student argument on claim, evidence, warrant, and backing component. The result of students' argumentation skill assessment on dynamic electric concept is presented as follows:

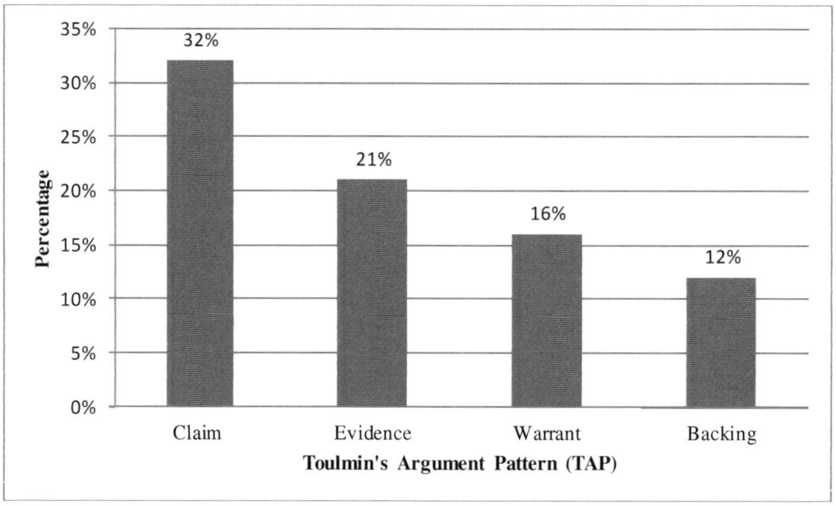

Figure 2. Profile of students' argumentation skills to class XI of Dynamic Electric Concepts

Figure 2 shows that argumentation skills student is low. Students still have difficulty in arguing scientifically. It is known that the argumentation component claims is 32%, evidence is 21%, warrant is 16%, and rebuttal is 12%.

Students' argumentation skills in claim are still less accurate to answer the problem or phenomenon. Students argue briefly, without explaining the intent of the answer. Then the student in providing the evidence that does not support the claim that has been filed, so the data does not provide appropriate evidence to justify the claim. Then the data presented there is no analysis that supports the claim, so the data is just writing to answer questions. Students can only provide non-theory-oriented data, then the student must have other alternative evidence to justify the claim. But students are accustomed to theory and then look for evidence [12]. The student in filing a justification or warrant does not adequately explain the exact relationship between data and claims. The warrant does not support the claim, so the data becomes incorrect. Then backing does not underlie warrant and thus does not support the claim. The proposed support is not based on the facts or theories applicable, so it is the same as the claim already delivered.

Therefore students' scientific argumentation skills have not been correct, so it is still low. Following up on this argumentation skill profile is expected to improve the quality of better learning in schools.

4. Conclusion

The conclusions obtained based on the results of the study, showed that students' understanding ability and argumentation skills are still low. This is evidenced by 42% an explanation, an interpretation of 36%, summarizes by 32%, compares by 26%, classifies by 28%, and inferr by 38%. While the argumentation claim score of 32%, evidence of 21%, warrant of 16%, and rebuttal 12%. These findings will serve as a basis for further research on innovative learning models and strategies that can improve students' comprehension and argumentation skills.

Acknowledgments

The author would like to thank Dr. Johar Maknun and Dr. Lilik Hasanah for his guidance and time in making the students' ability to understand and argumentation skills of dynamic electrical concept. The authors would like thank Dr. Aloysius Rusli and Dr. Winny Liliawati, M. Si for his willingness to be the instrument validator that the author made. The writer also expressed his willingness to the students of senior high scholl in South Tangerang for his willingness to be a sample of this research.

References

[1] Kuhn, Deanna and Wadya Udell 2003 The Development of Argument Skills. *J. Society for Reaseach in Child Development* 1245-1260
[2] Ford, M. 2008 Disciplinary authority and accountability in scientific practice and learning. *Science Education* 404-423
[3] Berland, Leema Kuhn and Katherine L. McNeil 2009 Using a learning progression to inform scientific argumentation in talk and writing. *LeaPS Conference* 4
[4] Acar, Omer and Bruce R. Patton 2012 Argumetnation and Formal Reasoning Skills an Argumentation Based Guided Inquiry Course. *Social and Behavioral Scienc* **46** 4756-4760
[5] Cepni, S 2009 Effect of Computer supported Instructional Material (CSIM) in Removing Students Misconceptions about concepts" "Light, light source, and seeing", *Energi Education Science and Techbnology Part B: Social and Educational Studies* vol **1** 5
[6] Sadler, T. D. and Zeidler, D. L. 2009 The Significance of Content Knowledge for Infromal Reasoing Regarding Sosioscientific Issue: Applyng Genetics Knowledge to Genetic Engineering Issue. *Science and Education* 71
[7] Pardhan, H. and Bano, Y., 2001 Science teachers' alternate conceptions about direct-currents, Int. J. Sci. Educ. **23** 301-318
[8] Lee, Y. and Law, N., 2001 Explorations in promoting conceptual change in electrical concepts via ontological category shift, Int. J. Sci. Educ. **23** 111-149
[9] Anderson, L.W. and Krathwohl, D.R 2001 *A Taxonomy for Learning, Teaching, and Assesing: A Revision of Bloom's Taxonomy of Educatioanl Objectives.* (New York: Addison Wesley Longman, Inc.)
[10] Campbell, Todd and Brooke Roberthaw 2013 Constructing arguments: Investigating pre-service science teachers' argumentation skills in a socio-scientific context *Science Education International* **24** 195-211
[11] Ranne, Carina S and Samuli Kolari 2003 Promoting the Conceptual Understanding of Engineering Students through Visualisation *Global Journal of Engineering Education* **7** 197
[12] Kuhn, D. 1993 Science as argument: Implications for teaching and learning scientific thinking. *Science Education*, p 319-337

IOP Conf. Series: Journal of Physics: Conf. Series **895** (2017) 012105 doi:10.1088/1742-6596/895/1/012105

Pre-Service Physics Teachers' Problem-Solving Skills in Projectile Motion Concept

S Sutarno[1,2,*], A Setiawan[1], I Kaniawati[1] and A Suhandi[1]

[1]Program Studi Pendidikan Ilmu Pengetahuan Alam, Sekolah Pascasarjana, Universitas Pendidikan Indonesia, Jl. Dr. Setiabudhi No. 229, Bandung 40154, Indonesia
[2]Program Studi Pendidikan Fisika, Fakultas Keguruan dan Ilmu Pendidikan, Universitas Bengkulu, Jl. W.R. Supratman, Kandang Limun, Bengkulu 38371, Indonesia

[*]sutarno_bengkulu@student.upi.edu

Abstract. This study is a preliminary research aiming at exploring pre-service physics teachers' skills in applying the stage of problem-solving strategies. A total of 76 students of physics education study program at a college in Bengkulu Indonesia participated in the study. The skills on solving physics problems are being explored through exercises that demand the use of problem-solving strategies with several stages such as useful description, physics approach, specific application of physics, physics equation, mathematical procedures, and logical progression. Based on the results of data analysis, it is found that the pre-service physics teachers' skills are in the moderate category for physics approach and mathematical procedural, and low category for the others. It was concluded that the pre-service physics teachers' problem-solving skills are categorized low. It is caused by the learning of physics that has done less to practice problem-solving skills. The problems provided are only routine and poorly trained in the implementation of problem-solving strategies.The results of the research can be used as a reference for the importance of the development of physics learning based on higher order thinking skills.

1. Introduction

Problem solving is one of the important elements in physics learning. Problem solving defined as a series of procedures that must be completed to solve the problem [1], is a complex cognitive activity that represents mental construction [2], and a movement process in purposing aim in case the way is not possible to be determined immediately [3]. Furthermore, problem-solving skills are defined as someone's capability to discover solution through a process that involves the information acquisition and organizing [4]. Problem solving depends on experience and people's perspective according to the problem being faced [5]. The structure of knowledge and problems character being faced are some of the factors that affect problem solving ability [6]. The presentation of the complex and contextual problems that involve laboratory activity and giving structured tasks with variously represented models will help students in practicing problem-solving skills [4, 7].

The research to explore students' problem-solving skills in physics has been widely practiced [8,9,10]. In general, there are two important parts representing investigative procedures in these studies, which are problem characteristics and problem-solving strategy steps that used to describe problem-solving skills. In general, the problems used are only textbook, in the form of mathematical problems, non-contextual, in the form of routine issues, and in a less appropriate format of presentation. While the steps in implementing the problem solving-strategies used are generally less detailed, usually involving only three to four aspects of problem solving, such as understanding problems, planning solutions, implementing solutions, and evaluating solutions.The characteristics of the problem used are believed to be less able to encourage students to think more deeply, while the steps applied are less able to describe students' problem-solving skills comprehensively. In contrast, in this study the problem used is contextual, related to the phenomenon of everyday life and presented in context-rich problems format. At the end of the problem presentation, students are directed to use a more comprehensive problem-solving strategy including useful description, physics approach, specific application of physics, physics equation, mathematical procedure, and logical progression.

The study to explore the profile of pre-service physics teachers' problem-solving skills needs to be done. The results obtained can be used as a reference in developing a physics learning strategy that can train problem-solving skills, both in class and through laboratory activities. Furthermore, the results of the research can encourage teachers to present a variety of physics learning activities that are capable of developing higher order thinking skills.

2. Method

The aims of this study is to explore the problem-solving skills of pre-service physics teachers. The participants of the study were 76 undergraduate students (male 34 and female 43) ranging in age from 19 to 22 years. This study was conducted at Study Program of Physics Education of a college in Bengkulu Indonesia in the even semester of 2016-2017 academic year. Pre-service physics teachers'problem-solving skills are identified using essay test questions on the concept of projectile motion. This concept was chosen on the consideration that all students in the even semester had already acquired the material's learning. Problems in the matter are related to the life context related to the principles of physics. At the end of the problem description, pre-service teachers are directed to solve problem using problem-solving strategies. The stages of problem-solving strategies analyzed using the rubric of physics problem solving [11].

3. Result and Discussion

3.1. Result Description

Pre-service teachers' problem-solving skills in this research are explored through one quantitative problems with the longest allocation completion time of 60 minutes. The answers are analyzed with problem solving-strategies rubric. Their problem-solving skills then identified based on each problem-solving strategies used. The physics problem used in this study is given in the following context rich problem: in a war game, a soldier was asked to fire cannonball off with velocity 100 m/s to the enemy's ship that moves closer with 20 m/s of velocity. The position of the cannonball provided with distance detector is on the cliff with 20 meters height from the sea level (as shown in figure 1). Unfortunately the soldier is confused, when should he start to fire the cannonball off so that it exactly hits the ship in P position? The soldier is asking your help to solve the problem as a reference in improving his shooting ability. To help the soldier, you must take these steps: (a) draw a physical sketch that indicates the situation, (b) determine the most appropriate physics concept and principles, (c) determine the specific situations that must be implemented so that the solution gained effectively, (d) according to physical sketch, the specific situation as well as the physics principles, formulate the proper physics equation, (e) do a mathematical procedure using known equation and units, (f) does your solutions clear, focus, and one part to each other is not contradicting? does the mathematical procedure accurate? does your answer rational?

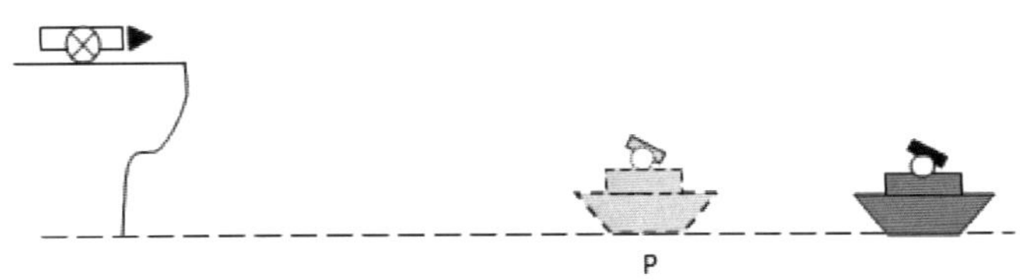

Figure 1. The cannonball and the ship position

IOP Conf. Series: Journal of Physics: Conf. Series **895** (2017) 012105 doi:10.1088/1742-6596/895/1/012105

The first stage being explored is a useful description, consists of pre-service teachers skills in drawing a physical sketch that illustrates the problem situation. The physical sketch that represents the problem shown in figure 2. The physical sketch includes: (a) the representation of cannonball position, the cliff is used as horizontal reference point, ship position when the projectile starts to shot (e.g in C), and ship position when it's shoot (in P), (b) the cannonball track uses dotted line forming parabolic trajectory, (c) vector of the initial cannonball velocity (not forming a specific angle with horizontal line) and vector of ship velocity, (d) physical quantity sign for the cliff height (h_o), cannonball velocity (v_p), ship velocity (v_k) with value and units, and (e) the representation of ship's distance from the bottom of the cliff when the cannonball starts to fired off that shows distance that must be determined. Target quantity sign can use alphabet such as x along with emphasizing sentence to indicate that this distance has to be determined, he sentence used could be: what is distance x so the cannonball hit the ship at P?

According to the answer analysis, it can be concluded that generally the physical sketch doesn't illustrate the problem situation exactly. One of the physical sketches produces by the student is shown in figure 3. Based on the analysis of the physical sketch, there are some errors identified, such as: (a) the initial velocity of cannonball curves a particular angle to the horizontal line, (b) analyzing the vector force doesn't needed, such as the description of weight and normal force acting on the cannonball, the description of weight and floating force acting on the ship, as well as the potential energy and kinetic energy acting on the ship in point P, (c) doesn't represent the ship distance that must be determined, (d) doesn't give the emphasizing sentence related to the targeted the quantity that will be searched, (e) the cannonball motion doesn't create parabolic trajectory, (f) the cannonball trajectory illustrated with solid line, not with the dotted line, (g) misinterpreting the ship's distance to be determined. Students interpret the distance to look for is the distance when the ship is hit by the cannonball, (h) the ship velocity vector illustrates longer than the cannonball velocity vector, whereas the value of the vector is vice versa.

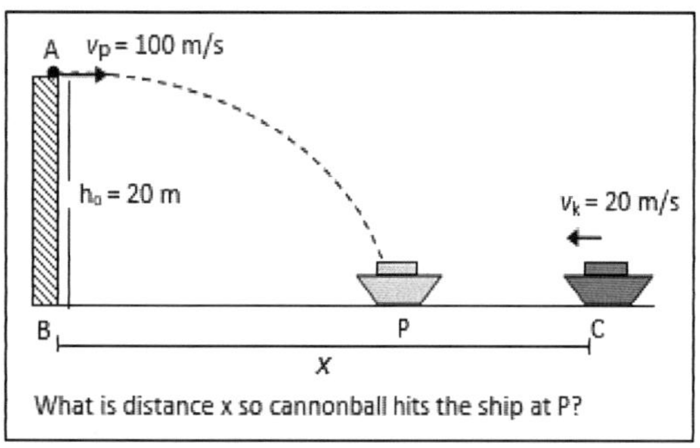

Figure 2. Physical sketch according to problem situation.

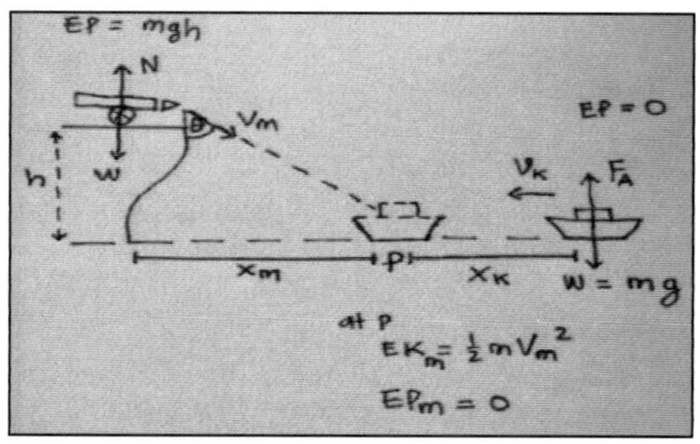

Figure 3. One of the physical sketches produces by the pre-service teacher.

The ability to implement problem-solving strategy steps in the second stage is physics approach, that contains student's ability in determining physical concept and principles accordance with the physical sketch made. The appropriate physics concept and principles in accordance with the problem are: (a) parabolic trajectory, it occurs when the cannonball is started to fire off until it hits the ship, and (b) linear motion, when the ship hasn't been shot, during the ship goes from C to P. Based on the answer identification given in Physics Approach stage, there are some errors done by the student. The physics concept and principles expressed are mostly unimportant and not directed to the solution. For example, the principle of energy conservation, Archimedes' Law, the principle of resultant vector, and the principle of free fall.

The ability to implement problem-solving strategy steps in the third stage is specific application of physics, consists of student's ability to identify and specify specific conditions that must be implemented based on physics principles. The specific conditions that must be implemented in the given problem are: (a) the time the cannonball takes to hit the ship in point P has the equal time for the ship to move from point C (when the cannonball shoot) to point P, (b) the projectile velocity in the horizontal axis is always constant, has the same value as the initial projectile velocity, and (c) when the projectile hits the ship, the last projectile height (h) equals zero. Based on the result of the answer analysis, it can be obtained that the illustration of specific conditions applied by the student is generally not important since it doesn't lead to the problem solution. Student defines specific condition by stating that the time needs by the projectile to hit the ship is equal to the time when the projectile is dropped to the bottom of the cliff, the ship initial velocity is equal to the projectile velocity in free fall. Some of those statements are true, but they're not related directly to the context of the problem. The statement given doesn't identify the specific conditions that able to direct into the solution attainment of the problem.

The ability to implement problem-solving strategy steps in the fourth stage is physics equation, consists of student's ability to define and determine the proper physics equations according to the principle and the implemented specific conditions. The written physics formulation must be relevant to physics approach aspect and drawn the physical sketch in useful description aspect. The suitable physics formulation according to the problem shown in figure 4. Based on the answer analysis, it can be resulted that some student's error descriptions are: (a) written physics formulation is not suitable with the physical sketch, (b) misinterpretation, the distance counted by most of the students is the horizontal distance when it was hitted by the projectile, and (c) physics formulation written is not complete yet.

in projectile motion from A to P:

$$h = h_0 - \frac{1}{2}gt^2$$

with h = 0 when the projectile hits the ship, so that:

$$h_0 = \frac{1}{2}gt^2 \text{ and } t = \sqrt{\frac{2h_0}{g}}$$

Projectile horizontal distance (x_{BP}) is :

$$x_{BP} = v_p\, t = v_p \sqrt{\frac{2h_0}{g}}$$

traveled distance by the ship before it's hit (x_{CP}) is:

$$x_{CP} = v_k\, t = v_k \sqrt{\frac{2h_0}{g}}$$

so that the distance of the ship from the bottom of the cliff when the projectile is begin to shoot (x) is:

$$x = x_{BP} + x_{CP} = (v_p + v_k)\sqrt{\frac{2h_0}{g}}$$

Figure 4. The suitable physics formulation according to the problem.

The ability to implement problem-solving strategy steps in the fifth stage is mathematical procedure, consists of consists of student's ability to do mathematical procedure right after the student formulates the proper physics equation. The mathematical procedure expected from this stage shown in figure 5. Based on the pre-service teachers' answer analysis can be found some weakness, those are: (a) mathematical procedure used is not appropriate because of wrong physics formulation, (b) doesn't input the physics quantity units in every processes of the calculation, the student only writes the units in the final obtained value, and (c) the student doesn't give strengthening sentences upon the obtained solution.

$$x = (v_p + v_k)\sqrt{\frac{2h_0}{g}}$$

$$= \left(100\frac{m}{s} + 20\frac{m}{s}\right)\sqrt{\frac{2 \text{ x } 20\ m}{10\frac{m}{s^2}}} = \left(120\frac{m}{s}\right) 2s$$

$$= 240\ m$$

So, in order the projectile hits the ship exactly in point P, the projectile must be shot when the detector instrument in the cannonball shows the ship position in 240 m from the bottom of the cliff

Figure 5. The expected mathematical procedure

The ability to implement problem-solving strategy steps in last stages is logical progression, consists of student's ability to do reflection the overall obtained solution. There are three reflection questions given related to the review whether the solution given has met the explication, focus, and not contradictive; does the description given has been proper and beneficial; does the mathematical procedure done has been correct; and does the answer obtained rational. Based on the answer given in logical progression stage, identified the inconsistency between the answer given by the student related to reflection question with the written the reality in the answer. For instance, the student answered that all the solutions have been clear, focus and rational, while actually there spotted the discrepancy between physical sketch and the equation used. The student state that the description made has been proper and beneficial, but in fact, they wrote unneeded descriptions, such as writing force vectors acting on the ship, energy conservation law when the projectile hits the ship, and other errors were done.

3.2. Discussion

Based on the analysis of the stages in problem-solving strategies applied by the pre-service teachers in solving the problems, it can be seen that result has not been like what expected. The pre-service teachers' skills are in the moderate category for physics approach and mathematical procedures stages, and in low category for useful description, the specific application of physics, physics equation and logical progression stages. Their problem-solving skills are still low although the problems given are only related to the fundamental concepts of physics. The students generally only recognize the problem based on the surface feature. They are less able to link the various information contained in the problem description with other information required. Students fail to build relationships between known and unknown variables.

They are not able to deliver beneficial description as a starting point in problem-solving strategy. They still don't have a good ability in drawing a physical sketch to represent the problem. This inability affects the weakness of their ability to define key concepts or principles so that the formulas that are formulated do not show the right relationships. The formulated physics equation is less proper to the physical sketch, physics concept, and principles, as well as the set specific conditions. Some of the students are able to draw a physical sketch and determine the proper physics principles, but they are not able to formulate the equations that direct them to the expected solution.

Generally, the pre-service teachers haven't been able to view the problem completely, their comprehension is fragmented into separated pieces. They are not able to construct the relations in physics concepts. They tend to rely on the information contained problem as their asset in solving the problem. The pre-service teachers' characteristic is more identified as a novice rather than the expert. Novice doesn't need knowledge structure obtaining, their comprehension consists of random facts and equation with less conceptual meaning and they access the principle individually from memory [12].

Student's skills in solving a problem are affected by several factors, among others the problem characteristics and their prior knowledge [13]. The problem characteristics are referred to the problem structure given. Based on the structure, the problem differs into well-structure problem and ill-structure problem [2]. The well-structure problem represents all information needed in solving a problem, needs implementation some of ordered and perspective ways, convergent answer, and determined solution. While the problems in everyday life are more disordered, undefined, the problem elements are usually unknown, needs a various solution, has various criteria to evaluate solutions, as well as doesn't have the certainty in needed concept and principle for finishing and organizing the problem [14].

A person's prior knowledge also affects their ability to solve problems. A person who is able to keep the principles of physics in their memories as a piece of information that is inseparable and can be used together will tend to be more able to solve the problems faced [15]. Students' skills in solving a problem are affected by their prior knowledge and experiences in solving problems. Learning strategies that are deliberately designed to train problem-solving skills can improve students' problem-solving skills [16]. The activities both in the classroom as well as in the laboratory are obviously potential to get that skills. A learning strategies that can be used for those purposes are problem-based learning. In a laboratory environment, problem-solving skills can be given through practical activity based in problem solving laboratory [17].

The low student ability to solve problem can be caused by several factors, such as students are not accustomed to using problem-solving strategy steps, the exercises are not yet using new nonroutine problem and the application of problem solving strategy, the applied learning strategy has not been deliberately designed to train problem-solving skills both in class and in laboratory activities. The development of problem-solving skills will be successful when it involves specific skills in problem solving, contextual assignment, practical process in problem solving, discussion related to troubleshooting process, and teaches how to solve problems before students complete all the subject matter [18].

4. Conclusion

Based on the data analysis, it can be concluded that the pre-service physics teachers' skills in solving physics problems are still in the low category. The lack of student skills in applying the problem-solving strategy steps can be caused by the students are not accustomed to using the problem-solving strategy steps in solving the problem, the physics study has not yet presented various nonroutine issues that are rich in context and the practice of implementing its solving strategy, and the applied physics learning have not been deliberately designed to solve problem-solving skills. The results of this study can be used as a basis for the importance of the development of physics learning strategy that is able to train and provide higher-order thinking skills, either through classroom learning or laboratory activities.

Acknowledgment

This research is supported by the Ministry of Research, Technology, and Higher Education of the Republic of Indonesia through BPPDN Scholarship

References

[1] Polya G 1957 *How to Solve It* (Princeton NJ: Princeton University Press)
[2] Jonassen D H 2011 *Learning to Solve Problems* (New York and London: Roudledge)
[3] Sfard A, Nesher P, Streetland L, Cobb P and Mason J 1998 *For the Learning of Mathematics* **18**(1) pp 41-51
[4] Sujarwanto E, Hidayat A and Wartono 2014 *Jurnal Pendidikan IPA Indonesia* **3**(1) pp 65-78
[5] Schoenfeld A 1992 *Handbook of Research on Mathematics Teaching and Learning* ed D. Grouws (New York: Macmillan)
[6] Chi T, Feltovich P and Glaser R 1981 *Cognitive Sciences* **5** pp 121-52
[7] Ibrahim B and Rebello N S 2012 Representational task formats and problem solving strategies in kinematics and work *Physical Review Special Topics-Physics Education Research* **8** (*Preprint* gr-qc/010126)
[8] Mason A and Singh C 2011 Assessing expertise in introductory physics using categorization task *Physical Review Special Topics-Physics Education Research* **7** (*Preprint* gr-qc/020110)
[9] Shih Y L and Singh C 2013 Using an isomorphic problem pair to learn introductory physics: Transferring from a two-step problem to a three-step problem *Physical Review Special Topics-Physics Education Research* **9** (*Preprint* gr-qc/020114)
[10] Savelsbergh E R, de Jong T and Hessler F M G M 2011 Choosing the right solution approach: the crucial role of situational knowledge in electricity and magnetism *Physical Review Special Topics-Physics Education Research* 7 (*Preprint* gr-qc/010103)
[11] Docktor J and Heller K 2009 *Proc. of the NARST 2009 Annual Meeting*
[12] Reif F and Heller I 1982 *Educational Psychologist* **17** pp 102-27
[13] Metallidau P 2009 *Teaching and Teacher Education* **25** pp 76-82
[14] Gok T 2010 *Eurasian Journal of Physics and Chemistry Education* **2**(2) pp 110-22
[15] Larkin H 1997 *Engineering Education* **70** pp 285-8
[16] Gill-Perez D, Dumas-Carre, Caillot M and Martinez-Torregrosa J 1990 *Studies in Science Education* **18** pp 137-51
[17] Heller K and Heller P 2010 *A user's Manual* (United State: Departement of Physics University of Minnesota)
[18] Mayer 2008 *Learning and Intruction* (Upper Sandle River NJ: Prentice Hall)

IOP Conf. Series: Journal of Physics: Conf. Series **895** (2017) 012106 doi:10.1088/1742-6596/895/1/012106

Teaching Science Process Skills by Using the 5-Stage Learning Cycle in Junior High School

E S Safaah[1,3,*], **M Muslim**[2] **and W Liliawati**[2]

[1]Program Studi Pendidikan Fisika, Sekolah Pascasarjana, Universitas Pendidikan Indonesia, Jl. Dr.Setiabudi No. 229, Bandung 40154, Indonesia
[2]Program Studi Pendidikan Fisika, Universitas Pendidikan Indonesia, Jl. Dr.Setiabudi No. 229, Bandung 40154, Indonesia
[3]Madrasah Tsanawiyah Negeri 1 Sumedang

[*]evasitisafaah@student.upi.edu

Abstract. Science process skills are a set of skills used in scientific activities. Students with science process skills are actively involved in learning. Opportunities to be actively involved in learning can be obtained by students ifthe learning stage is designed in such a way. This article is about a lesson that is expected to teach students' science process skills and can help build their understanding of the concept of buoyant force. This learning can be carried out for students at the junior high school level. This learning process uses the 5-stage learning cycle consisting of observation, manipulation, generalization, verification, and application phase. Each activity is expected to facilitates students to develop science process skills such as observing, inferring, predicting, asking questions, constructing hypotheses, designing experiments, applying concepts, and communicating.

1. Introduction

The nature of science education involves students in scientific investigation. Scientific investigation includes the way thinking, attitude, and steps scientific activities to obtain products and knowledge of science. Students should be able to integrate between skills, knowledge and attitudes, to develop knowledge of conceptual understanding better [1]. One of the most important basic skills in scientific inquiry is the skill of the science process. Scientific process skills are a set of skills used in conducting scientific activities, producing and using scientific information, and problem solving [2].

Such skills can be a character in student's self if they have the opportunity to do so in either learning or laboratory activities. The opportunity can be given to students if the learning environment is arranged in such a way that they can engage in science activities to improve their science process skills [3]. Therefore, in a learning's plan teacher plays an important role in choosing an approach that can give students the opportunity to use these skills as widely as possible. So far, the results of the study of education practitioners are still lacking in incorporating lesson plans. The buoyant force learning plan by using this 5-stage learning cycle is structured in such a way that it can facilitate students to develop their science process skills.

2. Science Process Skills

Science process skills are process thinking skills using scientific processes and approaches [4]. Scientific process skills are very important for every individual, because they are almost always used in everyday life [5]. In science learning there are two groups of process skills, namely basic science process skills include observing, classifying, inferring, measuring, communicating, and predicting, while integrated process skills include identifying variables, constructing hypotheses, tabulating and graphing data, defining variables, designing investigations, and experimenting [4]. In this article the skills of the process of science include the skills of observing, inferring, predicting, asking questions, constructing hypotheses, designing experiments, applying concepts, and communicating [6].

These skills can be assessed through the tests. The skills indicators on the test include: (a) observing: taking into account the properties of objects and events using the five senses, or a description of what is perceived; (b) provides an explanation of a particular object or event; (c) predicting: forecasting future events based on past observations or patterns of data formed; (d) asking question: requesting an explanation of a phenomenon; (e) hypothesizing: expressing suspected interim answers from observations or conclusions but ultimately subject to direct testing or by one or more experiments; (f) designing experiments: determine the tool or material of inquiry, investigation steps, and variables to be observed in the investigation; (g) applying concept: using a pre-built concept to describe an event in a new situation; and (h) communicating: using words, symbols, or graphs to describe an object, action or event [6].

3. Buoyant Force Learning Uses The 5-Stage Learning Cycle

This learning uses the 5-stage learning cycle of the Wenning version as an approach to buoyant force learning [7]. Five stages of the learning cycle include observation, manipulation, generalization, verification and application. In the five stages of this learning cycle, students play an active role in conducting learning activities.

Here are the general learning activities in the five stages of the learning cycle: **Observation** - Students observe interesting phenomena and are able to engage interest and elicit their responses. Students explain in detail what they see, do analogies and provide other examples based on the phenomenon. While the teacher makes the right questions for observation activities. **Manipulation** - Students propose ideas of investigations and develop approaches that can be used to study the phenomenon. After a discussion, they make plans to collect qualitative and quantitative data and then implement the plan. **Generalization** - Students build a new principle or law based on the results obtained in advance and provide a plausible explanation of the phenomenon. **Verification** - Students make predictions and test it using common law from the previous stage. **Application** - Students make conclusions and then apply to new situations [7].

In this buoyant force lesson plan, the five stages of the learning cycle are integrated into the learning and are designed in such a way as to enable students to use their science process skills. Here are the learning steps:

3.1. Observation

The series of activities at the observation stage, is expected to train students' skills in observing, inferring, predicting, and asking questions. First, the teacher prepares some objects that can float or sink in water, such as plastic balls, pieces of wood, metal spoons, and others. Students are given the opportunity to use their senses maximally in observing activities, such as putting objects into aquariums of water, weighing objects immersed in water, or pressing floating objects into water and feeling the presence of buoyancy. Then the students are assigned to write their observations on the board. Examples of observations: 'there are submerged objects and those floating in water', 'Objects that can sink more lightly after being put in water', 'floating objects are difficult to drown', 'there is a reaction force when the ball is pressed into the water', and others.

After that, the teacher assigns students to inferr the results of his observations. If necessary, the teacher can guide the student through referral questions until the student can conclude that any object immersed in the liquid will experience a buoyant force which is directed upstream of the liquid surface. Then, students are assigned to predict the existence of buoyant force on other objects if immersed in water. If students make observations and interpretations correctly, students should be able to predict that either in objects of mass or small mass, if they are in liquid they will experience buoyancy, whether they can float or sink in liquids. The teacher then encourages students to ask research questions. The questions should not be limited, students are given the widest opportunity to practice their ability to ask questions. Of the many questions asked, teachers and students choose one of the questions to be investigated in an experiment, for example: 'why the ship can float in the sea?', 'How is the relationship between the volume of objects with buoyant force?'. At this stage the student should be aware of the quality of the inquiry question.

3.2. Manipulation

At the stage of manipulation, students are given the opportunity to formulate research hypotheses. Students should be aware of the relevance of the hypothesis with the experimental design to be carried out next. In designing the experiment, students are given the opportunity to determine the tools and materials used. If this is not possible, then the teacher can provide tools and materials that are easily obtained in the environment. In this lesson plan with the theme of this buoyancy force, the tools that can be used to investigate the relation of the volume of objects to the buoyant force are the soda cans that can sink in the water and the spring balance [8]. Although the cans do not resemble perfect cylinders, however, we can determine the volume of cans based on the volume of soda in them.

Students can determine the experimental steps that will be done by writing it systematically. For example, the first student observe the volume of beverages listed on the can and are considered to be the same as the can volume, for example, the volume of the can is 330 ml. Then student signs the volume on the outside of the soda can, for a quarter, a half and three quarters of a part using a marker as in Figure 1.

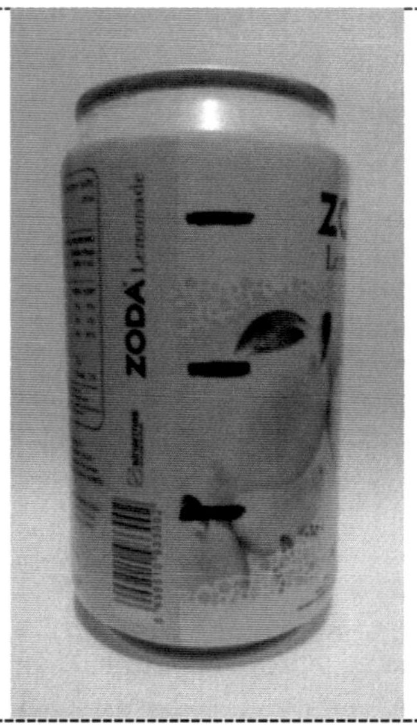

Figure 1. The soda cans that have been marked with volume boundaries to be immersed in water

Figure 1 shows that the mark on the can is made to provide a limit of the volume of the can to be immersed in water. The lowest mark indicates the volume of a quarter of a can, the middle mark showing the half-tin volume, and the topmost sign showing the volume of three quarters of a can.

Then the students measure the weight of the cans using the spring balance and get the weight of the object before dyeing 3.6 N (w_a) and then record the observations in a table. The next step students record the weight of the tin cans when dipped one quarter, half, three quarters, and all the parts into the aquarium containing the water as in Figure 2.

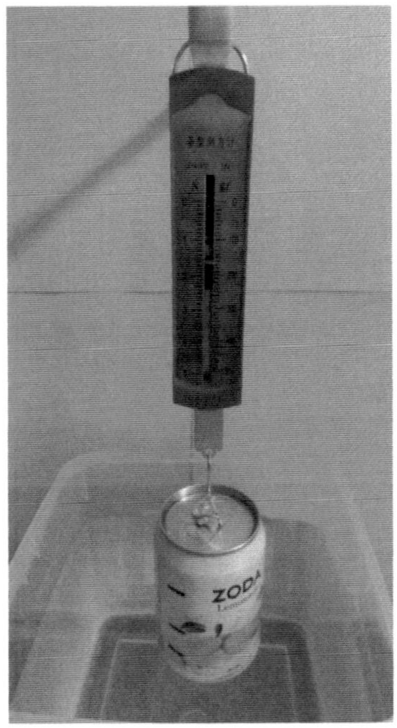

Figure 2. Measuring the apparent weight of a can by changing the volume of a dipped can

Figure 2 shows that by changing the volume of cans dipped in water, students can collect apparent weight data of cans. Based on the data of apparent weight cans, students can find that there is a weight difference before (w_a) and after dipped into water (w_f). The weight difference is a buoyant force (F_b). Thus students can create buoyant force formula that is $F_b = w_a - w_f$. Examples of observation table can be observed in Table 1.

Table 1. Data Table

Volume submerged (can)	Volume submerged (L)	Apparent weight can (N)	Buoyant Force (N)
0	0	3,6	0
¼ part	0,08	2,3	0,9
½ part	0,16	1,6	1,8
¾ part	0,24	0,8	2,7
all part	0,33	0	3,6

From the data collected as shown in Table 1, students practice communicating the results of their experiments by changing the form of presentation from the table into Graph. Examples of graphs based on data in Table 1 can be observed in Figure 3.

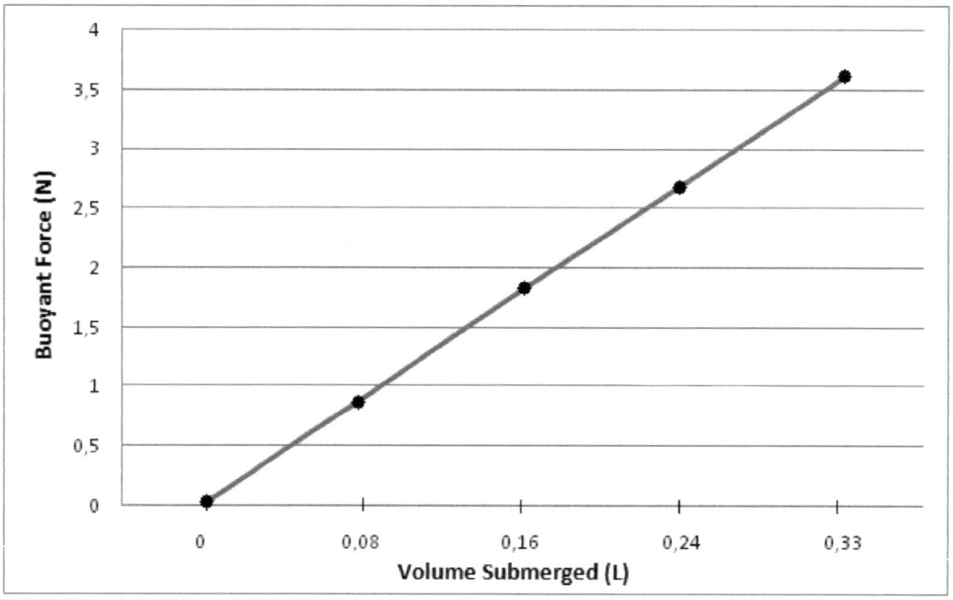

Figure 3. Graph of the relationship between buoyant force (F_b) and the objects volume submerged in water (V)

Figure 3 shows that the relationship between buoyant force and the volume of a submerged can form a linear equation. The results of this experiment are then used as materials to construct knowledge at the generalization stage.

3.3. Generalization

In the generalization stage, students build their understanding of the concept of buoyant force based on the experimental results. Experimental results show that the buoyant force is directly proportional to the volume of objects submerged in water ($\boldsymbol{F_b} \propto \boldsymbol{V}$).

3.4. Verification

In the verification stage, the teacher assisted students determine the relation of buoyant force to the volume of submerged objects using the concept of hydrostatic pressure difference at different depths. The diagram of forces acting on a floating object can be observed in Figure 4.

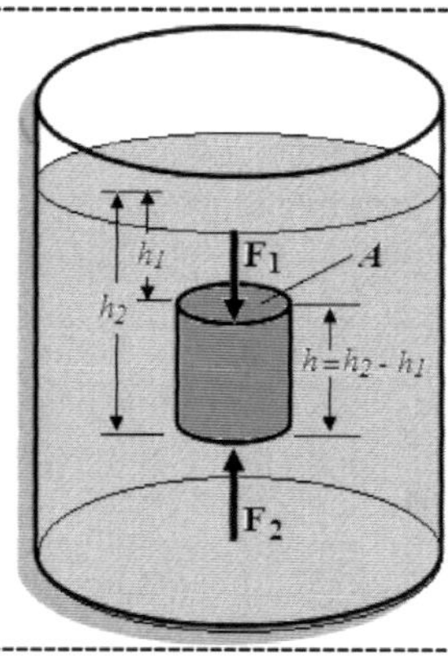

Figure 4. Diagram of forces acting on a cylinder immersed in a fluid

Figure 4 shows a force diagram on a cylinder floating in a fluid. A cylinder with an elevation h whose upper and lower edges have an area A and are immersed entirely in a fluid with a density of ρ_f. The fluid gives pressure $P_1 = \rho_f\ g\ h_1$ on the top surface of the cylinder. The force caused by the pressure at the top of this cylinder is $F_1 = P_1\ A = \rho_f\ g\ h_1\ A$, and heading down. In the same way, the liquid gives the upward force at the bottom of the cylinder equal to $F_2 = P_2\ A = \rho_f\ g\ h_2\ A$. The total force caused by the pressure of the liquid, which is the buoyant force (F_b) works upwards, the equation is :

$$\boldsymbol{F_b} = F_2 - F_1$$
$$\boldsymbol{F_b} = \rho_f\, g\, A\,(h_2 - h_1)$$
$$\boldsymbol{F_b} = \rho_f\, g\, A\, h$$
$$\boldsymbol{F_b} = \boldsymbol{\rho_f\, g\, V}$$

Thus, the relationship between buoyant forces proportional to the volume of objects immersed in the experimental activity can be verified through the equation.

3.5. Application

At the application stage, students are challenged to apply previously constructed concepts to new situations. For example, the teacher assigns students to make a plasticine boat design to accommodate marbles in large quantities. Students can design the project systematically and clearly. With prior knowledge built about the buoyant force, students should be able to explain the reasons for choosing the design of the plasticine boats they make.

4. Conclusion

Based on the literature review conducted by the author, this buoyant force lesson by using 5-stage learning cycle needs to be applied in learning and laboratory activities. This is done to get a picture of whether the learning can provide students with opportunities to use their science process skills.

References

[1] Zeidan A H and Jayosi M R 2015 Science Process Skills and Attitudes toward Science among Palestinian Scondary School Students *World Journal of Education* 5 (1)
[2] Aktamis H and Ergin O 2008 The Effect of Scientific Process Skills Education on Student's Scientifict Creativity, Science Attitudes, and Academic Achievements *Asia-Pacific Forum on Science Learning and Teaching* 9 (1)
[3] Akben N 2015 Improving Science Process Skills in Science and Technology Course Activities Using the Inquiry Method *Education and Science* (40) 179 111-132
[4] Rezba R J, Sprague C & Fiel R 2002 *Learning and Assessing; Scince Process Skills* (Iowa: Kendall/Hunt Publishing Company)
[5] Hirca N 2013 The Influence of Hands on Physics Experiments on Scientific Process Skills According to Prospective Teachers' Experiences *European Journal of Physics Education* 4 (1)
[6] Rustaman N 2009 *Pengembangan Butir Soal Keterampilan Proses Sains* (Bandung: FPMIPA UPI)
[7] Wenning C J 2011 The Levels of Inquiry Model of Science Teaching *Journal of Physics Teacher Education Online* 6 (2)
[8] Nelson J and Nelson J B 2015 Buoyancy Can-Can *The Physics Teacher* 53

Newton's Cradle Experiment Using Video Tracking Analysis with Multiple Representation Approach

A Anissofira[1,*], F D E Latief[2], L Kholida[2], P Sinaga[1]

[1] Program Studi Pendidikan Fisika, Sekolah Pasca Sarjana, Universitas Pendidikan Indonesia, Jl. Dr. Setiabudi No. 229, Bandung 40154, Indonesia
[2] Departemen Fisika, Fakultas Matematika dan Ilmu Pengetahuan Alam, Institut Teknologi Bandung, Jl. Ganesha 10, Bandung 40132, Indonesia

*almira_anissofira@student.upi.edu

Abstract. This paper reports a Physics lesson using video tracking analysis applied in Newton's Cradle experiment to train student's multiple representation skill. This study involved 30 science high school students from class XI. In this case, Tracker software was used to verify energy conservation law, with help from data result such as graphs and tables. Newton's Cradle is commonly used to demonstrate the law of energy and momentum conservation. It consists of swinging spherical bobs which transfers energy from one to another by means of elastic collisions. From the video analysis, it is found that there is a difference in the velocity of the two bobs of opposite ends. Furthermore, investigation of what might cause it to happen can be done by observing and analysing the recorded video. This paper discusses students' response and teacher's reflection after using Tracker video analysis software in the Physics lesson. Since Tracker has the ability to provide us with multiple means of data representation way, we conclude that this method could be a good alternative solution and might also be considered better than performing a hands-on experiment activity in which not every school have suitable laboratory equipment.

1. Introduction

Physics is sometimes considered as a subject which contains problem-solving work that mainly involves remembering formulas and doing some mathematical calculation. The phenomenon has becoming more frequent in physics education, which causes physics to be considered as abstract and hard subject [1]. Physics learning activity should be complemented by a proper experiment activity, which is fundamental in building an understanding that physics concepts are highly related to and can be applied in daily life. Nevertheless, the availability of suitable and sufficient laboratory equipment in every high school is sometimes till considered a luxury, especially in order to present a good science experiment. In the case of availability of the suitable laboratory equipment, the produced experimental data means nothing without a proper processing and analysis. Thus, it is essential for high school students to have a good ability to process and analyze experimental data. Experimental data can be presented in various ways (multiple representation), such as by tables, graphs, or diagrams [2]. Student should be proficient to make a cross-representation in order to have scientific reasoning which is fundamental for solving physics problems [3] [4] and doing physics experiments. Ironically, based on case study [5], it is found that the students' problem solving skill with multi representation approach is still low.

In this study, we use the video based learning analysis as an alternative tool for the students to practice their ability of multiple-representation skill. Advancement of technology has provided much better ways of teaching physics, one of the case is that physical quantities can be presented interactively in dynamic format [6]. Open Source Physics (OSP) is one of computation modeling software which is suitable to be applied in high school to improve students' exploration skill at the level of application in the learning cycle [7]. Tracker is one of video analysis and modeling software which is built by OSP Java Framework, which is designed as a proponent to fulfill various needs in physics education [8]. Previous studies have reported the use of Tracker video analysis in Physics education in order to solve various problems which are related to daily life such as pendulum [9], projectile [10], and free fall motion [11]. The results from the previous researches [10] [11] reported that, when the teacher deliver Physics lesson using Tracker video analysis and associated with particular Physics concept, the teacher received a lot of positive response from students. Tracker provides data analysis tool such as graph and table, curve fitting, etc., where we can understand, for example, the motion of a particle during an experiment.

Previous study, show us that a lot of physics experiment could be held thru video modeling. Besides, simplify experiment activity, Tracker video analysis shall be used to analyze physics law [12] such as Simple Pendulum. Those experiment utilize energy conservation law that is one of Physics concept that is frequently found in daily life. Numerous pendulum type, which often applied to the energy conservation law, however there is no study which discuss Newton's Cradle pendulum type in the physics classroom. Newton's Cradle is an apparatus that demonstrates the conservation of momentum and energy using a series of swinging bobs (sphere), in which energy is transferred through the stationary spherical bob and finally exerted to the last which pushes the last spherical bob to move upward. Newton's Cradle apparatus usually consists of 5 identical spherical bobs which is made of iron ball. According to those ideas, video based learning using Tracker video analysis on the Newton's Cradle experiment has been done to help student to understand energy conservation law.

The underlying concept which applied in the Newton's Cradle (assuming the spheres have the same mass) is

$$Energy_{initial} = Energy_{final} \quad (1)$$

$$\Delta EK = \Delta EK' \quad (2)$$

$$\frac{1}{2}m(v_f^{\,2} - v_i^{\,2}) = \frac{1}{2}m'(v_f'^{2} - v_i'^{2}) \quad (3)$$

$$\Delta v^2 = \Delta v'^2 \quad (4)$$

where the velocity of the spheres (v) was measured to verify the law of conservation of energy. The law can be verified by analyzing the difference between the velocity of the initial swinging sphere (Δv) at the left end and the velocity of the bouncing sphere ($\Delta v'$) at the right end.

2. Experimental Method

This study employ video based learning method using Tracker video analysis software. This research was preliminary study which aims to observe students' response and understanding for a certain physics subject, i.e., the energy and momentum conservation law which is applied in the Newton's Cradle experiment apparatus. The video was recorded using Canon Exilim EX-ZR200 with optical zoom of 12.5 times, 30 frame per second high definition mode. The video was recorded with a background of white paper and lighting directly from sunlight that make the video appear to be clear and bright. Tracker 4.61 was installed to all the students' laptop before the lesson activity.

Students are quite engaged, started form installing the Tracker software (off the school time) until tracking the motion of the Newton's Cradle by themselves on a video that already shared. For this experiment, students are not allowed to take the video of Newton's Cradle motion by themselves for time efficiency. The population in this study were 30 students of Class XI in one of the state high school in Bandung, Indonesia. Students are divided into 6 groups which are consists of 5 students. Each group were equipped with 2 laptops for operating Tracker. For the lesson plan, we used 45 minutes which is equal to one lesson session. The students' worksheet consists of a guide to operate the Tracker which is equipped with experimental questions related to table and graph analysis of video tracking. The students was given a pre-test at the beginning of the lesson using selected problem from TUG-K (Test of Understanding Graph-Kinematics) [13] which contains mostly questions related to graphs analysis. Pre-test activity was carried out in order to find out to what extend student understanding of multiple representation.

3. Result and Discussion

3.1. Pre-test result
Pre-test consist of 10 multiple choices which are adapted from TUG-K (Test of Understanding Graph Kinematics). TUG-K is considered as a representation to find out students' basic knowledge in graph analysis. From this pre-test, the average score was 28 out of 100. This score was very low and it indicates that overall students in this class did not have adequate understanding on how to solve problems related to multiple representation analysis.

3.2. Implementation of video based learning using Tracker video analysis.

In this lesson, we use Newton cradle apparatus as shown in Figure 1 which is consist of 5 series of pendulums with spherical bobs.

Figure 1. Newton's Cradle experimental device (series of sphere from the right to the left)

Subsequently, the motion of the swinging and bouncing sphere can be tracked. Figure 2 shows the tracking process of sphere 1 from the moment the ball was released from a certain position until it collided with the next sphere. The tracking process produce 10 traces of sphere 1 indicated with the red circle symbol. Newton's Cradle experiment is commonly used to demonstrate the law of conservation energy, where the kinetic energy of sphere 1 is transferred during the collision with the next consecutive spheres until the transferred energy deliver a force that pushed the last sphere to bounce upward. This last sphere (sphere 5) was also tracked and indicated with blue circles.

Figure 2. Tracking swinging Sphere 1 and pushes the Sphere 5 upward

After the tracking process, Tracker immediately provide graphical representation of the data, as well as the table that list the position vs. time. The data was presented as *x-t* (*x* position against time), *y-t* (*y* position against time), or even *v-t* (velocity magnitude against time) and many more variant of graph. We can adjust which graph that we want to analyze. For this experiment, we need velocity data, however, we present two other graphs as a comparison.

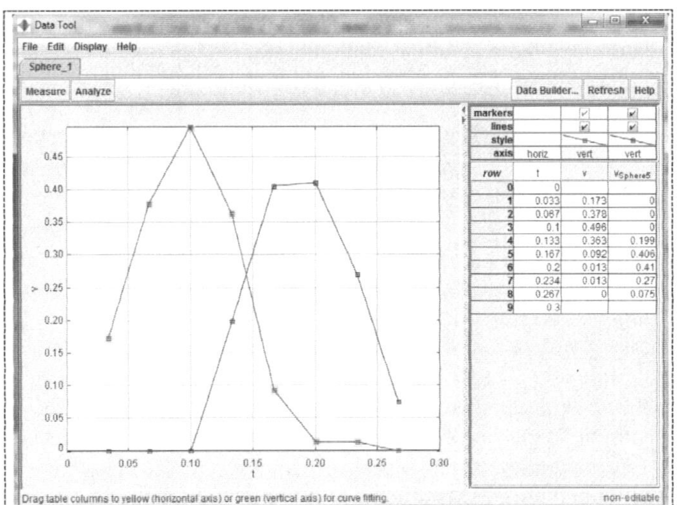

Figure 3. Visual display of analyse menu in the Tracker which show graph and table data of Sphere 1 and Sphere 5

From equation (4), we obtain the expression to prove the law of conservation energy, thus we should expect that $\Delta v = \Delta v'$ where velocity difference indicates the instantaneous velocity of the swinging sphere (Sphere 1) and the bouncing sphere (Sphere 5). In Figure 3, the experimental data result *v-t* was presented both in graph and table. Each data sequence of the table is related to the order of the sphere tracking.

Table 1 shows the summary data of Sphere 1 and Sphere 5, which are the time (s), *x* distance (m), *y* distance (m), and velocity (m/s). We substitute data from Table 1 to equation (4) and we obtain $\Delta v^2 = 0{,}0256$ m/s and $\Delta v'^2 = 0{,}0153$ m/s.

Table 1. Data Sphere 1 and Sphere 5

		Sphere 1			Sphere 5		
Data Number	Time (s)	Distance x (m)	Distance y (m)	Velocity (ms^{-1})	Distance x (m)	Distance y (m)	Velocity (ms^{-1})
1	0	0,147	0,031	-	0,042	0,010	-
2	0,033	0,146	0,030	0,173	0,042	0,010	0
3	0,067	0,138	0,023	0,378	0,042	0,010	0
4	0,100	0,125	0,016	0,496	0,042	0,010	0
5	0,133	0,108	0,011	0,363	0,042	0,010	0,199
6	0,167	0,102	0,010	0,092	0,029	0,012	0,406
7	0,200	0,102	0,010	0,013	0,016	0,017	0,410
8	0,234	0,102	0,010	0,013	0,004	0,023	0,270
9	0,267	0,102	0,010	0	0,001	0,027	0,075
10	0,300	0,102	0,010	0	0,001	0,027	-
		$\Delta v^2 = 0,0256$ m/s			$\Delta v'^2 = 0,0153$ m/s		

From the data analysis, the magnitude Δv^2 from Sphere 1 and $\Delta v'^2$ from Sphere 5 is not equal but close. Instantaneous velocity of Sphere 1 is greater than that of Sphere 5. This is due to the device of Newton's Cradle experiment is lack of precision. From the video we could observe that Sphere 2, 3, and 4 were not really stick together. The collisions between the spheres 2, 3, and 4 produced sound, therefore the transferred energy that reached Sphere 5 is decreasing, that is why $\Delta v'^2$ from Sphere 5 is less than Δv^2 from Sphere 1. This is what caused the swinging motion of the pendulums eventually comes to stop.

3.3. Student response and teacher reflection on using Tracker in the Physics lesson

During the lesson, students were really enthusiastic, especially because all of them were directly involved in data processing, not merely watching the process being demonstrated by the teacher. After the lesson, the students were asked about energy conservation law concept, and they admitted that they gained more understanding about the concept compared to before the lesson. They are very interested to do various kinds of Physics experiments using Tracker video analysis. At the same time, the Physics teacher admits that this method was really helpful for both the teacher and the students. This experiment activity will be arranged to be conducted in the class more often because by using Tracker video analysis, the students can still carry out experiments, even in the case of inadequate experiment equipment. The last, but not least, one of the most important thing is, with this Tracker video analysis, students are able to train their multiple representation skill, considering that the result of the experiment can be presented in various ways such as graph and table. The expected future outcome is that the students will be able to translate inter-multiple representation and to associate the representation to a contextual Physics concept.

4. Conclusion

Video based learning using Tracker video analysis is very good idea to be developed in Physics classroom. This method could be alternative solution despite of conducting hands-on experiment activity in the absence of full adequate equipment laboratory. Tracker software provides tools of multiple representation of the experiment data. The video tracking analysis using Tracker software should also be able to train the students' multiple representation skill in the context of Physics.

Acknowledgments
The author wish to acknowledge SMA Santa Angela Bandung, Indonesia and Mr. Eko Kusnanto for allowing us to conduct this study in the school.

References

[1] Funda O, William R and Mark P 2008 *Int J Environ Sci Educ* ; vol **3** (1):30–4.
[2] Alan V 2001 *Am J Phys* [Internet]. vol ;**69**(No. 11):1139–46. Available from: http://dx.doi.org/10.1119/1.1399043
[3] David R, Alan. V and Eugenia E 2009" *Phys. Rev. Spec. Top. - Phys. Educ. Res.*, vol. **5**, no. 1, pp. 1–13.
[4] Alan V 1991 *Am. J. Phys.*, Vol. **59** (10), p. 891–7.
[5] Almira A, Fourier D and Parlindungan S 2016. Seminar Kontribusi Fisika
[6] Zacharias Z and O. Roger A 2003 *Am. J. Phys.*, vol. **71**, no. 6, p. 618.
[7] Wolfgang C, Francisco E and Lyle. B 2011 *Open Source Physics*, *Science*, vol. 334, no. Icce, pp. 1077–1078, 2011.
[8] comPADRE 2012 *Tracker Video Analysis and Modeling Tool for Physics Education*, [Online]. Available: http://physlets.org/tracker/.
[9] Douglas B 2008, *AAPT 2008 Summer Meet.*, 2008.
[10] Loo K, Charles C, Giam H, Samuel T and Tat L 2012, *Phys. Educ.*, vol. **47**, no. 4, pp. 448–455
[11] Loo K, Kim K, Tze K and Ching T 2015 *Phys. Educ.*, vol. **50**, no. 4, pp. 436–442.
[12] Jyoti B, Vandana S, Arbind J and O.S.K.S Sastri, 2015 Conference Paper.
[13] Robert J 1994 *Am. J. Phys.*, vol. **62**, no. 8, p. 750, 1994.

The Kinematic Learning Model Using Video and Interfaces Analysis

T Firdaus [1,2,*], W Setiawan[2] and I Hamidah[1]

[1] Program Studi Pendidikan IPA, Sekolah Pascasarjana Universitas Pendidikan Indonesia, Jl. Dr. Setiabudi No. 229, Bandung 40154, Indonesia
[2] Program Studi Pendidikan Fisika STKIP Nurul Huda, Jl. Kotabaru Sukaraja, Buay Madang, OKU Timur, Sumatera Selatan 32161, Indonesia

*thohafirdaus@student.upi.edu

Abstract. An educator currently in demand to apply the learning to not be separated from the development of technology. Educators often experience difficulties when explaining kinematics material, this is because kinematics is one of the lessons that often relate the concept to real life. Kinematics is one of the courses of physics that explains the cause of motion of an object, Therefore it takes the thinking skills and analytical skills in understanding these symptoms. Technology is one that can bridge between conceptual relationship with real life. A framework of technology-based learning models has been developed using video and interfaces analysis on kinematics concept. By using this learning model, learners will be better able to understand the concept that is taught by the teacher. This learning model is able to improve the ability of creative thinking, analytical skills, and problem-solving skills on the concept of kinematics.

1. Introduction

Physics is based on experimental observations and is explained by quantitative measurement. Mathematics became the basic law, namely as a tool to build a bridge between theory and experiment. Cultivate the curiosity of students in learning physics into the main capital in pedagogy. However, in current physics education experiments in the laboratory alone is not enough in providing an authentic research experience. The skills needed in physics learning are the emphasis on critical thinking, creative thinking, analyzing and problem-solving skills. Moreover, providing opportunities for learners to participate in contextual based inquiry through direct observation and inquiry into the real world is very supportive in developing the ability of understanding the concepts of physics and scientific skills.

Current technological advances make it easy for educators and learners to learn science, Today's technology has developed many media such as video capture and video analysis [4] in physics learning, even today more affordable and accessible. Physics software like hers Vernier Logger Pro, a video analysis tools such as Videopoint and Tracker can be freely downloaded and used on a computer device.. Through such software, teachers can failitate learners in motion video capture and analsis [5]. Students simply insert a video file into the program and observe it from the motion of the object for analysis. Through this program students can capture video of a live event and analyze it very easily.

One of the new methods of creative teaching on the subject of physics, and what makes science more appealing to students is the video analysis using Logger Pro program. Logger pro is a learning tool that is widely used in the experimental activities especially in the field of physics. Software

Logger pro in the work requires the help of a device called the interface (LabQuest). Labquest is mounted and connected to sensor devices, such as voltage sensors, magnetic field sensors, oxygen sensors, carbon dioxide sensors, temperature sensors, etc. However, this software can also be used without the use of labquest, which is doing tracking video which is usually useful to calculate the speed of moving objects. This Logger Pro software can be accessed and downloaded through https://www.vernier.com/downloads/. This program is one that is able to make it easier to measure position coordinates (x and y) and time for a moving object. Logger Pro is also capable of displaying systematic data in the form of graphs and mathematical equations depending on the object under review.

The use of computer-based interactive media such as video and image modeling that are analyzed using Logger Pro program play an important role in teaching physics concepts related to dynamics and kinematics [6], and other physics concepts, because by using video recording and image modeling [7] can be observed physical phenomena of physics. Besides, practical use at low cost [8] making this learning more effective, creative and interesting [9] compared to using traditional learning [3]. Through an observation with video analysis and image modelling it significantly affects the level of student understanding [3], problem solving skills [10], Analyze, representational ability, predict, collect data and explain.

2. Experimental Method

There are six features in the analysis and experimentation activities in which students can act alone in the investigation. thereforethe criteria for activity is the level of independence of students associated with each characteristic. The following is a rubric of inquiry characteristics that has been developed and validated by [11] so that it is successfully applied in various scientific disciplines to determine the level of performance presented in Figure 1.

Characteristic	Level 0: Confirmation	Level ½: Structured inquiry	Level 1: Guided inquiry	Level 2: Open inquiry	Level 3: Authentic inquiry
Problem/Question	Provided	Provided	Provided	Provided	Not provided
Theory/Background	Provided	Provided	Provided	Provided	Not provided
Procedures/Design	Provided	Provided	Provided	Not provided	Not provided
Results analysis	Provided	Provided	Not provided	Not provided	Not provided
Results communication	Provided	Not provided	Not provided	Not provided	Not provided
Conclusions	Provided	Not provided	Not provided	Not provided	Not provided
	More structure				Less structure

Figure 1. Rubric of inquiry characteristics

From the description of the table above, the level corresponding to the level of competence of students physics teacher candidate is at level 2. At this level tutors only provide questions and theories to the students further stages undertaken by the students themselves without the guidance or mentoring of the tutor so that the ability and scientific skills can be explored and developed.

2.1. Teaching Strategy: Teaching kinematics in Higher Education

A very important skill owned by students in learning kinematics is the ability to create and interpret graphics. But the problem is that the ability is less controlled by learners, the problem they face is in connecting the concept of graphics and physics, graphics and the real world [12]. Reading and graphics require students to form abstractions from the real world, which involves individuals assuming that mathematics is a separate object [13]. This means it is difficult for students to connect the real world and graphics that are abstract representations of the same world. [12]

Experimental and demonstration methods are the right strategy to solve the above problems. By taking samples in nature, students are encouraged to observe the things that happen in everyday activities.

2.2. Video analisis dan Interface analisis

Video analysis is a learning tool that can be made as the right strategy to take samples in nature. Students in creative demands to take videos that occur in kinematics activities in everyday life, then they are in the race to understand the results of the video they get.

Interface analysis is a sensor program that can be used as a complement of video analysis. Through the analytics interface, this becomes an appropriate strategy for solving student problems. A teacher can practice this activity to the students. Video analysis and analysis interfaces have been built in a learning program called logger pro. Loger pro is one of the right programs to build analytical and problem-solving skills. In this program facilitates video analysis and interface analysis.

The model of video analysis and interface analysis can be packed with a level 2 inquiry study. The actual video analysis activities are performed by the students and data management through teacher guidance. While the activity of interface analysis is done entirely by the teacher, the students pay attention to the teacher in doing the demonstration.

3. Result and Discussion

In this connection PISA, divides the field of science applications into three groups, namely life and health, earth and environment, and technology. Problems and issues of science in these areas can be related to the individual child, as part of the community, and the citizens of the world.

Science teachers recognize that students should be prepared to get information related to technology issues of interest to the general public. While an important basic function of science teacher education is to prepare prospective teachers to connect science and technology optimally to the local community, to benefit the daily life of students, and wider social issues. [16,17]

Video analysis and interface analysis as a technology-based tool used to develop the ability to analyze and interpret the graph. These two activities lead students to study the science that exists in the natural environment. This learning model can be named as MIVA or abbreviation of Model Interface Video Analysis.

Capability development that can be done by student if using MIVA learning model is ability analysis, problem solving ability, and creative thinking ability. Here's the explanation in each development capability:

3.1. Creative Thinking Ability

In learning, students are grouped and then ordered to take video about the natural environment. Of course, taking videos according to the instructions and rules described by the teacher. So the indicators obtained by students, namely:

3.1.1. Increase creativity

After the teacher gives the correct instruction in taking the video, the student is given the flexibility to take the desired video, this makes a lot of creative thinking to make the video made as desired.

3.1.2. Active role of students

When the computer model learning process is supported by video analysis data, it can provide opportunities for students to engage in active inquiry-based learning. More importantly with video modeling when the object is driven by video analysis allows students to find better evidence and models for predicting and explaining bullet motion [14].

3.1.3. Development of teamwork and communication skills

The need for a video in doing this learning, making learners should think creatively in taking object. In addition to emphasizing to them to be able to work teams to produce the desired video, they can find good places and the correct technique in taking video.

3.2. Analytical skills

After the students have finished taking the video, they get guidance by the teacher to be able to manage the video results in the classroom. So this, get the following indicators:

3.2.1. Understanding the relationships between variable

After students are able to process the video into the application, they are required to understand the graph out of the relationship between variables that appear on the display.

3.2.2. Interpreting data

After students analyze the video, they are required to understand and process the data into an empirical data. Such as processing data to find mechanical energy.

3.2.3. Ability of representation

In this study [15] that analytical skills were able to apply student representation skills. This ability will be obtained when it has been able to understand the variables and interpret the data.

3.3 Problem solving skill

Problem solving skills, will be obtained by students when teachers demonstrate an interface activity analysis. Interface analysis is an additional program that uses sensors that are connected by programs contained on the computer. Here is an indicator of ability owned by students.

3.3.1 Tackling misconceptions

From the video analysis data obtained by students, then proven through sensor interface analysis to eliminate misconception.

3.3.2 Solve problems in everyday life

Data capture activities using video, this is the basis that this experimental activity is part of the application of everyday life.

3.3.3 Verification Hypothesis

By using an analytics interface, it is used to prove hypotheses for the work the learners get through video analysis.

3.3.4 Facilitate conceptual understanding

Activity using demonstration of sensor interface analysis tools, summarizes in student experiment activities, with this facilitate understanding the concept they get.

The analytical ability has an indicator of completion in representational skills, problems solving in everyday life, fostering creativity, making active students, and developing teamwork and communication skills. While the Interface analysis activities have an indicator of completion in tackling misconceptions, proving hypotheses, and facilitate understanding of concepts. This activity if united in the learning will be able to give understanding of relationships between variables, and interpret the data. More detailed description as described in Figure 2.

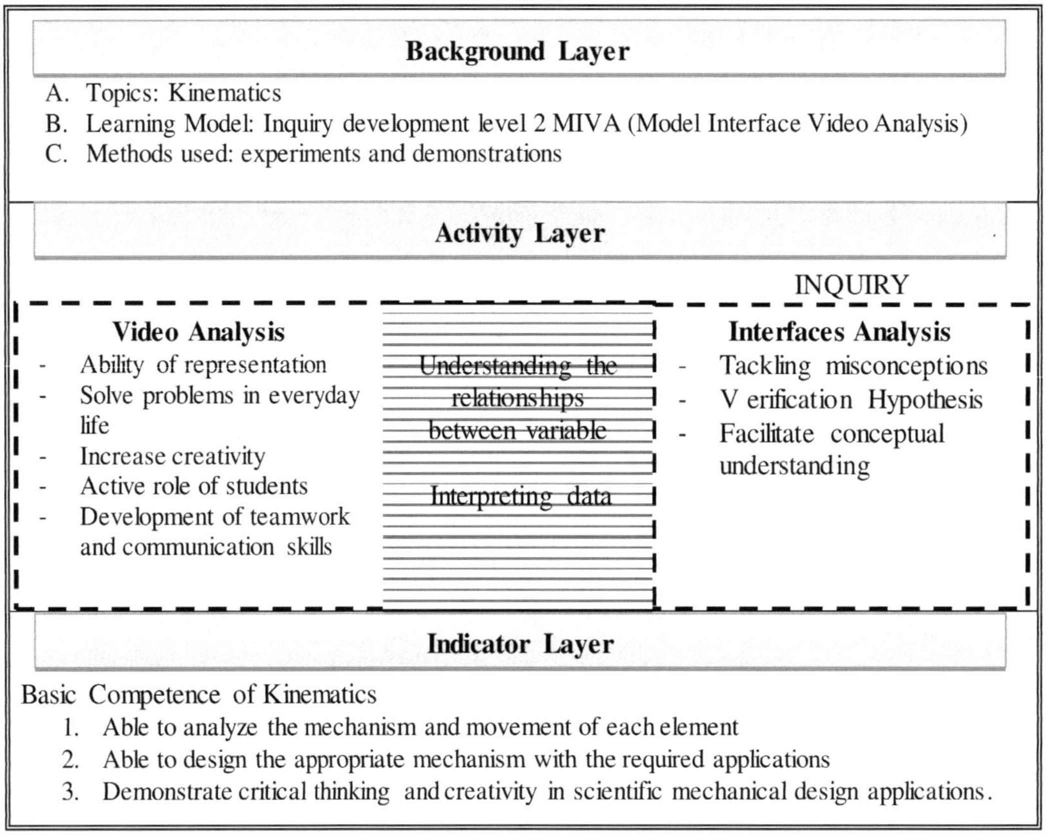

Figure 2. Layer of Learning Model MIVA

Kinematics learning has a basic competence that is consistent with that of the MIVA learning model. Basic competence in kinematics learning as described in the layer indicator in figure 2.

4. Conclusion

A science teacher, attempted to utilize the technology of learning activities as an effective learning tool and keep students away from the decline of the times. Kinematics is a material that takes many elements of real-world activities. The difficulty is how a teacher should be able to bring real-world activity into kinematics material. These two things can be bridged by technology-based learning media using video and interface analysis.

Acknowledgments

Thank you to the organizer of LPDP (Lembaga Pengelola Dana Pendidikan) who has provided financial support to the author so that we can finish this article easily..

References

[1] Ekici, E 2016 Why Do I Slog through the Physics? Understanding High School Students' Difficulties in Learning Physics. *Journal of Education and Practice* 7(7) 95-107

[2] Oon, P T & Subramaniam R 2011 On the declining interest in physics among students—from the perspective of teachers. *International journal of Science education 33*(5) 727-746

[3] Hockicko, P., Trpišová, B. and Ondruš, J. 2014 Correcting Students' Misconceptions about Automobile Braking Distances and Video Analysis Using Interactive Program Tracker *Journal of Science Education and Technology 23*(6) 763-776

[4] Laws P and Pfister H 1998 Using digital video analysis in introductory mechanics projects. *The Physics Teacher 36*(5) 282-287

[5] Wyrembeck E P 2009 Video analysis with a web camera. *The Physics Teacher 47*(1) 28-29

[6] Phommarach S, Wattanakasiwich P and Johnston I 2012 Video analysis of rolling cylinders. *Physics Education 47*(2) 189

[7] Rodrigues M and Carvalho P S 2013 Teaching physics with Angry Birds: exploring the kinematics and dynamics of the game *Physics Education 48*(4) 431

[8] Eadkhong T, Rajsadorn R, Jannual, P and Danworaphong S 2012 Rotational dynamics with Tracker *European Journal of Physics 33*(3) 615

[9] Dias M A, Carvalho P S and Vianna D M 2016 Using image modelling to teach Newton's Laws with the Ollie trick *Physics Education 51*(4) 045008

[10] Teiermayer A 2016 Problems based on phenomena and experiments in secondary school involving a digital camera *Physics Education 51*(6) 063002

[11] Buck L B, Bretz S L and Towns M H 2008 Characterizing the level of inquiry in the undergraduate laboratory *Journal of College Science Teaching 38*(1) 52

[12] Eshach H 2010 Re-examining the power of video motion analysis to promote the reading and creating of kinematic graphs. In *Asia-Pacific Forum on Science Learning and Teaching* (Vol. 11, No. 2, p. n2). Hong Kong Institute of Education. 10 Lo Ping Road, Tai Po, New Territories, Hong Kong.

[13] Hershkowitz R, Schwarz B B and Dreyfus T 2001 Abstraction in context: Epistemic actions *Journal for Research in Mathematics Education* 195-222

[14] Wee L K, Tan K K, Leong T K and Tan C 2015 Using Tracker to understand 'toss up'and free fall motion: a case study *Physics Education 50*(4) 436

[15] Chanijani S S M, Klein P, Bukhari S S, Kuhn J and Dengel A 2016, September Entropy based transition analysis of eye movement on physics representational competence. In *Proceedings of the 2016 ACM International Joint Conference on Pervasive and Ubiquitous Computing: Adjunct* (pp. 1027-1034). ACM.

[16] American Association for The Advancement of Science 1993 Benchmarks for Science Literacy (New York Oxford: Oxford University Press).

[17] National Research Coucil (1996) National Science Education Standards (Washington, D.C.: National Academy Press).

Pre-Service Physics Teachers' Concept Mastery and the Challenges of Game Development on Physics Learning

S Saprudin[1,3,*], **L Liliasari**[1] **and A S Prihatmanto**[2]

[1] Program Studi Pendidikan Ilmu Pengetahuan Alam, Sekolah Pascasarjana, Universitas Pendidikan Indonesia, Jl. Dr. Setiabudhi No. 229, Bandung 40154, Indonesia
[2] Sekolah Teknik Elektro dan Informatika, Institut Teknologi Bandung, Jl. Ganesha 10 Bandung 40132, Indonesia
[3] Program Studi Pendidikan Fisika, Fakultas Keguruan dan Ilmu Pendidikan, Universitas Khairun, Jl. Pertamina Kampus II Gambesi 97719, Indonesia

*saprudinunkhair@student.upi.edu

Abstract. This study is a survey that aims to describe pre-service physics teachers' concept mastery at a university in Ternate. Data were collected through test standard instrument for physics which used in the teacher certification program. Data were analyzed by using quantitative descriptive technique. Based on the results of data analysis, it was concluded that generally pre-service physics teachers' concept mastery can be categorized on low category (25.4%). The map of concept mastery will be used as a reference to developing game design in the physics learning context for pre-service physics teachers.

1. Introduction

Along with the development of science and technology, the use of digital games has been widely designed to facilitate students in learning physics concepts both at university level [1, 2, 3], high school [2, 4, 5, 6], junior high school [7, 8, 9, 10, 11, 12, 13, 14], Elementary school [15, 16] and also kindergarten [17]. Researchers have defined the game as an immersive, voluntary and enjoyable activity in which a challenging goal is pursued according to agreed-upon rules [18]. In the context of e-learning, a game is defined as an online environment that involves a competitive activity with a challenge to achieve a goal, a set of rules and constraints, and a specific context. Game features vary dramatically, including games of chance, games based on motor skills (also called twitch games), and games of strategy. Games for learning are called instructional games or serious games [19]. Educational game is competitive rule-based activities involving one or more players with an expressed goal of performing or meeting a goal at a superior level (i.e., winning) either in relation to a previous performance level (one player game) or in relation to the performance levels of other players. Success in the activity requires use of subject matter in some way [20]. In other literature, games can be viewed as tools for learning [8, 10, 12, 17, 21, 22, 23], tools for evaluation [7, 9, 16, 24] and can also be viewed as learning environments [11, 15].

Researchers have shown that integration of educational theory or implementation of strategies, models, learning methods has been done to optimize the use of digital games in facilitating physics learning such as cooperative/ collaborative learning strategies [4, 7, 13, 23], 5E learning cycles model [5], assessment [7, 9, 16, 24], meta learning/ meta-cognition [10], cognitive-affective interaction model and flow theory [15], conceptual physics [14], explanation-based approaches [11, 22], locus of control [1]. The results of papers riview show that the use of digital games in physics learning can increasing student engagement [2, 9], enjoyment [24], motivation [13, 16, 22], student perceptions [2, 7, 13, 16], conceptual change [14], cognitive elaboration of physical concept [12], divergent thinking [15], problem solving [16], which are ultimately can improving learning outcomes [1, 3, 4, 5, 7, 8, 9, 10, 11, 13, 16, 22] and student awareness [5].

Based on the paper analyzed, generally the use of games on physics learning at the middle school is more than the other level. Though the game has great potential to be developed for example in the context of physics learning in universities especially for pre-service physics teachers. As a first step, the researcher sees that it is very important to mapping pre-service physics teachers' concept mastery on the study population so that it can be used as a reference to identify characteristics of materials which were the weak achievement and to plan developing game design in the context of physics learning for pre-service physics teachers.

2. Method

This research is a survey that was focused to describe pre-service physics teachers' concept mastery at one of university in Ternate city. The survey involved 60 seventh semester students at one of the universities in Ternate city. The sample of the study was taken with consideration of the adequacy in taking the course of physics education. Data were collected through test standard instrument for physics which is used in the teacher certification program. The test instrument consists of multiple choice questions. Data analysis technique was done by using quantitative descriptive analysis technique.

3. Result and Discussion

The result of data analysis shows that the propile of pre-service physics teachers' concept mastery can be categorized as low (25.4%) as shown in figure 1.

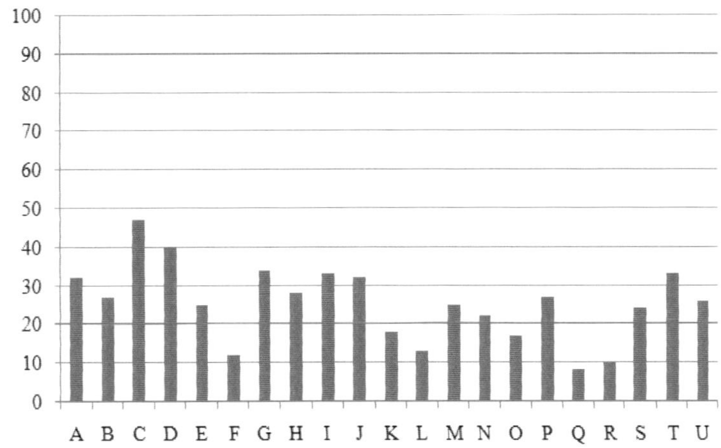

A; physics and measurement, B; motion in one dimension, C; motion in two dimensions, D; the laws of motion, E; static equilibrium and elasticity, F; universal gravitation, G; fluid mechanics, H; sound waves, I; superposition and standing waves, J; electromagnetic waves, K; electric fields, L; current and resistance, M; direct current circuits, N; magnetic fields, O; alternating current circuits, P; image formation, Q; interference of light waves, R; diffraction patterns and polarization, S; heat and the first law of thermodynamics, T; heat engines, entropy, and the second law of thermodynamics, U; modern physics

Figure 1. Profile of pre-service physics teachers' concept mastery

In the physics and measurement material, the concept mastery of pre-service physics teachers was categorized as low (32%). Physics and measurement material is a material that describes the standards of length, mass, and time, matter and model building, density and atomic mass, dimensional analysis, conversion of units, estimates and order-of-magnitude calculations, significant figures. Based on the test analysis, we found that physics teacher candidate students are still weak on indicators mentioning on length measurement results with vernier calipers & micrometer screw gauge, uses the rule of significant figure in calculating area or volume of an object based on vernier calipers measurement.

On the topic of motion in one dimension, the concept mastery of pre-service physics teacher was categorized as low (27%). In the population, we can identified the lack of concept mastery such as for indicator determines the average velocity, speed and acceleration of a moving object, describes the motion of an object based on the graph of $x = f(t)$, $v = f(t)$ and $a = f(t)$. Students also have difficulty in solving physics problems related to determining the time and place where both people meet if the distance, speed and initial time are known. While on the topic of motion in two dimensions, the concept mastery of pre-service physics teacher was categorized as medium (47%). One of the lacks of concept mastery is the indicator determines characteristics of projectile motion. If we look at the content of matter, these two topics involve quantities of physics to describe the motion of an object such as position, distance, displacement, velocity, speed, acceleration which were be the basic for next physics material achievement. This topic is closely related to the phenomenon that can be found in everyday life. On the other hand, to understand this material requires a mathematical understanding.

Pre-service physics teachers' concept mastery on the topic of the laws of motion can be categorized as low (40%). The lack of concept mastery was identified in the indicator on determining the effect of force to moving objects with certain speed and time. In the topic of static equilibrium and elasticity, Pre-service physics teachers' concept mastery was categorized as low (25%). Pre-service physics teacher have difficulty on indicators of determining appropriate experimental design to investigate one of the factors affecting the magnitude of constant of the spring.

Pre-service physics teachers' concept mastery on the topic universal gravity can be categorized as very low (12%). The force of gravity is an abstract concept, related to the interaction of one object with another object such as interaction on microscopic objects that can't be seen also between objects with macroscopic size that can't be presented in the class for example interaction between the sun and the planet or interaction between objects in the universe. In addition to understanding the force of gravity required understanding of mathematics. One of the weak concept mastery is the indicator determines speed of the satellite by applying the concept of gravity and centrifugal force.

Pre-service physics teachers' concept mastery on the topic fluid mechanics can be categorized as low (34%). Fluid mechanics is one of the physics topics that the application of the concept commonly was found in everyday life. However, not all application concepts can be presented directly into the class for students to observe [25]. Pre-service physics teachers' concept mastery on the topic fluid mechanics can be categorized as low (34%). Topics of fluid mechanics, is one of the topics that the application of the concept commonly was found in everyday life. In the population, the lack of concepts mastery was ideintified in indicator describes the graph of the relationship between the magnitude of the buoyant force and height, determine altitude from air pressure data or conversely.

The oscilations and mechanical waves in this survey are represented by topics of sound waves, superposition and standing waves, electromagnetic waves. Pre-service physics teachers' concept mastery can be categorized as low with the percentage of 28% on the topic of sound waves, 33% on the topic of superposition and standing waves and 32% on topic of electromagnetic waves. In the population, the lack of concepts mastery was ideintified in indicator determining the magnitude of wavelength and the speed of waves based on a wave equation, identifying the properties of electromagnetic waves, and determining frequency which was detected by an observer based on the Doppler effect.

Pre-service physics teachers' concept mastery on materials of electricity and magnetism can be categorized as low. Generally these materials contains abstract concepts, involving students in complex physical phenomena, requiring observation and interacting with three-dimensional phenomena, mentally difficult to transform 2D objects into 3D objects [8]. In addition, the aplication of concept on this material is very closely related to everyday life. In the population, the lack of concepts mastery was ideintified in indicator mentioning the result of voltmeter and ammeter measurement, determining the required resistance of an immersion heater that increases the temperature of water heated for a certain time while operating at certain potential differences, calculating the power delivered to the lamp which were connected to a certain voltage, determining the distance of a charge from the other charge if we known the resultant of the electric force is zero, determining the ratio of the magnitude of two charges at the point which the electric field is zero, and determining the number of primary or secondary windings in the step-up transformer.

The image formation topic is part of the geometrical optics. While the topic of interference of light waves, diffraction patterns and polarization is part of physical optics. On this topic, generally pre-service physics teachers' concept mastery was categorized as low. Mastery of concept on physical optics is the lowest achievement. The results of concept analysis show that the concept on this material was an abstract concept. This is supported by a number of studies showing that many students usually experience difficulty when conceptualizing abstract concepts in this material [26, 27]. In the population, the lack of concepts was ideintified in indicator determining the magnitude of the distance of ocular lens should be shifted so that the eye can be maximum magnification based on the observed object with a microscope, applying the concept of interference, diffraction and polarization in solving physics problems.

Pre-service physics teachers' concept mastery on thermodynamic materials can be categorized as low. In the population, the lack of concepts was ideintified in the indicators determining the rate of energy transfer on the event of conduction, determining the ratio of heat which necessary transferred to both types of objects with mass, specific heat and a certain temperature changes, calculating the energy transfer during phase change, using the concept of Carnot cycle to solve physics problems.

Pre-service physics teachers' concept mastery on modern physics materials can be categorized as low. In the population, the lack of concepts was ideintified in the indicators determine the photon energy with an certain wavelength and frequency, determining the wavelength for an electron using De Broglie hypothesis, interpretation the graphic of intensity and wavelength on blackbody radiation, identification of characteristics, advantages and limitations of various atomic models, determining the number of protons, neutrons and electrons of atoms and ions, predicting the behavior of alpha, beta, and gamma rays in an electric or magnetic field, determining radioactive half-life of an substance. These findings are reinforced by Gunawan's (2013) research which shows that students have difficulty in understanding the concepts in modern physics because generally this material contains abstract concepts so that need to be visualized [28].

On the several papers that have been studied, it has been shown that digital game has become an effective way of teaching physics concepts such as Newtonian dynamics [11, 13, 14, 22, 23, 29], Newton's playground [9], energy [7], electrostatics [4, 8], electrical energy [5], electromagnetic [3] and optics [2, 10]. However, generally the use of digital games in physics learning is mostly done at junior high school and high school. To be successful in studying concept of physics on the populations, it is very important to design learning activities that can organize and condition the learning situation so that more interesting, challenging, fun, can be improving student engagement in concept exploration activities, learning concepts that can't be practiced in the classroom, providing interaction experience with invisible and multidimensional objects, reinforcement of concepts, evaluating the concepts and provide for learning experiences related to the applications of concept. On the other hand, when teaching physics concepts, developing critical thinking and creative thinking becomes one of the challenges to planning further research. Games have the potential to be developed in the context of physics learning for pre-service physics teachers.

4. Conclusion

Based on the results of data analysis, it can be concluded that generally the profile of pre-service physics teachers' concept mastery was categorized as low (25.4%). The map of concept mastery can be used as a reference in developing the design of learning activities that can organize and condition the learning situation that is immersive, interesting, challenging, fun and improving student engagement in physics learning. On the other hand, developing critical thinking and creative thinking are important skills that must be possessed by pre-service physics teachers.

Acknowledgments
The authors would like to thanks for all the students who have participated in the data collection. The authors would also like to send his gratitude to the lecturers who have given the opportunity to conduct this survey.

References

[1] Yang J C, Lin Y L & Liu Y-C 2016 *Environmental Education Research* p 1-14
[2] Hamari J, Shernoff D J, Rowe E, Coller B, Asbell-Clarke J, Edwards T 2016 *Computers in Human Behavior* 54 170-179
[3] Tsai M-J, Huang L-J, Hou H-T, Hsu C-Y, Chiou G-L 2016 *Computer & Education* 98 115-129
[4] Echeverría A, Améstica M, Gil F, Nussbaum M, Barrios E, Leclerc S 2012 *Computers in Human Behavior* 28 1170-1177
[5] Dorji U, Panjaburee P, Srisawasdi N 2015 *Educational Technology & Society* 18 (1) 91-105
[6] Chen C-L D, Yeh T-K, Chang C-Y 2016 *Eurasia Journal of Mathematics, Science & Technology Education* 12 (5), 1379-1388
[7] Tsai F-H, Tsai C-C, Lin K-Y 2015 *Computers & Education* 81 259-269
[8] Anderson J L, Barnett M 2013 *Journal of Science Education and Technology* 22 (6) 914–926
[9] Shute V J, Ventura M, Kim Y J 2013 *The Journal of Educational Research* 106 423–430
[10] Verpoorten D, Castaigne J-L, Westera W, Specht M 2014 *Education and Information Technologies* 19 (2) 361–374
[11] Adams D M, Clark D B 2014 *Computers & Education* 73 149–159
[12] Sun C-T, Ye S-H, Wang Y-J 2015 *Computers & Education* 88 169–181
[13] Chen C-H, Wang K-C, Lin Y-H 2015 *Educational Technology & Society,* 18 (2), 237–248.
[14] Sengupta P, Krinks K D, Clark D B 2015 *The Journal of The Learning Sciences* 24 638-674
[15] Hsiao H-S, Chang C-S, Lin C-Y, Hu P-M 2014 *Journal of Computer Assisted Learning* 30 (4) 377-395
[16] Hwang G-J, Hung C-M, Chen N-S 2014 *Education Technology Research and Development,* 62 (2) 129-145
[17] Graaf J V D, Segers E, Verhoeven L 2016 *Computer & Education* 101 168–178
[18] Kinzie M B, Joseph D R D 2008 *Education Technology Research and Development* 56 643–663
[19] Clark R C, Mayer R E 2011 *E-Learning and the Science of Instruction* (San Fransisco: Pfeiffer)
[20] Rieber L P 2005 The Cambridge Handbook of Multimedia Learning Edited by Richard E Mayer Chapter: Multimedia Learning in Game, Simulations and Microworlds (New York: Cambridge University Press) p 549–568
[21] Shute V J, D'Mello S, Baker R, Cho K, Bosch N, Ocumpaugh J, Ventura M, Almeda V 2015 *Computers & Education* _86 224-235
[22] Killingsworth S S, Clark D B, Adams D M 2015 *International Journal of Education in Mathematics, Science and Technology* 3 (3) 162-186
[23] Eaton G V, Clark D B, Smith B E 2015 *International Journal of Education in Mathematics, Science and Technology* 3 (1) 1-13
[24] Kim Y J, Shute V J 2015 *Computers & Education* 87 *340–356*
[25] Nuzuliana A H, Bakri F, Budi E 2015 Prosiding Seminar Nasional Fisika (E-Journal) SNF2015 Vol IV (Jakarta: Jurusan Fisika, Fakultas MIPA, Universitas Negeri Jakarta) p 27
[26] Galili I, Hazan A 2000 *International Journal of Science Education* 22 (1) 57–88
[27] Horn M E, Leisner A, Mikelskis H F 2002 German Physical Society Spring Conference Leipzig 2002 (Beitrag 9.1 LOB Berlin 2002) p 1
[28] Gunawan, Setiawan A, Widyantoro D H 2013 Jurnal Pendidikan dan Pembelajaran 20 (1) p 25-32
[29] White B Y 1994 *Cognition and Instruction* 1 (1) 69-108

Training Students' Science Process Skills through Didactic Design on Work and Energy

S Ramayanti*, **S Utari and D Saepuzaman**

Departemen Pendidikan Fisika, Universitas Pendidikan Indonesia, Jl. Dr. Setia Budhi No. 229, Bandung 40154, Indonesia

*suci.ramayanti95@student.upi.edu

Abstract. Science Process Skills (SPS) has not been optimally trained to the students in the learning activity. The aim of this research is finding the ways to train SPS on the subject of Work and Energy. One shot case study design is utilized in this research that conducted on 32 students in one of the High Schools in Bandung. The students' SPS responses were analyzed by the development SPS based assessment portfolios. The results of this research showed the didactic design that had been designed to training the identifying variables skills, formulating hypotheses, and the experiment activity shows the development. But the didactic design to improve the students' predicting skills shows that the development is still not optimal. Therefore, in the future studies need to be developed the didactic design on the subject Work and Energy that exercising these skills.

1. Introduction

Science is a combination of processes and products that are related each other [1]. The process is a series of activities that needed to achieve the product [1]. This requires learning activities able to make the students' activeness in order to build up the students' knowledge through a series of activities that encourage students toward discovery process [2]. In order to find a knowledge is needed some skills such as observing, measuring, experimenting, interpret the data, communicate, and so on [2]. These skills can be used in the process of finding a knowledge which is then referred to as the Science Process Skills (SPS). The importance of SPS in science led to the various countries worldwide emphasis on SPS and developing higher skills through learning approach that centered on the student [3]. Therefore, the necessary means (didactic design) to exercising students' SPS.

However, this research shows that students' SPS are still not optimally trained in the classroom because of the experience lack in the learning activities investigation [4]. The similar results were found from preliminary studies through observation in the schools shows that the learning science is still dominated by the transfer of knowledge. The experimental activity does not start from the observation experiment, the students are not involved in building and designing experimental investigation questions. This is not counterpart with the government's recommendation to use the scientific approach in the learning activities [5]. According to the research findings, it could be suggested to a teacher requires in analyzing the students' characteristics. The students' learning capability will determine the role of a teacher conducting the learning in a class [6]. A learning process is one of the important aspects for the improvement of students' conceptual achievement [7]. Based on the observations result, it found that students' difficulty in the identifying variables based on the observations, constructing predictions, constructing hypotheses, and experiment. Several studies have tried to train students' SPS. However, not all of the students' skills be able to train optimally. Some results of these studies which are expressed by Lati that students' interpret data skill and construct the conclusion are still in fairly level [4]. Hodosyova, et al. reveals that the students' SPS is still low in terms of designing experiments and hypothesize [8]. It is also found by Akpokorie and Omajuwa expressed by Jack that of 15 SPS recommended curriculum, 70% of students having difficulty in mastering these skills [9].

The didactic design is the learning plans that include the ways to teach certain concepts through the students' responses on a draft action learning provided [10]. The aims of this research are trying to find the didactic design for training high school students' SPS on the Work and Energy through the SPS approach. Related to the subject of Work and Energy, the SPS will be trained to students can be seen in the SPS chart that designed by Rezba as shown in Figure 1 [11].

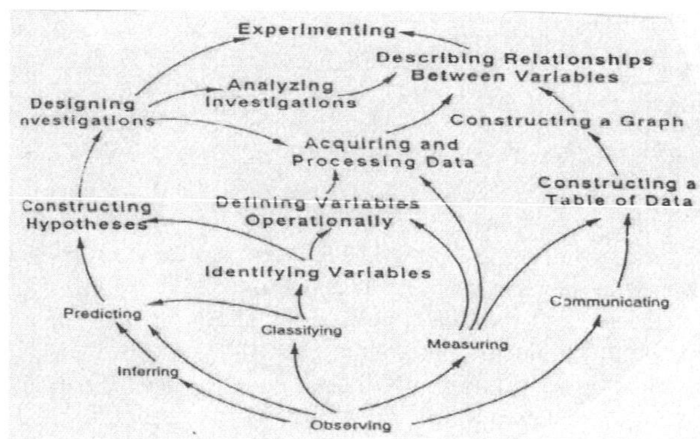

Figure 1. Diagram of Science Proses Skills

Based on Figure 1, the steps of SPS approach that utilized to design ways to train students' SPS that covering the aspects of observation to identify the variables and predicting, identifying variables to constructing the hypothesis, and then the students are invited to designing investigations to do the experimenting. Every aspect had been drilled through the questions that contained in the students' worksheets that were developed based on the SPS aspects. According to this description, the research will be able to describe the didactic design for training the SPS on the subject Work and Energy based on the analysis of portfolio assessment on students' worksheets.

2. Experimental Method

This study was conducted with the experimental study design using the one-shot case study to find the ways to train students' SPS on the subject of Work and Energy (didactic design). The study was conducted in one of the high schools in Bandung with a sample of 32 eleventh grade students were obtained by purposive sampling based on the observation of the students' SPS. The students' SPS responses were analyzed based on the increase of SPS from the assessment portfolio to the student's worksheets that developed from the SPS aspects. The students' worksheet is given to the students during three meetings. The scores in every SPS aspect obtained by the students during the three meetings compared to finding the students' SPS development. The percentage of student's score in the range of 0-50, 51-60, 61-70, 71-80, and 81-100 are interpreted as very poor, poor, fair, good, and excellent [4].

3. Result and Discussion

Based on the results of the research data acquisition, it obtained the description of the development students' SPS average score to identifying variables during the three meetings is shown in Figure 2.

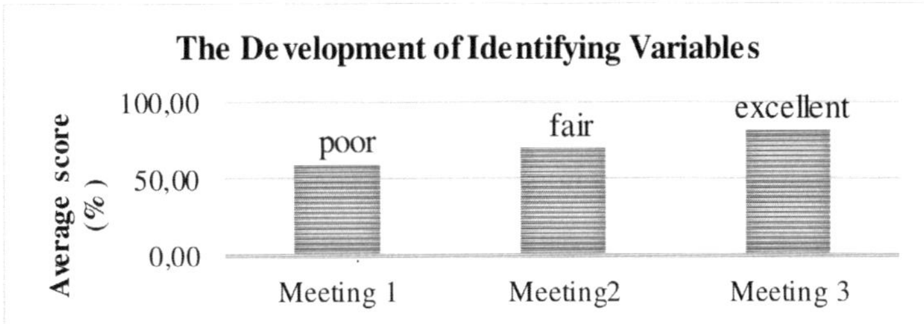

Figure 2. Graphics development of identifying variables during the three meetings

Figure 2 shows that the didactic design that had been designed could be training students' identifying variables skills during the three meetings. At the time of learning activities, the teacher begins the learning process by featuring the demonstrations directly in front of the class. Then the teacher asks the students to observe the demonstration that shown and ask the students to determine the quantities or the variables that observed based on the demonstration activity.

At the first meeting, the teacher did not use other media to aid the students doing the observation on the abstract things. For example, the velocity of the toy car, the teacher did not use a stopwatch to help students determine the fastest moving car. Teachers already planning a demonstration before assuming that students be able to identify the variables contained in the demonstration. However, students' difficulties in determining the variables that are abstract. Based on the findings of the problems at the first meeting, the teacher utilized the other media to aid in the students doing the observation for the next meeting.

In the second meeting, in order to know how much the energy of an object utilized the media such as the barrier boards. The further displacement of the barrier board from its first position after crushed by the toy car caused the energy on the toy car are also getting bigger. In observation requires a creative teacher activity by letting something abstract into something that can be observed directly [12].

At the third meeting, teachers utilized the video to help the students on doing the observation. This had been conducted so the students could observe the phenomenon more clearly. Johnston revealed that the utilized of the observation tools such as the projectors could assist students in the observation activity [12]. Rezba stated that the activities of the observations can be qualitatively and quantitatively [11]. The quantitative observations can be done with the help of measuring instruments. The goal is that the nature of the object can be observed by the students more closely.

On the Figure 3 displayed the tendency of students' answers on aspects of identifying the variables at the third meeting which is in conformity with the indicators that said by Rezba.

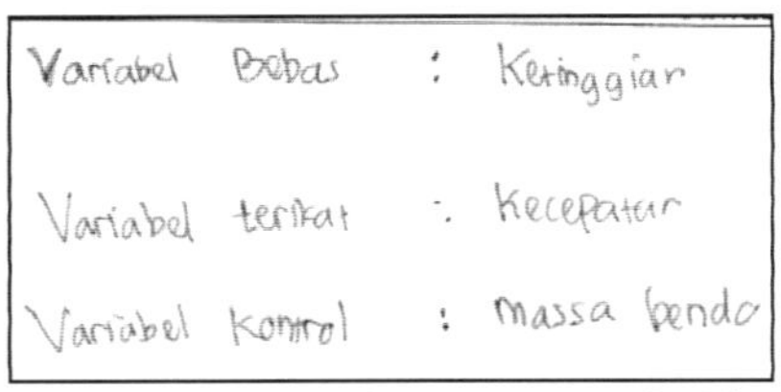

Figure 3. Photo the Example of the Tendency Students' Answer in Identifying Variables in the Third Meeting

The second aspect that measured is the predicting aspect. The development of the Predicting aspects during three meetings is shown in Figure 4.

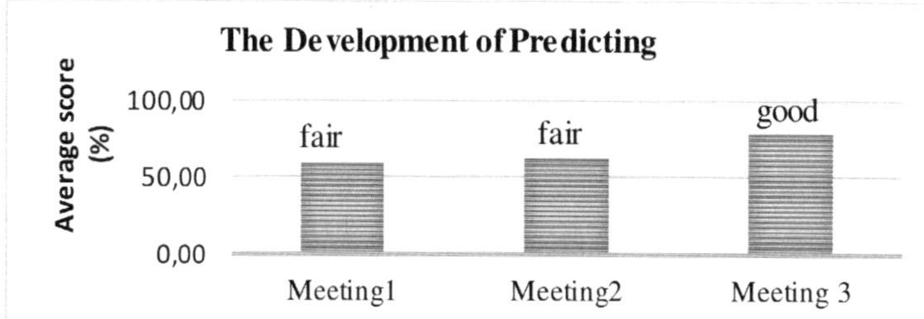

Figure 4. Graphics development of predicting during the three meetings

Figure 4 shows that students' predicting skill had been an increase in the first meeting to the second meeting, but the enhancement is not too great. At the third meeting, the students' predicting skills progressing into the Excellent category. According to Rezba, the predictions start from the observation activities, *inferring* to give an explanation for the observation, and predicting what will happen [11]. The predicting stage on the learning activities in the first and second meetings, the teacher asks the students orally to predict what would happen without the asked students' the reason why the predictions based on the observations. This is caused by the time limited to carry out all the stages in the learning process. Based on the phase expressed by the predicting activity requires *inferring* or provisional conclusions based on the observations from the teacher demonstrations [11]. This tentative conclusion that students use as a basis for making a prediction. However, at the first meeting and the second meeting, the teacher does not have students inferring or explaining the observations that have been made as an interim conclusion for students. The teacher immediately asked students to make a prediction because of the time constraints. This causes the students' predicting skill in the first and second meeting did not experience a large increase. But at the third meeting, the teacher asked some students to give an explanation for the prediction that had been delivered. The explanation that given with regard to the demonstration displayed the teacher. It could be training students' predicting skills so that at the third meeting they could achieve the good category.

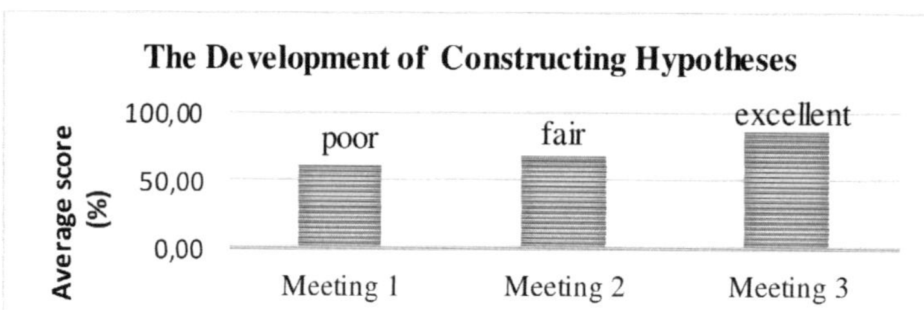

Figure 5. Graphics development of constructing hyphotheses during the three meetings

The Figure 5 shows that the skills to formulate hypotheses had been developed during the three meetings. The questions were developed to train students' constructing hypothesis skill is that students can answer the question of what will happen to the dependent variable with independent variables were deliberately altered.

Based on the comparing of the first and second meetings, the tendency of students' answers on the third meeting already shows that students can answer the questions with a good investigation. It can be seen from the students' answers to the question on the constructing hypotheses aspect on the students' worksheets that provided. At the third meeting, the students are able to construct the hypotheses as expected is to relate with what happened with the dependent variable due to the variables that deliberately tampered with. On Figure 6, shows the tendency of students' answers in answering questions on the constructing hypotheses aspect on the students' worksheets.

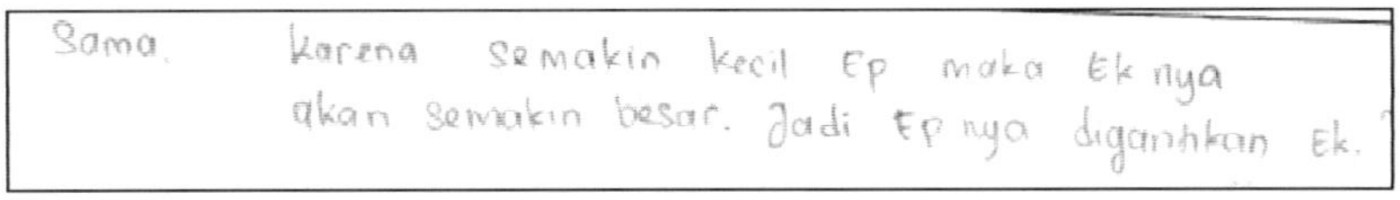

Figure 6. Photo the Example of the Tendency Students' Answer in Constructing Hypotheses in the Third Meeting

The constructing hypotheses activity starting from the identification of the variables that needed to make the phrase of the hypothesis on how the influence of the dependent variable due to the independent variable deliberately altered and fitted with an explanation for this hypothesis [11]. The identification of variables can be done by asking questions probe that display the issues that must be answered by the students [11]. This indicates that there are links between the students identifying variables skill with the constructing hypothesis skill. The improvement of students' constructing hypothesis skills also supported by the statement of Tomkins, that observation is the basis of constructing the hypotheses. The students can identify the variables through the observation of the demonstration that displayed as a basis the hypothesis [13].

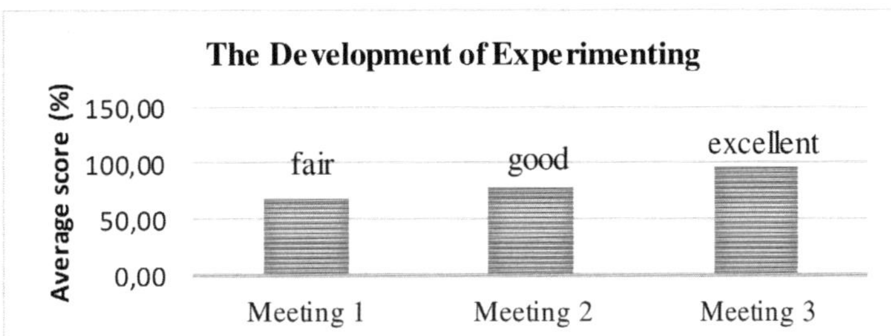

Figure 7. Graphics development of experimenting during the three meetings

Based on Figure 7, the students' experimenting skill due the progressing with the implementation of SPS approach in the learning process. The questions that developed is related to how students collect data, process and analyze the results of the experiment, and construct the conclusion of the trial. The tendency of students' answers on the third meeting, it has been able to collect the right data, process and analyze the data correctly, and make the right conclusions.

Rezba said that students' experimenting skill starting from the student's ability to identify the variables, constructing hypotheses, designing experiments, collecting data, explain and analyze the findings and make conclusions on experiments that have been conducted [11]. This suggests that the skills of identifying variables, predict, and hypothesized influencing students' experimenting skill. Correspondingly, the German revealed that the stage of experimentation done after designing the tables observation and in accordance with the procedures that have been designed [14]. The data obtained is then inserted into the table of observations. After the trial is completed, students are asked to analyze the results of experiments, construct the conclusions, provide an explanation for the conclusion and mention the mistakes that made in the study that could affect the study results. But at the first meeting and the second, the student does not provide an explanation for the conclusion and mention the mistakes made in the research. While at the third meeting, the students do so. This causes the students' experimenting skill at the third meeting greater than the two previous meetings. Another factor is the students begin to consult the results of their experiments to the teacher at all three meetings so the teacher can explain in accordance with the difficulties experienced by the students. This action is able to construct the development of students' skills. This is supported by Roth who stating that students should consult with friends and teachers at each meeting on the results of the experiment were found [15].

4. Conclusion

The results of this research showed that the didactic design on the subject of Work and Energy that are designed capable of developing students' SPS from the *poor* category into the *excellent* category. However, the study has not found ways to develop students' predicting skill to the excellent category. This is because the learning stage in training the predicting skills still not performing as it should.

Acknowledgments

The author would like to thank for all those who have helped in conducting research especially to the high school that have helped the investigation.

References

[1] Sheeba, NM 2013 An Anatomy of Science Process Skills in the Light of the Challenges to Realize Science Instruction Leading to Global Excellence in Education *J. Educationia Confab* **2** 108-123

[2] American Associates of Advancement of Science – AA AS 1968 *Science – A Process Approach* (New York: Xerox Corporation)

[3] Shulman L S and Tamir P 2004 *Research in teaching in the National Science. Second hanbook of Research teaching*: (rd) Rober M. W. T. (Rawera Chicago Rand McNally College Publishing Company)

[4] Lati W, Supasorn S, Promarak V 2012 Enhancement of learning achievement and integrated science process skills using science inquiry learning activities of chemical reaction rates *J. Procedia-Social and Behavioral Science* **46** 4471-4475

[5] Anies B 2016 Peraturan Menteri Pendidikan dan Kebudayaan Nomor 22 Tahun 2016 tentang *Standar Proses Pendidikan Dasar dan Menengah* (Jakarta: Kementerian Pendidikan dan Kebudayaan)

[6] Karim S, Prima EC, Utari S, Saepuzaman D, Nugaha MG 2017 Recostructing the Physics Teaching Didactic based on Marzano's Learning Dimension on Training the Scientific Literacies *Journal of Physics: Conference Series* **812**012102

[7] Karim S, Saepuzaman D and Sriyansyah SP 2016 The Learning Reconstruction of Particle System and Linear Momentum Conservation in Introductory Physics Course *Journal of Physics: Conference Series* **739** 012111

[8] Hodosyova M, et al 2015 The Development of Science Process Skills in Physics Education *J. Procedia-Social and Behavioral Science* 982-989

[9] Jack GU 2013 The Influence of Identified Student and School Variables on Student Science Process Skill Acquisition. *Journal of Education and Practice* **4**16-22

[10] Lia SA, Kusnandi, Yulianti K 2016 Desain Didkatis Konsep Luas Permukaan dan Volume Prisma dalam Pembelajaran Matematika SMP *J. Matematika dan Pendidikan* Matematika **1** 2502-5872

[11] Rezba JR, et al 1999 *Learning and Assessing Science Process Skills* (Kendall/Hunt Publishing Company)

[12] Johnston J 2009 Observation as an Important Enquiry Skill *Primary Science* **106** 15-17

[13] Tomkin PS, Tunnicliffe SD 2001 Looking for Ideas: Observation, Interpretation and Hypothesis-Making BY 12-Years-Old Pupils Undertaking Science Investigation *International Journal od Science Education* **23** 791-813

[14] Germann PJ, et al 1996 Identifying Pattern and Relationship Among the Responses of Seventh Grade Students to the Science Process Skill of Designing Experiments *Journal of Research in Science Teaching* **33** 79-99

[15] Roth WM 1993 The Development of Scence Process Skills in Authentic Contexts *Journal of Research in Science Teaching* **30** 127-152

IOP Conf. Series: Journal of Physics: Conf. Series **895** (2017) 012111 doi:10.1088/1742-6596/895/1/012111

Development of Teaching Material Oxidation-Reduction Reactions through Four Steps Teaching Material Development (4S TMD)

B S Syamsuri*, **S Anwar and O Sumarna**

Departemen Pendidikan Kimia, Universitas Pendidikan Indonesia, Jl. Dr.Setiabudi No. 229, Bandung 40154, Indonesia

*bebbysylvianasyamsuri@student.upi.edu

Abstract. This research aims to develop oxidation-reduction reactions (redox) teaching material used the Four Steps Teaching Material Development (4S TMD) method consists of four steps: selection, structuring, characterization and didactical reduction. This paper is the first part of the development of teaching material that includes selection and structuring steps. At the selection step, the development of teaching material begins with the development concept of redox based on curriculum demands, then the development of fundamental concepts sourced from the international textbook, and last is the development of values or skills can be integrated with redox concepts. The results of this selection step are the subject matter of the redox concept and values can be integrated with it. In the structuring step was developed concept map that provide on the relationship between redox concepts; Macro structure that guide systematic on the writing of teaching material; And multiple representations which are the development of teaching material that connection between macroscopic, submicroscopic, and symbolic level representations. The result of the two steps in this first part of the study produced a draft of teaching material. Evaluation of the draft of teaching material is done by an expert lecturer in the field of chemical education to assess the feasibility of teaching material.

1. Introduction

The learning process has three main components that are interrelated, that is teachers, students and teaching materials. The interactions of these three components form an integrated process entity, where there is a transformation of knowledge from the teacher to the learner so that the student gains the learning experience [1]. Therefore, the feasibility of a teaching material should receive serious attention, because teaching material is one of the main components in the learning process.

Teaching material for teachers serves to direct the learning activities as well as the substance of competence that should be taught to students. While for students, teaching materials have a function as a guide in the learning process and is a substance of competence that should be studied and as an evaluation tool of learning achievement [2].

Development of teaching material should pay attention to several aspects in order to help students learn independently and get mastery in the learning process, such as: giving examples or illustrations in accordance with the intended learning materials, giving practice questions or tasks as feedback for students to measure mastery of contextual, contextual based, and The use of simple language to make it easier for students to understand the material materials independently [3]. Therefore, the authors do the development of teaching materials through the method of 4S TMD (selection, structuring, characterization and didactic reduction), because through this method of teaching material that meet the criteria of a teaching material from the aspect of the suitability of the curriculum, true scientifically, has the value of usefulness for students , And the censoring of concepts [1]. This paper is the first part of the development of teaching materials that includes the selection and structuring steps to obtain a draft of teaching materials. Some studies suggest that students have difficulty in studying redox [4][5] in understanding the theory of oxygen binding and release, electron transfer, and changes in oxidation numbers. This is because the redox phenomenon is abstract and difficult to observe that it is difficult to convey both the concept and its application. It can be caused by the materials used. Teaching materials that connect theory with its application make it easier for students to understand the concept, by showing macroscopic, microscopic and symbolic [6][7]. Therefore, it is needed a teaching material that can facilitate user to understand it, so this research proposes to develop redox teaching materials through 4S TMD method. [8] The development of teaching materials is done by paying attention to aspects of concepts, linguistics and presentation that have an important role in realizing the teaching materials in accordance with the demands of education.

2. Experimental Method

Development of teaching materials is done through 4S TMD method. This article is the first part of the development of teaching materials that includes the selection and structuring steps. At the selection step, the development of teaching material begins with the selection of the concept of developing oxidation-reducing material (redox) based on curriculum demands, development of indicators, then development of basic concepts sourced from basic chemistry books, school chemistry books and international textbooks Related (concept identification), and the last is the development of values or skills that can be developed through redox material. The concept that has been selected is reviewed by an expert lecturer. The structural step includes concept maps, macro structures, and multiple representations of the oxidation-reduction reaction material that results in draft material.

This research uses the first type of Development Research design that is aimed not only to design and develop from a product but also to evaluate it [9]. This research procedure consists of three steps, namely design, production and evaluation steps. The Design step begins by conducting a literature review of the chemical concepts that students often have difficulty learning. Based on the results of literature review found that one such concept is redox. Literal analysis was then done to find out some of the causes of student learning difficulties. Based on the analysis it is found that one of the causes of students' difficulties in studying chemistry is the teaching materials used. At the production step, the procedure used is the teaching materials development procedure. In this research, the method chosen to develop teaching materials is Four Steps Teaching Material method known as 4S TMD. There are four steps: selection, structuring, characterization, and didactic reduction. But in this paper is the first part of the development of teaching material that includes selection and structuring steps. Evaluation of the draft of teaching materials is done by an expert lecturer in the field of chemical education to assess the feasibility of draft teaching material

3. Result and Discussion

3.1 Selection Step

Selection step included the development of indicators, concepts selection and value selection. At the selection step, the development of teaching materials covered the development of the scope of the oxidation-reduction reaction (redox) material based on curriculum demands and it was begun by selection of Basic Competence (KD). The most important aspect in designing the development of teaching materials, namely attention to the demands of the curriculum [10]. This means that teaching materials must be prepared in accordance with the applicable curriculum. In the redox subject there are two KD given by Depdiknas, that is:

3.9 Analyzing the development of the concept of oxidation-reduction reactions and determining the oxidation number of atoms in molecules or ions.

3.10 Implementing IUPAC rules for naming inorganic and organic compounds.

First step of selection step obtained that K.D 3.9 will be developed in teaching materials because the KD contain main concept of redox while KD 3.10 contain the nomenclature of a compound and it is not related to the concept of redox. After the curriculum analysis was done, the next step was the development of indicators. It aimed to measure the achievement of KD and as a guide in the development of teaching material. The development of indicators obtained are listed in Table 1.

Table 1. Development of reduction-oxidation reaction's indicators

KD	Indicator	Concept Label
Analyzing the development of the concept of oxidation-reduction reactions and determining the oxidation number of atoms in molecules or ions.	1. Explaining the Phlogiston theory 2. Explaining oxidation and reduction reactions based on the binding and release of oxygen 3. Explaining oxidation and reduction reactions based on the release and binding of electrons 4. Explaining oxidation and reduction reactions based on changes in oxidation numbers 5. Explaining oxidation number 6. Determining oxidation number of an atom in a molecule or ion 7. Determining oxidizer and the reducing agent in the redox reaction 8. Explaining autoredox reactions 9. Explaining application of the concept of redox in daily life	1. Phlogiston theory 2. oxidation and reduction reactions (Redox) based on : a. binding and release of oxygen b. binding and release of electron c. the changing of oxidation number 3. Oxidation number 4. OxdatordanReductor 5. Autoredox 6. application of the concept of redox in daily life

Table 1 shows that there are 9 indicators that developed in redox subject matter. The concept of the flogiston theory is included because it relates to the redox concept development as the beginning of the term oxidation. The results of this step of selection are the subject matter of the redox concept of the theory of flogiston, redox based on the binding and release of oxygen, redox based on the release and binding of electrons, and redox based on changes in oxidation numbers, oxidation numbers and applications in daily life. The next step was to identify the concept label in accordance with the indicators developed. The concepts were defined by textbooks so that the compilation of redox reaction material was obtained. Material selection begins by selecting the material source. In this study the source material in the form of basic chemistry books and international textbooks. Books that used as a reference are:

1. Bradi, James, Jespersen,Neildan Hyslop,Alison.2015.Chemistry Seventh Edition. New York: John Wiley & Sons.
2. Chang,Raymonddan Overby,Jason.2005.General Chemistry: The Essential Concepts, Sixth Edition.New York: The McGraw-Hill.
3. Davis, Raymond E., Metcalfe, H.Clark., Williams, John E., danCastka, Joseph E. 2002. Textbook modern chemistry.USA: Holt, Rinehart dan Winston
4. Gallagher,RoseMariedan Ingram,Paul.2009.Chemistry IGCSE Revision Guide. New York: Oxford University Press.
5. Norris, Roger., Ryan, LawriedanAcaster, David. 2011. Cambridge International AS and A Level Chemistry Coursebook. UK: Cambridge University Press

6. Glencoe.2002.Chemistry concepts and application.USA: The McGraw-Hill.

The next step is the development of values or skills that could be developed through redox reaction materials such as religious values, social care, tolerance, creative, environmental care and discipline. Table 2 show a sample of value linkage with concept

Table 2. value linkage with concept

Concept description	Related value	Value linkage with concept
 Picture of burning garbage Plastic base material such as ethene, vinyl chloride, ethan and others,If burned, ethene (C_2H_4) changes to CO_2 and H_2O $C_2H_4(g)+3O_2(g)\rightarrow 2CO_2(g) + 2H_2O(l)$ Ethene binds oxygen so that it undergoes oxidation	Environmental care	Household garbage piled up, if not burned or recycled, will certainly decay and cause major disasters. Because bacteria thrive and cause various diseases. In general, household waste is plastic, paper, leaves, and food scraps. If the waste is burned, then smoke and other harmful gases are formed. For example the burning of waste plastic produces CO_2 gas is very dangerous because it can raise the temperature of the earth. Therefore we must be more wise in using plastic

Table 2 show the value of environmental care in the redox concept based on the binding and release of oxygen. This teaching material is made oriented to SETS. Where SETS is a way of view that everything facing humans in this life contains aspects of science, environment, technology and society that affect the reciprocal [11][12].

3.2 Structuring Step

Structuring step is generated concept map as part of teaching materials that provide guidance on the relationship between the existing concepts on redox material; macro-structure which gives guidance on the systematic writing instructional materials; And multiple representations which are the development of teaching materials that connected between macroscopic, submicroscopic, and symbolic level representations. This step can facilitate students in connecting one concept with another concept, so that students' knowledge was more structured. [1] the students knowledge are structured in cognitive structure will facilitate students in considering new information received. [13] In addition to showing the relationship between concepts in teaching material, concept maps used as a first step to define planning (in the form of an outline) into writing.

Some of the benefits of structured teaching material are provide conformity between the needs of certain levels of learners with the content of teaching material, be able to describe the structure of science and helping students in describing cognitive structures in their brains [1]. In macro structures, there was mensions of progression and elaboration dimensions. The progression dimension is a vertical dimension in the form of a downward plot showing macro precision, while the elaboration dimension is the horizontal dimension of flow to the right as a way of reaching the clarity or explanation criteria of a subject matter. There are 4 core macro dimension, that is the phlogiston theory, redox reactions, oxidizing and reducing agents and the autoredox. Figure 1 show the part of the macro structure.

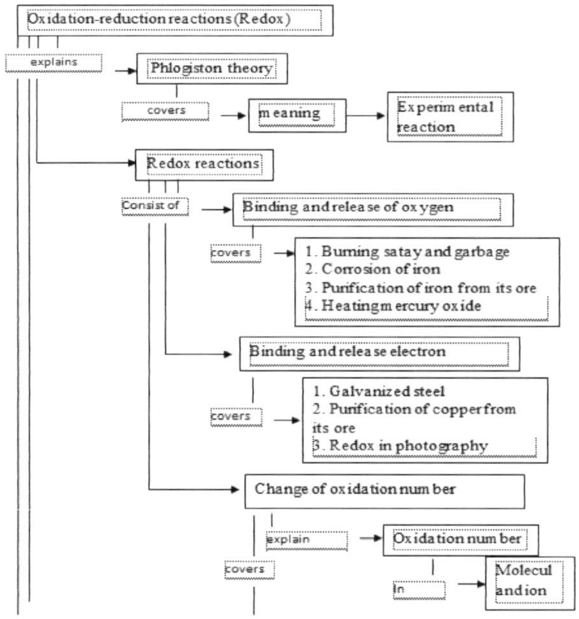

Figure 1. Redox macro structure

The characteristics of chemistry was based on the theory of the particulate nature of an object at the submicroscopic, macroscopic, and symbolic levels. [14] The chemical representations has three levels. The macroscopic level is a real and visible. The submicroscopic level is used for explanations in molecular form or symbolic explanations. The latter is a symbolic level consisting of different types of image representations and symbols. The success of chemistry learning involves the construction of mental associations between the macroscopic, submicroscopic and symbolic levels of the representation of chemical phenomena by using different representations [15]. Therefore it is necessary to integrate the three chemical representations to facilitate students in understanding a material.

Table 3. Redox multiple representation

Concept	Macroscopic	Submicroscopic	Simbolic
Redox based on the release and binding of electrons		Fireworks is one example of a redox reaction. One of the substances contained in it is magnesium which gives a bright white color to the fireworks. Magnesium (Mg) reacts with oxygen (O2) to produce MgO. In this reaction Mg releases an electron which means Mg is oxidized and Oxygen binds an electron indicating it is undergoing a reduction	$2Mg(s) + O_2(g) \rightarrow 2MgO(s)$ (Oksigen bond the electron) 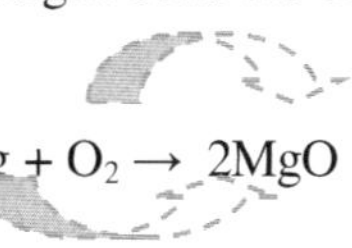 $2Mg + O_2 \rightarrow 2MgO$ (Magnesium released the electron) OxidaMagnesium (MgO) is the ionic that arranged from Mg^{2+}dan O^{2-} ions. $2Mg \rightarrow 2Mg^{2+} + 4e^-$ $O_2 + 4e^- \rightarrow 2O^{2-}$

IOP Conf. Series: Journal of Physics: Conf. Series **895** (2017) 012111 doi:10.1088/1742-6596/895/1/012111

Then the compilation of the matter was made in the draft of teaching material. The evaluation of the draft of the teaching material is done by an expert lecturer in the field of chemical education to assess the feasibility of teaching material. Views of teaching material obtained can be seen in Figure 2.

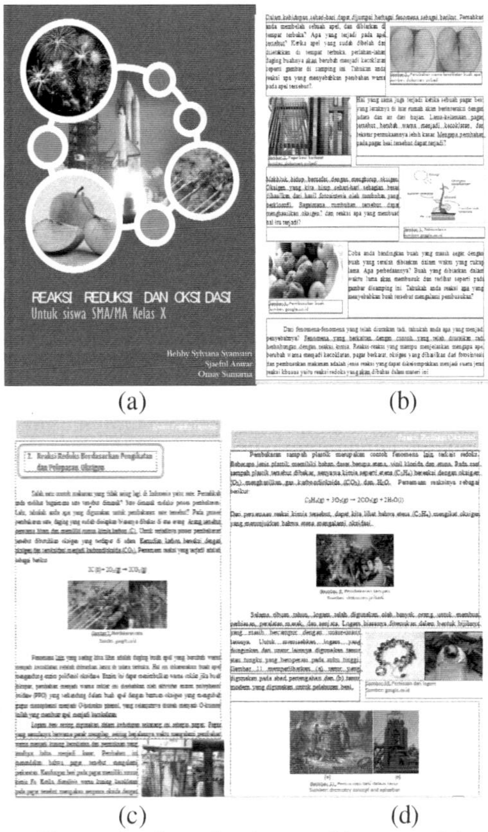

Figure 2.View of redox teaching material

4. Conclusion

The subject matter that developed on teaching materials was a development of KD 3.9 with 9 indicators. These indicators include the concept of the phlogiston theory, redox based on the binding and release of oxygen, redox based on the release and binding of electrons, redox based on changes in oxidation numbers, oxidation state and its application in daily life. At the structuring step was developed concept map as a part of teaching materials that provide the relationship between redox concepts; Macro structure that guide systematic on the writing of teaching material; And multiple representations which are the development of teaching materials that connection between macroscopic, submicroscopic, and symbolic level representations. The result of the two steps in this first part of the study produced a draft of teaching material.

Acknowledgments

The author would like to thank Mr. Sjaeful Anwar and Mr. Omay Sumarna for providing guidance during this research.

References

[1] Anwar S 2015 *Pengolahan Bahan Ajar (4 Steps Teaching Material Development)* [HandoutPerkuliahan] (Bandung: Unpublished Manuscript)

[2] Lestari I 2013 *Pengembangan Bahan Ajar Berbasis Kompetensi* (Padang: Akademia Permata)

[3] Widodo C S and Jasmadi 2008 *Panduan Menyusun Bahan Ajar Berbasis Kompetensi* (Jakarta: PT Elex Media Komputindo)

[4] Chiang W W, Chiu M H, Chung S L, and Liu C K 2014 Survey of High School Students' Understanding of Oxidation-Reduction Reaction *Journal of Baltic Science Education* vol 13(**5**) pp 596-607.

[5] Brandiet A R and Bretz S L 2014 Measuring meta-ignorance through the lens of confidence: examining students' redox misconceptions about oxidation numbers, charge, and electron transfer *Chemistry Education Research and Practice* **15** 729-746

[6] Suchocki J A 2013 Conceptual Chemistry 5th Edition (New Jersey: Prentice Hall)

[7] McClary L M and Talanquer V 2011 College students' mental models of acids and acid strength *Journal of Research in Science Teaching* 48(**4**) pp 396-413.

[8] Abidin Y 2014 *Desain Sistem Pembelajaran Dalam Konteks Kurikulum 2013* (Bandung: PT.Refika Aditama)

[9] Richey R C and Klein J D 2005 Developmental Research Method: Creating Knowledge From Instructional Design and Development Practice *Journal of Computing in Higher Education* 16(**2**) pp 23-28

[10] Depdiknas 2008 *Panduan pengembangan bahan ajar* Direktoral Jendral Manajemen Pendidikan Dasar dan Menengah

[11] Binadja A 2006 *Pedoman Pengembangan Silabus Bervisi dan Berpendekatan SETS (Science, Environment, Teaching, and Society) atau (Sains, Lingkungan, Teknologi dan Masyarakat)*. Laboratorium SETS: Program Pascasarjana UNNES

[12] Poedjiadi A 2010 *Sains Teknologi Masyarakat* (Bandung: PT Remaja Rosdakarya)

[13] Sinaga P 2014 Pengembangan Program Perkuliahan Fisika Sekolah III Untuk Meningkatkan Kompetensi Menulis Materi Ajar Calon Guru Menggunakan Multi Modus Representasi. Disertasi. Program Studi IPA Sekolah Pascasarjana UPI.

[14] Johnstone A H 1982 Macro and Micro Chemistry *School Science Review* 227(**64**) p 377-379

[15] Cheng M and Gilbert J K 2009 Towards a Better Utilization of Diagrams in Research into the Use of Representative Levels in Gilbert J.K &D.F.Treagust (eds) 2002 *Multiple Representation of Chemical Education* (Boston:Springer)

How to Develop Electrochemistry SETS-Based Interactive E-Book?

M Munawwarah[*]**, S Anwar and Y Sunarya**

Departemen Pendidikan Kimia, Universitas Pendidikan Indonesia, Jl. Dr. Setiabudi No. 229, Bandung 40154, Indonesia

[*]munawwarah@student.upi.edu

Abstract. This study aims to develop SETS-based interactive e-book teaching material through 4S TMD methode. The research methode in this study is the Development Research (RD) Richey and Klein that consists of design, phase, and evaluation. The design step was to analyze and plan the types of teaching materials instructional developed. There are 12 indicators from 3 standard competences that produced in selection step based new curriculum, the compatibility subject matter and indicators, and the relations between value and subject matter. Structuring steps yield concept map, macro structure, and multiple representation that were arranged to be first draft of teaching material that was used for develop the instruments for characterization step. Chatacterization test have been done to students in 12nd grades with 68 texts. Characterization results indicated that there were some texts included to difficult text. Difficult texts have been reduced with the ways back to qualitative steps and particulation. The second draft of teaching material was arranged based the results of didactic reduction of difficult texts. This draft was used for arranged interactive e-book. The characteristics of this SETS-based interactive e-book that developed were mention about the connection between science with environment, technology, and society. This interactive e-book consists of animation, task, and quizes that taken the interaction of students directly.

1. Introduction

Electrons in an electrochemical cell flows from the anode to the cathode. Electrons from the anode to the cathode will continue to flow until the electron's potential energy in the two half-reactions is equal to one another. An example of a spontaneous reaction is a reaction that occurs between the zinc metal and the copper (II) sulfate solution. The reaction can be written as follows:

$Zn\ (s) + Cu^{2+}\ (aq) \rightarrow Zn^{2+}\ (aq) + Cu\ (s)$

The electrolysis cell is a reverse reaction of voltaic cells. Electric current needed to generate the redox reaction (electric energy into electrochemical) [1] [2].

Teaching materials is anyone part of the learning resources. Through the good teaching materials, students can be master the competencies in accordance with the speed of learning that they have. Therefore, it was expected that the compiled teaching materials can make students become more active in learning activities to achieving the competences. The guidance about the development of teaching materials said that the instructional materials that were compiled must pay the attention to the demands of the curriculum, that's meaning the manner of teaching material that we will develop must be in accordance with the curriculum [3]. In the new curriculum, the core competencies and basic competencies have been established by the government, so that all schools can achieve the same competencies at the end of a particular subject. Therefore, the instructional materials were expected to be used in various schools that implement that curriculum. The good teaching materials are that can be understood by all students with different backgrounds and developments.

There were some method that can be use for develop the teaching material. Anyone of that is 4 steps teaching material development or commonly called with 4S TMD. The steps of development the teaching materials on 4STMD provided clear direction for the researchers in the process of developing the material teaching. Four important steps of this method are the selection, structuring, characterization, and didactic reduction. The steps performed from curriculum analysis to testing the feasibility of teaching materials. Each steps focuse on a particular aspect and mutually supports the next stage. After going through a step to next step, it was expected that the next stage will improve the quality of teaching materials that was developed. As the results [5] there was increasing in the level of legibility of teaching materials after the reduction step compared with the results of comprehension test at the characterization step.

The development of teaching materials based on SETS can improve students' concept comprehension because the presentation of the material was arranged based on daily life [5]. In addition, this approach can also be used in developing teaching materials that present integrated lessons and in accordance with standards of content and competence of graduates [6]. It explains that the teaching materials depend on the current curriculum and the conclusions made from that context [7]. Therefore, through this approach, it was expected that the development of teaching materials can bring students closer to their real life. so that students can solve the problems that were expected to appear in their daily life.

In this study, an interactive e-book was compiled with SETS-based developed using the 4STMD method. The development of this resource have done because most of the existing e-book is just a transfer of printed book into the software format without any interactive aspects in it [8]. Therefore, through the development method, it was expected that the teaching materials in the form of interactive e-book can be compiled to meet the criteria of e-learning materials and criteria set by the National Education Standards Board consisting of the feasibility of content, language, presentation, and graphic [9][10]. Besides that, the interactive e-book composed presents electrochemical material related to science, environment, technology, and society aspects that were directly related to daily life. So that, the purpose of this study is to produce a SETS-based Interactive e-book resource developed by using 4S TMD.

2. Experimental Method

The research method that used in this research is *development research* method Richey and Klein [11]. This research was used to develop and validate educational products [4]. Figure 1 is a brief overview of Development Research stages in general.

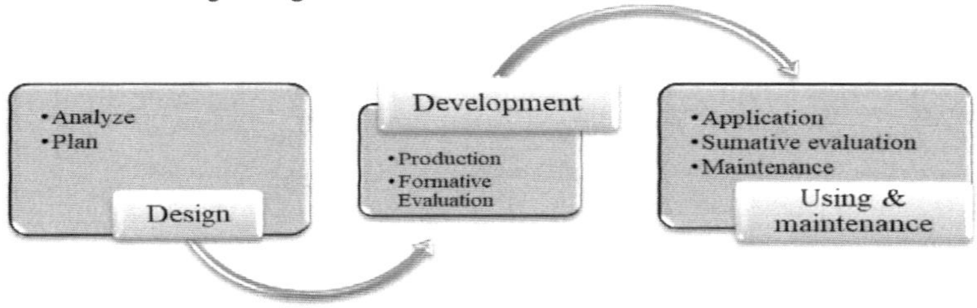

Figure 1. The steps of development research by Richey and Klein

The development process of teaching materials using procedures developed is 4S TMD (Four Step Teaching Material Development). The teaching materials development procedure consists of four steps: Selection, Structuring, Characterization, and Didactic Reduction [12] in accordance with Development Research procedures by Richey and Klein.

Some instruments that used in this study are the selection review instrument, the structuring review instrument, the characterization instrument, and the didactic reduction reviews instrument. The characterization test was conducted in anyone school in Sumedang on 40 students of 12nd grade.

3. Result and Discussion

3.1. The development of teaching material SETS-Based interactive e-book

Teaching material that developed in this study is teaching materials in the form of interactive e-books on SETS-based teaching materials using the steps of development of 4S TMD teaching materials. The four stages consist of the stage of selection, structuring, characterization, and didactic reduction. Each stage has been through the stage of review and evaluation phase. Interactive e-book development is done twice after the characterization stage and after the didactic reduction stage. The e-book produced after the characterization stage is the first draft of the teaching material used for the characterization stage. The e-book produced after the didactic reduction stage is a product of the development of teaching materials.

The steps of interactive e-book development through 4S TMD are:

a. Selection step

Selection step was divided into two part: the first part consisting of the contents standard on the curriculum related to the development indicators of matter, subject matter sources related to the identification of the concepts with the indicators, as well as the value related to chemicals. The second step of selection consists of conceptual analysis with value, compilation of teaching materials, and instruction material revisions. Selection step is more specific on the material delivered on the teaching materials that is related to the grid on the teaching material, the collection of material, and the linkage between competences and indicators to be achieved. The three basic competencies were selected in this selection process which developed into 12 indicators used as the basis in the selection of appropriate concept labels on teaching materials. The results of the formulation of these indicators can be seen in the table below.

Tabel 1. Indicators of KD-3 in Electrochemistry matter

Basic Competencies	Indicator
3.3 To evaluating phenomenon or process that occurs in an electrochemical cell sample (voltaic cells and electrolysis cells) that used in daily life.	1. Explain the meaning of redox and electrochemical reactions
	2. Write down the steps of equalizing the redox reaction equation by using half-reaction and the change of oxidation number
	3. Mention the component of voltaic cell
	4. Write down the reaction on the voltaic cell, cell notation, and determine the standard cell potential value of a redox reaction
	5. Determine the spontaneity of redox reactions by using reduction potential data
	6. Mention the voltaic cell utilization in daily life
	7. Describe the components and types of electrolysis cells
	8. Mention the electrolysis cell utilization in daily life
3.4 Analyze the factors that influence the occurrence of corrosion and propose ideas / ideas to overcome them.	9. Describes the reaction that occurs in the corrosion process
	10. Analyzing the factors that cause corrosion
	11. Propose ideas on how to solve / prevent corrosion.
3.5 Applying the law / rules in the relevant calculation electrochemical cell.	12. Resolving issues related to the stoichiometry of the electrolysis and the voltaic cell by using Faraday's law

Selection step was conducted to ensure the fulfillment of the criteria of the teaching materials in accordance with the criteria of the teaching materials consisting of the criteria of conformity to the applicable curriculum, the truth of the concept of the scientific aspect, the usefulness of the students, the suitability of the available time and the delivery of important concepts studied by the students [12]. Some of these criteria are related to the depth and breadth of material to be presented on teaching materials.

b. Structuring step

This step was conducted to determine the systematic preparation of a resource and a first draft teaching materials consisting of a concept map, the macro structure, and multiple representations. Concept maps are arranged to show the relationship of a concept with another concept clearly. Concept map was arranged from general concept to the specific concept [13]. Therefore, through the whole concept maps that students can develop the knowledge, explore and explain their views, connects the concept of their thinking, to achieve better results [14]. The concepts that were connected through a concept map is based on three basic competencies, which has been mentioned in the earlier stages in the selection. The three sub themes consist of redox reactions, voltaic cells, and electrolysis cells.

Macro structure shows a macro concept (subject matter) in a material equipped with micro (subordinate) concept of macro concept. The macro structure is used to control the writing of the text of the material as a whole. Therefore, each subject in the teaching materials were analyzed and compiled with the text representation model in the form of macro-structure of the text [15]. Multiple representation is the presentation of chemicals in teaching materials consisting of macroscopic, submicroscopic, and symbolic aspects [16][17]. Not only understand on representation level, but also students can conducted every single steps.[18][19][20][21].

Teaching materials draft was prepared and through the steps of review and validation of supervisors. Draft teaching materials then used as a reference for preparing the storyboard in preparing an interactive e-book developed with the cooperation of media experts. The e-book of the first draft teaching material was developed and used for the characterization step.

c. Characterization step

The characterization step was performed after the drafting of the first teaching materials developed in the form of an interactive e-book. The characterization test via e-book provides more effective and efficient process in terms of paper use through technology utilization. The interactive e-book compiled based on the first draft of the teaching material is accompanied by a column writing down the main idea and the choice of selecting the characteristics of a text in the teaching material. Not only composed by containing the text only, in this interactive e-book also presented some phenomena that were displayed in the form of images and video that can support the explanation of a text.

The structured characterization step of the text was tested for its understanding. Each student gives an assessment as much as the student's understanding on each text is measured by the text characteristic level (easy, medium, difficult) and the writing of the main idea of the text. The result of the recapitulation of text characterization test of each student was presented in PDF format which is then processed by the researcher. There are 68 total texts tested for this resource and 7 text are classified as text. The following is the text analysis of the characterization step. The difficult texts continue at the didactic didactic reduction step.

Table 2. Percentage of score results of characterization per text

Characteristic	Amount	Percentage (%)
Easy	61	89,70%
Difficult	7	10,30%

d. Didactic reduction

This step aims to reduce the level of difficulty in teaching materials so that can be easily to understood by students. At this step the draft of teaching materials that have been prepared as a result of the step of structuring was reduced by the grid obtained by the characterization stage. Reduction was done on every difficult text according to the type of difficulty that the text has. The reduction of the difficulty level of the text on the teaching material was done in several ways including: (1) return to the qualitative step; (2) particulate. Draft teaching materials that were prepared after going through the didactic reduction stage were a draft of SETS-based teaching materials (Science, environment, technology, and society).

3.2. Characteristics of an interactive e-book

SETS-based Interactive e-book materials on electrochemical matter is teaching materials that presented in electronic form. Electrochemical material was delivered by linking aspects of SETS aspects of science, environment, technology and society. The SETS pattern used in the development of interactive e-books was connecting science (concept, theory, and law) with SETS as a supplement to the topic in accordance with indicators that have been developed [22][23]. The good chemistry education is that give the students the value of daily life [24]. It shows that a material text presented in this interactive e-book was not directly linked to the four aspects of SETS. In a text only highlighted one or two aspects of SETS while the other aspects are implied.

The interactive aspect of this e-book teaching material is the presentation of the material and its application in daily life in the form of video that will be observed by the students as information in understanding the electrochemical matter. In addition, there were also tasks with feedback given directly to the answers given by the students. Another interactive aspect is the animation and quiz given at each end of the sub-material and the results can be directly known by the students when they have successfully completed the given quiz. In addition to knowing the results of the quiz score, students can also find out the correct answers from the quiz questions they have been working on.

SETS-based interactive e-book on electrochemical material presents the material in an integrated way between matter of chemistry and daily life. The most important aspect of SETS in the development of this interactive e-book is to bring students into situations to utilize the concept of science to the form of technology for the benefit of society [25].

4. Conclusion

Based on the results of analysis and data processing research that has been done, obtained the following conclusions "Interactive E-book" materials on SETS-based electrochemical materials are developed using Four Steps Teaching Material Development method with four steps: selection, structuring, characterization, and didactic reduction. SETS-based Interactive E-book Teaching material on electrochemical matter" developed by 4S TMD method have the characteristic of simulation, video, task, and quiz display with automatic processing of results.

Acknowledgments
The author would like to thank Mr. Sjaeful Anwar and Mr. Yayan Sunarya for providing guidance during this research. The authors also thanked the Director General of Kemendikbud who has provided excellent scholarships used during the research.

References

[1] Bradi, James. Et al. 2015. *Chemistry, seventh edition*
[2] Chang, Raymond. 2005. *Kimia dasar*. Jakarta: Erlangga.
[3] Depdiknas 2008 *Panduan pengembangan bahan ajar*. Direktoral Jendral Manajemen Pendidikan Dasar dan Menengah.
[4] Arifin, Z. 2011 *Penelitian pendidikan metode dan paradigma baru*. Bandung: PT Remaja Rosdakarya.
[5] Binadja A 2006 *Pedoman Pengembangan Silabus Bervisi dan Berpendekatan SETS (Science, Environment, Technology, and Society) atau (Sains, Lingkungan, Teknologi dan Masyarakat) Bahan Pembelajaran Penerbitan Khusus Media MIPA UNNES*. Semarang: Laboratorium SETS. Universitas Negeri Semarang.
[6] Budiharti R, Ekawati E Y and Pujayanto 2015 *Pengembangan Modul IPA Terpadu berbasis SETS dengan Tema "Pelestarian Lingkungan" (Dalam Tinjauan Validitas Isi)*. Surabaya: Prodi Pendidikan Fisika JPMIPA FKIP UNS.
[7] Kansanen P and Matti M 2015. *Didactic relation in the teaching-studying-learning process*. Finland: University of Helsinki Finland.
[8] Kong SC, Chan TW, Huang R and Cheah H M, A review of e-Learning policy in school education in Singapore, Hong Kong, Taiwan, and Beijing: implications to future policy planning. *Journal Computer Education*. Vol. 1 (1-3): pp 187-212
[9] Kozma R B and Russell J 1997 Multimedia and understanding: expert and novice responses to different representations of chemical phenomena. *Journal of Research in Science Teaching, 34(*9) 49-68
[10] Pietzner V 2014 Computer based learning in chemistry classes *Eurasia Journal of Mathematic, Science, and Technology Education* Vol 10 (4) pp 297-311
[11] Richey R C, Klein J D and Nelson W A *Develomental research : studies of instructional design and development*. http://aect.org/edtech/41.pdf
[12] Anwar S 2014 Pengolahan bahan ajar [*hand out perkuliahan*]. *Unpublish manuscript*. Bandung: Universitas Pendidikan Indonesia.
[13] Arifin. 2015. Pengembangan bahan ajar IPA terpadu pada tema udara berbasis nilai religius menggunakan 4 steps teaching material development [Tesis]. *Unpublish manuscript*. Bandung: Universitas Pendidikan Indonesia.
[14] Okonkwo I 2014 Effect of concept mapping teaching strategy on students' achievement in environmental concepts in chemistry. *International Journal Of Scientific Research*. Vol. 3 No. 2277-8179.
[15] Novak J D 1990 Concept maps and vee diagram: two metacognitive tools to facilitate meaningful learning *Instructional science* 19: 29-52.
[16] Gwen Lawrie & Trevor Appleton. 2011. Using multiple representations to enhance understanding of molecular structure: a blended learning activity. *International Journal of Environmental & Science Education*. Vol 8, 67-71.
[17] Jaber L D and BouJaoude S 2012 A macro–micro–symbolic teaching to promote relational understanding of chemical reactions. *International Journal of Science Education 34* (7) 973-998

[18] Johnstone A H 2000 Chemical education research: Where from Here? *Chemistry Education 4* (1) 34-48

[19] Winnie Sim Siew L I and Mohammad Yusof Arshad 2014. Application of Multiple Representation Levels in redox Reactions among Tenth Grade Chemistry Teachers. *Turkish Science Education*. Vol 11 issue 3

[20] Treagust D F, Chittleborough G D and Mamiala T L 2003. The role of sub-microscopic and symbolic representations in chemical explanations. *International Journal of Science Education, 25* (11), 1353-1369.

[21] Tan K C D, Goh N K, Chia L S and Treagust D F 2009 Linking the macroscopic, submicroscopic and symbolic levels: The case of inorganic qualitative analysis. In Gilbert, J.K., & Treagust, D.F. (Eds.), *Multiple representations in chemical education* Springer, pp 137- 150.

[22] Pudjiadi A 2007 Pendidikan sains teknologi masyarakat. *Rujukan filsafat, teori, dan praksis ilmu pendidikan.* Bandung: UPI Press.

[23] Bridstock M, Burch D, Forge J, Laurent J and Lowe I 1998 *Science, technology, and society.* UK: Cambridge University Press

[24] Lobo M D, Novell J M 2015 How to prepare didactic experiments related to chemical properties related to chemical properties for primary, secondary and high school. *International Journal of Advanced Research in Chemical Science* (IJARCS) 2 (5) pp 41 – 49

[25] Humairoh F and Wasis 2015 Pengembangan *E-Book* Interaktif Berbasis Salingtemas (Sains, Lingkungan, Teknologi, Masyarakat) pada Materi Fluida Dinamis untuk Meningkatkan Pemahaman Konsep Siswa dan Penerapannya. *Jurnal Inovasi Pendidikan Fisika (JIPF),* Vol. 04 No. 02, Mei 2015, 69-75.

IOP Conf. Series: Journal of Physics: Conf. Series **895** (2017) 012113 doi:10.1088/1742-6596/895/1/012113

Construction of Context-Based Module: How OLED can be used as a Context in High School Chemistry Instruction

I R Anugrah*, **A Mudzakir, O Sumarna**

Departemen Pendidikan Kimia, Sekolah Pascasarjana Universitas Pendidikan Indonesia, Jl. Dr. Setiabudi No. 229, Bandung 40154, Indonesia

*indah.rizki.anugrah@student.upi.edu

Abstract. Teaching materials used in Indonesia generally just emphasize remembering skill so that the students' science literacy is low. Innovation is needed to transform traditional teaching materials so that it can stimulate students' science literacy, one of which is by context-based approach. This study focused on the construction of context-based module for high school using Organic Light-Emitting Diode (OLED) topics. OLED was chosen because it is an up-to-date topic and relevant to real life. This study used Model of Educational Reconstruction (MER) to reconstruct science content structure about OLED through combining scientist's perspectives with student's preconceptions and national curriculum. Literature review of OLED includes its definition, components, characteristics and working principle. Student's preconceptions about OLED are obtained through interviews. The result shows that student's preconceptions have not been fully similar with the scientist's perspective. One of the reasons is that some of the related Chemistry concepts are too complicated. Through curriculum analysis, Chemistry about OLED that are appropriate for high school are Bohr's atomic theory, redox and organic chemistry including polymers and aromatics. The OLED context and its Chemistry concept were developed into context-based module by adapting science literacy-based learning. This module is expected to increase students' science literacy performance.

1. Introduction

The common problems faced in science education today are how to motivate students and increase their interest in learning science. One of the most widely used solutions is through a context-based learning approaches [1]. Context-based approaches are approaches that adopted in science learning where context and application of science are used as starting points for the development of scientific ideas [2]. Context-based learning can bring science closer to student daily life so that they are more interested in learning it and ultimately the understanding of science itself increases. The development of context-based teaching materials projects in the UK (Salter's Advanced Chemistry), USA (Chemistry in the Community) and Germany (Chemie im Kontext) is one of the application of context-based approaches.

The application of context-based teaching materials is based on the results of several studies such as students 'lack of interest in science lessons (Ramsden, 2003), declining students' interest in science in secondary school (Reiss, 2004; Simpson and Oliver, 1990) and student assumption that science doesn't have relevance to their lives (Reiss, 2000) [3]. Reported by de Jong, meta-analysis has been done by Bennett, Hogarth and Lubben on 66 studies of the effects of context-based learning approaches. The results showed that this context-based approach motivate students in their science lessons and enhance more positive attitudes to science more generally [4].

Meanwhile, most of the teaching materials that have existed and developed in Indonesia focus only on dimension of content [5]. The application or context of the knowledge content itself is not much explored. One effort to improve science literacy performance of Indonesian students is to construct context-based module. The topic to be used as a context may refer to the selection principles of science learning-based on PISA: relevant to real-life situations that will still be relevant for at least a decade ahead and closely related to the competence of the process [6].

One interesting topic is Organic Light-Emitting Diode (OLED). This is because OLED is an up-to-date research theme with widespread application and can be made from local resources. An OLED is a light emitter with a thickness of about 100–200 nm [7]. The uniqueness of OLED is the use of organic materials that are easier and cheaper than inorganic materials (silicon). Even, mangosteen, turmeric and carrots extract can be used as an OLED emitters [8]. OLED is widely used in lights, mobile phones, televisions display, etc.

This study focused on the reconstruction of science content structure of OLED into content structure for instruction and developing it to become context-based module for high school students. The reconstruction is required because not all OLED related Chemistry concepts can be delivered to high school students. This context-based module was developed by adapting the sequences of the Science and Technology Literacy (STL) learning stages proposed by Nentwig et al [9]. By following these STL learning stages, it is hoped that students can develop their thinking process and be able to integrate some Chemistry concepts through the OLED context. Furthermore, through OLED context based-module, it is expected that students' science literacy performance will be increased.

2. Method

The research design used in this research is Model of Educational Reconstruction (MER) developed by Duit [10]. The fundamental idea of this model is that the content structure for instruction can not be taken directly from the science content structure. It should be reconstructed in particular by paying attention to the educational objectives as well as the students' cognitive and affective perspectives. Therefore, besides science content structure, students' conceptions and educational goals must be equally considered [11].

MER consists of three components, namely: 1) analysis of content structure, 2) research on teaching & learning, and 3) development and evaluation of instruction – and their close interplay [12]. The first stage is conducted through literature analysis of OLED books and journals to obtain scientific perspectives on OLED including its definitions, components, characteristics and working principles. Later, an analysis literature of related Chemistry concepts related to OLED conducted using high school and university chemistry textbooks as references. The second stage is conducted through clinical interviews of high school students to explore preconceptions about OLED. Interview guideline adapted from the Laherto format [12]. The interview questions consist of 13 unstructured-format items. These items were divided into six sections that exploring: 1) general knowledge of OLED; 2) emission process of light and Chemistry concepts that explains the process; 3) OLED structure and its related Chemistry concepts; 4) characteristics of OLED and its related Chemistry concepts; 5) basic principles of OLED and its related Chemistry concepts and 6) student's interest in OLED-context modules. The third stage is the module construction process through the didactic reduction.

The module that have been constructed are tested for its readability and feasibility. Readability test is conducted using Fry formula. The feasibility assessment is undertaken by Chemistry and Indonesian teachers, including language, presentation and graphics.

3. Result and Discussion

3.1. Scientist's perspective about OLED

OLED is a light emitter with a thickness of about 100–200 nm [7]. The advantages of OLED compared to inorganic LEDs and LCDs are better color resolution and lower power consumption to operate. In addition, with its thickness OLED can be used to create thin and flexible displays [13]. The basic structure of OLED is the organic material positioned between the cathode and anode made of transparent conductive glass, usually indium tin oxide, ITO (indium tin oxide).

Light emission on OLED occurs by making the organic molecule in an excited state. This process is then followed by the decay of electron pairs and holes (electron defects) back to their ground state resulting light emission. However, excitation in organic molecules is difficult. Excitation of organic molecules will only occur if the molecule exhibits an intrinsic conductivity both for electrons and holes and an energy level alignment as found in inorganic semiconductors. The key to electrical conductivity in the polymer is the existence of a conjugated system [14]. The conjugated system is a system in which the position of the double bond and the single bond are alternating with each other.

The overlap of π orbitals of the conjugated system between organic molecules induce an unfolding in the HOMO and LUMO level, originating narrow bands that consist in energetic steps called bandgap (of organic semiconductors that act like the similar bandgap of inorganic ones) that usually range from 1 eV to 4 eV. This bandgap is corresponding to the minimum energy to generate an excited state. This bandgap usually decreases with the increase of the delocalization, because the π electronic system increases [7]. The smaller the bandgap, the lower the bandgap energy, the easier the organic molecule will be excited.

Although the key to the conductivity of the polymer is the existence of a conjugated system, it does not necessarily make the polymer electrical conductor. An interference is required through the removal of electrons (oxidation) and the addition of electrons (reduction). This process is called doping. Through the redox principle, electrons will move easily and flow along the molecules to create an electric current [15].

Light emission on OLED takes place through a process called electroluminescence. Electroluminescence is a non-thermal generation of light resulting from the application of an electric field to a substrate. Emissions on OLED devices are explained through the principle of electron-hole recombination. When electrons and holes recombine, a high-energy quasiparticle is formed. These quasi-particles are called exitons, which are similar with single molecules, but with more high energy. This exciton produces light when it decays [13]. The basic principle of the electroluminescence process occurring in OLED is illustrated in Figure 1.

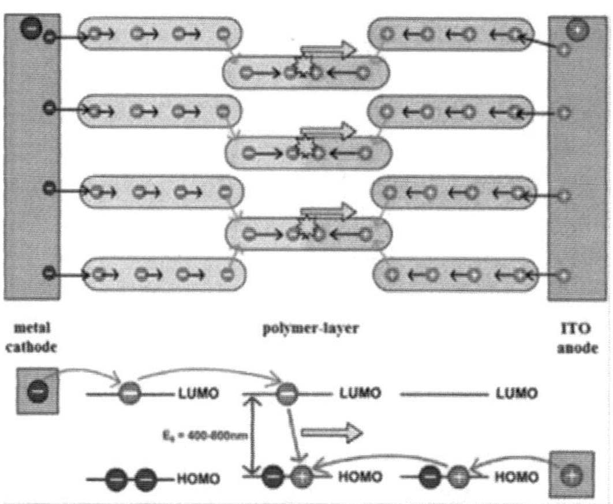

Figure 1. Principle of electroluminescence in an OLED (Banerji, Tausch & Scherf, 2013).

Based on Figure 1, the electroluminescence process can be described as follows: 1) Charge injection: using a power supply, the electrons are drawn from the anode to form a hole, then the electrons are forwarded to the cathode so that the electrons are excessive; 2) Charge transport along the polymer layer from one molecule to another in opposite directions and 3) Charge recombination (electrons and holes) to give an exciton that emitting light when it decays [14]. The wavelength of the light emission produced depends on the exciton energy. Therefore, we can control the color produced by regulating the energy through the selection of the organic material. This is the advantage of OLED displays [16].

3.2 High school Chemistry concepts related to OLED

Science content structure of OLED can not be used directly as content structure for High School Chemistry instruction. Therefore, it is important to analyze the relevance of chemistry concepts related to OLED to the ongoing curriculum. The purpose is that the instruction experienced by students is appropriate with the level of cognitive for high school students.

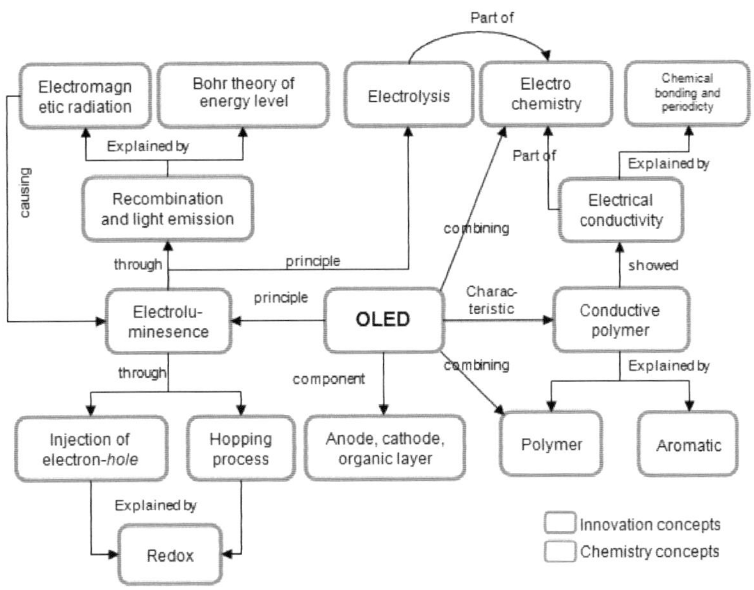

Figure 2. Scheme of context OLED and its related Chemistry concept.

Figure 2 illustrates the linkage between the OLED concept as an innovation concept with High School Chemistry concept. Those related High School Chemistry concepts are:

1. Bohr's Theory about energy level, to describe light emission process
2. Periodicity, to describe electrical conductivity of semiconductor
3. Organic compound, related to basic material of OLED
4. Polymer, especially conductive polymer
5. Alkene and aromatic compound to describe conjugated system
6. Redox and electrolysis, to describe principle of OLED that makes conductive polymer can emitt lights.

3.3 Student's preconceptions about OLED and its related Chemistry concepts

Student's preconceptions about OLED are explored through a clinical interview which adapting the Laherto format. The preconceptions include general knowledge of OLED technology, Chemistry concept explaining the characteristics and principles of OLED and student's interest in using context-based modules that use OLED topic. Overall, these student's preconceptions are not fully similar with the perspective of science. One reason is that some of those related Chemistry concepts are not taught in high school.

Table 1. Student's preconceptions about OLED and its related Chemistry concepts

	Student's Preconceptions
General knowledge of OLED (before picture of OLED sample is given)	• Students do not know OLED
General knowledge of OLED (after picture of OLED sample is given)	• Light, flexible and eco-friendly light-emitting technology
Emitting light process and it's related Chemistry concepts	• Students do not know the mechanism of light emitting • Students know Bohr's atomic theory, but do not know that the process of light emitting can be explained from this theory
The basic structure of OLED and it's related Chemistry concepts	• It's similar with electrochemistry, but only some students anwer correctly, that is electrolysis
OLED characteristics and it's related Chemistry concepts	• Organic compounds are naturally electric insulators • The conjugated double bond is where the double bond with the single alternates • Students do not understand the delocalisation of electrons in compounds that have conjugated systems (e.g. benzene)
OLED working principle and it's related Chemistry concepts	• Some students are able to identify some of the processes that occur in electroluminescence (electron transfer, charge movement), but do not recognize the recombination process
Student interest in the OLED-context module	• Students are interested in using the OLED-context module because the material presented is more useful and not boring

These preconceptions shown in Table 1 are used to determine the starting point of conveying the OLED context of the constructed module. In this case, the development of television display is used to deliver OLED because most students recognize OLED as a television technology. In addition, students' preconceptions become one of the inputs in the preparation of the text. It is intended that the module is constructed conveniently read by students.

Most students complain about so many theories being taught in schools today. Context-based learning can prevent students from overloaded curricula [17]. Students are interested in using context-based module because it's more relevant with student's need. The relevance of science to the needs of students is directly proportional to the interest of students studying science [18]. Thus, the findings in this study reinforce the results of previous studies on the role of context-based approaches in improving student's positive attitudes (including motivation and interest) to science.

3.4 Construction of OLED-context module

Didactic reduction is conducted to reduce the difficulty level of the concept to be written on the module, making it easier for the students to understand. Didactic reduction is gone through four ways: back to qualitative phase, neglecting process, using picture to help explanation and using historical development. The module is designed based on *teaching and learning sequence of Science and Technology Literacy (STL) that adopted from Chemie im Kontext [9].* The constructed module is divided in two part, The Blue Part contains explanation about OLED and The Green Part that explores high school Chemistry concept related to OLED. By adapting STL learning phase stage, it is expected that student's thinking processes and integrating ability increase. The sequence of the module is:

1. Contact phase
   Television display development is used in this phase as an introduction in The Blue Part that contain explanation about OLED.
2. Curiosity phase
   In this phase, student's curiosity is stimulated by question about why organic compound that generally isolator can conducting electricity like in television and hand phone.

3. Elaboration phase
   This phase is focusing on OLED explanation, including characteristics and how OLED works.
4. Decision making phase
   The purpose of this phase is to make students gets right and obvious answer for question in curiosity phase. In the constructed module, this phase is represented in form of summary.
5. Nexus phase
   This last phase is the decontextualizing phase from all of content that has been submitted. In this module, nexus phase is presented by giving example of some organic compound to be analysed by students whether it can be used as OLED raw material or not.

3.5 Review result of the constructed OLED-context module

The feasibility test was conducted by three High School Chemistry teachers and one Indonesian teacher. Referring to the feasibility score percentage of teaching materials according to Slavin, it can be concluded that in terms of language, presentation and graphics as shown in Table 2 the constructed module can be declared feasible.

Table 2. Review of content feasibility

No.	Criteria	Assesment				Average
		1	2	3	4	
1	Language	80 %	85 %	80 %	80 %	81,25 %
2	Presentation	85 %	80 %	80 %	85 %	82,5 %
3	Graphics	80 %	80%	85 %	80 %	81,25 %

Readability test of the constructed module used Fry formula. Three representative samples are selected to be tested. The grade reading level is determined from the intersection of the average number of sentence and the average number of syllables per hundred words by plotting it to the Fry graph, which is in this study is level 11. This means that the module is appropriate for high school.

4. Conclusion

The current teaching learning trends are shifting towards a context-based approach. OLED as an up-to-date and relevant topic to students' daily life is an appropriate theme to be used as a context, including in Chemistry instruction. The reconstruction process is needed to adjust the OLED science content structure to be appropriate for high school instruction so that it meets the criteria for being a context. In this study, reconstruction of OLED content was performed using MER by combining scientist's perspective, student's preconceptions and curriculum objectives. The module is constructed by adapting the STL learning sequence, divided into two parts: Blue Part (contains explanation about OLED) and The Green Part (explores high school Chemistry concept related to OLED). Interview result shows that context-based modules help improve students' positive attitudes toward Chemistry and emphasize the relevance of Chemistry in everyday life. On this basis, it is hoped that students' science literacy performance can increase.

Acknowledgments

We would like to show our gratitude to all of lecturers who gave suggestion for improvements and did triangulation. We also thank all of the students who voluntarily became respondent for inteview session and the teachers who did feasibility test.

References

[1] Rannikmae M, Teppo M and Holbrook J 2010 Popularity and Relevance of Science Education Literacy: Using a Context-based Approach *Science Education International* **21** 116-125
[2] Bennett J, Lubben F and Hogarth S 2006 Bringing Science to Life: A Synthesis of the Research Evidence on the Effects of Context-based and STS Approaches to Science Teaching *Sci Ed* **91** 347-370.

[3] Otter C 2011 Context based learning in post compulsory education: Salters Advanced Chemistry project. *Educació Química EduQ* **10** 11-17
[4] De Jong O 2006 *Context-based chemical education: How to improve it?* (Seoul, Korea)
[5] Nugraha D A, Binadja A and Supartono 2013 Pengembangan Bahan Ajar Reaksi Redoks Bervisi SETS, Berorientasi Konstruktivistik *Journal of Innovative Science Education* **2** 27-34.
[6] Hayat B and Yusuf S 2010 *Mutu pendidikan* (Jakarta: Bumi Aksara)
[7] Pereira L 2012 *Organic Light Emitting-Diodes: The use of rare-earth and transition metal* (Taylor & Francis Group)
[8] Praditya A 2011 *Fabrikasi dan karakterisasi material gel pemancar cahaya menggunakan elektrolit berbasis kristal cair ionik fatty imidazolinium dan emitter berbahan baku sumber terbarukan lokal* (Bandung: FPMIPA UPI)
[9] Nentwig P, Parchmann I, Demuth R, Gräsel C and Ralle B 2002 *Chemie im Context-From situated learning in relevant contexts to a systematic development of basic chemical concepts* (Kiel, Jerman)
[10] Duit R, Gropengießer H, Kattmann U, Komorek M and Parchmann I 2012 The model of educational reconstruction – A framework for improving teaching and learning science *Sci. Educ. Res. and Pract. in Europe: Retrospective and Prospective* **5** 13–37
[11] Duit R 2007 Science educational research internationally: Conception, research method. *Domain research. Eurasia jurnal of mathematics, science & technology education* **3** 3-15
[12] Laherto A 2012 *Nanoscience education for scientific literacy: Opportunities and challenges in secondary school and in out-of-school settings* (Faculty of Science of the University of Helsinki)
[13] Mitschke U and Bäuerle P 2000 The electroluminescence of organic materials *J.Mater. Chem.* **10** 1471-1507
[14] Banerji A, Tausch M W and Scherf U 2013 Classroom experiments and teaching materials on OLEDs with semiconducting polymer. *Educ. quím.* **24** 17-22
[15] Nordén B and Kruimeijer E 2000 *Advanced Information: The Nobel Prize in Chemistry, 2000: Conductive Polymers* (The Royal Swedish Academy of Sciences)
[16] Tsujimura T 2012 *OLED Display: Fundamentals and Applications* (John Wiley & Sons Inc.)
[17] Pilot A and Bulte A M W 2007 Why do you 'need-to-know': context-based education *International Journal of Science Education* **28** 953-955
[18] Holbrook J and Rannikmae M 2009 The Meaning of Scientific Literacy *International Journal of Science Education* **4** 275-288

IOP Conf. Series: Journal of Physics: Conf. Series **895** (2017) 012114 doi:10.1088/1742-6596/895/1/012114

Pre-Service Science Teachers (PSTs)' Creative Thinking Skills on Atoms, Ions and Molecules Digital Media Creation

R R Agustin[1,2*], **L Liliasari**[1,3)], **P Sinaga**[1,4)] and **D Rochintaniawati**[1,2)]

[1] Sekolah Pascasarjana, Universitas Pendidikan Indonesia, Jl. Dr. Setiabudi No. 229, Bandung 40154, Indonesia
[2] International Program on Science Education, Universitas Pendidikan Indonesia, Jl. Dr. Setiabudi No. 229, Bandung 40154, Indonesia
[3] Departemen Pendidikan Kimia, Universitas Pendidikan Indonesia, Jl. Dr. Setiabudi No. 229, Bandung 40154, Indonesia
[4] Departmen Pendidikan Fisika, Universitas Pendidikan Indonesia, Jl. Dr. Setiabudi No. 229, Bandung 40154, Indonesia

*rikarafikah@upi.edu

Abstract. Atoms, ions and molecules are considered as abstract concepts that often lead to students' learning difficulties. This study aimed at providing description of pre-service science teachers (PSTs)' creative thinking skills on atoms, elements and compounds digital media creation. Qualitative descriptive method were employed to acquire data. Instruments used were rubric of PSTs' digital teaching media, open ended question related to PSTs' technological knowledge and pre-test about atoms, ions and molecules that were given to eighteen PSTs. The study reveals that PSTs' creative thinking skills were still low and inadequate to create qualified teaching media of atoms, ions and molecules. PSTs' content and technological knowledge in regard with atoms, ions and molecules are the most contributing factors. This finding support the necessity of developing pre-service and in-service science teachers' creative thinking skill in digital media that is embedded to development of technological content knowledge.

1. Introduction

Atoms, elements and compounds are abstract concepts that commonly cause learning difficulties [1]. Appropriate technology in the term of digital teaching media can be used to help students' comprehend the concepts more easily. Nevertheless, it is thought that students' learning achievement is influenced more by teachers' practice than the technology itself [2]. Thus, study on teachers' competence in creating and implementing digital teaching media about atoms, ions and molecules is urgently required.

Various studies on in-service and pre-service science teachers (PSTs)' competence to integrate technology in learning has been conducted. Especially for PSTs, it is found that their competence in integrating technology is significantly improved during the two years master program [3, 4]. PSTs of undergraduate program are not totally ready to effectively integrate technology into science teaching [5]. In addition, description of PSTs' competence in integrating technology were also explored in various contexts such as inquiry-based learning [3], case-based learning [6] and internet use [7]. PSTs' technology competence were also explored on various constructs and other factors, such as, gender, teaching experience and age [8]. Other factors affecting technology competence are the ICT learning experience [9], experience using ICT [10, 11], and fulfilling content representation [12]. However, none of the studies investigating PSTs' technology competence related to creative thinking skill.

The investigation of PSTs' creative thinking skill on digital science teaching media creation has not been performed. Therefore, this study aimed at investigating PSTs' creative thinking skill on Atoms, Elements and Compounds digital teaching media creation. In this study, aspects of PSTs' creative thinking skill was captured by means of analyzing digital teaching media of atoms, ions and molecules. Technology and content knowledge of PSTs were analyzed from their pre-test and open-ended question of technology integration in science learning.

2. Method

This study employed descriptive method to explore how PSTs' creative thinking skill performed on digital teaching media about atoms, ions of molecules. Eighteen PSTs were involved in the study. The study was conducted in the context of school science course that aims to prepare PSTs to develop science learning material as well as the learning media for teaching science in junior high school.

Every science topics was carried out through eight following stages: 1) Identifying big ideas to assess content knowledge 2) identifying teaching procedures to assess pedagogical knowledge 3) identifying the use of technology to assess technological knowledge 4) planning to integrate appropriate teaching approach 5) planning to integrate appropriate technology for teaching Atoms, Elements and Compounds 6) planning to integrate technology for certain teaching method 7) Creating digital media for teaching Atoms, Elements and Compounds. PSTs' creative thinking skills were investigated by analysing the digital media they made.

This study used multiple data sources which are rubric of digital teaching media, open-ended question on technological knowledge and pre-test about atoms, ions and molecules. The rubric of digital science teaching media consists of creative thinking skills indicator which are fluency, flexibility, originality and elaboration [13]. Open-ended questions about PSTs' technological knowledge were embedded into course assignment. The questions asking about tools of technology that is appropriate for teaching atoms, ions and molecules as well as ways to overcome absence of the technology. Meanwhile, pre-test was given before the implementation of the course. The pre-test was arranged in the form of concept map task about atoms, ions. Data were analysed using qualitative approach. Creative thinking aspects performed in digital science teaching media were identified based on creative thinking skill indicators. Open-ended question exploring PSTs' technological media was used to triangulate the data.

3. Result and Discussion

This section provides description of PSTs' creative thinking skills appearing in digital science teaching media about atoms, ions and molecules. Moreover, the association of PSTs' creative thinking and other variables which are content knowledge and technological knowledge are also discussed.

3.1. Pre-service Science Teachers(PSTs)' Creative thinking skill on Fluency aspect

Fluency are indicated by a number of big ideas that the PSTs derived for the topic of Atoms, Elements and Compounds. The ideas are: 1) Classification of element 2) The development of atomic theory 3) Nomenclature of compounds 4) Periodic table of element and 5) Ions and Molecules. These ideas are fluently presented in the teaching media in the form of definition. This type of knowledge indicates PSTs' content knowledge [14]. The example of digital teaching media slide presentation is shown in Figure 1. It is shown that PSTs fluently derive certain big idea into several information. However, it is found that mostly PSTs present the ideas by means of delivering one direct information.

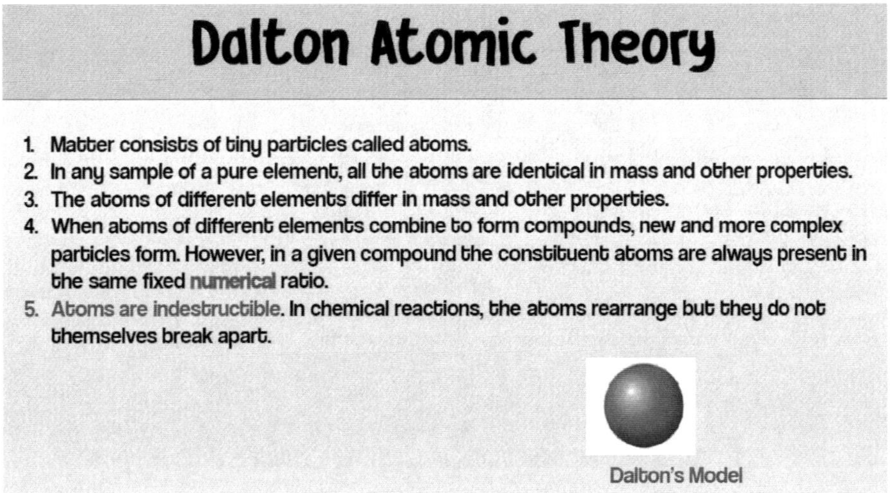

Figure 1. Example of one direct information in digital science teaching media

The PSTs are also fluently generate teaching strategies appropriate in delivering the topic of Atoms, Elements and Compounds. Those strategy are lecturing with analogy, the use of elements' characteristics cards and games. Games is one learning strategy that can be used to enhance creative thinking skills [15].

3.2. Pre-service Science Teachers' Creative thinking skill on Flexibility

Flexibility can be identified when PSTs are able to perform the following ability: 1) generate various ideas, answers, questions as well as view problems from different perspectives (2) Search numerous alternative solution 3) change approach and thoughts [13]. It is found that all PSTs use Power Point TM. This finding implies that PSTs' creative thinking in flexibility aspect is still low. The use of certain software in this study echoes the lack of PSTs' technological knowledge. It is also found that PSTs do not insert any more sophisticated aid to deliver the topic. This fact indicates the need of ICT course infused into teacher preparation program [10]. The only visual aid inserted in the digital media is pictures. Figure 2 shows the use of picture to endorse the delivery of Atoms, Elements and Compounds topic. This phenomenon is associated with less technological knowledge which is knowledge about hardware and software for enhancing the learning [14].

Based on figure 2, it is found that PSTs use static figures to deliver Atoms, Elements, Compounds and Mixtures. It implies the PSTs' low ability to insert more sophisticated media which currently provided freely in internet. Thus, it is important to develop intervention so that the PSTs are fluently able to insert pictures, video as well as simulation into their digital science teaching media. The intervention can be in the form of instructional design based on technological pedagogical content knowledge (TPACK) [16].

Data from pre-test reveals that PSTs' content knowledge do not guarantee their creative thinking skills. As a consequence, PSTs' creative thinking skills need to be developed along with the program developing content knowledge and technological knowledge.

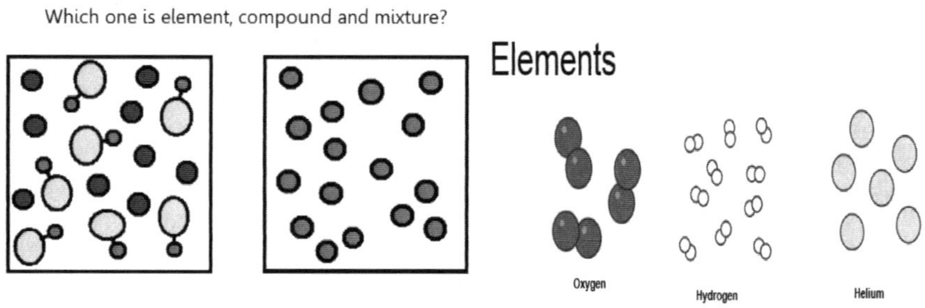

Figure 2. Example of the use of picture in digital science teaching media

3.3. Pre-service Science Teachers' Creative Thinking Skill on Originality and Elaboration

Originality and elaboration have not appear in the digital science teaching media. No one of the digital science teaching media contains new and unique statements. All the digital media contains common description and explanation delivering the topic of Atoms, Elements and Compounds. In contrast, originality refers to the ability to generate unusual combinations [13]. According to the digital media created, there is no feature performing pre-service science teachers' originality to present unique software as well as digital application to teach Atoms, Elements and Compounds. Regarding this, an effort developing pre-service science teachers' originality in building digital science teaching media is necessary. Specific program or training needs to be implemented in order to prepare PSTs in creating digital media. This finding support view about the necessity of technological pedagogical content knowledge within PSTs preparation program [9].

Elaboration is not performed in PSTs' technological knowledge. Elaboration is indicated by the skill to enrich and develop an idea or product, add or derive details of certain object, ideal or situation to be more interesting. This skill is also associate with Association between PSTs critical thinking skill and technological knowledge can be bridged by program teaching technological abilities. One of the ability is coding [17].

4. Conclusion

Based on the study it is concluded that PSTs' creative thinking skills were still low and inadequate to create qualified teaching media of atoms, ions and molecules. Creative thinking skill aspects that appears is limited to fluency and flexibility. The PSTs have not been developed originality and elaboration. PSTs' content and technological knowledge in regard with atoms, ions and molecules are the most contributing factors. Meanwhile, content knowledge about atoms, ions and molecules does not guarantee PSTs' creativity to create digital science teaching media. This finding support the necessity of developing pre-service and in-service science teachers' creative thinking skill in digital media that is embedded to development of technological content knowledge.

Acknowledgments

The authors would thank to Dr. Phil. Ari Widodo, M.Ed. as the head of school science course, Ikmanda Nugraha, M.Pd. and Dr. Eka Cahya Prima as lecturers team of the course.

References

[1] Agustin R R 2013 Pengembangan keterampilan generik sains melalui penggunaan multimedia interaktif *Jurnal Pengajaran MIPA* **18** 253-7

[2] Howard S K, Chan A, Mozejko A and Caputi P 2015 Technology practices: Confirmatory factor analysis and exploration of teachers' technology integration in subject areas *Computers and Education* **90** 24-35

[3] Maeng J L, Mulvey B K, Smetana L K and Bell R L 2013 Preservice Teachers' TPACK: Using Technology to Support Inquiry Instruction *Journal of Science Education and Technology* **22** 838-57

[4] Hofer M and Grandgenett N 2012 TPACK development in teacher education: a longitudinal study of preservice teachers in a secondary M.A.Ed. program *Journal of Research on Technology in Education* **45** 83-106

[5] Agustin R R and Liliasari L 2016 Pre-service science teachers' readiness to integrate technology (an exploration toward tpack in preliminary practical context) *Jurnal Pengajaran MIPA* **22**

[6] Mouza C and Karchmer-Klein R 2013 Promoting and Assessing Pre-Service Teachers' Technological Pedagogical Content Knowledge (TPACK) in the Context of Case Development *Journal of Educational Computing Research* **48** 127-52

[7] Sahin I, Celik I, Oguz Akturk A and Aydin M 2013 Analysis of Relationships between Technological Pedagogical Content Knowledge and Educational Internet Use *Journal of Digital Learning in Teacher Education* **29** 110-7

[8] Bilici S C, Guzey S S and Yamak H 2016 Assessing pre-service science teachers' technological pedagogical content knowledge (TPACK) through observations and lesson plans *Research in Science & Technological Education* **34** 237-51

[9] Alayyar G M, Fisser P and Voogt J 2012 Developing technological pedagogical content knowledge in pre-service science teachers: Support from blended learning *Australasian Journal of Educational Technology* **28** 1298-316

[10] Koh, Woo H-L and Lim W-Y 2013 Understanding the relationship between Singapore preservice teachers' ICT course experience and technological pedagogical content knowledge (TPACK) through ICT course evaluation *Educ Asse Eval Acc* **25** 321-39

[11] Lin T-C, Tsai C-C, Chai C S and Lee M-H 2013 Identifying Science Teachers' Perceptions of Technological Pedagogical and Content Knowledge (TPACK) *Journal of Science Education and Technology* **22** 325-36

[12] Agustin R R and Liliasari L 2016 Investigating Pre-Service Science Teachers (PSTs)' Technological Pedagogical Content Knowledge Through Extended Content Representation (CoRe). In: *MSCEIS,* (Indonesia: Journal of Physics)

[13] Torrance E P 1969 *Creativity, What Research Says toThe Teacher* (Washington, DC: National Education Association)

[14] Pringle R M, Dawson K and Ritzhaupt A D 2015 Integrating Science and Technology: Using Technological Pedagogical Content Knowledge as a Framework to Study the Practices of Science Teachers *Journal of Science Education and Technology* **24** 648-62

[15] Navarrete C C 2013 Creative thinking in digital game design and development: A case study *Computers & Education* **69** 320-31

[16] Lee C J and Kim C M 2014 An implementation study of a TPACK-based instructional design model in a technology integration course *Educational Research Technology and Development* 437-60

[17] Dufva T and Dufva M 2016 Metaphors of code—Structuring and broadening the discussion on teaching children to code *Thinking Skills and Creativity* **22** 97-110

IOP Conf. Series: Journal of Physics: Conf. Series **895** (2017) 012115 doi:10.1088/1742-6596/895/1/012115

Integrating Students' Understanding of Prerequisite Concepts into Capability to Synthesize Organic Compounds

L N Amsad[1,2*], Liliasari[3], and A Kadarohman[3]

[1] Sekolah Pascasarjana, Universitas Pendidikan Indonesia, Jl. Dr. Setiabudi No. 229, Bandung 40154, Indonesia
[2] Program studi Pendidikan Kimia, Jurusan Pendidikan MIPA, FKIP Universitas Cenderawasih, Jl. Abepura, Jayapura, Papua 99351, Indonesia
[3] Departemen Pendidikan Kimia, Universitas Pendidikan Indonesia, Jl. Dr. Setiabudi No. 229, Bandung 40154, Indonesia

*lusia_amsad@upi.edu

Abstract. The capability to synthesize organic compounds requires understanding of organic chemical prerequisite concepts. Students need to integrate organic chemical concepts to analyze stages for synthesizing organic compounds. This study aims at understanding students' organic chemical prerequisite concepts and students capability to combine their prerequisite concepts in synthesizing organic compounds using descriptive method. The respondents of this research were students from chemistry education departement. The study used multiple choice with reasons test and essay test analyzed using non parametric correlation statistics. The result showed a relationship between student understanding of prerequisite concepts with conclusion of organic compounds synthesis.The result found that students have a good understanding on prerequisite concepts such as: type of reaction, starting material and reagent. Conversely, students prerequisite concepts were very lacking in the concept of carbocation ions, reaction products, electrophile formation of C-C, and disconnection of compounds.

1. Introduction

The concept of synthesis organic compound required by students in Advanced Organic Chemistry learning. Based on the implementation of Organic Chemistry learning, it is known that students still depend on familiarity with the related reaction. Futhermore, they have lack of problem solving test because they cannot reminiscence the answers or they just seek to available answers [1]. According to early research, students do not have the skills and capability to create a fit reaction mechanism [2]. Consequently, some of them tend to learn by roting or resolving problems by making an analogy. Even previous research stated that the success of an individual who resolve chemical problems begins by describing the initial structure to discover ways to solve the problems [3]. In addition, there is recognition to present a constant connection between the concepts. The integration between the concepts will have an impact on the students' capability in completing the assignment of organic synthesis.

It also found that student assignments involved only the explanation of general term and linked it to the mechanism of reaction and the whole process transformation of a compound [4]. These examines are only part of how students answer an organic synthesis problem. These examination do not emphasis on how to solve an organic synthesis problem completely. Students have sense in advance exertion on the part that should merely be the initial stage of sequence part of the organic compound synthesis resolution. The difficulties of students accomplish the assignment is affected from students capability to resolve more complex problem that related to organic synthesis compound. Moreover, if students are given a suitable assignment in resolving organic chemical problems, then they will find no difficulties to resolve the problems. According to early research, students must given organic synthesis problems by using mechanistic solution [5]. Through this assignment, students will be more mindful to resolve organic chemistry problems mechanically. Students who learn organic chemistry requisite to practice synthesis of organic compounds in new molecular environments and different intermediates. It will make them to think about the conventional organic synthesis mechanisms. The assigments that given were out of the ordinary or simple context in which they were first taught. It is also recognized the important for an approach that connected to learning process in science with students confidence [6]. Thus, it can provide a meaningful learning for the students [7]. Meanwhile, it pedagogically provides information about how students thinking process, along with how the students resolve the problems of synthesis of a target molecule compound.

Similar with other research found if students depend on solely heuristic ability, there will be effortless or no process skill from the students to achieve the right answer [8]. Sometimes students use practical directions that help them to simplify the task and solve problems. But it has been proven regularly lead students cannot understand the current concepts [9], [10] . Even in learning chemistry, studens can have misconceptions likewise in other lessons, so the misconceptions need to be identified from the beginning in order not to impede the process of student acceptance in learning [11], [12]. However, if students do not use the strategy and skills, they will not be able to achieve the intended learning outcomes. Thus, they may be capable to answer the given questions. A mental model of student can effect to student'understanding because a mental model emphasizes function rather than form[13]. It is founded that students serve as subjects in the study who are not yet experts in organic They will not demonstrate consistent behavior with problem solving strategies which were successfully used by experts. Complexity of compounds considered as a factor that affected how students work with the problems. For that purpose, this paper will present about the understanding of organic chemistry concept from students and how they integrated it to resolve organic synthesis problem.

2. Method

The research used descriptive statistic method then analyzed by using nonparametric correlation test to know the integration between the level of understanding of the concept possessed by the students by way of student completion on the problem of organic synthesis. This study was conducted at one of the universities in Papua. This research involved students who were taking Advanced Organic Chemistry courses in the fifth semester of the third year. The number of students involved was 14 students in the third. The course has a prerequisite course, namely Organic Chemistry courses 1 and 2.

The instrument used was multiple choice with reasons test and essay test. Questions were related concepts in organic chemistry prerequisites such as: the type of reaction, the structure of the compound, the reaction product, electrophiles in the formation of C-C. These problems considered the connection to resolve organic synthesis problems focus on aldol condensation, benzene acylated and tertiary amine benzene. Figure 1 shows an example of a given problem for the students:

IOP Conf. Series: Journal of Physics: Conf. Series **895** (2017) 012115 doi:10.1088/1742-6596/895/1/012115

A. An example quetions of understanding of prerequisites concepts:

1. From these benzene compounds, which one have the para offering groups ....

a) 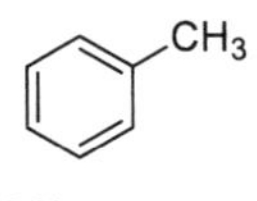 b) 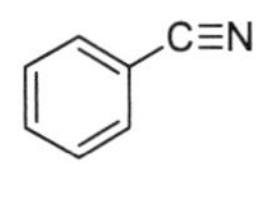 c) 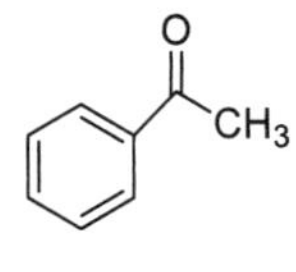 d) 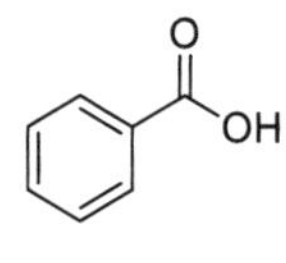

Reason:..

2. Based on the structure of the following compounds that would be appropriate to the tertiary amine compound is ....

a) 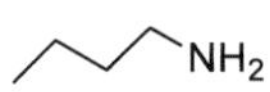 b) 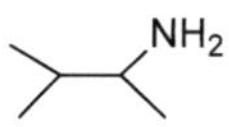 c) 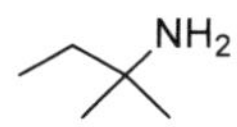 d) 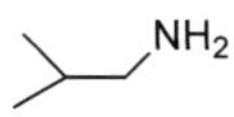

Reason:..

B. Examples of organic synthesis problems:

Write down the synthesis of the target molecule of the starting compound of the following materials:

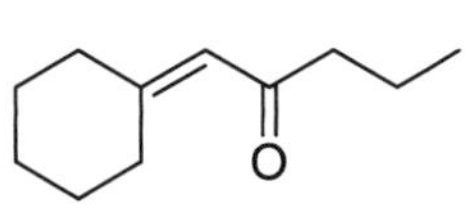 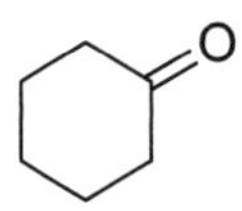

Target molecule Starting material

Figure 1. The example questions of understanding of prerequisite concepts and organic synthesis problems

The problems given to students is based on the synthesis problem of target molecular compounds. Data from retrieval process came from students answers. The questions were 20 items about the multiple choice with reasons used to find students understanding of the prerequisite concept. From students answer classification based on student answer. It is classified that students responded correctly (if the answer and the reason are true), and classified wrong (if the answers and reasons are wrong, or answered correctly but the reason is wrong or no reason, or no answer at all for each item). From 20 items of the questions for students' understanding of prerequisite concepts divided into 7 items for aldol condensation compound (35% of all items), 7 items for acylated benzene (35%) and 6 items for tertiary amine benzene (30%).

The percentage outcome of understanding the concept of student prerequisites is seen as related to the percentage of the number of students who can use the prerequisite concept at the stage of each synthesis of organic compounds. The statistics used for Spearman rank correlation at significance level $\alpha = 0.05$. Spearman rank correlation coefficient value [14]. The students' understanding of prerequisite concepts in organic chemistry owned by the students associated with subjects such as: type of reaction, the structure of the compound, starting material, the reaction product, electrophiles used, reagents, and disconnection that is then associated with a way resolve the problems of organic synthesis.This study was directed to the synthesis of aldol compounds, acylated benzene, and tertiary amine benzene.

3. Result and Discussion

The results of students undersanding of prerequisite concepts and then viewed its integrity to resolve the synthesis compounds problem can be seen in table 1:

IOP Conf. Series: Journal of Physics: Conf. Series **895** (2017) 012115 doi:10.1088/1742-6596/895/1/012115

Table 1. Percentage of level of understanding of student prerequisite concepts and percentage of use of prerequisite concepts in answering synthesis of organic compound.

No.	*Concepts*	*Students who understand the concept(%)*	*Students who used of the concept of prerequisite in answering synthesis of organic compounds (%)*
A. Aldol condensation compound			
1	Type of Reaction	85,72	42.86
2	Structure of the compound	42.88	14.28
3	Starting material	85,72	28.57
4	The result of the reaction (Product)	14.28	14.28
5	Electrophile on C-C formation	14.28	0
B. Benzene acylated compound			
6	Reaction of Nucleophilic Substitution	14.28	14.28
7	Reagent of acylated compound	71.43	14.28
8	Reagent function	71.43	42.86
9	Carbocation ions	7.14	0
C. Tertiary amine benzene compound			
10	Steering Group	28.57	0
11	Structure of Compound	50.00	28.57
12	Reduction Reactions	57.14	14.28
13	Properties of Compound	47.62	14.28
14	Disconection of compound	14.28	0

Based on the result of the percentage of students' understanding of the prerequisite concept with the completion of organic matter as a whole found the existence conection. But the views of specific integrity for each compound we can seen in figure 2,3 and 4.

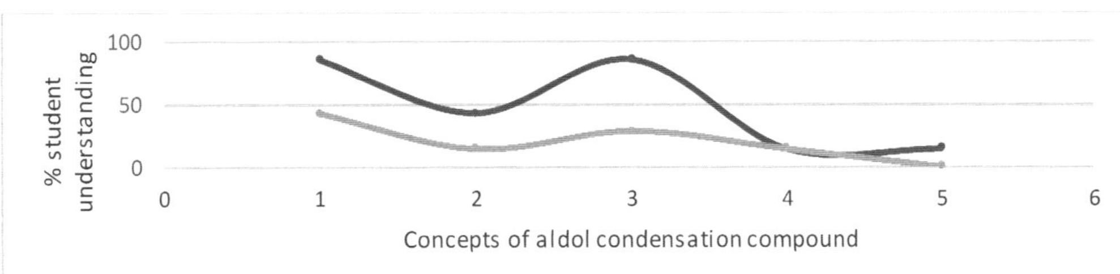

Figure 2. Graphic integration level of student understanding of prerequisite concept with the use of the concept to solve the problem of synthesis of aldol condensation compounds.

Based on figure 2, it can be seen a relationship between students understanding of prerequisites concepts with the percentage of the use of prerequisite concepts on the completion of organic compounds. In aldol condensation synthesis was found that the students percentage understanding on the concept of the type of reaction and the starting material is equal to 85.72%, on the concept of compound structures 42, 88%, and on concepts of the reaction product and electrophiles in the same formation on acquiring C-C percentage of 14.28%.

In the synthesis of benzene acylated compound found the students understanding is found on the concept of type and function of reagents that have the same percentage of 71.43%, on the concept of nucleophilic substitution reaction reactions and carbocation ions are 14, 28% and 7.14%.

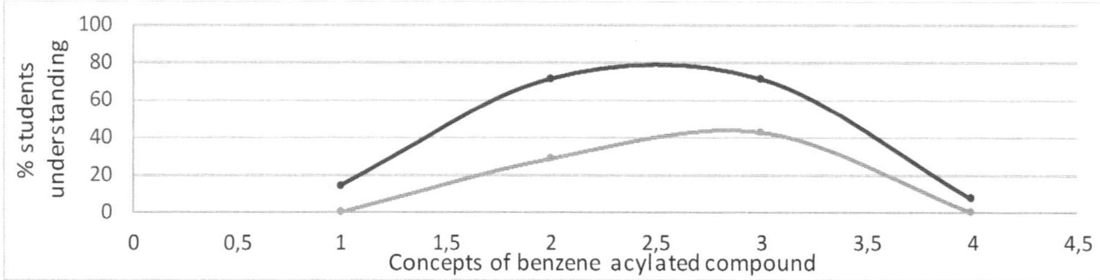

Figure 3. Graphic integration level of student understanding of the prerequisite concept with the use of the concept of prepay to solve the problem of synthesis of benzene acylated compound.

From figure 3, it can be seen that the absence of the meeting point is indicated if the level of understanding of the concept of student pre-requisite does not have integrity with the way of completion on this compound. These results suggest that there is no correlation between student understanding of prerequisite concepts with the percentage of the use of prerequisite concepts on the completion of organic compounds.

In synthesis of tertiary amine benzene compound, there was a sufficient level of understanding of the concept of student prerequisite on the concept of compound structure, reduction reaction and compound properties of 50.00%, 57.14% and 47.62%, respectively. The concept of prerequisite of the steering group and the disconnection of each compound 28.57% and 14.28%.

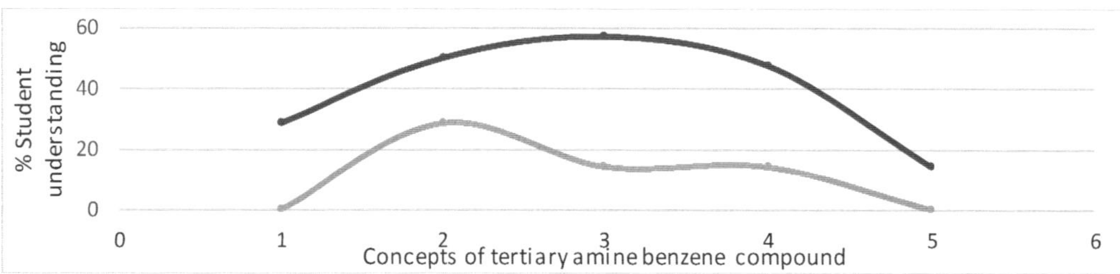

Figure 4. Integration level of students understanding of the prerequisite concept with the use of the concept of prepay to solve the problem of synthesis of tertiary amine benzene compounds.

Figure 4 suggest that there is no correlation between understanding the concepts of student prerequisites with the percentage of the use of prerequisite concepts on the completion of organic compounds. The findings obtained show that the students understanding of concepts prerequisite kind of reaction, the starting material the same percentage of 85.72%. Meanwhile, the prerequisite concepts of carbocation ions 7,14%, reaction result, electrophile forming of C-C, and disconnection of compound with equal percentage equal to 14,28%.

According to previous research shows that the question of some kind of prerequisite concept of reaction type and structure of the compound is a question which is only a repetition [15]. It is known that mostly students only give answers related to cognitive aspects of knowledge or rote just like the concept of the type of reaction, the starting materials and reagents. The student's tendency to memorize a set of rules and previous reactions in organic chemistry has been found to inhibit student learning [16], [17]. There is a lack of understanding of concepts from the majority of students so that they can not apply organic chemical concepts to solve problems related to synthesis organic compounds. In accordance with the findings on the synthesis of aldol condensation compounds students are only able to answer questions related to alcohol reactions. It is also possible to relate to students' inadequacies, leaving them with no other way to solve the problem. It is seen in the synthesis of benzene acylated compounds found that students cannot connect between the concept of reagents used and the reaction of benzene compounds. Similarly in synthesis of tertiary benzene amine compounds, students were found unable to relate the concept of benzene compound structure with benzene reactions.

But if the views of its specific integrity, it is found an integration between understanding the concept of student prerequisites and how students apply it to the synthesis of aldol compounds. But it is different in the synthesis of benzene acylated compounds and tertiary amine benzene compounds that show no connection. This suggests a gap between the knowledge possessed by students and the skills to achieve synthesis and retrosynthetic analysis. Students make a study based on authentic and tangible problems that are similar to the complex problems facing scientists that have an impact on improving the ability of students to think critically and learn independently and gain learning experience. This is a process of assimilation that must be passed by students to be transformed into practitioners or scientists in the field of organic chemistry [18]. Meaningful learning can be done through the implementation of the organic compound synthesis experiment in laboratory. It is known that lab work performed by the student in the laboratory can help improve the affective, psychomotor and cognitive can even evaluate students on these aspects [19–21]. In fact, through the implementation of experimental synthesis of organic compounds in the lab[22], the students involved in an authentic problems within a specific domain and not just laboratory work in the laboratory that verification [23].

4. Conclusion

Inclusive, there is a relationship between students understanding of prerequisites concepts with the completion of organic synthesis. When it is viewed, every synthesis of organic compounds found that in the synthesis aldol condensation students can apply the concepts of preference to the organic synthesis of the synthesis but different from the synthesis of terylated benzene compounds. In Tertiary amina benzene, it is found that students can not combine the concept of prerequisite with the completion of synthesis of organic compounds. It can be suggestion to Organic Chemistry lecture, for more comprehensive teach organic prerequisite concept that corelated with synthesis organic compound. Especially when they teach about benzen compound. For more futher research, the learning process about this subject matter must done in inquiry and project based learning by using laboratory.

Acknowledgement
We give gratitude to KEMENRISTEK-DIKTI that funded the BPPDN scholarship for author and author wish to thank organic lectures team from one state University in Papua for their help to collected data from the students.

References

[1] Flynn, Allison B 2014 *Chem. Educ. Res. Pract.* **15** pp 747-762
[2] Bhattacarya, G 2013 *J. Chem. Educ.* **90** pp 1282-1289
[3] Bodner, GM & Domin, DS 2000, *Univ. Chem. Educ.* **4 (1)** pp 24-30.
[4] Strickland AM, Kraft & Bhattacharyya G 2010 *Chem. Educ. Res. Pract.* **11 (4)** pp 293-301
[5] Anderson, JP 2009 Learning the language of organic chemistry: How do students develop reaction mechanism, problem-solving skills? Doctoral dissertation, Retrieved from *ProQuest*, UMI Dissertations (Purdue : Purdue University)
[6] Gunes, P, Katircioglu, H, & Yulmaz, M 2015 *J. Turkish Sci. Educ.* **12** pp 71-83
[7] Hakim, A; Liliasari; Kadarohman, A 2016 *J. Turkish Sci. Educ.* **13 (2)** pp 27-36
[8] Aydin, A & Altuk, Y G 2013 *Int. Educ. Stud.*, **6 (5)** pp 104–116
[9] Maeyer, J & Talanquer, V 2010 *Sci. Educ.* **94 (6)** pp 963-984
[10] McClary, L &Talanquer, V 2011 *Int. J. Sci.Educ* **33 (10)** pp 1433-54
[11] Harman, G 2014 *J. Turkish Sci. Educ.* **11 (4)** pp 81-106
[12] Ozdemir, Ő 2015 *J. Turkish Sci. Educ.* **12 (1)** pp 29-42
[13] Heiser, J & Tversky B 2006 *Cognitive Sci.* **30 (3)** pp 581-92
[14] Spearman, C 2009 "Spearman ' s rank correlation coefficient," in *Royal geographical society*, **15** pp. 72–101
[15] Raker, J R & Holme, T A 2014 *J. Chem. Educ.* **90** pp 1437-1442
[16] Anderson, T L & Bodner, G M 2008 *Chem. Educ. Res. Pract.* **9 (2)** pp 93-101
[17] Grove, N P & Bretz, S L 2011, *Chem. Educ. Res. Pract.* **13 (3)** pp 201-208
[18] Flynn, Allison B & Biggs, Robbyn 2011 *J. Chem. Educ.* **89** pp 52-57

[19] Bretz, L S, Fay M, Bruck, Laura B, & Towns, Marry H 2013 *J. Chem. Educ.*, **90** pp 281-288
[20] Galloway, Kelli R & Bretz, S L 2015 *Chem.Educ. Res. Pract.* **16** pp 705-706
[21] Cịbik, A., S & Yalcịn, N 2013 *J. Turkish Sci. Educ.* **10** pp 108-135
[22] Bhattacharyya, G & Bodner, G 2014 *J. Res. Sci.Teach.* **51 (6)** pp 694-713
[23] Hakim, A; Liliasari, Kadarohman, A & Syah, Y M 2016 *J. Chem. Educ.* **93 (1)** pp 193-196

IOP Conf. Series: Journal of Physics: Conf. Series **895** (2017) 012116 doi:10.1088/1742-6596/895/1/012116

The Binary System Laboratory Activities Based on Students Mental Model

A Albaiti [1,2,*], S Liliasari [1], O Sumarna [1] and M A Martoprawiro [3]

[1] Universitas Pendidikan Indonesia, Jl. Dr. Setiabudi No. 229, Bandung 40154, Indonesia
[2] Universitas Cenderawasih, Kampus Uncen Waena Jayapura, Indonesia
[3] Institut Teknologi Bandung, Jl. Ganesha No.10, Bandung 40132, Indonesia

*albaiti@unicen.ac.id

Abstract. Generic science skills (GSS) are required to develop student conception in learning binary system. The aim of this research was to know the improvement of students GSS through the binary system labotoratory activities based on their mental model using hypothetical-deductive learning cycle. It was a mixed methods embedded experimental model research design. This research involved 15 students of a university in Papua, Indonesia. Essay test of 7 items was used to analyze the improvement of students GSS. Each items was designed to interconnect macroscopic, sub-microscopic and symbolic levels. Students worksheet was used to explore students mental model during investigation in laboratory. The increase of students GSS could be seen in their N-Gain of each GSS indicators. The results were then analyzed descriptively. Students mental model and GSS have been improved from this study. They were interconnect macroscopic and symbolic levels to explain binary systems phenomena. Furthermore, they reconstructed their mental model with interconnecting the three levels of representation in Physical Chemistry. It necessary to integrate the Physical Chemistry Laboratory into a Physical Chemistry course for effectiveness and efficiency.

1. Introduction

Physical chemistry is a branch of chemistry that is studied in college. Physical chemistry can be interpreted as a study based on the principles of physics and mathematical language that determines the properties and behavior of chemical systems. It can be studied at the level of macroscopic, sub-microscopic and symbolic. Students when studying physical chemistry that refers to the phenomena observed (macroscopic), then they can explain the phenomenon through sub-microscopic explanation. It explains what was observed at the macroscopic level, using things like the movement of electrons, molecules, atoms, or ions in sub-microscopic ways. Explanation at sub-microscopic level strongly supports by symbolic explanation that can be derived as formula or equation of phenomena observed. Students need to interconnect the phenomenon of macroscopic, sub-microscopic and symbolic in studying Physical Chemistry. It is intended that the students have a holistic understanding of the chemical [1]. The ability of students to interconnect those levels reflects their mental models [2]. Conception is often viewed as a mental model therefore it also describes someone conception [3]. Researches on the students conception or comprehension in studying Physical Chemistry describes their mental models are considerably varied. Several studies identified that students had alternative conceptions, misunderstandings and misconceptions about some topics of Physical Chemistry [3-14]. Binary system is part of phase equilibrium that studied in Physical Chemistry. The results of research concerning student mental model about phase equilibrium showed that most of the students had three mental models of vapor pressure. It showed an alternative conceptualization of the entity or the properties of vapor pressure and connectedness in the system [12]. There was misunderstanding of the concept of evaporation, melting, the effect of temperature and pressure on the transitions phase. Although students understanding were inadequate [5].

The analysis of research result about binary system showed that all researches were focused on the study in the laboratory [15-16]. These researches focused on investigating the behavior of binary systems with emphasis on the symbolic level of the data obtained in the laboratory. In thermodynamics, closed system is an equilibrium condition if it fulfills the criteria of thermal, mechanical, and material equilibrium. We suppose the closed system which is in thermal and mechanical equilibrium in *P-V* work only has not yet reached phase equilibrium. A two-phase system is exemplified α phase and β phase. In certain conditions, if $T^\alpha > T^\beta$, heat flows spontaneously from α phase to the β phase until the temperature in both phase is the same ($T^\alpha = T^\beta$). If $P^\alpha > P^\beta$, "work" flows spontaneously from α phase to the β phase until $P^\alpha = P^\beta$. Similarly, if $\mu_i^\alpha > \mu_i^\beta$ then *i* component in the system flows spontaneously from α phase to the β phase until $\mu_i^\alpha = \mu_i^\beta$ [17]. Two component systems that may exhibit two liquid phases are often discussed in relation to Gibbs phase rule [18]. Students regularly have difficulty with interpretation of a *T-X* diagram and the implication of the Gibbs phase rule. At a temperature and bulk composition within the immiscible region, coexisting liquids, Liq_A and Liq_B, are connected by tie line at their compositions (x_1 and x_2 respectively). The Lever rule is used to determine their relative amounts (because the Gibbs phase rule says nothing about quantities), nothing that length of Lever arm inversely correlates with amount [19]:

$$\frac{Liq_A}{Liq_B} = \frac{x_2 - x_0}{x_0 - x_1}$$

Binary system practices based on students mental model using hypothetical-deductive learning cycle is one of the efforts that have been made to learn them on this matter by interconnect three levels of representation. Students should be examined their hypothesis against their scientific concepts and determine the appropriate results of the investigation in the laboratory through hypothetical-deductive reasoning and experimentation. This approach required that students had to meet their intellectual needs in improving understanding of the concept and improve thinking skills [20]. Based on this, students activities in laboratory facilititate the increasing of GSS. GSS are important for learners to develop skills in their discipline area. Some GSS indicators are logical self-consistency, concept formation, direct observation, mathematic modeling, and symbolic language [21]. Generic skills are "employability" skills used in the application of knowledge. The degree to which students develop these skills determine solving problems, writing reports, functioning in teams, self-assessing and performing reviews of others, learning new knowledge, and managing stress in confronted with change [22]. Overall, the purpose this study was to determine the improvement of students GSS through the binary system labotoratory activities based on their mental model using hypothetical-deductive learning cycle. However, research on how to learn and improve GSS of students by interconnect the three levels of representation in the binary system has never been done previously.

2. Methods

This study was a mixed methods embedded experimental model research design [23]. This research involved 15 students in chemistry education department of a university in Papua, Indonesia. They were divided into six groups (2-3 students per group) based on similar students initial mental models. The students initial mental model was seen from the individual hypothesis formulation on their worksheets. Two students groups worked on the same binary system. Binary system was used in this study were acetic ethyl and water, 1-butanol and water, also cyclohexane and water systems. Those students designed laboratory activities collectively in their own group and consulted it to their lecturer. Lecturer observed students activities during investigation in laboratory and providing guidance if necessary. Students made discussion about anything they found in the investigation. Lecturer facilitated the class discussion at the next meeting. During the class discussion, students of other group were given the opportunity to give their response. Otherwhile the stages of binary system practices based on mental model to improve students GSS was presented in Table 1.

Table 1. Stages of Binary System Practices Based on Students Mental Model

Stages of learning	Students Activity
Engagement phase	Students were given binary system phenomena to create cognitive conflict and motivated them to conduct practical activities.
Exploration phase	Students developed hyphotheses (if.....then......) based on the issues contained in the passages individually and evaluated each hyphoteses, selecting hyphoteses to be tested.
Explanation phase	Students designed a binary system laboratory activities to test the hyphoteses that had been formulated. Having decided upon the experimental method to be used. They made observations and measurements to collect data.
Elaboration phase	Students made discussion about anything they found out in the investigation. Analysis of the laboratory results allows for hyphoteses was rejected or retained. Lecturer helped in facilitating class discussion in the next meeting. During the in-class discussion, students of group should be responded.

Written test which used to analysis the improving of students GSS was essay form. It was consists of 7 items. Each questions and answers was designed to interconnect macroscopic, sub-microscopic and symbolic levels. The results showed the validity of question number 2 and 6 were revised while five others were valid. The reliability of this test was 0.78 and was considered high [24]. The Increase of students GSS could be seen from their N-Gain. The N-Gain value formula as $\langle g \rangle$= (post-test –pre-test)/(max –pre-test) x 100% [25]. N-Gain value adapted in percentages [26]. The results were then analyzed descriptively. The GSS indicators which were enhanced in each stage of learning in the binary system were presented in Table 2.

Table 2. GSS Indicators on Each Stage of Learning

Stages of Learning	GSS Indicators
Engagement and exploration phase	Logical self-consistency
Explanation phase	Concept formation Indirect observation
Elaboration phase	Mathematical modelling Symbolic language Logical self-consistency

3. Results and Discussion

3.1. Results

Students were grouped by similarity of their initial mental model based on their hypothesis formulation in their worksheets. Each group designed laboratory activities of binary systems by themselves. They did investigation in the laboratory to prove their hypothesis created. All students of six groups managed to obtain data on each model of liquid-liquid equilibrium of two components, i.e.

the liquid-liquid equilibrium system of acetic ethyl and water, 1-butanol and water, and cyclohexane and water systems. Students skills in deciding temperature system from cloudy mixture turned clean through indirect observation influenced the accuracy of the data. These skills can only be trained to students through laboratory activities (hands on). Students mental model was improved from this study. Example reconstruction of student initial mental model through this model was presented in Table 3.

Tabel 3. Reconstruction of Student Initial Mental Model

Student Initial Mental Model		Reconstruction of Student Initial Mental Model	
Macroscopic Level	Symbolic Level	Interconnection of macroscopic and sub-microscopic Levels	Symbolic Level
Adding water into 1-butanol will yield two layers.	It will form asymetric curve : 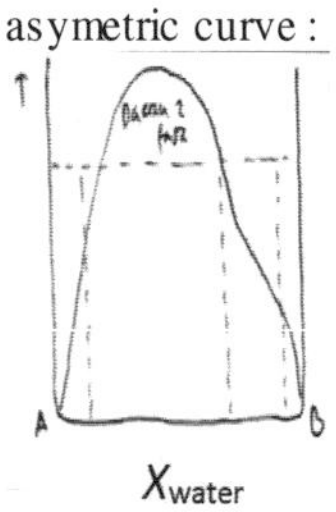	Adding 3.2 ml water into 5 ml 1-butanol yield two layers. Water soluble in the 1-butanol at top layer. There is interaction between 1-butanol molecule and water molecule and interaction between water molecules through hydrogen bonding in top layer. 1-butanol soluble in the water at bottom layer. There is interaction between 1-butanol molecule and water molecule and interaction between 1-butanol molecules through hydrogen bonding in bottom layer. 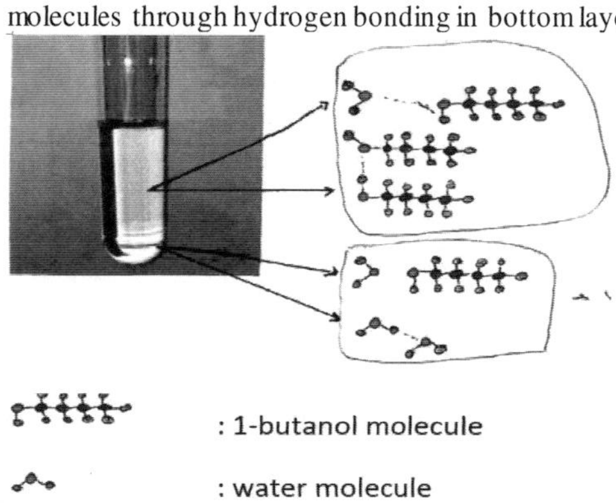	Example, top layer is symbolized as α and bottom layer as β, so that $\mu^{\alpha}_{water} = \mu^{\beta}_{water}$ and $\mu^{\alpha}_{1-butanol} = \mu^{\beta}_{1-butanol}$

Based on Table 3, at first, student only interconnected macroscopic and symbolic levels to explain binary systems phenomena. Furthermore, student reconstructed their mental model with interconnecting the three levels of representation. The results of the increasing students GSS was presented in Table 4.

Table 4. The Increasing of students GSS in binary system

No.	GSS Indicators	Item	N-Gain, %
1.	Logical self-consistency	1	44.70
		5	12.89
		6	43.89
2.	Concept formation	2	34.13
3.	Indirect observation	3	47.67
4.	Mathematic modelling	4	56.84
5.	Symbolic language	7	38.89

According to Table 4, all students GSS in binary system was improved. Logical self-consistency indicator that increased in this research was to find out logical relation between two rules or more. After the implementation of this learning model, there were an increase of the N-gain for the logical self-consistency of 44.70% for item 1 (moderate category), 12.89% for item 5 (low category), and 43.89% for item 6 (moderate category). Concept formation develops further the idea of an object or process in order to understand a phenomenon that cannot be done with daily conversational language. The results showed an increase of 34.13% (moderate category). Indirect observation in this study was enabling the students to determine the temperature system by using thermometer (item 3). Students also trained to be able to explain at the sub-microscopic level about the increasing of temperature system until cloudy mixture dissepeare. The N-gain increase after the implementation of this learning model was of 47.67% (moderate category). Mathematic modeling is a formula that describes the law for the natural phenomena using mathematic language. Being able to formulate with this language is an application of mathematical modeling. The N-gain increase for the mathematical modeling indicator was of 56.84% (moderate category). Symbolic language is the language used to talk about natural phenomenon that cannot be expressed with plain conversational language. The indicator of the symbolic language that increased in this study was when determining the upper absolute temperature of n-hexane and water system based on the curve gained (item 7). The rise of the N-gain of the symbolic language was of 38.89% (moderate category).

3.2. Discussion

A number of approaches to generic skill development have been formulated and proven to be effective in science and engineering education [27]. Similarly, the result has shown that there was an increase in students GSS for five indicators which focused of this research. The indicators mentioned are logical self-consistency, concept formation, indirect observation, mathematical modeling, and symbolic language. Specifically, the highest increase of students GSS occurred on the mathematic modeling with the result of 56.84% (moderate category). While, the lowest one was 12.89% (low category) occurring on the logically self-consistency. These indicate that students mental model based learning of binary system via hypothetical-deductive learning cycle can enhance students GSS. Such learning accommodates students in conducting experiments at laboratory by interconnecting the levels of macroscopic, sub-microscopic, and symbolic for their meaningful knowledge and skill gains. The understanding chemistry on the macroscopic, sub-microscopic, and symbolic levels is necessary for the development of students skills like the capacity for abstraction, logical reasoning, and the understanding of natural phenomena and themselves in the world [28]. The researches show the importance of discussing this topic and for researchers to help teachers improve their class work.

In addition, most of students prefer conducting experiments because they can obtain valuable skills such as a skill to solve contrary data. Such laboratory activities can improve the mental process of learners dealing with science. They can examine their assumptions towards the scientific concepts and identify them based on findings of the research via deductive hypothesis considerations and experiments [29]. The primary emphasis of laboratories should not be limited to learning certain scientific methods or laboratory techniques, but rather laboratories should allow students to investigate phenomena by using the methods and procedures of science thereby enabling them to solve problems [30]. Learners can learn to think better if they are learning by doing. In this respect, the majority of science programs assume that a laboratory study is a stepping stone due to students' active learning engagement. Moreover, this laboratory activity can be categorized into the student-learning centered strategy [29]. The development of any skills is best facilitated by giving students practice and not by simply talking about or demonstrating what to do [27].

This study indicates that the enhancement of students GSS on logically self-consistency indicator was the lowest. In this respect, students understanding about Lever rule to determine each component into two phases and their understanding about intermolecular forces influenced their generic science skills achievement. Students have alternative conceptions about intermolecular forces [11]. So, this concept was a difficult concept. It is abstract concept. Besides that, students must become able to not only interpret symbolic expressions and perform mathematical manipulations but they must also develop an understanding of the physical significance of the symbolic representations. As instructors of chemistry, we hope that students become able to interpret and work with mathematical expressions such as this with understanding, not just routinely apply algorithms and formulae. However, instructors should be aware of how these symbolic forms contribute to student reasoning in various contexts if they are to meaningfully facilitate student's reasoning with such inscriptions [6].

4. Conclusion

Students mental model and generic science skills were improved from this study. They were interconnect macroscopic and symbolic levels to explain binary systems phenomena. Furthermore, they reconstructed their mental model with interconnecting the three levels of representation in Physical Chemistry.

Acknowledgments

The researchers show gratitude to the chairperson of Chemistry Education Department, the head and assistants of Chemistry Laboratory, Universitas Cenderawasih in Jayapura Papua for their facility supports towards the process of this study. The appreciation is also forwarded to the students who were willing to be the subjects of this research.

References

[1] Albaiti Liliasari and Sumarna, O. 2016 *Indonesian Journal of Science Education* **5 (1)** p 6.
[2] Chittleborough G D 2004 *Ph. D. Thesis* (Perth: Curtin University of Technology).
[3] Nilsson T and Niedderer H 2014 *Chem. Educ. Res. Pract.* **15** pp 336-353.
[4] Smith K C and Nakhleh M B 2011 *Chem. Educ. Res. Pract.* **12** pp 398-408.
[5] Yalcin F A 2012 *Chem. Educ. Res. Pract.* **13** pp 369-377.
[6] Becker N and Towns M 2012 *Chem. Educ. Res.Pract.* **13** pp 209-220.
[7] Turanyi T and Toth Z 2013 *Chem. Educ. Res. Pract.* **14** pp 105-116.
[8] Sreenivasulu B and Subramaniam R 2013 *Chem. Educ. Res.Pract.* **35(4)** pp 601-635.
[9] Cheng M M W and Gilbert J K 2014 *Int. J. Sci. Educ.* **36 (8)** pp 1373-1407.
[10] Hernandez G E Criswell B A Kirk N J Sauder D G and Rushton G T 2014 *Chem. Educ. Res. Pract.* **15** pp 354-365.
[11] Sendur G 2014 *Chem. Educ. Res.Pract.* **15** pp 689-719.
[12] Tumay H 2014 *Chem. Educ. Res.Pract.* **15** pp 366-379.
[13] Stains M and Sevian H 2014 *Res. Sci. Educ.* **45(6)** pp 807-840.
[14] Haglund J Andersson S and Elmgren M 2015 *Chem. Educ. Res. Pract.* **16** pp 537-551.
[15] Dai M and Chao J P 1985 *Fluid Phase Equilib.* **23** pp 315-319.
[16] Alessi P Fermeglia M and Kikic E 1989 *J. Chem. Eng. Data* **34 (2)** pp 236-240.
[17] Levine I R 2009 *Physical Chemistry* (New York : McGraw-Hill).
[18] Logam S R 19989 *J. Chem. Educ.* **7(3)** pp 339-342.
[19] Erikson J P 2016 *J. Chem. Educ.* **94(1)** pp 75-78.
[20] Lawson A E 1988 Three Types of Learning Cycles: A Better Way To Teach Science. Paper Presented at The Annual Convention of The National Association for Research in Science Teaching, Lake Ozark. MO.
[21] Liliasari 2007 *Proceeding of The First International Seminar on Science Education* (Bandung: Indonesia University of Education).
[22] Kamsah M Z 2004 *Developing generic skills in classroom environment: Engineering Student's Perspective*.
[23] Creswell JW and Clark VLP 2007 *Designing and conducting mixed methods research* (California: Sage Publication).
[24] Guillford J P 1986 *Fundamental statistics in physychology and education* (New York: McGraw-Hill Book co. inc).
[25] Hake R R 1998 *American J.Phys.* **66(1)** pp 64-74.
[26] Fynewever H 2008 *Chem. Educator.* **13** pp 264-269.

[27] Woods D R Hrymak A N Marshall R R Wood PE Crowe C M Hoffman T W Wright J D Taylor P A Woodhouse K A and Bouchard C G K 1997 *J. Eng. Educ.* **86 (2)** pp 75-91.
[28] Santos V C and Arroio A 2016 *J. Turkish Sci. Educ.* **13 (1)** pp 3-18.
[29] Marasigan A C and Espinosa A A 2013 *Journal of Education Research and Behavioral Sciences* **2 (11)** pp 206-228.
[30] Hofstein A and Lunetta V N 2004. *Sci. Educ.* **88(1)** pp 28-54.

IOP Conf. Series: Journal of Physics: Conf. Series **895** (2017) 012117 doi:10.1088/1742-6596/895/1/012117

Inquiry-Based Laboratory Activities on Drugs Analysis for High School Chemistry Learning

I Rahmawati[1*], H Sholichin[2], and M Arifin[2]

[1] Program Studi Pendidikan Kimia, Sekolah Pascasarjana, Universitas Pendidikan Indonesia, Jl. Dr. Setiabudi No. 229, Bandung 40154, Indonesia
[2] Departemen Pendidikan Kimia, Universitas Pendidikan Indonesia, Jl. Dr. Setiabudi No. 229, Bandung 40154, Indonesia

[*]irahmairma@gmail.com

Abstract. Laboratory activity is an important part of chemistry learning, but cookbook instructions is still commonly used. However, the activity with that way do not improve students thinking skill, especially students creativity. This study aims to improve high school students creativity through inquiry-based laboratory on drugs analysis activity. Acid-base titration is used to be method for drugs analysis involving a color changing indicator. The following tools were used to assess the activity achievement: creative thinking test on acid base titration, creative attitude and action observation sheets, questionnaire of inquiry-based lab activities, and interviews. The results showed that the inquiry-based laboratory activity improving students creative thinking, creative attitude and creative action. The students reacted positively to this teaching strategy as demonstrated by results from questionnaire responses and interviews. This result is expected to help teachers to overcome the shortcomings in other laboratory learning.

1. Introduction

Chemistry lab activities in high school, especially in acid-base titration is commonly used cookbook procedure with not real problems. Experimental activities only use materials in chemical laboratories. Whereas, many daily use materials and chemical phenomena can be used [1]. The cookbook practice will only encourage student passivity (creeping passivity) so that the level of student involvement is low [2]. Learning chemistry with conventional lab should be converted into activity that can make students actively seek their own knowledge required for competence development. In accordance with the rationalization of the development curriculum in Indonesia, namely the improvement of passive learning patterns into active-discovery learning [3].

With the result that students are not given the opportunity to have high creativity to solve various problems. Even though this concept requires the ability to think creatively in solving the problem. This will make the student a difficulty on the concept of titration, because this concept emphasizes instruction on numerical problem solving [4], and requires creative thinking skills [5]. Creativity is a high-level thinking ability that is very important for high school students. Because without creativity, students will have difficulty in solving problems and linking principles between productive and chemistry as adaptive subjects. With creativity, students will demonstrate the skills and dispositions which required to generate new ideas according to the problems faced by [6]. In addition, Paradigm of Education section of XXI Century [7], explained that one of the competencies or skills that must be owned by 21st century students is creativity and innovation skills. Where students are able to develop their own creativity to produce innovative breakthroughs.

The learning model that is very suitable for enhanching creativity high school students is Inquiry-Based Laboratory (ILAB), in accordance with the results of previous study by Feldhusen and Treffinger [8], Dehaan [9], Yakar and Baykara [10], Amalina [11], and Sari [12]. However, study on acid-base titration using inquiry-based laboratory has not much discussed in the topic of drugs analysis. Whereas it is important to investigate, this is because the chemical properties of drugs that will not be separated from the concept of acid and base. In this study we aim to investigate whether using the inquiry-based laboratory can improve high school students' creativity in acid based titration of drugs analysis.

1.1. Inquiry-based Laboratory and Creativity

The learning step with Inquiry-Based Laboratory (ILAB) according to Hofstein and Walbreg [13]: (1) problem identification; (2) formulating hypotheses; (3) designing experiments; (4) collecting data; (5) data analysis; and (6) make a conclusion. Creative thinking is measured following the creative thinking component of Guilford and Torrance [14], theres fluency (generating many ideas), flexibility (shifting perspective easily), originality (conceiving of something new), and elaboration (building on other ideas). While the components for creative attitude that is curiosity, imaginative, challenged by plurality, dare to take risks, and appreciative [15].

1.2. Quantitative Analysis of drugs with acid-base titration

Acid-base titration can be taught by using different learning plans. Thus, to work with ILAB, each group of students will have a real problem of drug analysis. They should get an initial volume of acid/base to be titrated and determine the percentage content of the drugs, wherein each compound are classified as acidic, basic, neutral or zwitterion [16]. An acid and base drugs can be used to determine the levels of the active substance by acid-base titration method. Because the basic principle of acid and base is the neutralizing reaction, between H^+ and OH^- ion, as shown in Table 1 [17].

Table 1. Chemical Reactions of the Acid–Base Titration.

Titration Type	Before the Equivalent Point	Equivalent Point
Strong acid – strong base	$HCl + H_2O \rightarrow H_3O^+ + Cl^-$	$HCl + NaOH \rightarrow H_2O + NaCl$
Weak acid – strong base	$HOAc + H_2O \rightarrow OAc^- + H_3O^+$	$OAc^- + H_2O \rightarrow HOAc + OH^-$
Strong acid – weak base	$B + H_2O \rightarrow BH^+ + OH^-$	$BH^+ + H_2O \rightarrow B + H_3O^+$

2. Experimental Method

In this study, we adopt mixed-methods by combining descriptive and experimental research with Pretest-Postest, Nonequivalent Control Group Design. In the experimental class, the inquiry-based laboratory was used in chemistry experiment teaching. On the other hand, the students in the control class received the conventional teaching. The design of the study can be diagrammed as follows:

$$G_1 \qquad O_1 \text{------} X_1 \text{------} O_2$$
$$G_2 \qquad O_1 \text{------} X_2 \text{------} O_2$$

with G_1 eksperiment class; G_2 control class; O_1 pretest; O_2 posttest; X_1 inquiry-based laboratory; and X_2 conventional lab activity.

The subjects sampled in this study were 73 students at XI grade from pharmaceutical high school in Sumedang Indonesia. There were 34 students in the experimental class and 39 students in the control class, taken through purposive sampling technique. This study was conducted during four meetings on the acid-base titration concept, including pretest, implementation of inquiry-based laboratory, postest, answer questionnaires and interviewed.

The following tools were used to assess the students'creativity: (1) creative thinking test on acid base titration, an essay test are used to collect information on students' creative thinking skills before

IOP Conf. Series: Journal of Physics: Conf. Series **895** (2017) 012117 doi:10.1088/1742-6596/895/1/012117

and after the implementation Inquiry-Based Laboratory models; (2) creative attitude and action observation sheets; and (3) questionnaire and interviews of inquiry-based lab activities. Instruments used have been through expert validation and revision process to produce the right instruments to measure the research problem. Content validity ratio (CVR) of creative thinking test, showing that instruments can be used (CVR=1).

3. Result and Discussion

3.1. Inquiry-based Laboratory Design

Chemistry learning with ILAB requires a worksheet that will guide students in activities. Real problem that must be solved is to determine the percentage of drugs that are easy to find in drugstores around students. Tabel 2 shows the design of ILAB worksheet for drugs analysis.

Table 2. Design of Inquiry-based Laboratory Worksheet.

ILAB Activity	Task	Activity	Purpose
1. Problem identification	Acid-base on daily lives	Students are faced with a real-life problem, in this activity the problem will be presented is gastritic acid in ulcer.	Elicit ideas about pH, concentration, strength of acid-base, and how to overcome the pain with acid-base aplications. And to determine students motivation.
2. Formulating hypotheses	Neutralization of acid-base solutions	Students were were asked to explain their prediction about: Case#1: When you mix a strongly acidic with weakly acidic solution, does the pH of the two solutions after the mixing change? Case#2: Still with the same solution, but we will mix it with different volumes. Which component will cause the pH change to be more neutral? Case#3: How many tablets of drugs are needed to neutralize stomach acid?	To determine which basic principles students will use. To show how the various concepts are connected, and as a basis for further questions. And to determine students creative thinking skills.
3. Designing experiments	a. Acid-base titration b. Indicators on acid-base titration	Students design an experiments procedure by determining control variables, independent variables and their own dependent variables	Elicit ideas about indicators of acid and base titration, tools, materials, procedure and drugs sampel to be determined. And to determine students elaboration and creative thinking skills.
4. Collecting data	Acid-base titration	Students collect data from their design experiment, as the activity of capturing the information needed to test the hypothesis.	Elicit ideas about titration type, sampel preparation, volume, concentration, changing colors, end point, and equivalent point
5. Data analysis	a. Stoichiometric reaction b. Accuracy of titration results	Students organize and analyze the data, connect to the hypothesis, predict, select findings are consistent with what is already known.	Elicit ideas about stoichiometric calculation in titration, concentration of solution, percentage active component of drugs and to determine students elaboration, and their creative thinking skills.
6. Conclusion	Substitute for chemical medicine	Students describe and interpret the findings obtained based on the results of hypothesis testing.	To determine students elaboration, and their creative thinking skills

In ILAB activity, students are required to be able to create something new, prepare and then assemble it in accordance with self-designed procedures. Besides that students are expected to make the procedure different from others. The results of the student experiment design are presented in the Table 3.

Table 3. Students experimental design of acid-base titration.

Group	Experimental Purposes	Active component	Sample
1	Determine content of the active component in salicyl powder	salicylic acid	salicyl powder
2	Determine content of the active component in vitamin C	ascorbic acid	vitamin C
3	Determine content of the active component in aspirin	asetil salicylic acid	aspirin
4	Determine content of the active component in ibuprofen tablet	ibuprofen	ibuprofen tablet
5	Determine content of the active component in naspro tablet	asetil salicylic acid	naspro tablet
6	Determine content of the active component in bodrex	asetil salicylic acid	bodrex

3.2. Students Creativity in Drugs Analysis

Improvement of students creative thinking obtained from the results of pretest and posttest data before and after inquiry-based laboratory implementation. Pretest and posttest results data then processed statistically by calculating the average value of the gain is normalized. Recapitulation scores creative thinking indicated on Table 4.

Table 4. Students creative thinking score in elaboration component.

No	Indicator	Control Class		Gain	N-Gain	Category	Experiment Class		Gain	N-Gain	Category
		Pretest	Posttest				Pretest	Posttest			
1	Neutralization of acid-base solutions	2,03	2,21	0,18	0,09	Low	2,03	3,53	1,50	0,76	High
2	Application of acid base titration	2,95	3,38	0,44	0,41	Moderate	2,79	3,06	0,26	0,22	Low
3	Indicators acid base titration	1,74	2,05	0,31	0,09	Low	2,09	2,97	0,88	0,30	Moderate
4a	Designing experiment	1,64	1,90	0,26	0,19	Low	1,09	2,38	1,29	0,68	Moderate
4b	Accuracy of titration results	1,05	1,51	0,46	0,24	Low	0,97	2,09	1,12	0,55	Moderate
5	Substitute for chemical medicine	3,64	4,38	0,74	0,55	Moderate	3,53	4,35	0,82	0,56	Moderate
	Results				0,26	Low				0,51	Moderate

Table 4 shows the result of students' creative thinking in generally improved. But, experimental class (0,51) has N-gain higher than control class (0,26). This proves that learning with inquiry based laboratory can improve students' creative thinking skill. Creative thinking ability of experimental and control class produces very different results on question number 1, about neutralization of acid-base solutions. Students in the control class did not correct the previous error on pretest. This is seen from the control class students who wrote the same answer with the pretest as much as 56.76% and the students whose score decreased by 29.73%. This causes the N-gain value to be only 0.09.

In question number 1, student are given predictive questions about two different cases of acid base neutralization, $CH_3COOH(aq)$ and $NaOH(aq)$ with same concentration but different volume ratio. Student should have predicting which case will be close to neutral solution. In cases A, they are adding 8mL $CH_3COOH(aq)$ with 2mL $NaOH(aq)$, and case B shows that they are adding 6mL

$CH_3COOH(aq)$ with 4mL $NaOH(aq)$. Student in experiment class have been able to correcting prediction errors (91,18%) in the case after learning through inquiry-based laboratory activities. The situation is the same as the previous study [18], which states that inquiry based laboratories may improve assessment outcomes.

Table 5. Students creative attitude at the learning phase of Inquiry-Based Laboratory.

	Inquiry-Based Laboratory Activity	Mean	Percentage	Category
1	Problem identification	1,417	70,83	Good
2	Formulating hypotheses	1,472	73,61	Good
3	Designing experiments	1,250	62,50	Good
4	Collecting data	1,867	93,33	Very Good
5	Data analysis	1,417	70,83	Good
6	Conclusion	1,250	62,50	Good
	Result	1,530	76,48	Good

The next component of creativity is the creative attitude. The creative attitude aspect examined is curiosity, that is the student's wish to ask many questions in every step of Inquiry-Based Laboratory activity. Table 5 shows that generally students creative attitude is in good result (76,48%), this proves that in Inquiry-Based Laboratory activity, students have high curiosity attitude. Many students ask questions during this lesson compared to ordinary learning. The highest percentage is in the data collection stage (93,33%), because at this step the students test the concentration of the medicine active substance by the procedure they design themselves. So the curiosity of the lab results makes many students ask their fellow questions through the discussion or to the teacher. The same as the previous study [19], whereas students who design their own experiments, spent relatively more time discussing the experiment and revisions of their work. The non-design students focused their efforts mostly on the mathematical procedure.

3.3. Student Responses to Inquiry-Based Laboratory

Inquiry-based Laboratory was first conducted in the pharmaceutical high school in Sumedang, but feedback from students who attended the class was very positive. The result show that chemistry learning before Inquiry-based laboratory implementation, mostly (52,84%) using methods that do not require student creativity. Even though, 100% students agree with statement that creativity is very important to owned by the vocational high school student. Whereas with ILAB students 100% stated that on learning ILAB their creativity is really used. This is supported by interviews stating that:

"*All this time, chemistry class is only done regular learning, where students only discuss and practice with procedures like cooking recipes. In my opinion, my creativity is not needed during learning. Different from ILAB, here I am required to make my own procedures, I think this is a very interesting and fun thing. I can't wait to try the experiment with my own procedure*"

Furthermore Inquiry-based laboratory has many advantage, that is improving curiosity, students become more active, learning chemistry becomes more fun and easy to understand, making students think more deeply, correct the students mistake and helping students find new ideas.

4. Conclusion

In this paper, we have reported the implementation Inquiry-based Laboratory in drugs analysis with acid base titration improving pharmaceutical high school students creativity. Moreover, the students reacted positively to this teaching strategy as demonstrated by results from questionnaire responses and interviews.

Acknowledgments
The author would like to thank the Bhakti Sejahtera Jatinangor Pharmaceutical High School in Sumedang who have allowed the author conducted research. To the teachers who are willing to take the time to follow the research was also delivered thank you profusely.

References
[1] Tatsuoka T, Shigedomi K, Koga N 2015 *Journal of Chemistry Education* **9** 1526–1530
[2] Schroeder J, Greenbowe T 2008 *Chem. Ed. Res. & Prac.* **9** 149-156
[3] Departemen Pendidikan dan Kebudayaan 2013 *Salinan Lampiran Peraturan Menteri Pendidikan dan Kebudayaan Nomor 70 Tahun 2013 Tentang Kerangka Dasar dan Struktur Kurikulum Sekolah Menengah Kejuruan/ Madrasah Aliyah Kejuruan* (Jakarta: Depdikbud)
[4] Sheppard K 2006 *Chem. Ed. Res. & Prac.* **7** 32-45
[5] Chijioke U, Offiah F 2013 *Universal Journal of Education and General Studies* **2** 336-53
[6] O'Neil H, Lang J, Perez R, Escalante D, Fox F 2014 What Is Cognitive Readiness? *Teaching and Measuring Cognitive Readiness* ed O'Neil H, Perez R, Baker E (New York: Springer) pp 3-23
[7] BSNP 2013 *Paradigma Pendidikan Abad XXI* (*Buletin BSNP* Vol 8) (Jakarta: Badan Standar Nasional Pendidikan) pp 3-5
[8] Fasko D 2000 *Creativity Research Journal* **13** 317–27
[9] Dehaan R 2009 *CBE—Life Sciences Education* **8** 172–181
[10] Yakar Z, Baykara H 2014 *Eurasia Journal of Mathematics, Science & Technology Education.* **10** 173-183
[11] Amalina I 2014 *Dampak Pengembangan Pembelajaran Inkuiri Laboratorium Terhadap Kemampuan Inkuiri, Berpikir Kreatif, Dan Penguasaan Konsep Siswa Pada Materi Sifat Koligatif Larutan* (Thesis) Postgraduated School, Universitas Pendidikan Indonesia, Bandung
[12] Sari 2014 *Pengembangan Kreativitas Mahasiswa Calon Guru Kimia Dalam Pembelajaran Dengan Metode Inkuiri Laboratorium Melalui Tema Minuman Kemasan* (Thesis) Postgraduated School, Universitas Pendidikan Indonesia, Bandung
[13] Hofstein A, Shore R, Kipnis M 2004 *International Journal of Science Education* **26** 47–62
[14] Smith J K, Smith L F 2010 Educational Creativity *The Cambridge Handbook of Creativity* ed Kaufman J, Stenberg R (New York: Cambridge University Press) p 256
[15] Munandar S 1999 *Pengembangan kreativitas anak berbakat* (Jakarta: Rineka Cipta) p 193
[16] Charifson, P.S, Walters, W.P, 2014 *J. Med. Chem.* **23** 9701–9717
[17] Gomez, D.G, Rodríguez, D.A, Cañada, F.C, Jeong, J.S 2015 *Journal of Chemistry Education.* **5** 860–861
[18] Maria B, Salim S, Marjan G, Joe S, Lisa S 2013 *International Journal of Innovation in Science and Mathematics Education* **21** 42-53
[19] Etkina E, Karelina A, Ruibal M, Rosengrant D, Jordan R, Hmelo C 2010 *The Journal of the Learning Sciences* **19** 54-98

IOP Conf. Series: Journal of Physics: Conf. Series **895** (2017) 012118 doi:10.1088/1742-6596/895/1/012118

Enhancement of Self Efficacy of Vocational School Students in Buffer Solution Topics through Guided Inquiry Learning

Ardiany M*, Wahyu W and Supriatna A

Departemen Kimia, Universitas Pendidikan Indonesia, Jl. Dr. Setiabudi No. 229, Bandung 40154, Indonesia

*mimy.ardiany16@gmail.com

Abstract. The more students who feel less confident in learning, so doing things that are less responsible, such as brawl, drunkenness and others. So researchers need to do research related to student self efficacy in learning, in order to reduce unwanted things.This study aims to determine the effect of guided inquiry learning on improving self-efficacy of learners in the buffer solution topics. The method used is the mixed method which is the two group pretest postest design. The subjects of the study are 60 students of class XI AK in one of the SMKN in Bandung, consisting of 30 experimental class students and 30 control class students. The instruments used in this study mix method consist of self-efficacy questionnaire of pretest and posttest learners, interview guides, and observation sheet. Data analysis using t test with significant $\alpha = 0{,}05$. Based on the result of inquiry of guided inquiry study, there is a significant improvement in self efficacy aspect of students in the topic of buffer solution. Data of pretest and posttest interview, observation, questionnaire showed significant result, that is improvement of experimental class with conventionally guided inquiry learning. The mean of self-efficacy of student learning there is significant difference of experiment class than control class equal to 0,047. There is a significant relationship between guided inquiry learning with self efficacy and guided inquiry learning. Each correlation value is 0.737. The learning process with guided inquiry is fun and challenging so that students can expose their ideas and opinions without being forced. From the results of questionnaires students showed an attitude of interest, sincerity and a good response of learning. While the results of questionnaires teachers showed that guided inquiry learning can make students learn actively, increased self-efficacy.

1. Introduction

Until now still found students who are not active when learning takes place. This is shown from students who are lazy to ask or answer. Students lack confidence and are not sure to be able to ask or answer questions, and do not want to compete. Such conditions indicate the low self-efficacy of the student. Research with guided inquiry learning and self-efficacy is still required in education world. Research has been done and can produce a positive self-efficacy in learning activities. It turns out that the success can be influenced by self efficacy such as the increase of understanding, development of high-level thinking and others. According to Tenaw 2013 [1] based on his research on the relationship between self efficacy and academic achievement, he suggested to innovate the learning process through the implementation of various strategies that can trigger student self-improvement to be better. This is suitable with national educational goals which is increating confidence, and create a bright young generation

Guided inquiry learning according to experts can influence student mind set and process skills in lab practice, so students can expose ideas/ thoughts and be challenged in learning chemistry / science. Guided inquiry learning is a learning based on students steps to find their own learning problems, through teacher guidance and aims to encourage students to become more courageous and have creative imaginary, which guided to discover ideas and thought or discoveries and tools that have never existed before Anam & Khairul 2015 [2]. This becomes the motivation for students to learn.

Research on self efficacy with cognitive abilities is interconnected. The higher someone's belief in learning, the more cognitive ability that can increases. In the current learning process of education is not enough only to teach (transfer of knowledge) but also educate (transfer of value). According to Lickona's 2013 [3] opinion, character education process involves various aspects of students, which is cognitive, affective, and psychomotor as an aspect of wholeness in the context of cultural life. It is also mentioned in Permendikbud No. 54 of 2013 [4], to achieve the competency standards of graduates, students need to have 3 dimensions that is knowledge, attitude, and skills.

In the learning process that can determine high student self-confidence is required in various affective, cognitive and psychomotor aspects. Where the students' graduation competence is influenced by the hard skill and soft skill aspects, which determine the quality of the graduation, One of soft skill that influences student learning motivation is self efficacy in the student. Self efficacy plays an important role in the advancement of education, Self efficacy can help students feel confident in their abilities, and able to effectively handle the difficulties they foced in the learning experience. According to Bendura 1994 [5] self efficacy is one's self-belief that it can handle situations and produce positive outcomes. This is consistent with the theory of Behaviorism which emphasizes behavioral change as a basis for the success of the desired learning process. Suyono and Hariyono 2011[6].

Guided Inquiry learning was first developed by Richad Suchman in 1996 [7] which views the nature of learning as a thinking exercise through questions. Then in the book Edi Hendri, Suchman (1962), put forward the core idea of inquiry learning: (1) Students will ask (inquiry) if they find confusing, unclear or discrepant events, (2) students have the ability to Analyzing their thinking, (3) the way of thinking can be taught and added to students, and (4) inquiry can be more meaningful and effective when done in group learning. Thus, this inquiry learning can develop student mindset more creative, logical and analytical.

Research conducted on the role of self efficacy can improve process skills in chemistry learning, and biology is developed by research Uzuntriyaki & Aydin 2009 [8].The meaningful learning of guided inquiry exists in the very close relationship between guided inquiry learning and laboratory activity. Thus, laboratory-based inquiry learning is expected to improve students' understanding and even reduce students' misconceptions, improve student laboratory skills, and students' attitudes toward chemistry and laboratory work.

Guided Inquiry learning is a learning which will make students to be creative, active, critical and able to develop their ideas freely and deeply, so that they can express their imaginations and thoughts. Guided Inquiry learning has advantages according to Bruner 2008 [9], namely: (1) Students will understand the basic concepts and ideas better; (2) Assist in using memory / cognitive abilities and transfer in new learning process situations; (3) Encourage students to think of initiatives and formulate their own hypotheses; (4) provide intrinsic satisfaction; (5) The learning process becomes more stimulating.

So in particular this study aims to analyze the influence of guided inquiry learning on the increase of self efficacy on the topic of buffer solution. Based on the description above it is necessary to do research "The Influence of Guided Inquiry Learning to Improve Self Efficacy".

2. Experimental Method

Target of this research is student of SMK class XI SMKN 13 Bandung counted 60 student, class XIAK3 and XIAK4. The research design is a mixed method research. Mix Method is a research

IOP Conf. Series: Journal of Physics: Conf. Series **895** (2017) 012118 doi:10.1088/1742-6596/895/1/012118

method that combines quantitative and qualitative methods according to the research problem and to answer the problem formulation Creswell, 2012 [10].

According Creswell (2012) there are 4 types of mix methods are: embedded, explanatory, exploratory, and triangulation. Research type chosen by the researcher is embedded exprimental that is research based on quantitative and qualitative data collection. The data were analyzed by the researcher separately.

The design of this study is the two group pre test post test design, which is described as follows:

G1:	O1	X1	O2
G2:	O1	X2	O2

Figure 1. Research design
Wiersma and Jurs, 1990 [11].

Information:

G1 Show Experiment Class, G2 Show Control Class, X1 Show Guided Inquiry Learning, X2 Show Conventional Learning, O: menunjukkan Pre test and Post test

Learning devices used in the study are: (1) Syllabus; (2) RPP; (3) LKS. The research instruments used are: (1) Observation Sheet of the implementation of Guided Inquiry Learning; (2) Self Efficacy Questionnaire Sheet; (3) Student Question Sheet; (4) Master Question Sheet; Interview Sheet.

Data collection methods used are the method of observation and test methods. The observation method is used to observe the inquiry of guided inquiry learning at the time of practicum learning in laboratory and the result of students' self-efficacy abilities during the guided inquiry learning on the buffer material. The test is given at the beginning of the lesson and the end of the lesson.

The study was conducted 3 times. The first to third meetings are held for learning. While the initial and final tests outside the lesson. The teacher divided the students into 6 groups, each group consisting of 5 heterogeneous student.

Data collection techniques conducted in this study is to fill the self efficacy questionnaire. Students fill out the questionnaire test before learning and after learning. Grid of Self-Efficacy Self-Questioner

Table 1. Table Quisioner Self Efficacy

Item Quotion	Poorly 1	2	Very Poorly 3	4	Average 5	6	Well 7	Very Well 8	9
1.To what extent can you explain chemical laws and theories?									
2.How well can you choose an appropriate formula to solve a chemistry problem									
3.How well can you establish the relationship between chemistry and other sciences ?									
4.How well can you describe the structure of an atom ?									
5.How well can you work with chemicals ?									
6.How well can you describe the properties of element by using periodic table ?									
7.How well can you read the formulas of elements and									
8.To what extent can you propose solutions to everyday problems by using chemistry ?									
9.How well can you interpret chemical equations ?									
10.How well can you explain the particulate nature of matter ?									
11.How well can you construct laboratory apparatus?									
12.To what extent can you explain everyday life by using chemical theories ?									
13.How well can you collect data during the chemistry?									
14.How well can you use the equipment in the chemistry laboratory ?									
15.How well can you understand the news/documentary									
16.How well can you interpret graphs/charts related to chemistry ?									
17.How well can you write a laboratory report summarizing main findings?									
18.How well can you interpret data during the laboratory?									
19.How well can you solve chemistry laboratory?									
20.How well can you recognize the careers related to chemistry?									
21.How well can you carry out experimental procedure in the chemistry laboratory									

Answer honestly
v Answer according to your opinion Hint: Put a check mark (v) on the appropriate column with your opinion Description: SS: Strongly agree S: Agree TS: Disagree STS: Strongly Disagree.
Table 2 Distribution of guided inquiry learning is by SS = 4, S = 3, TS = 2 and STS = 1, then summed. After that it is then inserted into MSI (Method Survivef Interval) to get the value that can be processed with SPSS.

Table 2. Distribution of Guided Inquiry Learning Inquiry Questions

No	Task
1	I can identify the buffer solution components in guided inquiry learning
2	This learning makes it easy for me to interact with other friends
3	I am always keen to learn to understand the material of the buffer solution
4	I tried to put forward an opinion in the discussion
5	In the discussion, all group members are entitled to propose ideas / ideas
6	I became challenged to answer questions from other frie
7	I am excited to join practicum activities.
8	I feel happy to attend classes and labs
9	Guided inquiry learning makes learning chemistry easier with buffer material.
10	Guided inquiry learning facilitates students to ask questions and communicate these issues with other student
	Average

Table 3 shows that self efficacy of the experimental students are better than the control class through guided inquiry learning on the topic of buffer solution through lab workshop. In Lickona's (2013) opinion, character education process involves various aspects of students, cognitive, affective, and psychomotor aspects Wholeness in the context of cultural life. It is also mentoined in Permendikbud No. 54 years 2013, to achieve the competency standards of graduates, students need to have 3 dimensions as knowledge, attitude, and skill.

3. Results and Discussion

Table 3 Self Efficacy Student Score and Statistical test results

Data	Self efficacy Pretest		Self Efficacy Posttest	
	Class Experiment	Control Class	Class Experiment	Control Class
N	30	30	30	30
Average	4,65	405	7,13	6,00
SD	13,89	10,30	7,80	9,72
Normalitas (Sig)	0,503	0,244	0,991	0,126
Homogenitas (Sig)	0,11	0,11	0,311	0,311
Uji t (Sig)	0,179	0,179	0,002	0,002

Trial Criteria
Reject H_0 if significance obtained < α. Base on the Table *Independent Samples Test*
Value obtained significance *t-test for Equality of Means* is $0.002 < 0.05 = \alpha$. Therefore H_0 is rejected. There is a significant difference in self-efficacy between the experimental class and the control class. Self efficacy has an important role in improving students' confidence. Self-efficacy characters known as self-beliefs can improve the ability of individuals to form behavior that is relevant to a specific task or situation (Pervin, 1984)
Table 4 . The correlation between inquiry learning is guided towards the improvement of self efficacy of SMK students on the topic of buffer solution
In this study the authors do a correlation analysis that is technically correlation analysis

This is assisted with SPSS 22.0 for Windows software. The result of correlation analysis can be seen in table 4 below:

Table 4 Correlations

		Guided Inquiry Learning	Student Self Efficacy
Guided Inquiry Learning			.737
	Sig. (2-tailed)		.047
	N	30	30
Student Self Efficacy	Pearson Correlation	.337	1
	Sig. (2-tailed)	.047	
	N	30	30

From the above table it is found that the correlation value between guided inquiry learning to the increase of self-efficacy of students is 0.737 with the value of Sig. (2-tailed) is 0.047. Criteria Testing reject Ho if Sig. (2-tailed) $<\alpha$, where $\alpha = 0.05$ is the level of significance. Since the value of Sig. (2-tailed) between inquiry learning toward self-efficacy of students is 0.047 <0.05 then Ho is rejected.

This means that there is a relationship between inquiry learning to the improvement of students' self efficacy.To see how the relationship between inquiry learning on the improvement of students' self efficacy is seen from the value of the coefficient of determination categorized in the following table 5.

Table 5. Determination Coefficient Interval

Determination Coefficient Interval	Classification
0,000 – 0,199	very low / Weak can be ignored
0,200 – 0,399	Low/ Weak
0,400 – 0,599	Medium
0,600 – 0,799	High / Strong
0,800 – 1,000	Very high / very Strong

Based on the output of SPSS 22.0 for Windows. Obtained coefficient of determination is presented in the table 6 below:

Table 6 Model Summary

Model	R	R Square	Adjusted R Square	Std. Error of the Estimate
1	.737[a]	.543	.734	8.13324

a. Predictors: (Constant), Guided Inquiry Learning

Determined coefficient value of 0.543. The value is at the interval of 0.400 - 0.599. This means the relationship between inquiry learning and self-efficacy in moderate students.

R Square or coefficient of determination has result of calculation equal to 0,543 has meaning that the increase of self efficacy of student influenced by inquiry learning equal to 54.3%, while the rest that is 100% - 54.3% = 45.7%, influenced by other factors not examined by writer.

Next is a simple regression analysis. Simple regression analysis technique used to find out how the dependent variable (Y) improvement of self efficacy of students can be predicted through independent variables (X) guided inquiry learning.

Table 7 Coefficients[a]

Model	Unstandardized Coefficients B	Unstandardized Coefficients Std. Error	Standardized Coefficients Beta	T	Sig.
(Constant)	55.043	10.156		5.420	.000
Guided inquiry learning	.165	.335	.037	.195	.047

a. Dependent Variable: student self efficacy

From the data processing using the existing facilities in the software SPSS 22.0 for windows obtained the estimated value of regression coefficient as follows:

a = 55.043

b = 0.165

Of these results obtained by regression equation as follows :

Y = 55.043 + 0.165 X

Based on the above equation the regression model can be interpreted that the constant is 55.043, meaning that if the inquiry learning is guided (X) the value is 0,165 then the student self efficacy improvement (Y) is 55.043. The regression coefficient of inquiry learning variable (X) is 0.165, meaning that if the inquiry learning is guided to increase 1 unit, then the student self efficacy (Y) will increase by 0.165. Coefficient of positive value means there is a positive relationship between guided inquiry learning on the increase of self-efficacy of students, the more up inbuilt learning is guided then the increased self-efficacy students. From the above table found that the correlation value between guided inquiry learning toward the students' cognitive improvement is 0.543 with the Sig score. (2-tailed) is 0.047.

Testing Criteria

Reject Ho if Sig. (2-tailed) <α, where α = 0.05 is the level of significance. Because the value of Sig. (2-tailed) between inquiry learning is guided against increasing efficacy

Student self of 0.047 <0.05 then Ho rejected. According to Hasheminasab, et al 2014[12], stated that there is a correlation between discipline with self efficacy and achievement academic.

Table 8. N-gain Independent Samples Test

		Levene's Test for Equality of Variances		t-test for Equality of Means						
									95% Confidence Interval of the Difference	
		F	Sig.	t	Df	Sig. (2-tailed)	Mean Difference	Std. Error Difference	Lower	Upper
GainScore	Equal variances assumed	.675	.415	.482	58	.032	.19667	.40832	-.62068	1.01401
	Equal variances not assumed			.482	55.537	.032	.19667	.40832	-.62145	1.01478

Trial Criteria

Reject H_0 if significance obtained < α. Base on the Table Independent Samples Test

Value obtained significance t-test for Equality of Means adalah 0.032 < 0.05 = α. Therefore H_0 is rejected. There is a significant difference in self-efficacy between the experimental class and the control class

Table 9. Table showing the conclusions of the influence of Guided inquiry learning in self-efficacy of SMK students in of buffer solution the topic

Average t Guided Inquiry learning (*N*=30,*sig*=0,05)	Sig	Average t self efficacy (*N*=30, sig=*0,05*)	Sig
Post test	0,00	0,737	0,047
N-Gain	0,00	0,487	0,032

Table 9 shows that there is an influence of guided inquiry on the improvement of self efficacy of SMK students on the topic of buffer solution. According to Tenaw (2013) based on his research on the relationship between self efficacy and academic achievement he suggested that innovation in the learning process through the implementation of various strategies that can trigger student self-improvement to be better.

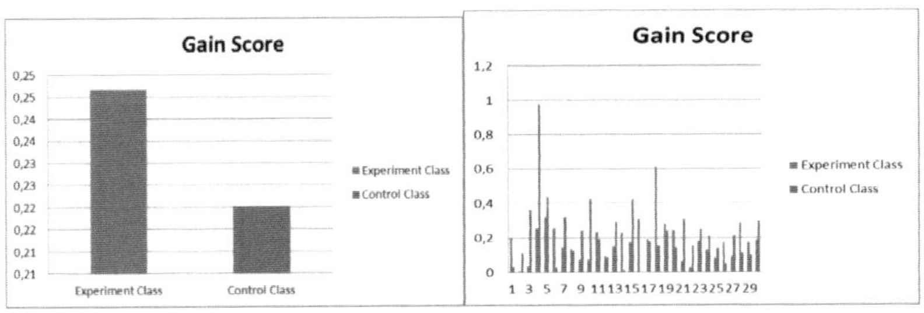

Figure 2. shows that there is an influence of guided inquiry on the improvement of self efficacy of SMK students on the topic of buffer solution by looking at the experimental N-gain class grade compared to the control class.

$$G = \frac{\text{posttest score} - \text{pretest score}}{\text{maximum score} - \text{pretest score}}$$ Hake, 1999 [13]

Research on improvement of concept and self efficacy through model of Orinted Guided Inquiry Learning by Oktarina (2016). The results of this study can improve the mastery of concepts and self efficacy through this inquiry learning with the concept of making colloids,

Figure 3 Show there are 2 meeting of Practicum at the Laboratory by filling out the student questionnaire in conducting the guided Inquiry learning activity. The1st meeting is explaining the nature of the buffer solution The second meeting is measuring the pH of the buffer solution

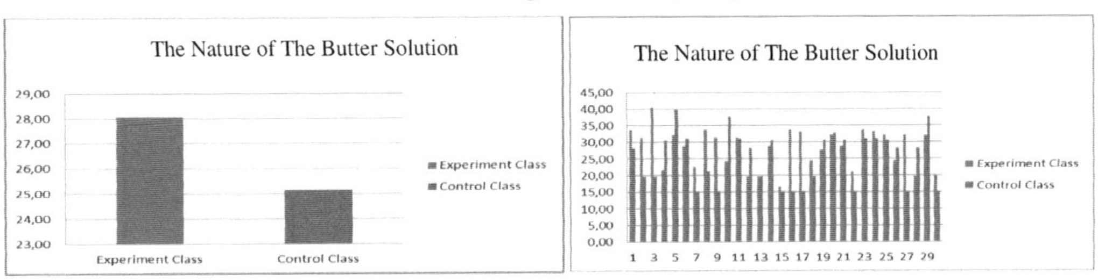

Figure 3. The Nature of The Butter Solution (One meeting)

The graphic shown above shows the guided and conventional inquiry teaching activities in the experimental class and control class with the material explaining the nature of the buffer solution.

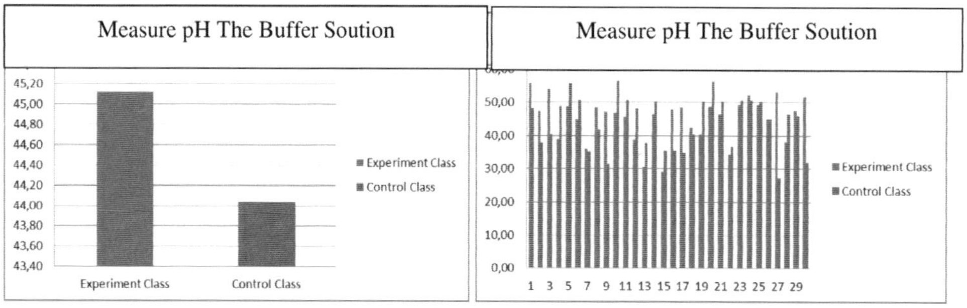

Figure 4 The experiment class shows better activity than the control class Measure The pH of The Buffer Soution

Figure 4 Show Measure The pH of The Buffer Solution (second meeting)
The graphic shown above Guided and conventional guided inquiry teaching activity in the experimental class and control class with the material Measuring the pH of the Buffer Solution. The experiment class shows better activity than the control class
Figures 3 and 4 show guided inquiry learning through practicum on the buffer topic at the first meeting between the control class and the experiments there are differences, in which the experimental class is better than the control class at the first and second meeting. This shows the guided inquiry learning succeess in this study. Guided inquiry learning is a learning based on students' steps to find their own learning problems, through teacher guidance and aims to encourage students to become more courageous and creatively imaginative, guided to discover ideas and ideas or discoveries and tools that have never existed before Anam & Khairul 2015 [14]

In Guided Inquiry Learning, it is seen show the teacher's activity in the laboratory pursuit activities is collaborated in the buffer solution topic.

Table 10. Show data on the implementation Guided Inquiry Learning Guidance of Buffer Solution Practicum

	hoImplementation of Guided Inquiry Practicum		
	Psychomotor	cognitive	Affective
Pert I	82,8	83,3	82,5
Pert. 2	83,6	83,7	83,5

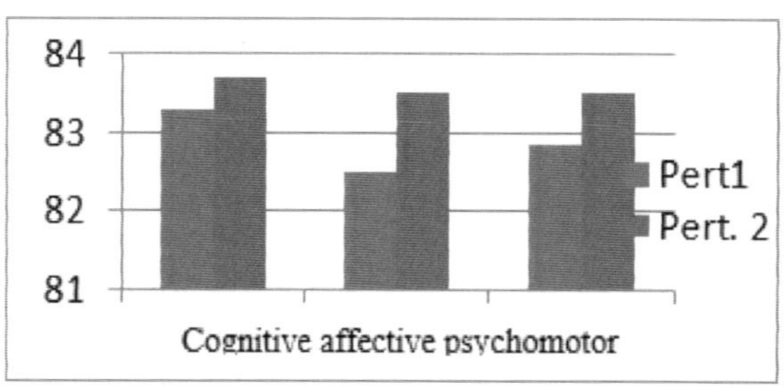

Figure 5 Show grafic on the implementation Guided Inquiry Learning Guidance of Buffer Solution Practicum

Figure 5 Demonstrates the guided Inquiry learning experience on the topic of the laboratory buffer solution in the laboratory, an increase from the average of the second meeting of meeting 1

Pert 1 Explains the nature of the buffer solution
Pert 2 Calculates the pH of the Solution

4. Conclusion

Based on the result of research and discussion, it can be concluded that there is a significant improvement in self efficacy of students in guided inquiry learning with buffer topic from experimental class compared with control class with mean value of 0,047. While the influence of inquiry learning is guided towards self efficacy students of the experiment class compared with control class is 0,737 with regression equation for guided inquiry relationship with self efficacy:

$Y = 55{,}743 + 0{,}165X$

Acknowledgments

The authors would like to thank all those who have assisted the authors in examining the effect of guided inquiry learning on improving self-efficacy of students on the topic of buffer solutions, so that the authors can produce the expected papers, although there is still a sense of desire to further examine (the relationship better from the authors to produce). Because the authors feel confident that this inquiry study is challenging for creative students and who are willing to issue their ideas, so that they can run maximally.

The authors also feel that in guided inquiry learning is accompanied by this experiment takes quite a long time to do, and requires hard work for teachers to create a fun learning for students.

References

[1] Anam, Khairul 2015 *Pembelajaran Berbasis Inkuiri Metode dan Aplikasi*. (Yogyakarta: Pustaka Belajar).
[2] Bandura A 1994 *Self Efficacy* (New York: Academic Press).
[3] Cresswell J W 2012 Educational Reasearch, *Plannin, Conducting, and Evaluating, Quantitative and Qualitative Reasearch*, 4th edition, (Boston: Pearson).
[4] Hake R R 1999 *Analizing Change Gain Score* Dept of Physics. Indiana University.
[5] Hasheminasab M, Zarandi Z G, Azizi J,and Zadeh M S 2014. Investigating the Relationshop between Self Efficacy with Academic Achievment, Dicipline, Urban-Rural and Order Bith of High School Student if Rafsanjan, *International J of Pssichology and Behavioral Research*. 3(4). Hlm 258-264.
[6] Lickona T 2013 *Mendidik Untuk membentuk Karakter*.Alih bahasa, Kartadinata. (Jakarta: Bumi Aksara).
[7] Oktarina H 2016 *Peningkatan penguasaan Konsep Dan Efikasi Diri Siswa SMA Melalui Model Process Oriented Guided Inquiri Learning (POGIL) Pada Konsep pembuatan Koloid*, Bandung: Program Pendidikan PascaSarjana UPI Bandung.
[8] Peraturan Menteri Pendidikan dan kebudayaan, 2013 Standar Kompetensi Lulusan Siswa SMA/SMK,Jakarta 54.
[9] Pervin 1984 Personality: *Theory and Reseach Fourth Edition*. Canada: John Wiley and Sons, Inc.
[10] Suchman J R. 1996. *The Inquiry Developmant Program-An Inquiry Session*, Philedelphia: Science Research Assosiates Inc.
[11] Suyono dan Hariyono 2011 *Belajar dan Pembelajaran* Bandung: PT Remaja Rodakarya
[12] Tenaw Y A 2013 Relationship Between Self Efficacy, Academic AchivementAnd Gender In Analytical Chemistry *At Debre Markos College Of Teacher Education ACJE*, 3(1), hlm. 3-28
[13] Wiersma and Jurs 1990 *Educational Measurement and Testing*, Boston, London, Sydney
[14] Uzuntiryaki E and Aydin YA 2009 Development and Validition of Chemistry Self-Efficiency Scale for College Students, *Res Sci Educ*, 39 hlm. 539-551.

IOP Conf. Series: Journal of Physics: Conf. Series **895** (2017) 012119 doi:10.1088/1742-6596/895/1/012119

A Study of Cognitive Load for Enhancing Student's Quantitative Literacy in Inquiry Lab Learning

E Nuraeni*, T Rahman, D P Alifiani and R S Khoerunnisa

Departemen pendidikan Biologi, Universitas Pendidikan Indonesia, Jl. Dr. Setiabudi No. 229, Bandung 40154, Indonesia

*eninuraeni@upi.edu

Abstract. Students often find it difficult to appreciate the relevance of the role of quantitative analysis and concept attainment in the science class. This study measured student cognitive load during the inquiry lab of the respiratory system to improve quantitative literacy. Participants in this study were 40 11th graders from senior high school in Indonesia. After students learned, their feelings about the degree of mental effort that it took to complete the learning tasks were measured by 28 self-report on a 4-point Likert scale. The Task Complexity Worksheet were used to asses processing quantitative information and paper based test were applied to assess participants' concept achievements. The results showed that inquiry instructional induced a relatively low mental effort, high processing information and high concept achievments.

1. Introduction

In the 21st century mathematical ability is needed in various activities such as when working, studying, even while doing daily activities. Current demands lead to the ability to understand, to apply and integrate numerical matters and data analyses in everyday as a habit of mind called quantitative literacy. Quantitative literacy is the ability to understand, to read and to describe quantitative information in everyday life [1] and the comfort of a person in operating and or processing numerical data [2]. QL involves responding to quantitative information presented verbally, graphically, or symbolic form [3]. Some researcher claimed that quantitative literacy is the most important skill for individuals [4,5,6]. Even [7] confirmed that quantitative literacy is more important than even high-level mathematics courses. Mathematics is The basic knowledge for mastering quantitative, but it does not mean quantitative literacy is imposed into the mathematics curriculum only. Students need to learn quantitative literacy from various other subjects.

Unfortunately, the quantitative literacy of high school students, Biology students, Biology education students and biology teachers in Indonesia is relatively weak [8,9]. This weak quantitative literacy is due to the fact that teachers have not developed quantitative literacy consciously in biology learning. In order to develop quantitative literacy, students need to get interesting experience and clear experimental activities so that students can feel the benefits of literacy Quantitative in the real world [10] .Students also need to be directly involved in generating quantitative data that can be interpreted [8], get more opportunities to make decisions that involve information-gathering and assessment, quantitative analyses, and communications about quantitative topics [10]. The students need to participate actively in scientific processes through research to acquire necessary knowledge in various fields [11].

An individual who has strong quantitative literacy skills has more ability to think and solve quantitative problems from various authentic contexts and situations in everyday life. Thus quantitative literacy is a tool of thought. Students can practice to solve quantitative problems by doing research in inquiry activities. In Inquiry Learning approach students can integrate laboratory activities, where there are initial activities before the learning is to identify the problem of investigation, determine the purpose of an investigation, conduct the investigation in accordance with the problems made and make a scientific question [12]. Inquiry studies have a positive impact on the students' independent in formulating and conducting an inquiry, able to measure the skills of the science process, make formulas and hypotheses, identify variables, experiment, interpret data, observe, measure, ask questions, communicate, And concluded [13], training students to overcome challenges, solve problems, test hypotheses, explain phenomena, or answer questions that are a method of a scientist [14]. These data are then analyzed to find relationship among variables. In inquiry lab Teachers helping students' difficulties, but the primary responsibility for designing an experiment, collecting data, analyzing and interpreting data, and communicating the results is borne by the students.

Applying certain learning strategies to develop the quantitative literacy is affected by students' working memory. Working memory goes consciously and it is able to process information quickly with its limited capacity [15]. The capability to analyze information (information processing) constitutes the process of information to be processed in order to produce learning result patterns [16]. Each person has limited and distinctive capacity for receiving information. That limited capacity will make someone hold something heavy and burdensome when he has to receive a lot of information. Such a issue has been described in the Cognitive Load Theory (CLT). Cognitive load theory aims at predicting the learning result by considering the capability and limitation of human cognitive conception [17].

According to the cognitive load theory, there are three types of cognitive load namely intrinsic cognitive load, extraneous cognitive load, dan germane cognitive load. It depends on a number of elements and its interactivity. Extraneous or ineffective cognitive load is generated by an ineffective learning design . All the irrelevant information processing with the learning objective illustrates the extraneous cognitive load. The relation between two cognitive loads elaborated by [15]. If intrinsic cognitive load is low, the level of extraneous cognitive load became less important since the total of cognitive loads does not burden the working memory capacity. Germane cognitive load is generated while the students have some deep cognitive processing and used for construction and automation of schemes.

Inquiry lab learning can be applied to improve the students' capabilities for quantitative literacy. However, it is necessary to consider whether the addition for quantitative literacy in this inquiry lab learning will bring about cognitive loads for the students or not. Therefore, there should be some study concerning the cognitive loads that the students have when they implement the inquiry lab learning in an effort to increase their quantitative literacy.

2. Method

This descriptive research aimed to identify the cognitive loads consisted of intrinsic cognitive load, extrinsic cognitive load and Germane cognitive load in the inquiry lab learning that develops the Senior High School (SMA) students' quantitative literacy. The participants in this research were 40 senior high school students of Grade XI for Natural Science (IPA) in Bandung, Indonesia (at the age of 16-17 year). The participants were the students studying by using inquiry lab learning.

All the students attended two-week inquiry lab learning in group. The class began by watching video regarding free diving, then continuing by formulating the reasearch questions in the experimental activity. The theme of first-week practical work was respiratory tidal volume in a human adult. Every group decided the research variables and experimental designs in order to answer their questions. The variable observed by the students had to be in the form of quantitative data. Subsequently, they made research in lab guided by the teacher. As a result of their activities, the resultant quantitative data were could be used to practice the students' quantitative literacy. At the end of every learning, each student was given 8 questions functioned to guide them developing their quantitative literacy, namely: 1) How much the average of vital capacity and complementary air as well as the frequency does the insect breathe in the practical work? 2) How is the determined factors effects for the lung vital capacity ? Add the supporting quantitative proof! 3) How is the determined factors effects for the lung supplementary air? Add the supporting quantitative proof. 4) Is there any relation between the determined factors and the lung vital capacity volume? Explain! 5) Is there any relation between the determined factors and the lung supplementary air volume? Explain! 6) Are the determined factors the most influential ones? Compare with other groups results using other factors. 7) Based on the observation, what does it means with the change of water volume contained in the bottle? 8) Based on the obtained observation data, change the table data into the graphical form. On the second-week learning activity, the students were doing inquiry lab activities with the theme of insect respiratory frequency.

The data collecting of cognitive load component is implemented during and after every inquiry lab learning taking place. Intrinsic Cognitive Load (ICL) is assessed using *Task Complexity Worksheet* given in the form of worksheet contained 16 questions to measure the students' capabilities of receiving and processing information during the learning process in the subject of respiratory system. The *Subjective Rating Scale* questionnaire is given every time the learning process to measure the mental efforts (ECL) completed during the learning process. There are 28 questions in the questionnaire sheet illustrating the compatibility between the learning strategy and the teaching material and also the students' difficulties in processing information (ICL) at the week 1 and 32 questions for the week 2 using 4-point Likert scale. Germane Cognitive Load (GCL) is measured by 20 multiple choice concerning the respiratory concept and 10 essay regarding the quantitative literacy with indicator by [18]. A non-parametric correlation test (Kendall's tau) taken to figure how strong the relation between each cognitive load component. Data of Germane Cognitive load collected after and before inquiry lab learning.Normalized test-gain (N-gain) of GCL calculate abd categorize by Hake formula [19].

3. Result and Discussion

3.1 Instrinsic Cognitive Load dan Extrinsic Cognitive Load

Table 1 shows the ICL presented by mental effort score and the ECL presented by the students'processing information score in inquiry lab learning 1 and 2.

Table 1. Mean score of ICL and ECL in inquiry lab 1 and inquiry lab 2

Cognitive Load Component	Inquiry1	Sd	Inquiry 2	Sd	Mean	Sd
Processing Information (ICL)	83,61	6,38	85,27	7,49	84,44	5,63
Mental effort (ECL)	46,16	8,47	44,45	9,14	45, 81	8,19

Based on Table 1, it is found out that the illustrated ICL by the capability to process information during the learning process fall into the high with the score of 83.61 (inquiry lab 1) and 85.27 (inquiry lab 2). This result presents the evidence that the inquiry lab is suitable for the quantitative information processing. ICL can be changed by natural change from what has been studied or by their own learning action [20]. It means that the changing that the students naturally have themselves will change the cognitive scheme in their memory. Therefore, the high capability to process information (ICL) is not enough to proof the low cognitive load that the students have, and it should be followed by the results of mental effort (ECL). If the students feel more difficult in or tired of the implemented learning process, it means *Extraneous Cognitive Load* will be higher. In this research, it is found out that the Mental Effort score (ECL) that the students have is low, and the students tend to be able already to cope with the experienced difficulties and able to accept the learning process by using inquiry lab (44.45). A good instructional design is learning design that provides tasks that can achieve sufficient and not excessive levels of ICL, can decrease ECL, and improve GCL [21].

In Table 1, the mental efforts (ECL) are not distinctive from inquiry lab 1 (44.16) to inquiry lab 2 (44,45). The considerable mean of the students' mental efforts based on Table 1 is 44.80 (relatively low). This finding shows that the implemented inquiry lab does not burden the students' working memory. The National Research Council has recommended the effort to improve the quantitative literacy for biology curriculum through interdisciplinary approach including stressing for the use of mathematics and statistics [22]. In the learning attended by the students, they make their own data, then make calculation, representation, interpretation, and communicate the data in order to understand the respiratory concept [13]

The important finding from this research is that there are ICL data inversely proportional to the ECL. It means that, there is no cognitive load, the inquiry lab learning strategy helps the students in obtaining and processing information (teaching materials). The mean of students' cognitive load shows that the students' learning achievment and the processing information capability are higher than the mental efforts. Strategy of learning is effective when the extraneous cognitive load as low as possible during the learning process [15]. The students' learning results are directly affected by the capabilities to receive and to process information, so that if the students have less burden in receiving and processing information, then they will tend to have less burden also in constructing the cognitive scheme. So, the good learning results from the students shows that they have less burden in learning the respiratory system using the inquiry lab learning strategy. However, the mental effort remains to have contribution to forming the cognitive scheme.

ECL that presented through the students' mental efforts has a little more mean compared to the ICL that presented through processing information capability (Table 1). Good GCL is obtained as a result of the contribution of more capability to Receiving and Processing Information (RPI) by the students compared to the Mental Efforts (ECL) [23]. In this case, based on Table 1, the ICL score is higher than ECL so that it enables the students to produce the good Learning Results (GCL) for them too.

3.2 Germane Cognitive Load

Table 2 demonstrate the score of the Germane Cognitive Load component

Table 2. The Germane Cognitive Load score

Component of GCL	Mean	Sd	Normalized gain
Content	67	9,05	0,62
Quantitative literacy	90	9,05	0,55

In Table 2, there is content mastering score for respiratory system and the ability of quantitative literacy. The high *germane cognitive load* can be affected by ICL or ECL. *Intrinsic cognitive load* has direct effect to the GCL forming, whereas ECL does not directly affect to the GCL forming. Normalized gain content achievement score and quantitative literacy fall into the moderate score category [19] that shows that the students still have the difficulties in understanding the two aspects of GCL.

Table 3. The percentage of students correctly answer the Quantitative literacy test

No	QL Indicator	%
1.	Interpretation	73
2.	Calculation	74
3.	Analysis	73
4.	Representation	89

Based on Table 3, it seems that from the measured four indicators during the inquiry lab learning, the students' representation capabilities tend to be higher (89%) of the students answering right compared to the other three quantitative indicators that is interpretation, calculation and analysis. The students' calculation capability score also has the adequate percentage value (74%). The calculation capability involved is the ability to perform average calculations.. The students are capable of answering the calculation questions easier because this capability has already been taught since the elementary school. On the other hand, the indicator for the interpretation and analysis quantitative data has the lowest percentage compared to other two indicators (73%). The interpretation and analysis capabilities is at the lowest in percentage where the students answering right though. In order to find out the cognitive loads that the students have, the measurement is not adequate just from the high or low score for every cognitive load components obtained by the students, it has to observe also the relation among the cognitive loads respectively [24].

Table 4. Normality test of ICL, ECL adn GCL

Data	N	Mean	Sd	Sig. value	Summary
ICL	43	84,44	5,57	0,513	Normal
ECL	43	44,80	8,20	0,876	Normal
GCL	43	74,65	6,30	0,110	Normal

Table 5. The result of correlation test

Variabel	R	sig. Value
ECL terhadap ICL	-0,097	0,372> 0,05
ICL terhadap GCL	0,065	0,576> 0,05
ECL terhadap GCL	-0,033	0,772> 0,05

Based on Table 5., it can be described that the correlation results from each components of students' cognitive loads. The variable value for Mental Effort (ECL) to the capabilities in Receiving and Processing Information (ICL) is negative; not significant (r=-0,097). It means that there is not any contribution from the mental efforts to the capabilities to receive and to process information clearly, but capable of describing that the reducing of mental efforts can improve the capabilities to receive and to process information. The students still have cognitive loads during the learning process even it is a little. On the other hand, the correlation of information processing capability (ICL) to the Learning Result (GCL) is positive; not significant (r=0.0065). It means that the contribution (ICL) to the Learning Result (GCL) tends to be uncertain. Apart from it, the correlation of Mental Effort (GCL) shows negative value; not significant (r=-0.033). It can mean that the contribution for the reducing of mental effort to the learning result is uncertain, it is possible that the students still have cognitive loads even it is a little. Cognitive load theory is concerned with techniques for managing working memory load in order to facilitate the changes in long term memory associated with schema construction and automation [15]. The results of the study show that students do not yet have quantitative literacy as the ability stored in long-term memory.Based on the overall correlation results, that the relation among the components respectively can be confirmed uncertain, it is shown by the unavailability of significant result from the correlation test. The important finding from this research is that the inquiry learning can practice the students' quantitative literacy shown by the students' percentage which is relatively high in answering right the quantitative literacy test. In this case, the measurement resulting quantitative data constitutes the initial step in developing quantitative literacy [25].

4. Conclusion

Inquiry lab learning improved the quantitative information processing with relatively-low mental efforts. In the applied inquiry lab, the determination of quantitative variables by the students, continued by making measurement and analysis for quantitative data succeed in training the students to develop their quantitative literacy capabilities especially representation. However, there are still cognitive loads found in developing it. Unfortunately, the limited research is carried out in two sessions. Quantitative literacy is habits of mind, it is need to continoues practice until it becomes a habit.

References

[1] Richardson R R and McCallum W G 2003 *The Third R in Literacy* (In Madison and Steen 2003) pp 99-106

[2] Rhodes T L and Finley A 2013 *Using the VALUE Rubrics for Improvement of Learning and Authentic Assessment* (Washington DC: Association of American Colleges and Universities)
[3] Frith V and Gunston G 2011 Towards Understanding the Quantitative Literacy Demands of a First-Year Medical Curriculum *AJHPE* **3** (1) pp 19-23
[4] Kemp M 2003 Critical numeracy: helping people to decide *Proceedings of the International Conference The Decidable and the Undecidable in Mathematics Education* (Brno Czech Republic) pp 144-148
[5] Skalicky J 2004 Quantitative Literacy in a Reform-based Curriculum and Implications for Assessment *AARE 2004 International Education Research Conference Paper Abstract* (Melbourne)
[6] Sevgi L 2006 Speaking with Numbers: Scientific Literacy and Public Understanding of Science. *Turk J Elec. Engin.* **14** (1) pp 33-40
[7] Steele B and Kilic-Bahi S 2010 Quantitative Literacy: Does it Work? Evaluating of Student Outcomes at Colby-Sawyer College *Numeracy* **3** pp 1-16.
[8] Nuraeni E 2016 Program perkuliahan anatomi tumbuhan berbasis kerangka instruksional dimensi belajar marzano untuk mengembangkan literasi kuantitatif mahasiswa *Dissertation* (Not published)
[9] Nuraeni, E, Rahmat A, Redjeki S, and Riandi 2014 Profil literasi kuantitatif mahasiswa calon guru biologi *Prosiding Mathematics and Sciences Forum 2014* (Semarang) pp 375-380
[10] Taylor C H 2009 Assessing Quantitative Reasoning *Numeracy* **2** (2) pp 1-5
[11] Duncan S I, Bishop P, and Lenhart S 2010 Preparing the "New" Biologist of the Future: Student Research at the Interface of Mathematics and Biology *CBE—Life Sciences Education* **9** pp 311–315
[12] Wenning C J 2005 Levels of Inquiry: Herarchies of Pedagogical Practices and Inquiry Processes *J Phys. Teach. Educ. Online* **2** (3)
[13] Wenning C 2010 Levels of inquiry: Using inquiry spectrum learning sequences to teach science. *Journal of Physic Teacher Education Online* **5** (3)
[14] Linden D and Madison W 2005 *The Wisconsin Program for Scientific Teaching* From http://cst.yale.edu/sites/default/files/Uncook_handout.pdf
[15] Paas F, Renk and Sweller J 2003 Cognitive Load Theory and Instructional Design: Recent Development *Educational Psychologist* **38** (1) pp 1-4
[16] Yovan 2008 *Memori dan Pembelajaran Efektif* (Jakarta: Yrama Widya)
[17] Kalyuga S 2011 Cognitive Load Theory: How Many Types of Load Does It Really Need? *Educ. Psychology Rev.* **23** pp 1–19
[18] Association of American Colleges and Universities 2009 *Quantitative Literacy Value Rubric* from http://www.aacu.org/value/rubrics/pdf/QuantitativeLiteracy.pdf
[19] Hake R R 1999 Analizing Change/Gain Scores from http://www.physics.indiana.edu/~sdi/AnalyzingChange -Gain.pdf
[20] Moreno R and Mayer R E 1999 Cognitive Pronciples of Multimedia Learning: the Role od Modality and Contiguity *Journal of Educational Pscychology* **91** (2) pp 358-368
[21] Meissner B and Bogner F X 2013 Towards Cognitive Load Theory as Guideline for Instructional Design in Science Education *World of Journal Education* **3** (2) pp 24-37
[22] National Research Council 1996 *National Science Education Standard* (Washington D.C: National Academy Press)
[23] De Jong T 2010 Cognitive Load Theory, Educational research, and instructional design. *International Science* **38** (2) pp 105-134
[24] Rahmat A and Hindriana A F 2014 Beban Kognitif Mahasiswa Dalam Pembelajaran Fungsi Terintegrasi Struktur Tumbuhan Berbasis Dimensi Belajar *Jurnal Ilmu Pendidikan* **1** (20) pp 66-74
[25] Mayes R L, Peterson F, and Bonilla R 2013 Quantitative Reasoning Learning Progressions for Environmental Science: Developing a Framework *Numeracy* **6** (1) pp 1-30

IOP Conf. Series: Journal of Physics: Conf. Series **895** (2017) 012120 doi:10.1088/1742-6596/895/1/012120

Development of Virtual Laboratory Based on Interactive Multimedia on Planting and Painting Bacteria

A Yuniarti[1*], L F Yeni[2] and Yokhebed[2]

[1] Departemen Pendidikan Biologi, Universitas Pendidikan Indonesia, Jl. Dr. Setiabudhi No. 229, Bandung 40154, Indonesia
[2] Departemen Pendidikan Biologi, Universitas Tanjungpura, Jl.Prof. Dr. H. Hadari Nawawi, Pontianak 78124, Indonesia

[*]anisyah.yuniarti@student.upi.edu

Abstract. The Aim of this research is to produce Virtual Laboratory based on interactive multimedia on sub material of planting and painting bacteria. Knowing eligibility of the media and knowing the student's respond that has been taught of using Virtual Laboratory. The method of this research is development research. This research consists of two stages are preliminary (preparation and design) and formative evaluation (self evaluation, expert reviews, small group and fields test). Data collection techniques used in this research are questionnaires and tests. The subject of this research is 37 students. Based on validation result from validator expert, Virtual Laboratory is categorized valid. This value is obtained because Virtual Laboratory can be installed in various hardware and software, easy to use, the program works well, and Virtual lab has a unique shape, interactive and good animation. Meanwhile, students give the positive respond, it means student can operate and simulate the experiment of planting and painting bacteria with Virtual Laboratory and helps students to understand the material of planting and painting bacteria easily. So it can be concluded that Virtual Laboratory has legibility to be used as learning media.

1. Introduction

Virtual Laboratory (VL/Virtual lab) is a set of laboratory tools in the form of interactive multimedia based computer software that operates with computer hardware and can simulate activity in the laboratory as if the user is in the actual laboratory (1,2). Interactive multimedia means a device equipped with user-operated controller tools, so users can choose what they want for the next process. Interactive multimedia combines and synergizes all media consisting of text, graphics, audio, and interactivity (3).

On the previous research Virtual lab is able to provide an overview of the International Retrofit Energy Trial at Cork Institute of Technology, Ireland. As well as providing an example of this type of analysis can be done by using this platform (4). Another benefit like a web-based Virtual lab on a frequency modulation experiment helped students to be able to carry out experimental work and gain hands-on experience of using an expensive spectrum analyzer according to their own schedule and pace on an individual basis (5). The Virtual lab also has interactive screen experiments and be innovative media for distance learners (6). However, Virtual lab for planting and painting bacteria experiment is not exist yet.

The aims of this paper is to development of Virtual lab, knowing eligibility of the media and knowing the student's respond that has been taught of using Virtual lab. Virtual lab development starts from the preliminary stage including content preparation and Virtual lab design. Then proceed to the formative evaluation stage including self-evaluation, expert reviews, small groups, and field tests (7). The result of this research is Virtual lab can be developed and utilized by students to do planting and painting of bacteria experiment. The advantages when using a Virtual lab is so effective because it does not need much time to do the simulation lab and can be repeated if students do not understand. So it helps students more easily understand the planting and painting bacteria material.

2. Experimental Method

The method of this research uses development research with type formative research. Figure 1 give illustration for the formative research design:

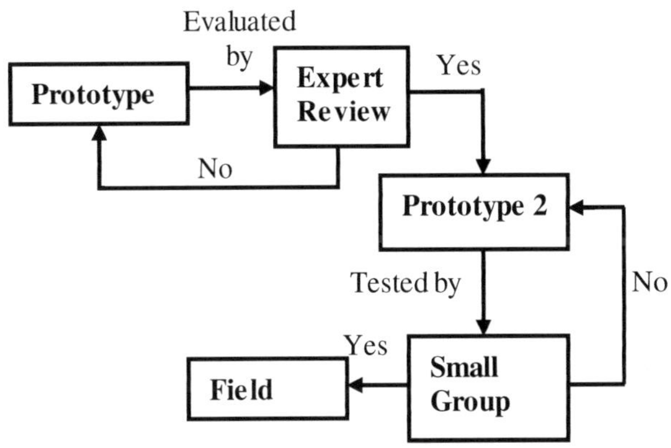

Figure 1. Flowchart of Formative Research Design

The development research focus on two stages are preliminary consists of preparation and design. Preparation stage works for analyzing of content, while the design for a device of learning material and media. Furthermore formative evaluation, first Self Evaluation: On this stage all learning devices that have been made will be checked again before validation. Then, Virtual lab must be measures by the researcher to know how does the condition and function of the program. The second Expert Review: All learning devices are validated by the validator. Validator consists of two are material and media. Virtual lab that is validated on this stage is prototype 1. Comment and suggestion from the validator uses for revision and fixed learning devices. The repair of Virtual lab of this research calls prototype 2. Third, Small Group Evaluation: On this stage the Virtual lab measurement prototype 2 on small group of student (6 students). Fourt, The real field test uses all learning devices that has been revised and validated.

The subject of this research is students of class X for 37 students at Senior High School (SMAN 1 Pontianak). The technique of collecting data on this research is questionnaires of eligibility Virtual lab and student's respond. The eligibility of Virtual lab media is valid by the experts 3 from media and 3 from material expert. The research instrument is questionnaires that have been validated by 2 lecturers Biology education and 1 biology teacher witth the result validation is valid.

The questionnaire eligibility of Virtual lab uses formula as follows: $RTV_{TK} = \frac{\sum_{i=1}^{n} A_i}{n}$ (1)
with notes RTV_{TK} (total mean) validity, Ai (mean aspect i), i(aspect) and n (more acpects). The questionnaire eligibility of Virtual lab that is validated by the media expert consists of 10 questions which divided into two aspects are engineering software and audio visual communication. The student's respond analysis using likert scale consists of 4 options answer, such as very agree, agree, disagree and very disagree. There are 30 questions are 15 positive questions and 15 negative questions.

3. Result and Discussion

3.1. Development and Eligibility of Virtual lab Based on Interactive Multimedia on Sub Material of Planting and Painting Bacteria.

Development of Virtual lab presents the simulation of planting and painting bacterial eksperiment. The validation result can be seen on Table 1:

Table 1. Recapitulation of Validation Result of Media Expert through Virtual lab based on Interactive Multimedia on Sub Material of Planting and painting Bacteria

Aspect	Criteria	Validator score on-			(Ki)	(Ai)
		1	2	3		
Software Enginering	Compatibility (it can be installed/run by hardware and software).	3	3	4	3,33	3,42
	Reliability (the program can have good function).	4	3	4	3,67	
	Usability (it is easy and simple to be used and operated).	3	3	4	3,33	
	Maintainability (it can be protected/managed easily).	3	3	4	3,33	
Communication Audio Visual	Creative on idea and concept.	3	4	4	3,67	3,44
	Visual (layout, design and color).	3	4	4	3,67	
	Audio (narrative, sound effect, back sound and music).	4	3	4	3,67	
	Animation.	3	3	3	3	
	Interactive Layout (navigation pad).	3	3	4	3,33	
	Interactivity.	3	3	4	3,33	
∑Ai						**6,86**
RTV_{TK}						**3,43**
Conclusion : Valid						

Information:
Ki = Average Criteria -i
Ai = Average Aspect -i
RTV_{TK} = Average total of validation both two aspects

According to Table 1 it can be known that Virtual lab is valid and eligible to be used as learning media with the score 3.43. If the average scores up to 3, so the learning media can be claimed valid [8]. Engineering Software Aspect, this aspect can be valid with average score 3.42. Virtual lab can be installed or run in the existing variety of hardware and software, the Virtual lab program is able to run good function, easy and simple. Audio Visual Communication Aspect, communication aspect of audio visual gets average score of 3.44 and valid. A creative Virtual lab on making ideas, visual (layout design and colour), audio (narrative, sound effect, back sound and music), animation, interactive layout (navigation knob) and interactivity of Virtual lab has fulfilled the eligibility standard on learning media.

Based on the explanation above, Virtual lab has eligibility as learning media. Nonetheless, there are suggestions from validator to be good prototype 1, so it needs revision (Tabel 2).

Table 2. The Change Media Before and After Revision (Prototype 1)

Before Revision	After Revision
The application is still on fla format.	App has been gathered on one compitable exe.
Instrumentation is not appropriate with user command.	The instrument move based on user command.
The sound on evaluation question is still crowded	The sound on evaluation question is neat and clear.
The object media target is small so it is hard to drag and drop.	The target object media has been large to be dragged and dropped.
The icon knob is small and it is hard to be clicked.	The size icon has been large and it is easy to be clicked.

This is the recapitulation result from media expert through Virtual lab based on interactive multimedia can be seen on Table 3.

Table 3. The Recapitulation Result from Material Expert through Virtual lab Based on Interactive Multimedia on Sub Material of Planting and Painting Bacteria

Aspect	Criteria	Validator Score On-			(Ki)	(Ai)
		1	2	3		
Format	Accord with sound, animation and color.	2	3	3	2,67	3,33
	Use size, shape, font color that is easy to be read.	2	3	3	2,67	
	Appropriate as learning media in the class/laboratory.	4	4	4	4	
	Be an alternative for doing practicum of planting and painting bacteria.	4	4	4	4	
Content	The material formulas based on the basic competition, indicator and the aim of learning syllabus.	4	4	4	4	3,67
	The using appropriate for sub material of planting and painting bacteria.	4	4	4	4	
	The clear concept of sub material for planting and painting bacteria on Virtual lab.	3	3	3	3	
Language	The language easy to understanding.	4	4	4	4	4
	The santance was right based on EYD	4	4	4	4	4
∑Ai						**11**
RTV_{TK}						**3,67**
Conclusion : Valid						

The validation material expert shows that Virtual lab based on interactive multimedia on sub material of planting and painting bacteria is valid and eligible to be used as learning media with 3.67. The result of validation data by the material expert shows that Virtual lab is valid with 3.67. It is displayed that Virtual lab can be used as learning media. Based on the aspect below: Format Aspect, format aspect gets the average score 3.33 and it belongs to valid. Virtual lab has accord with sound, animation and good color. Using of size, shape and color shape that has been read. The Virtual lab becomes an alternative to be practiced on practicum on planting and painting bacteria. Content Aspect, the content aspect gets score 3.67 and is valid. Virtual lab arranges with material formulas based on basic competition, indicator and the aim of learning syllabus. Virtual lab based on sub material. Language Aspect, the language aspect gets the average score 4 and it belongs to valid. Virtual lab uses easily language to be understood. But, validator gives suggestion to fix the content of Virtual lab.

Table 4. The Change Material Before and After Revision (Prototype 1)

Before Revision	After Revision
Wrong *pronounciation* on words "*Virtual lab*" dan inconsistent intonation.	Record again the words and sentences on *Virtual lab* with good *pronounciation* and intonation.
Kind, color and font are hard to be read.	Observe clearly about color, font which are easy to be read.
Wrong sterilization.	The right sterilization concept.
There is no explanation about source of bacteria on *streak plate* stage.	The additional explanation about source of bacteria on *streak plate* stage.

3.2 Student Respond through Learning on Planting and Painting Bacteria using Virtual lab

The result of student respond through Virtual lab based on interactive multimedia on sub material of planting and painting bacteria can be seen on Table 5.

Table 5. Questionnaire Analysis Result of Student's Respond on Field Test Stage

Criteria	**Indikator**	**Interval (%)**	**Respon**
Respond	Respond Virtual lab that is used on learning activities.	89, 19	Very Positive
	Respond through learning material.	83,78	Positive
	Respond through language.	84,12	Positive
	Respond through presentation (visual).	87,84	Positive
	Respond through relevance uses *Virtual lab*	81,59	Positive
Reaction	Attention through *Virtual lab* on learning.	83,61	Positive
	Confidence uses *Virtual lab* on learning.	78,89	Positive
	Satisfaction through *Virtual lab* on learning	83,11	Positive
	Student's Respond Through *Virtual lab*	**84,15**	**Positive**

The student's respond on this research are their respond and reaction through Virtual lab. According to analysis data, student's respond through Virtual lab amounted 84.15 % with positive respond. There are student's respond that consists of 2 criteria's are respond and reaction. Respond, respond through Virtual lab that is used on learning is very good because the operation of Virtual lab is easy. One of good learning media is the easy on usage, operation and it is not hard to be operated [9]. Most of students are happy and interest because they able to operate practicum simulation for planting and painting bacteria with Virtual lab. Students give positive respond through learning material of Virtual lab. This is important because media must appropriate with characteristic of fact, concept, principle and procedural to help the effective learning process [9]. The student respond seems from their comment that states through Virtual lab is very easy to understand for direct simulation practicum."

Virtual lab uses language, so student can understand the meaning of the words. Using picture and animation on Virtual lab is good. The good media must be clearly and neatly on layout, format, sound, written, and picture illustration or animation [9]. Virtual lab based on interactive multimedia made due to the student's need on sub material of planting and painting bacteria, the simple media on layout is focused as basic material to be learnt by the student. Reaction, Virtual lab able to make students interested. It can be seen from student's comments that like the visual (picture and animation), the animation of Virtual lab is good and interesting, and the display of laboratory animation and the tools seems like the real and the simulation practicum on Virtual lab like the real practicum".

Visual is able to develop the interest in a learning subject. In addition, the interest can improve student's motivation with taking, creating and involving students on learning process [10]. The messages through simulation will more interesting, because with attention students will be encouraged on their self motivation to learn.

Student's reaction on self confidence criteria use Virtual lab on learning is categorized positive. But, the presentation is less positive on confidence than other indicator. It is because some of students still disbelieve on their own selves to do simulation practicum of planting and painting bacteria with using Virtual lab based on interactive multimedia.

Regarding to this research, it has been known that students feel satisfied and get benefit after learning to use Virtual lab based on interactive multimedia on sub material of planting and painting bacteria (positive reaction). Some of students suggests that learning with Virtual lab based on interactive multimedia is interesting experience because use picture, animation and simulation.

4. Conclusion

Based on the result of the research, it can be concluded that Virtual lab based on interactive multimedia on sub material of planting and painting bacteria is valid and eligible to be used as learning media with RTV_{TK} 3.55. Furthermore, student's respond through Virtual lab based on interactive multimedia on sub material and painting bacteria belongs to positive with average presentation of 84.15%.

Acknowledgments

I want to say thankyou for my Lecturer Laili Fitri Yeni, S. Si, M. Si and Yokhebed, S. Pd, M.Pd for guidance and advice so that I can finish this thesis. Thankyou Desti and Cita who have helped me finish the software development program of learning media of Virtual lab. Student and Biology teacher SMAN 1 Pontianak for being my research object.

References

[1] Santoso H 2009 *The effect of real laboratory use and Virtual lab on physical learning judging from the critical thinking skill of students* (Surakarta: Tesis Program Pasca Sarjana Universitas Sebelas Maret) p 26

[2] Nazipah 2013 *Effect of Virtual lab usage on practicum of class XI buffer solution IPA SMA Islam Al-Falah Jambi* (Jambi: Pendidikan Kimia FKIP Universitas Jambi) p 3

[3] Green T D and Brown A 2002 *Multimedia project in the classroom* (USA: Corwin Press, Inc) p 2-6

[4] Murphy M D, Breen , O'Donovan A, Delaney F, Hill M, and Sullivan P D O "A Virtual lab for the simulation of sustainable energy systems in a low energy building: A case study" International Conference on Advances in Renewable Energy and Technologies (ICARET 2016) IOP Conf. Series: Earth and Environmental Science Vol 32, pp 1-5, 2016.

[5] Ko C C, Chen B M, Hu S, Ramakrish V, IEE, Cheng C D, Zhuang Y and Chen J "A Web-Based Virtual lab on a Frequency Modulation Experiment" IEE Transaction On System, Man, Ans Cybernetics- Part C: Aplications and Review Vol 31 No 3, pp 302, 2001.

[6] Hatherly P A, Jordan S E, and Cayless A "Interactive screen experiments innovative virtual laboratories for distance learners" IOP European Journal of Physics Vol 30, pp 751-762, 2009.

[7] Tessmer M 2005 *Planning and Coducting Formative Evaluation: Improving The Quality of Education and Training* (Oxon: British Library) p 15

[8] Yamasari Y 2010 *Development of qualified ICT-based mathematics media* (Surabaya: Seminar Nasional Pascasarjana X ITS) p 2

[9] Asyhar R 2012 *Creative developing learning media* (Jakarta: Referensi Jakarta) p 23

[10] Smaldino S E, Lowther D L and Russel J D 2012 *Instructional technology and media for learning* (Jakarta: Kencana) p 73

E-Beam Capture Aid Drawing Based Modelling on Cell Biology

T Hidayat[1,2*], A Rahmat[1], S Redjeki[1], and T Rahman[1]

[1] Program Studi Pendidikan IPA, Sekolah Pascasarjana, Universitas Pendidikan Indonesia Jl. Dr. Setiabudhi No. 229, Bandung 40154, Indonesia
[2] Program Studi Pendidikan Biologi, Universitas Muhammadiyah Bengkulu, Jl. Bali Kota Bengkulu, Bengkulu 38119, Indonesia

*tomihidayat@umb.ac.id

Abstract. The objectives of this research are to find out how far Drawing-based Modeling assisted with E-Beam Capture could support student's scientific reasoning skill using Drawing-based Modeling approach assisted with E-Beam Capture. The research design that is used for this research is the Pre-test and Post-test Design. The data collection of scientific reasoning skills is collected by giving multiple choice questions before and after the lesson. The data analysis of scientific reasoning skills is using scientific reasoning assessment rubric. The results show an improvement of student's scientific reasoning in every indicator; an improvement in generativity which shows 2 students achieving high scores, 3 students in elaboration reasoning, 4 students in justification, 3 students in explanation, 3 students in logic coherency, 2 students in synthesis. The research result in student's explanation reasoning has the highest number of students with high scores, which shows 20 students with high scores in the pre-test and 23 students in post-test and synthesis reasoning shows the lowest number, which shows 1 student in the pretest and 3 students in posttest. The research result gives the conclusion that Drawing-based Modeling approach assisted with E-Beam Capture could not yet support student's scientific reasoning skills comprehensively.

1. Introduction

The Drawing Based Modeling (DBM) approach is an effective way to teach scientific thinking skills. Drawing Based Modeling (DBM) is a learning approach where students themselves build the models of scientific phenomena [1]. The model is a representation of the reality used by scientists as a means to understand the observed phenomena. Approaches, strategies, and models of learning in the teaching and learning activities in the classroom cannot be separated from the teaching media. The selection and use of teaching media should be appropriate, especially in the approach of Drawing-Based Modeling (DBM) which uses computer media that is E-Beam Capture.

Using E-Beam Capture in the teaching-learning process is useful for students and teachers especially for facilitating in describing abstract phenomena. Computer models allow students to manipulate experimental variables with ease and produce learning products that are more predictable and easier to visualize [2]. In the E-Beam Capture, students draw objects, assign characteristics and behaviors to objects and observe how the interaction between objects takes place by running a simulation. This practice is called drawing based modeling.

The Drawing Based Modeling (DBM) approach using E-Beam Capture is an application of computer-based technology that has many advantages over manual approach or paper and pencil.

Some of the drawing functions in education system are as follows: (1) enhancing and activating learners' motivation, (2) learning for the representation and for providing a deeper understanding of scientific concepts for students; (3) increasing scientific reasoning, for example in creating a series of drawings (4) as a Learning Strategy. Helping learners build their knowledge, and (5) as a means of communication with others [3].

In organizing the class in general or in presenting the lesson material in particular, the teacher must be able to determine the appropriate strategy, model or approach so that the learning objectives can be achieved. The choice of learning model should include the function of giving encouragement, the expression of the growing interest in learning, the delivery of learning materials, the creator of a conducive learning atmosphere, the energy to generate creativity, the impetus for self-assessment in the learning process and outcomes, and the incentive to complete the learning outcomes. The models that are developed to achieve three important learning objectives, namely academic learning outcomes, acceptance of diversity, and the development of social skills, are cooperative learning models [4]. Which means, that the process of involving students in learning activities or make students as an active learning subject and the interaction between teachers or lecturers with students to stimulate, support or improve the ability of scientific reasoning of learners.

Good classes such as observation, experimentation, and even scientific discussion should enable students to become active learning subjects, enabling such activities to trigger students' thinking processes and also provide meaningful experiences. Student involvement in cognitive processes as well as class organizing is essential to train students so that they are capable of science literacy. Science literacy as a learning objective, encompasses students' understanding of the nature of science as well as the scientific reasoning of students [5].

In the teaching-learning process in the classroom, in addition to involving students in learning and using models, strategies, learning approaches, the educators should also be able to choose and use the appropriate teaching media. In schools or campuses that are lacking in terms of infrastructure, computer-aided image and modeling tools are the most appropriate medium to use in classroom learning.

In order to support the students' thinking ability especially scientific reasoning ability in Cell Biology course, the chosen approach is Drawing Based Modeling. Drawing Based Modeling in accordance with the demands of the curriculum and able to support the thinking process especially the students' scientific reasoning abilities is the answer to all the above problems.

The Drawing Based Modeling (subject drawing based modeling), this approach is more effective in achieving the conceptual and operational understanding that can stimulate scientific reasoning skills rather than other learning approaches [6]. Modeling can also provide an opportunity for students to think and discuss in a scientific way about students' ideas [7]. The description of the above background is the basis of the researcher to know the extent to which Drawing-based Modeling approach assisted with E-Beam Capture can support students' scientific reasoning abilities, especially in Cell Biology course.

Theoritical material on Cell Biology about mechanisms of transport across the cell membranes consists of several subjects, namely active transport, passive transport, endocytosis and exocytosis [8]. Cell membrane it is thin, flexible, lipid barrier that separates the contents of the cell or organelles from its surroundings (9].

1.1 Problem Identification

The formulation of this research are: 1) Is Drawing-based Modeling approach with E-Beam Capture able to support students' scientific reasoning ability? 2) How is the students' scientific reasoning ability using the Drawing-based Modeling approach assisted with E-Beam Capture?

1.2 Research Objectives

This research aims to determine the extent to which Drawing-based Modeling approach assisted with E-Beam Capture can support students' scientific reasoning ability and figure out the percentage of students' scientific reasoning ability using the Drawing-based Modeling approach assisted with E-Beam Capture.

2. Method

2.1. Research Method

The method used in this research is the experiment method with the one group pre-test and post-test design [10]. Instruments used include: questionnaires, observations, interviews, and tests. The experimental class is given a learning treatment with Drawing Based Modeling approach assisted by computer application in the form of E-Beam Capture. The technique of data analysis and processing used during the research is descriptive analysis and percentage technique.

2.2 Research Design

This research uses the one group pretest and posttest design. One group pretest and posttest design can be fully seen in figure 1 below:

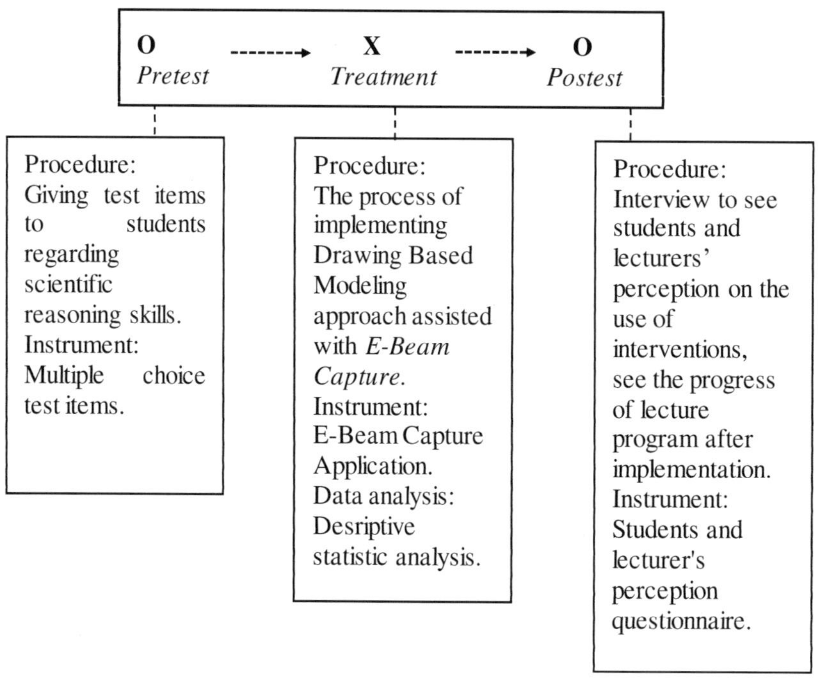

Figure 1. One group pre-test and post-test design

2.3 Research Subject and Location

The subjects of this research are 25 students of the Biology Education University of Muhammadiyah Bengkulu, consisting of 17 female students and 8 male students. This research involves validators consisting of: material experts, media experts, teaching tool specialists, 3 high education practitioners. Experiment I, is a test of the implementation of drawing based modeling approach and the use of E-Beam Capture application in supporting students' scientific reasoning capability, Trial II, is to test the functioning of lecturing programs, instruments, drawing based modeling approach by using E-Beam Capture application in supporting students' scientific reasoning ability. The location of the research was conducted in Biology Education Study Program of Muhammadiyah University of Bengkulu located in the city of Bengkulu.

2.4 Research Procedure

Implementation of research was done by doing things that are necessary in the execution of research such as preparing the assessment rubric format of scientific reasoning ability [11], making a schedule of research activities, making syllabus and semester course plan for Cell Biology course, and arrange the instrument in the form of learning observation, pre-test and post-test. Data collection through interviews, observation and documentation and test result data. The selection of the classes is done in consideration of the number of students, characteristics, and capabilities of individual students who are considered equal.
The stage of approaches of learning in Drawing Based Modeling in Cell Biology consists of four steps, namely: The first step, Expressive, students express ideas about the Structure and Cell Organelle topic. Then describe their initial idea by using a computer application in the form of eBeam Capture. In the second step, Experimental, students collect empirical data to validate and improve their image models through books and lecturers' explanations. In the third step, Evaluative, students evaluate the images they make by comparing their image models with their peers and then exchanging critical ideas with each other. In the fourth step, Exploratory, the lecturer gives a new phenomenon on each subject of Cell Biology in hope of facilitating the students to improve the image they have made. After observing the new phenomena given by the lecturers, the students then explore to create new images using eBeam Capture application.

2.5 Data Collection
Scientific reasoning skill data is collected by giving multiple choice questions before and after learning activities and viewing the learning recordings and discussions on drawing-based modeling implementation using the eBeam Capture computer application and assessed in the rubric of scientific reasoning [11].

2.6 Data Analysis
Scientific reasoning test is performed by giving test items. The test consists of 30 items written test of multiple choice with 5 choices. The results of the test are divided into six indicators of scientific reasoning skills, namely generativity, elaboration, justification, explanation, logic coherency, and synthesis.

3. Result and Discussion
Scientific reasoning tests are performed using the written pre-test and post-test, this test consists of 30 items written test of multiple choice with 5 choices. The results of the test are divided into six indicators of scientific reasoning skills, namely generativity, elaboration, justification, explanation, logic coherency, and synthesis. Based on those reasoning indicators, the research results can be seen in Figure 2 below:

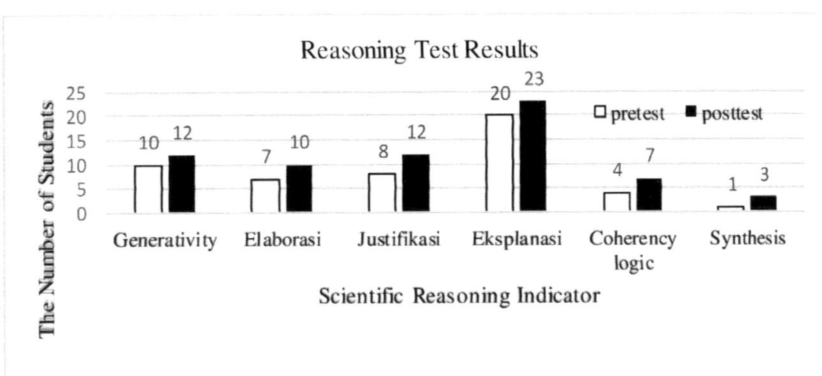

Figure 2. Scientific reasoning test results data

The graph 2 shows the number of students in each scientific reasoning which achieve high scores in pre-test and post-test of the indicator of the implementation of drawing based modeling approach assisted with eBeam Capture in Cell Biology. In general, it appears that the students' scientific reasoning skills in every indicator show improvements from pre-test to post-test, this is shown in generativity reasoning which has 2 students achieving high scores. Then in elaboration reasoning consists of 3 students, justification consists of 4 students, explanation consists of 3 students, coherency logic consists of 3 students, synthesis consists of 2 students. This research result also shows that in explanation reasoning consists of the most number of students with high scores, which is 20 students in the pre-test and 23 students in post-test, this is because of the students' ability in making and explaining pictures with the explanation of parts of the picture completely. While synthesis reasoning consists of the least number of students achieving high scores, which is only 1 student in the pre-test and 3 students in post-test, this is because the students could not synthesize image explanation that they made, therefore, the answers did not form a systematic theory structure.

The students' scientific reasoning skill is still categorized low in Biology Cell using drawing based modeling approach assisted with eBeam Capture, because the use of eBeam Capture application in assisting drawing based modeling is the first time ever used in Biology Cell. The use of a computer application for the first time should be tested frequently and designed insomuch to the students with hope for easier use in a learning activity [11].

Students' scientific reasoning skill is still low because in the process of teaching and learning especially in group discussion, there are still many students that could not cooperate with others. Scientific reasoning skill is gained by the process of thinking logically individually and collaborative cognition through a problem-based teaching and learning model [12]. Besides, based on the journal article analysis result that has been conducted, reasoning skill can be improved through the application of some inquiry teaching and learning model [13], Problem Based Learning [14], ICT-based [15], playing and group discussion [16]. In other context, besides the application of teaching and learning model to the improvement in reasoning skill, it can be in form of reasoning correlation to conceptualization [17], presentation [18], logical thinking [19].

Based on the progress of students' scientific reasoning skill both on pre-test and post-test can be seen in table 1 below:

Table 1. Result data of pre-test and post-test of scientific reasoning

Test	Student	N	Total Test Score	
			SD	G-Gain
Pretest	All	25	1,45	g=0,05
Postes	All	25	1,50	

The table 1 above shows that the score of post-test (SD = 1.50) is statistically higher than the score of pre-test (SD = 1.45), with N-Gain of 0.05. This number shows that drawing based modeling approach assisted with E-Beam Capture still gives contribution although it is not so huge or not too significant.

From the score, it can be seen that the lecturer using the Drawing-Based Modeling approach assisted with E-Beam Capture could not yet support students' scientific reasoning skill in Cell Biology. This is in accordance with the explanation that the role of teachers in learning using a drawing approach with the help of computer media should be tested initially and re-designed in order to improve a scientific thinking skill of students [20].

4. Conclusion

Based on the research that has been conducted above, it can be concluded that Drawing-based Modeling approach assisted with E-Beam Capture can not yet support students' scientific reasoning skill comprehensively, this can be seen from the very small and the number of students achieving high scores in every reasoning indicator is still very small. E-Beam Capture serves as a medium in the Drawing-based Modeling Approach which can not yet support scientific reasoning skill comprehensively, as seen in the post-test score and pre-test score with a very small. The Drawing-based Modeling Approach with E-Beam Capture Assistance in Cell Biology course should be simulated or tested first. Trials are conducted as an effort to anticipate the lack of implementation of the learning steps due to the limited time. Through the trial, it can be known how much the time should be allocated at each stage of learning and the weaknesses contained in the scenario that has been prepared. Thereby can be done improvement effort before applying to the actual research.

Acknowledgments
That is all that the author can convey the material that has been discussed in this article. Acknowledgments to all those who have assisted in conducting this research. I want to say thankyou for my Lecturer Dr. rer. Nat. Adi Rahmat, Prof. Hj. Dr. Sri Redjeki, M.Pd and Dr. Taufik Rahman, M.Pd for guidance and advice so that I can finish this disertation. The authors also thanked to the Directorate General of Higher Education who has funded this preliminary study.

References
[1] Louca, L T and Zacharia, Z C 2011 Modeling-based learning in science education cognitive, metacognitive, social, material and epistemological contributions (*Educational Review* vol 64) pp471–492
[2] Rutten N V, Joolingen W R and Van Der Veen J T 2012 The learning effects of computer simulations in science education (*Computers & Education* vol 1) pp 136–153
[3] Ainsworth S, Prain V, and Tytler R 2011 Drawing to learn in science (*Representations 333)* pp 1096–1097
[4] Rusman 2010 *Model-model Pembelajaran* Jakarta PT Raja Grafindo Persada pp 27-29
[5] Lawson A E 2009 *J. Science and Mathematic Education,* 2 307
[6] Harrison A G and Treagust D F 2000 *J. Science Education* 22 9 1011
[7] Rouwette E, Vennix J and Thijssen C M 2000 Group Model Building A Decision Room Approach (*Simulation and Gaming* vol 3) pp 359–379
[8] Alberts B, Bray D, Johnson A, Lewis J, Raff M, Roberts K and Walter P 1998 Essensial Cell Biology (*Garland Publishing)*
[9] Karp G 1999 Cell and Molecular biology : Concept and experiment 2nd Ed (*John Wiley and Sons, Inc*) chapter 8
[10] Creswell J W 2007 *Desgning and Conducting Mixed Methods Research.* University of Nebraska-Lincoln pp 67-68
[11] Heijnes D 2015 *Stimulating scientific reasoning with drawing-based modeling* Utrecht University
[12] Kuhn D2015 Thinking Together and Alone (*Educational Research* vol *1)* pp 46-53
[13] Kilic D 2014 *J. Biological Education* 2 63-70
[14] Jensen L J 2011 Effects of collaborative group composition and inquiry instruction on reasoning gains and achievement in undergraduate biology (*CBE—Life Sciences Education* vol 10) pp 64–73
[15] Muthivhi A 2012 J. *Psychology of South African* 42 SSNI 0081-2463
[16] Zheng W 2015 *J. PLoS ONE* 10 0120080.
[17] Knight J 2013 Understanding clicker discussions: student reasoning and the impact of instructional cues (*CBE Life Sciences Education* Vol. 12) pp 645–654
[18] Chen, Chun-Ting and She, Hsiao-Ching 2014 *J. Science and Mathematics Education* 20 1Y20.
[19] Mackey A 2015 *J. PLoS ONE* 10 e0137627
[20] Gijlers H, Weinberger A, Van D A M, Bollen L and Van J W 2013 *J. Computer-Supported Collaborative Learning* 810.1007/s11412-013-9180-5

Traditional Knowledge of Local Wisdom of *Ammatoa* Kajang Tribe (South Sulawesi) about Environmental Conservation

H K Surtikanti*, A Syulasmi and N Ramdhani

Departemen Pendidikan Biologi, FPMIPA UPI, Jl. Dr. Setiabudhi no 229 Bandung 40154, Indonesia

*hertien_surtikanti@yahoo.com

Abstract. Education may improve the knowledge how to build the people attitude especially environmental aware surrounding it. The aim was to study about environmental education of the local wisdom people in conserving their environment. The method was qualitative descriptive using second document, questioner/interview instrument and field observation. This research is done in local wisdom of *Ammatoa* Kajang village (South Sulawesi). The respondens were eldery people (*tetua adat*), local governmentand people(15 adult couplesand 15children). The majority of local people was educated at elementary school. Environmental education is studied in school, however informal education is heritated from eldery people. The field study showed that the people keep the environment wisely. It can be proved with the presence of sacred forest, waste recycle, moor (*tegalan*) maintainance, mutual cooperation, no natural resource exploitation, keep clean, etc. The people submissive customs rules and believe that people will get punishment form environment itself. In conclusion, traditional knowledge from community is implemented in caring the environment

1. Introduction

The increasing number of people have an increasing number of human activities in exploiting the environment. People exploited the environment in purpose to fulfill their needs. It affected to the living things [1]. People attitude in caring the environment is influenced by their value beliefs. Their beliefs were reflected in environmental ethics which is a guide life for people how to treat the environment. Their understanding and implementation of environmental ethics is valuable in contributing the environmental conservation [2]. Furthermore, environmental education learning in the school may enhance their understanding in environmental conservation.

Environmental ethics in modern people is replenishing. It can be seen with the environmental damage due to human activities. Having benefit from environment was inbalance with saving the environment. Therefore, environment maynot supply continuously for future generation [3]. Environmental education play important role for people. Because with education, people may enhance and improve their quality life. Conservation and saving the environment is one example that can be learnt in environmental education.

Since long time ago, local wisdom people sustained culture value and environmental ethics, especially in saving the environment and environmental care. Some studies have been proved it. Other study showed that Aboriginal traditional knowledge influenced in saving the environment in Canada [4]. Mount Salak traditional people have belief in maintaining and preserving the environment [5]. Timorese Community in Timor Island appreciated in environmental management [6]. Anak Dalam people in Jambi get benefit from the environment and save the forest [7]. It showed that the local people still follow their environmental ethic. Mostly local wisdom have traditional knowledge in improving environmental knowledge [8].

IOP Conf. Series: Journal of Physics: Conf. Series **895** (2017) 012122 doi:10.1088/1742-6596/895/1/012122

There are numerous local wisdom in Indonesia having 34 provincies. Each provincies may have some tradition local wisdom. Some of them were community of Saham village in West Kalimantan [9], community of *Anak Dalam* in Jambi [6], community of Dompu County in West Nusa Tenggara [10], local wisdom in Karimunjawa, *Dayak Tunjung* community in East Kalimantan [11], *Moronene* community in North East Sulawesi [12], and Dukuh wisdom Village in West Jawa [13]. However, there were limited studies investigating about environmental education of local people wisdom in preserving the environment. Therefore, study about *Ammatoa* Kajang in caring their environment is done.

2. Method

This study is done in Tana Toa, Kecamatan Kajang, Kabupaten Bulukumba, South Sulawesi. Their community is called *Ammatoa* Kajang people. This research used qualitative method. Collecting data technique is done using secondary document, questionnaire/intervieuw instruments, and field observation (Table 1). Secondary document is provided from local government office. Questionnare and intervieuw are done at the same time through leader of the community, 15 parents and 15 children. Those number of respondents is obtained using *purposive sampling*.The number of questionnare and intervieuw (30) are used to obtain data on environmental knowledge, society, belief, culture, livelihood, education and attitude of *Ammatoa* community (Table 2). Field observation is done by investigating all activities of the people in their village.

Table 1. Collection data methods

No	Methods	Data	Source
1.	Secondary data	Geography, peopleand government system	Local government office and Literature studies
2.	Questioner and intervieuw	Background Environmental knowledges of people formal and informal	Teacher, parents, children and leader of community
3.	Field observation	People activities related to environmental conservation	Field

Table 2. Items in questioner and intervieuw

No	Subject	Data
1	Environmental knowledge and education	a. Environmental knowledge in school b. Environmental knowledge in society c. People attitude in environmental conservation
2	Public view in caring environment	a. Society b. Belief c. Culture d. Livelihood e. Education
3	Attitude of *Ammatoa* community to environment	a. Attitude of *Ammatoa* community to environment b. Effort in environmental conservation c. Environmental management Rule and customary law in environmental conservation

3. Result and Discussion

3.1. Study location
Local wisdom of Kajang is one of the local wisdom in Tana Toa village, Kabupaten Bulukumba, South Sulawesi Provinci. Kajang sub-district is about 129.1 km^2, while the area of Tana Toa village is about 5.25 km^2[14]. The village of Tana Toa consists of 8 hamlets, including Dusun Sobbu, Benteng, Pangi, Tombolo, Luarayya, and Balambina located within the *Ammatoa* customary area.

The location of Tana Toa village is $5^0 20'$ LS dan $120^0 22'$ BT. There are 4 rivers surrounded this village, Limba, Doro, Tulia, and Sangkala river. The distance from the capital city to that village is about 270 km.

3.2. Background of Ammatoa Kajang community (government, religion, culture, social and economy)
The Kajang have their own customary government system. They are led directly by an *Ammatoa* as a traditional elder. *Ammatoa* in traditional duties and ceremonies was accompanied by two *Anrongs* and some traditional leaders called *Galla*.

Almost the *Ammatoa* community are Muslim. Although Islam is recognized by the *Ammatoa* community as the only religion within the *adat* area, but in their life are still mixing up with the teachings of their ancestors (*patuntung*). Traditional ceremonies and rituals are widely held in the Kajang tribe. One of the special ceremonies and rituals concerning the environment is the ceremony held after harvest. The ceremony is intended as a form of gratitude to God for what has been given. Rituals of *panrobel* and *tunu passau tunas* are also performed if there are people who break the rules such as cutting trees without the permission of *Ammatoa*.

The culture adopted by Kajang society is the culture of *kamase-kamasea*. The culture teaches the Kajang people to live a simple life. According to them, excessive life can lead to conflicts between communities that will result in disharmony in the community.

Ammatoa's livelihoods are farming, fishing, weaving, trading in the market. While young people and *Ammatoa* community generally choose to live outside the area to study and work. *Ammatoa* food ingredients are generally obtained from farming. The types of plants are tomatoes (*Solanum lycopersicum*), pepper (*Piper nigrum*), rice (*Oryza sativa*), corn (*Zea mays*), vegetable such as green beans (*Vigna radiata*), soybean (*Glycine max*) and others.

As social creatures, family relations among Kajang people are very strong. The *Ammatoa* community also greatly appreciates and respects to other communities. People know their first name, nickname, occupation, and number of families (children and followers). Houses have a big role in social relationships between neighbors and fellow nuclear families because it is used as a gathering place for community. The good relationship between *Ammatoa* Kajang community is formed from routine activities in various activities, such as clean up the environment, help each other when planting rice and harvest together and then the result is divided equally to all *Ammatoa* Kajang community.

3.3. Education process about environment
Education is seen as a lifelong learning process. Learning process may take place anywhere, anytime, anyone for the whole lifelong. The lifelong learning that occurs for every individual in obtaining values, attitudes, skills and knowledge is not only derived from school but also from family, daily experience or from other sources around the environment [6].

Based on the findings in Table 3 it is known that *Ammatoa* Kajang community environmental science is obtained from parents (60%), traditional elders (56.6%), nature (86.6%), education in schools (63.3%), (53.3%) and the community (80%). The environmental science that *Ammatoa* communities gain in schools is tailored to the applicable curriculum with a wider and more structured material review of the environment. While the environmental knowledge obtained by the Kajang Tribe community is based on pairs of Kajang and *patuntung* trust which is more to keep the forest as a natural resource in Kajang custom area. The educational process is largely informal education in which the community at the time of gathering (*A'borong*) with *Ammatoa* in Baruga or during routine activities, discussing the *Ammatoa* Kajang environment, customs and tribes in the Kajang and discussing environmental issues which is rife in the Kajang tribe.

Table 3. Environmental education sources in *Ammatoa* Kajang

No	Sources	Percentages
1.	Parents	60
2.	Elder people	56.6
3.	Environment	86.6
4.	School	63.3
5.	Local law	53.3
6.	*Ammatoa* Kajang community	80

Ammatoa parents and communities provide environmental science to children by way of lectures, direct practice, demonstrations and story telling. For example, when looking for wood for home making, home furnishings and looking for plants to color the weaving thread, parents will take children and teach them verbally and practice directly. The results also show that parents teach their children not only about cultivating and weaving, but taught little by little about Kajang's *pasangri* tides to cut trees in the forest indiscriminately without the permission of *Ammatoa*. Types of trees that are felled are also not just any but only certain trees such as *asa* , *nyato* and *pangi* wood.

The inheritance of knowledge is passed down orally from generation to generation. Their traditional knowledge has formed their attitude in caring the environment [15]. The way of delivering knowledge to the community through lectures is a way considered easily accepted and understood by the community [16]. In this study the results of education from parents and from school, applied in everyday life (Table 4). Similarly, that environmental knowledge obtained by children from their parents, can be understood and implemented to maintain and preserve the environment in Kampung Naga, Tasikmalaya [17].

Table 4. The implementation of environmental education

No	Statements	Percentages
1	The implementation of environmental education from school	76.6
2	The implementation of environmental education from community	83.3

In the sense of *Ammatoa* society (Table 5) contain more meaning than just a message. Each message conveyed is a mandatory mandatory to be obeyed and implemented. When the mandate is not implemented it will lead to negative impact, *i.e.* there is no balance between social and ecological systems. Meanwhile, *patuntung* trust views the function of the forest ecosystem as a source of rain and water springs (*tumbusu*) [18]. Therefore, the *Ammattoa* society treat the environment in maintaining the function of the ecosystem.

Table 5. *Pasang-pasang* of the *Ammatoa* Kajang tribe on forest preservation

No	*Pasang-pasang*	Meanings
1	*Jagai linoa lollongbonenakammayya tompa langika siagangrupa taua siagang boronga*	Keep the earth and its contents (sky, people, and forest)
2	*Nikasipalliangngi ammanra'-manrakia borong*	Prohibited (*kasipalli*) and abstinence
3	*Anjo boronga iya kontaki bosiya nasaba konre mae pangairangnga iaminjo boronga nikua pangairang*	Destroying the forest
4	*Punna nitabbangngi kajuariborongnga,nunipappi rangnga*	If the wood in the forest is cut down, the rain will decrease and the spring will disappear (dry up)
5	*Narie' kaloro battu riborongnga, narie' timbusu battu rijakua na battu ri kalelengnga*	The existence of streams derived forest, the spring comes from trees and lianas
6	*Boronga pangallui nitallasi, erea battu ri kaloro lupayya*	Forests need to be conserved because water comes from small rivers
7	*Iyamintu akkiyo bosi anggenna erea nipake a'lamung pare, ba'do appa'rie'timbusia Anjo*	The forests bring rain water that can be used to plant rice, corn, etc.
8	*Tugasa'na Ammatoa nolarangngi annabbang kaju ri borongnga*	The duty of an *Ammatoa* people is to prohibit logging in the forest
9	*Iyaminjo nikua ada'tana Iyaminjo boronga kunne pusaka Talakullei nisambei kajua, iyato' minjo kaju timboa*	Thus the law that applies here, that the forest is our heirloom. The wood has to grow and it should not be replaced
10	*Talakullei nitambai nanikurangi borong karama, nilarangngi tauwa a'lamung-lamung riborongnga, nasaba se're hattu larie' tau angngakui bate lamunna*	Not allowed to change the other type of wood in customary forests. That's all wood grows naturally. It can not be added, reduced, and prohibited the existence of planting in customary forest, because a time will arise recognition of plantation rights

The ban on cutting timber in the forests is focused on the *borongkaramaka* (sacred forest) bund, while *Ammatoa* communities are allowed to cut trees in *borong batasayya* (border forests). The type of wood that can be cut by the *Ammatoa* community is limited such as timber (*Tamarindus indica*), *nyatoh*, *pangi* wood (*Pangium edulu*). The wood harvested by the *Ammatoa* community for building houses and making furniture.

The real condition that is illustrated through the environment of the cool Kajang tribe and the circumstances outside the region gives its own understanding for the *Ammatoa* community. They are aware that vehicle pollution, household waste and arbitrary acts against forests will destroy the environment.

3.4. The role of education in determining public caring attitude towards environmental conservation.
The community realizes that today the environment tends to be increasingly damaged. The environmental degradation is largely due to human activities and human behaviour that are not environmentally sound. However, despite the environmental damage caused by the lack of care of the

community, there is still hope to improve it is by raising human awareness of the importance of maintaining the environment.

The respondents responses in Table 6 indicates that Ammatoa as elders of *adat*, parents and education in school are equally instrumental in forming the attitude of Kajang society to the environment. Education undertaken by the community at school may increase the insight of the *Ammatoa* community about the importance of protecting the environment. Findings on other research also known that both education in schools and education in the Aga Medaha and Metruna tribes alike affect the behavior change of Aga Tenganan Bali [19].

Table 6. The Role of Education in Caring attitude towards Environmental Conservation

No	Statements	Percentages
1	*Ammatoa* community's attitude towards the environment is influenced by parents	76.6
2	*Ammatoa* community's attitude towards the environment is influenced by education in schools	56.6
3	Kajang's attitude towards the environment is influenced by *Ammatoa*	80
4	The attitude of the *Ammatoa* community to the environment is influenced by people's self-awareness	40

The environmental knowledge that Kajang communities gain from parents and the teachings of the Kajang tide of *Ammatoa* is supported by the knowledge gained from the school will certainly increase the community's insight and further strengthen the community to care about the environment. Education can be utilized as a means of forming awareness of the environment effectively because through education can be realized mental readiness and tendency to behave positively toward a certain object such as environment [20].

In line with that in the history of human life, education is done through an integrated process of informal in everyday life [21]. Humans learn language, behave, values to become effective members of society and through individuals in society such as parents and community leaders. Therefore education such as informal education through family and environmental education is needed to ensure in the management and conservation of the environment in a certain region or tribe such as *Ammatoa* Kajang.

3.5. The role of education towards the psychomotor attitude of Ammatoa society to the environment

Based on this research, it is known that polluting the environment and cutting down trees is one form of prohibition that should not be done by the *Ammatoa* community. Because it is included in custom rules and can cause damage to nature. In addition to the necessities of life, nature is also respected in order to keep the Kajang community from all forms of threat.

The attitude shown by the community in maintaining the environment is reflected in the daily behavior of the community (Table 7). More than 63.33% of the people have done positive activities in managing waste, preserving the forest and utilizing the yard. It is shown that, both traditional and formal education may build their character in caring the environment. Similar study also showed that moral character of Karimunjaya society reflected with the high conservation of wildlife and sacred places [21]. Based on this, it is appropriate if the environment is well managed by the community for the welfare of mankind. The purpose of environmental management is to achieve the harmony of the relationship between humans and the environment as a goal to build a whole human, controlled resource utilization wisely and manifestation of human as environmental coach [22].

Table 7. Psychomotor Attitudes Ammatoa Society Against the Environment in the Kajang Tribe

No	Statements	Percentages
1	The *Ammatoa* Kajang community do not dispose of the garbage	83.3
2	*Ammatoa* Kajang people do not justify to dispose of garbage	86.6
3	The *Ammatoa* Kajang people do not throw garbage in the river	80
4	*Ammatoa* Kajang communities sort out organic and inorganic waste	73.3
5	The garbage that is in the neighborhood by the Ammatoa Society is collected and then buried	63.3
6	The *Ammatoa* Kajang community utilizes the surrounding yard for a useful crop	83.3
7	Cut down indiscriminately forest trees is not allowed in the Kajang tribe	83.3
8	The *Ammatoa* Kajang community replanted after cutting down trees	90
9	To rebuke people who do not take care of the environment needs by *Ammatoa* Kajang Community	96.6

The role of rules and customary law related to environmental conservation in Kajang tribe
Based on the responses of respondents in Table 8 shows that 76.6% of respondents think they do not object to the customary rules applicable in the Kajang. 93.3% of respondents feel the rules to protect the environment and maintain the forest is very beneficial to the life of the community in the sense that the customary rules that notabennya have a purpose to maintain and preserve the environment, especially the forest is not felt burdensome *Ammatoa* community. They believe that the rule will not bring harm to their lives but to give good luck in the world and in the hereafter.

Table 8. Traditional law related to environmental conservation

No	Statements	Percentages
1	The *Ammatoa* community do not object to customary rules regarding preserving the forests	76.6
2	Rules (preserving the environment and preserving the forest) are beneficial to *Ammatoa* Kajang community life	93.3

4. Conclusion

Ammatoa people have learnt environmental science from school and community. Young generation received their traditional knowledges mostly from their parents, eldery people and their environment. Their knowledges are heritated through young generations. The learning process took place with giving a lecture and field exercise during routine daily life actifities. This learning method was effective and easily to understand. Their knowledges affected to their positif attitudes and efforts in saving the environment and keeping the natural resources (forests) from high exploitation. It is good learning for ordinary people to implement it in their daily life.

Acknowledgements
Thanks to *Ammatoa* community for their valuable information on this research. This work is supported by UPI through competitive grants (PPKBK-UPI 2017) that funded this research.

References

[1] Affandi R 2013 Integrasi pendidikan lingkungan hidup melalui pembelajaran IPS di sekolah dasar sebagai alternative menciptakan sekolah hijau. *Pedagogia,* pp 98-108
[2] MKDU PLH 2014 *Pendidikan lingkungan hidup*. Semarang: Universitas Negeri Semarang
[3] Junianto B 2007 *Persepsi, Sikap dan Perilaku Masyarakat Sekitar Terhadap Keberadaan Hutan Penelitian Haurbentes (Studi Kasus di Desa Jugalaya, RPH Jasinga, BKPH Jasinga)*. Skripsi Fakultas Kehutanan IPB Bogor
[4] Hobson G 1992 *Traditional Knowledge is Science*. CARC Northern Perspective. **20** (1)
[5] Royyani M F & Waluyo E B 2012 The Perspective Environment Basedon Local Beliefs and Ritual Tradition: A case study of communities in Mount Salak. *Journal of Biology Indonesia*. **8** (1) pp 71-84
[6] Ardan A S, Ardi M, Hala Y, Supu A & Dirawan G D 2015 Needs Assessment to Development of Biology Textbook for High School Class X-Based the local Wisdom of Timor. *International Education Studies* Vol.8 No.4.
[7] Andhika R R, Hariyadi B, Saudagar F 2015 Etnobotani Penghasil getah oleh Suku Anak Dalam di Taman Nasional Bukit Duabelas Kabupaten Soralangun, Jambi. *Jurnal Ilmu Pertanian Indonesia* (JIPI).
[8] Ardan A S 2016 The Development of Biology Teaching Material Based on the Local Wisdom of Timorese to Improve Students Knowledge and Attitude of Environment in Caring the Perservation of Environment. *International Journal of Higher Education* **5**, No 3.
[9] Winda G W, Budhi S & Sisillia L 2015 Etnobotani Masyarakat Desa Saham. *Jurnal Hutan Lestari*. **4** (1) pp 1-8
[10] Nurrahmah N, Zamroni Z & Sumarno S 2016 An Etnographic Study of Elementary Education in the Rural Area of Dompu County, the Province of West Nusa Tenggara. *Research and Evaluation in Education*. **2**. 1 pp 79-91
[11] Setyowati FM 2010 Etnofarmakologi dan pemakaian tanaman obat Suku Dayak Tanjung di Kalimantan Timur. *Media Litbang Kesehatan* **20** (3) pp 104-112.
[12] Setiawan H & Qiptiyah M 2014 Kajian Etnobotani Masyarakat Adat Suku Moronene di Taman Nasional Rawa Aopa Watumohai. *Jurnal Penelitian Kehutanan Wallacea* **3**(2) pp 107-117
[13] Hidayat S 2009 *Kajian etnobotani masyarakat Kampung Adat Dukuh, Kabupaten Garut Jawa Barat*. Skripsi. Institut Pertanian Bogor
[14] Madasir H 2010 *Statistik Kabupaten Bulukumba (Statistical yearbook of Bulukumba)*. Badan Pusat Statistik Kabupaten Bulukumba.
[15] Riggs E M 2004 *Indigenous fields based education and knowledge of Essential Companents of Geoscience Education for Native American Community*. Wiley Periodicals Inc. Sci (ed) pp 1-18
[16] Saputro FBE 2012 *Pelestarian kearifan local masyarakat desa pakramantenganan pegringsingan dalam pengelolaan hutan*. [Online]. Diakses dari http://repository.unej.ac.id/bitstream/handle/123456789/59077/Fajar%20Bagus%20Ekanda%20Saputro.pdf?sequence=1
[17] Pertiwi S E 2016 *Pendidikan pengetahuan lingkungan masyarakat lokal Kampung Naga Tasikmalaya mengenai konservasi lingkungan*. Skripsi. Universitas Pendidikan Indonesia
[18] Istiawati NF 2016 Pendidikan karakter berbasis niali-nilai kearifan lokal adat ammatoa dalam menumbuhkan karakter konservasi. *ISSN*. ***10*** (1). pp 1-17
[19] Fatimah T 2016 *Pendidikan Lingkungan masyarakat Suku Aga Tenganan Pegeringsingan Bali mengenai Konservasi Lingkungan*. Skripsi Program Studi Pendidikan Biologi FPMIPA UPI.

[20] Hamzah S 2013 *Pendidikan lingkungan.* Bandung: Rafika Aditama.
[21] Sudarmin & Pujiastuti S E 2013 Scientific knowledge based culture and local wisdom in Karimunjaya for growing soft skills conservation. *International Journal of Science and Research* **4,** pp 598-604
[22] Pramudya S 2001 *Melindungi lingkungan dengan menerapkan ISO 14001.* Jakarta: PT Grasindo.

IOP Conf. Series: Journal of Physics: Conf. Series **895** (2017) 012123 doi:10.1088/1742-6596/895/1/012123

Effects Total Solar Eclipse to Nasty Behaviour of the Several Legume Plants as a Result Student Research

S Anggraeni*, S Diana and B Supriatno

Departemen Pendidikan Biologi, Universitas Pendidikan Indonesia, Bandung, Indonesia

*anggraeni _said@upi.edu

Abstract. Some group students of plant Physiology course have given task to do free inquiry. They investigated of the nasty behaviour of several legume plants in response to changes in light during the partial solar eclipse that occurred at March 9, 2016. The investigation carried out in UPI Bandung, West Java, Indonesia, which is in the penumbra region of a total solar eclipse with the location coordinates of latitude: -6.86105, longitude: 07.59071, S 6057' 37.53553 "and E 107035' 24.29141". They were measuring the movement of opening leaves every ten minutes at the beginning of the start until the end of the eclipse compared with the behaviour without eclipsing. Influence is expressed by comparing the leaf opening movement (measured in the form of leaf angular) at the time of the eclipse with a normal day. Each group was observed for one plant of the legume, there are: Mimosa pudica, Bauhinia purpurea, Caesalpinia pulcherrima, and Arachis pintoi. The results showed that the changes in leaf angular in plants Mimosa pudica, Caesalpinia pulcherrima, and Arachis pintoi differently significant, except for Bauhinia purpurea. In conclusion, the total solar eclipse in the penumbra area affects the movement of some nasty legume plants. It is recommended to conduct a study of the nasty behaviour of legume plants in the area umbra in the path of a total solar eclipse.

1. Introduction

The total solar eclipse that crosses Indonesia at March 9, 2016 is a very rare phenomenon. The eclipse not only crosses Indonesia but also Asia, Australia and the Pacific. Especially in Indonesia, the eclipse across Sumatra, Borneo, Sulawesi and Maluku islands [1] tracking the umbra's shadow from the moon on the earth is called the Totality path. It is usually 10,000 miles long but only about 100 miles wide. It covers less than 1% of the entire earth's surface area. To see the Sun being completely obstructed by the moon, the observer must be somewhere in the narrow path of totality. The shadow of the solar eclipse also swept the island of Java [2].

There was an atmospheric anomaly in the British Isles caused by a partial solar eclipse on March 20, 2015. Decreased temperatures indicate consistent cooling and increased humidity associated with eclipse events [3].

Solar eclipses affect animals too; birds will stop singing when the moon passes in front of the sun. The darkness of clogged sunlight triggers the behavior of nocturnal birds. Nocturnal animals become alert and livestock, pets looking for a place to sleep. Insects, moths and butterflies are affected by the change of light. A total solar eclipse will affect the way moths and butterflies, because they fly use light for navigation [4]. While [5] studied at a farm in Ternate, apparently during a total eclipse on

March 9, 2016 several species of birds (Columbiformes, Galliformes, Anseriformes) were silent and immobile. While some fish species (Claridae, Cypridae) take shelter in shallow waters. This is due to the influence of environmental factors such as light, temperature, and humidity. Behavior of animals is observed by the method of continuous recording during the eclipse.

How about plants? [6] An observation to circadian rhythm of Mimosa pudica plants heve been conducted. Observations were made since the total solar eclipse on March 9, 2016. The results showed that the period of total solar eclipse was very short, but it has influence on showing by behavioral change of Circadian rhythm of this plant.

Students that participate in the Plant Physiology course at Unversitas Pendidikan Indonesia, each approaching the end of the lecture is always given the task to conduct free inquiry. This task is intended to explore physiological phenomena around the campus or the daily life. Along with that total solar eclipse event, this triggered the spirit of students to search the topic of research on the physiology of plants associated with solar eclipse.

There are many tropical plants that showed phenomena influences by sunlight. Thus, it is very interesting to examine these behavioral changes during a total solar eclipse. The motion of chloroplast a photo nastic phenomena or even at the rate of photosynthesis and respiration are examples. In tropical plants, many physiological phenomena are affected by light, temperature or humidity. For example, red beans (Phaseolus vulgaris) have circadian rhythmic behavior or biological clocks that are influenced by infrared and far red rays. Bending up and down of the leaves in red beans are influenced by phytochrome that is sensitive to changes in sunlight [7, 8].

Students make some research topics about plant physiology relating to solar eclipse. Almost groups of students proposed about nastic behavior of plant and others about plant metabolism. How they do and what their results, will share in this paper, especially about what effects of total solar eclipse to nasty behavior of the several legume plants. So, the aims of this study to describe some of the physiological behavior of tropical plants primarily nastic movement the legume plants during a total solar eclipse in the penumbra area that occurred on March 9, 2016 in Bandung city.

2. Method

Four group students proposed research as a free inquiry for nastic movement of the leaf from the legume plants. Each group worked with one of these plants, that is Mimosa pudica, or Bauhinia purpurea, or Arachis pintoi, or Caesalpinia pulcherima. Their planning was evaluated by lecturer before they did the work.

They conducted the research at the place of the occurrence of partial solar eclipse on campus Universitas Pendidikan Indonesia in Bandung, West Java with coordinates: latitude: -6,86105, longitude: 07.59071, S 6057'37,53553 "and E 107035'24,29141". On Wednesday (March 9, 2016) morning in Bandung starts at 06.19 and ending at 08.32 when a partial solar eclipse occurs and at the same time on Thursday morning (March 10, 2016). They were using tools like hygrometer, light meter, thermometer, hand phone camera, Canon camera, standing camera and computer.

The effect of partial solar eclipse on the legume plant as a nastic movement is expressed by measuring the change in the magnitude of the movement of the opening and closing of the legume leaves during the solar eclipse. The size of the leaf angle is measured using Corel Draw based on the leaf motion picture. The magnitude of the angular is expressed in degrees.

The behaviour of opening the closing of the leaves is followed by photo recorder every ten minutes from the beginning of the solar eclipse to completion. Measurements were also performed on other days of normal circumstances beyond the day of a solar eclipse. Their observations were recorded using observation sheets by several observers and recorded using a camera. Meanwhile, the changes in light radiation will be measured using flux meters, air temperature and humidity using a Hygrometer per ten minutes.

First, normal distribution of the data calculated to determine using of parametric or nonparametric tests. The data of leaf angular between the partial solar eclipse and the normal angular calculated using t tests.

IOP Conf. Series: Journal of Physics: Conf. Series **895** (2017) 012123 doi:10.1088/1742-6596/895/1/012123

Figure 1. Showed the leaf angle and how measured by corel draw

3. Result and Discussion

They have worked hard and enthusiastic. Their lecturer was give guiding and controlling for ensure the investigated enough valid. They have hypothesis that there was no significant difference in opening movements of the plant leaf legume between partial solar eclipse and normal day. They results have been resume and report here.

Table 1. Normal distribution and independent sample t test of angular leaf from legume plants

Legume plants	Sig. Normal distribution test of the Kolmogorov Smirnov		independent sample t - test
	Non solar eclipse	Solar eclipse	
Mimosa pudica	.003	.129*	.011*
Bauhinia purpurea	.678*	.678*	.403
Caesalpinia pulcherrima	.445*	.973*	.000*
Arachis pintoi	.174*	.982*	.000*

Because data on the four plants have normally distributed so we continued with independent sample t-test to examine the difference. It turns out that from the four legume plants has significant difference except *Bauhinia purpurea*. The results showed that there was a significant difference of the leaf angular between the partial solar eclipse and the normal condition (without solar eclipse). The difference was seen in three legume plants tested, that is *Mimosa pudica*, *Caesalpinia pulcherrima* and *Arachis pintoi*. The leaves opening at a partial solar eclipse slightly smaller than at normal day, but for *Bauhinia purpurea* shown the angle of the leaf look coincident at the both conditions.

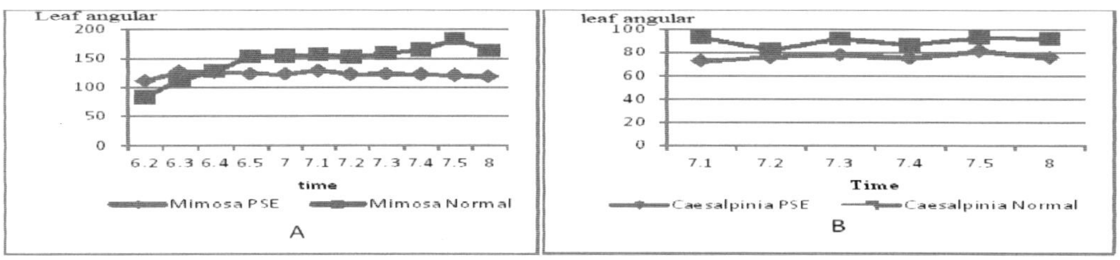

Figure 2. Leaf angular difference of Mimosa pudica (A) and Caesalpinia pulcherrima (B) between Partial Solar Eclipse (PSE) and normal day

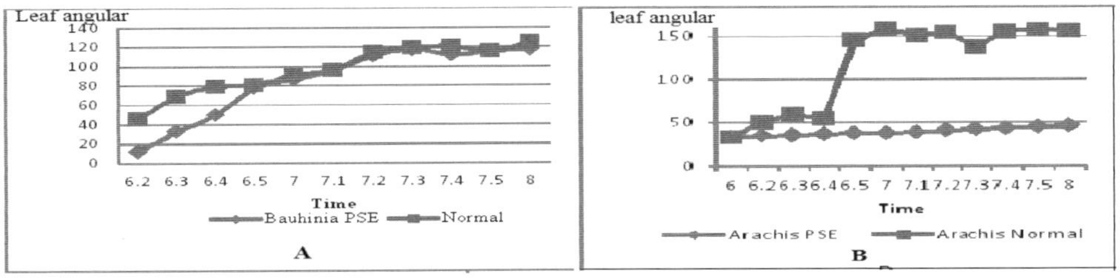

Figure 3. Leaf angular difference of Bauhinia purpurea (A) dan Arachis pintoi (B) between Partial Solar Eclipse (PSE) and normal day.

Mimosa pudica, Caesalpinia pulcherrima and *Arachis pintoi* opening of the leaf at a partial solar eclipse is smaller angular than during normal day (non PSE). This phenomenon was corresponded to environmental factors as light intensity, temperature and humidity. Partial solar eclipse influenced environmental factors, both on the intensity of light reaching to the earth and the temperature of the environment. This can be seen from the following graph, the intensity of light at the time of the eclipse decreased compared to normal times. Reduced light intensity will lower the environmental temperature, but it will not automatically affect the humidity. This is evident from the following graph:

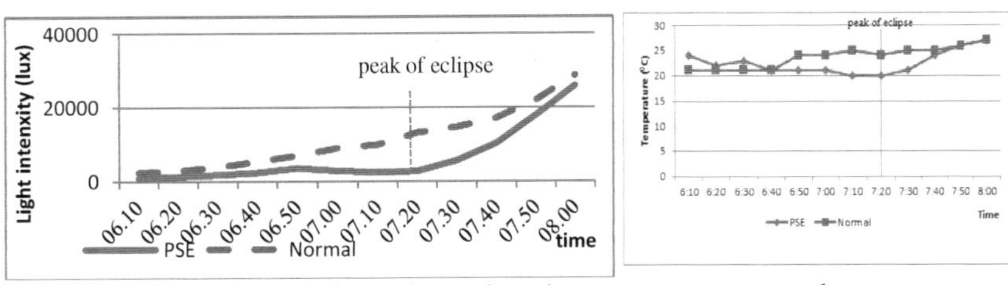

Figure 4. Light intensity and environment temperature between Partial Solar Eclipse (PSE) and normal day

The phytochrome system provides information for plants about the quality of light. The sensing mechanism allows plants to adapt to changes in lighting conditions [7].

Light affects the blue pigment (Phytochrome) present in plants. In the morning the phytochrome should absorb the red light wave as much as possible to transform itself into the pythochrome far Red (Pfr). Sunlight in the morning contains a lot of red light, so converting to Pfr is faster than converting to Pr, therefore the ratio of Pfr to Pr increases in sunlight. If the sunlight is weak, so the formation of Pfr will decrease this will affect the metabolism of plants, so the energy to implement the K + pump was reduced. This will affect the movement of nasty because the plants in the morning will open the leaves with perfect energy. According to [7] most plant processes, such as the transpiration and synthesis of certain enzymes, rise and fall throughout the day. Some of these cyclical variations are responses to changes in the intensity of light, temperature, and humidity that accompany the 24-hour day-to-night cycle. If there were more energy will more actively the K + pump works and the more leaf will open perfectly. Conversely, if the light intensity is less, the temperature decreases it will decrease the activity of metabolism so the energy decreases. That is, which causes the difference of leaf angle when legume plants *Mimosa pudica*, *Caesalpinia pulcherrima* and *Arachis pintoi* starting opening in the morning when a solar eclipse comparing with normal day.

The changes leaf angular of *Bauhinia purpurea* not significant because the plant not more sensitive than to the other plants in light changes, although *B. purpurea* is often seen to have closed and leaf-opening movements, especially if the temperature decreases or the humidity rises high enough. The fact in heavy rain or late afternoon this plant will close its leaves. So the changes in temperature and light intensity at the time of the eclipse cannot be responded well or less affect the daily activity of this

B. purpurea. Other possibilities *B. purpurea* is not sensitive to light and temperature drop, but more sensitive to moisture enhancement. While based on data, the humidity in both conditions is not too different.

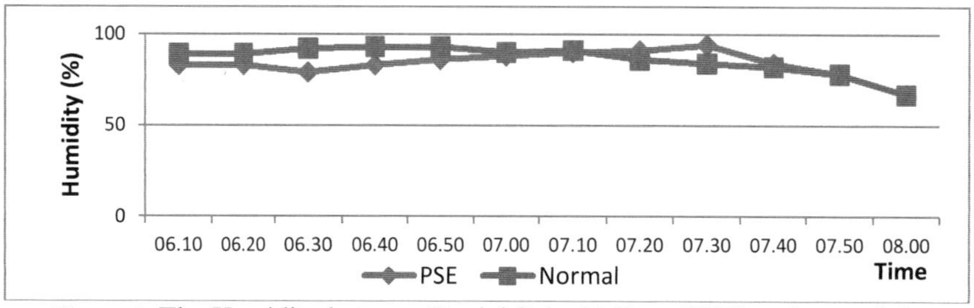

Figure 5. The Humidity between Partial Solar Eclipse (PSE) and normal day

The cycle of opening and closing leaves or flowers in frequency of 24 hours and not directly controlled by environmental variables is called circadian rhythms. Circadian rhythms are common present in all eukaryotic organisms. In the legume plant, this rhythm is shown with nastic movement that is the closing and opening the leaves are driven by motor cells in the pulvinus region [8]. The wide of leaf angle when it opening, it can be influenced by the amount of Pythochrome which absorbed more red rays of light. If more the red light can be absorbed in the morning so the leaves will ability to open perfectly.

The factor that sets the biological clock up to 24 hours a day is light. Phytochrome and blue light receptors can adjust the circadian rhythms in the growing nature; the interaction between phytocrom and biological clocks causes plants to measure the passage of night and day [7].

Some legume plants are able to show the mechanism of this circadian movement clearly as in *Phaseolus vulgaris* plants. Starting from 12 pm, this plant and upright (bending up) leaves slowly until at 12 am really open perfectly. After that, the petiole of leaves began to bending down slowly and completely down (hyponastic) at 12 pm and this happens rhythmically [7, 8].

4. Conclusion

Based on student data of the difference in the opening and closing the leaves of the legume, it can be stated that their research hypothesis "There is no significant difference in the movement of legume leaves between a partial solar eclipse and normal day" is rejected, especially for *Mimosa pudica*, *Caesalpinia pulcherrima* and *Arachis pintoi*. So, partial solar eclipse gives influence to nastic movement on some legumes plants. While the influence of the total solar eclipse (penumbra) on nastic movement in *Bauhinia purpurea* is not proven.

It is interesting to examine the nastic movement on the *Bauhinia purpurea*. This plant does not provide a positive response to the solar eclipse phenomenon so it should be further investigated whether including nyctinastic or photonastic, thermonastic, or hygronastic movement.

Another suggestion is to examine the nastic behaviour of legume plant actually in the umbra line that is in the path of total solar eclipse. It is also interesting to examine other physiological behaviour of plants during solar eclipses, such as photosynthesis, respiration and transpiration.

Acknowledgments

Thanks to our colleagues for their contribution, also to the Department of Biology Education and Faculty of Mathematics and Natural Sciences Education (Universitas Pendidikan Indonesia) for special support. Thanks a lot to students as participants in Plant Physiology course at 2016 and for the data contribution.

References

[1] Espenak F and Anderson J 2016 *NASA Eclipse Bulletins*
[2] Espenak F 2009 Solar Eclipse for Beginners *Mr.Eclipse.com*
[3] Smith J 2015 Solar Eclipse and Its effects *Guardian Liberty voice*
[4] Gray S L and R Harrison 2016 *Phil.Trans.R.Soc.* A 374:20150224
[5] Kusumo H 2016 *Proc. International Symposium on Sun, Earth, and Life* (Indonesia: Institut Teknologi Bandung)
[6] Syamsudin T S 2016 *Proc. International Symposium on Sun, Earth, and Life* (Indonesia: Institut Teknologi Bandung)
[7] Campbel *et al.* 2005 *Biology*. 7th ed. (San Fransisco : Pearson Benyamin Cumming)
[8] Taiz L and Zeiger E 1990 *Plant Physiology* 2th.ed (Sunderland : Sinauer Ass. Inc. Publ.)

Efforts to Handle Waste through Science, Environment, Technology and Society (SETS)

D Rahmawati*, T Rahman and A Amprasto

Program Studi Magister Pendidikan Biologi, Universitas Pendidikan Indonesia, Bandung, Indonesia

*raahmaawaatiidiinaa@gmail.com

Abstract. This research to identify the attempt to deal with the waste through a learning SETS to facilitate troubleshooting and environmentally conscious high school students. The research method is weak experiment, with the design of the study "The One-group pretest-Posttest Design". The population used in this study is an entire senior high school class in Ciamis Regency of Indonesia many as 10 classes totaling 360 students. The sample used in this study were 1 class. Data collected through pretest and posttest to increase problem-solving skills and environmental awareness of students. Instruments used in this research is to test the ability to solve the problem on the concept of Pollution and Environmental Protection, in the form of essays by 15 matter, the attitude scale questionnaire of 28 statements. The analysis N-gain average showed that the SETS problem-solving skills and environmental awareness of students in the medium category. In addition, students' creativity in finding out pretty good waste management by creating products that are aesthetically valuable and economic appropriately.

1. Introduction

The problem of environmental pollution effect on human life and human life at this time to come [1]. It is known that the number of industrial tahu (tofu) in Ciamis Regency of Indonesia as many as 103 companies with a production capacity ranging from 6 tons to 66 tons, and 1,500 tahu up to 90 million tahu [2]. Total industry tahu each year continues to increase along with demand for the tahu product. Of industrial waste treatment process tahu this is still using the wastewater treatment plant (IPAL) is simple. For waste tahu no impact on the environment, it needs special handling of the waste or waste water tahu into a more useful product. Education plays an important role in preventing environmental pollution, environmental destruction, and preserve the environment. Conserve natural resources, developing thinking about environmental issues, building a future clean and healthy obtained from study [3].

The activities carried out in the school environment is expected to become a habit and become a character that is strongly attached to the students [4]. The learning experience can be powered directly through the learning of science, environment, technology, society (SETS). Students are guided to have a sensitivity to the problems in the community and play an active role to contribute to finding solutions [5]. According to one study, learning SETS designed and developed to interpret science and technology as social complexity, and to develop critical faculties, scientific, decision-making, able to act and be responsible for those decisions [6].based learning SETS effective to increase creativity in solving problems of environmental pollution [7]. Determine what the students do and can do is an important part of an effective classroom practice and is an important component of learning [8].

In this study, students were invited to direct observation of tofu factory near their homes so that students can immediately identify the problems that occur in their real environment that knows untreated waste into products that are more useful. After the observation, students design a wastewater treatment product out. Products made of students, namely, chips, nuggets and kecap from the waste out. This research includes elements of science, environment, technology, and society (SETS) more integrated and students to implement what they have learned in their real life. Learning so that SETS this research can facilitate problem-solving skills and environmental awareness of students. Based on the description above, it is important to examine more deeply and more about problem-solving skills and environmental awareness of students by using learning SETS who raised the issue in a real environment such as students or in community waste out.

2. Experimental Method

The research method used is the method of weak experiment/pre experiment. Subjects were all students of senior high school class in Ciamis many as 10 classes totaling 360 students. Samples are first class numbered 34 students, 18 girls and 16 boys. Measurement increase problem-solving skills and environmental awareness of students is carried out through the pretest and posttest, so the design of the study is "The One-group pretest-Posttest Design"[9]. Instruments used in this research is to test the ability to solve problems on the concept of Pollution and Environmental Protection, in the form of an essay of 15 questions, the questionnaire scale of attitude as much as 28 statements with positive statements and negative are required of students at one of the columns consisting of columns yes, hesitated and did not. The data in this study using a support program Microsoft Excel 2010 and SPSS version 24.

3. Result and Discussion

3.1. *Students' ability to solve the problem*

The statistics student's ability to solve problems before (pre-test), after (post-test), and N-gain during the learning process *SETS* can be seen in table 1.

Table 1. Data statistics student problem solving ability

Statistics	value		N-gain
	pretest	posttest	
Average	15.79	31.29	0.55
Variance	9.68	18.64	0.02
At least	10	21	0.19
Maximum	21	39	0.83
Standard deviation	3.11	4:32	0.15
Median	15	32	0.58

Table1 shows that there is a difference between pretest to posttest results and the value of N-gain indicates how much influence learning after learning SETS. The result of more increases than pretest posttest, so that the N-gain medium category. Learning SETS managed to make a difference and increase the number of frequencies on the posttest results. The learning process teachers try to make learning more active and fun. At first studied teachers to motivate students to make observations to the factory know, students were very enthusiastic about the learning activities outside of school and a real touch with their surroundings.

3.2. Students' environmental awareness

Based on the attitude likert scale questionnaire showed students experience a change in consciousness and environmental awareness as evidenced by the percentage difference in attitude scale questionnaire results before and after the learning SETS. Students are directly involved with the problem that is happening is the waste out. Innovation and creativity of students also increased by making a product refined from waste out, the students get a certain satisfaction when they succeeded in designing a product and can make the product. Because of the limited learning time, products made of students is not so perfect, but the cracker products, nuggets and ketchup from the waste out can be consumed and waste signify know can be processed into a very useful product.

3.3. Student activities during the learning SETS

Figure 1 shows the learning activities *SETS* that have been done student, that student successfully treat waste out into a food product that is suitable for consumption.

Figure 1. Figure product crackers and nuggets of waste out

Currently tofu factory observation activities, students carry out the observation of that waste will processed back into a useful product so that students can find solutions to overcome and control the environment caused by waste pollution know. Thus the ability to solve problems and increase the environmental awareness of students learning *SETS*. This research has managed to achieve the main goal of learning SETS namely the development of a scientific attitude of students, conceptual understanding, critical thinking, and problem solving skills through engagement with real-world problems [9].

4. Conclusion

Learning SETS can facilitate problem-solving skills and environmental awareness of high school students in an effort to deal with waste on the concept of Pollution and Environmental Conservation. Any student creativity in making a product can increase. Students can tackling the problem of environmental pollution. However, it takes additional time to actually be able to increase significantly the ability to solve problems and environmental awareness of students, the current regulatory research school schedule is not enough to do more research and more deeply.

Acknowledgments
The authors would like to thank all reviews those who have provided financial and moral support for this research.

References

[1] Cetin and Seda 2010 Enhancing Awareness Students' environmental *Elsevier* **2** pp 1830-1834
[2] Dinas Perindustrian dan Perdagangan 2016 *Company Directory Food, Small Industries, and Medium* (Ciamis)
[3] Kiraz A and Firat A 2016 Analyzing the Environmental Awareness of Students According to Their Educational Stage *Researchers World* **7** (2) pp 2231-4172
[4] Ernst and Elizabeth 2015 Redigned High Schools for STEM Learning Transformed: Pilot Performance Assessment Outcome *Journal of STEM Education* **16** (4) pp 1-10.
[5] Binadja and Ward 2006 *High School Chemistry Teaching Quality Improvement Through Application CBC Visionary* SETS *Approach*. Proposed Research Grants Program Grant A2 Chemistry Department Semarang.
[6] Pedretti 2003 *Teaching Science, Technology, Society and Environment (STSE) Education*
[7] Sugiarto and Djukri 2015 SETS based learning as Efforts to Improve Creativity in Problem Solving Environmental Pollution Innovation *Journal of Science Education* **1** (1) pp 1-11
[8] Walker and Jeffrey 2014 Students Performance Driving Assessment: What Students Can Do Learning *National Council Of Teachers Of Mathematic* **19** (8) pp 468-476
[9] Tal, Dori and Keiny 2010 Assessing Conceptual Change of Teachers Involved in Education and Curriculum STES Development the Stems Project Approach *IJSE* **23** (3) pp 247-262

IOP Conf. Series: Journal of Physics: Conf. Series **895** (2017) 012125 doi:10.1088/1742-6596/895/1/012125

Mastery of Content Representation (CoRes) Related TPACK High School Biology Teacher

W R Nasution*, S Sriyati, R Riandi and M Safitri

Program Studi Pendidikan Biologi, Sekolah Pascasarjana Universitas Pendidikan Indonesia, Bandung, Indonesia

*wildarizkinst@student.upi.edu

Abstract. The purpose of this study was to determine the mastery of Content Representation (CoRes) teachers related to the integration of technology and pedagogy in teaching Biology (TPACK). This research uses a descriptive method. The data were taken using instruments CoRes as the primary data and semi-structured interviews as supporting data. The subjects were biology teacher in class X MIA from four schools in Bandung. Teachers raised CoRes was analyzed using a scoring rubric CoRes with coding 1- 3 then categorized into a group of upper, middle, or lower. The results showed that the two teachers in the lower category. This results means that the control of teachers in defining the essential concept in the CoRes has not been detailed and specific. Meanwhile, two other teachers were in the middle category. This means that the ability of teachers to determine the essential concepts in the CoRes are still inadequate so that still needs to be improved.

1. Introduction

Teachers today are growing according to its function to foster students to achieve educational goals [1]. In the era of globalization, not only the students who are required to have a lot of skill and compete in the technological progress that is increasingly sophisticated. Teachers also need to update their knowledge and skills, both in the fields of technology and science [2]. The rapid development of technology today makes many researchers feel that the technology should become an important part of the learning process, as was done by Koehler and Mishra that add technology into PCK concept introduced by Shulman becoming TPACK [3][4]. They assume that good teaching is teaching not only use components of content and pedagogy alone, but there must be a technology component in it. So that teachers is required to have seven essential components derived from TPACK components, namely TK, PK, CK, TPK, TCK, PCK, TPACK.

With the demands of procurement of component technology and pedagogy in learning, teachers often end up ignoring the importance of the content because it is too focused on the selection or the use of technology and pedagogy. As revealed by Shulman in research Van Driel, Verloop, and De Vos that teachers should not ignore the content in the learning process [5]. The use of technology and pedagogy basically is a tool that can be selected teachers to facilitate the delivery of concepts to students. That is why these components become crucial in the learning process and should be considered by the teacher. Another problem in the classroom is that many of the teachers are not in agreement with the educational background [6]. So that teachers will have difficulty in delivering the right content to the students, even the teachers could be transferred her misconception to students that will raise the conceptual difficulties students [7].

Teachers are expected to select the content to be taught to students, mapping the core material that must be learned by students, where the additional material, and the material is not the time studied by the students [8]. Teachers are also expected to identify misconceptions or errors that may occur in the understanding of their students, as well as overcome the difficulties students in learning concepts to be taught. It certainly can be realized if the teacher has a good knowledge of the content, so it will have a positive influence on the teachers themselves and will have an impact on effective teaching [9]. Besides, the ability of teachers to identify the content to be taught influenced by curriculum factors, knowledge of the subject to be taught, and assessment standards to be used by the teacher [10]. Many studies of CoRes has been done, one of which is used to improve teacher PCK [11]. Research on CoRes also been done on the Concept of Human and Environment which reveals that PCK experienced teachers are in the medium category and prospective teachers in the high category [12]. Based on the explanation above, this study aims to see mastery Content Representation (CoRes) related teacher TPACK in learning biology.

2. Experimental Method

This study is a descriptive study which will give an idea of knowledge Content Representation (CoRes) teacher related TPACK on learning Biology. Researchers did not give the treatment of research subjects and only retrieve data without any changes. The subjects were four teachers biology class X MIA from four different schools in Bandung. Data collected through the document CoRes developed by Loughran [13]. CoRes consisted of 10 questions, namely: (1) What do you intend to the students to learn about this material? (2) Why is it important for students to master these concepts? (3) Related to the concepts to be taught, what concept that need to be explained and untimely known by students? (4) What are difficulties in teaching these concepts? (5) What is error of understanding or students misconceptions that is common in the concept? (6) What factors are likely to influence the teaching of the concept to the students? (7) What is local potential anything that can be used to help convey the concept? (8) What are the procedures or the order of teaching the concept? (9) How do you know the students already understand or not the concepts they are studying? (10) What are source learn that you used in this material?. These CoRes were analyzed using a scoring rubric CoRes with coding 1-3 developed from Hadiyanti [12] then categorized based on the number of coding score into the category of upper, middle, or lower. (See Table 1).

Table 1. Categories of coding results

Number of coding score	Categories
24-30	Upper
17-23	Middle
10-16	Lower

3. Result and Discussion

Data used in this study is the result that obtained from CoRes document which developed by Loughran [13] and filled by participants before learning. All answers from the teacher at CoRes documents described below.

3.1. What do you intend to the students to learn about this material?

Concepts which students need to learn is the essential concepts that are important to know to develop their understanding of this topic [13]. Total concept or big idea raised by the teacher was never determined, but is generally 5-8 teacher raises important concept [13]. Based on the responses of teachers in the document CoRes about Kingdom Animalia, MI 8 raises important concept, RI 5 important concept, EI 6 important concept. As for the material Kingdom Plantae, CI raises 13 important concept. But from the results of interviews conducted, all the teachers said that the essential concept to be taught to students of this topic is only about general characteristics, classification, and role. The exact number of essential concepts that emerged from the interviews could not be determined. Differences in the number of important concepts raised by the teachers influenced by the curriculum, time allocation, and teaching experience [12] [14].

3.2. Why is it important for students to master these concepts?

Two of the three teachers who teach the material Kingdom Animalia say that this concept is important for students so that they can distinguish animals from the different phylum. Students can also identify the animals in the surrounding environment better. In other words that consideration of the teacher to teach this concept is based on curriculum and benefit in student's daily life. However, one teacher can not express the importance of the concepts taught and said that this concept is already listed in the book and have to be taught. This is because teacher consideration in choosing concept is just textbooks.

As for the teachers who teach the Kingdom Plantae argued important concept is that the students should be able to distinguish the types of plants from different families and classes, as well as can identify plants in the neighborhood. A successful teacher is a teacher who can convey the concepts and relate them with student's daily life and create the meaningful learning [13].

3.3. Related to the concepts to be taught, what concept that need to be explained and untimely known by students?

Almost all teachers have difficulties and confusion in answering this question. All the teachers have not been able to determine the limits of the concept. Clearly, even MI said that all concepts must be explained. RI and CI said that the scientific names of animals and plants as well as families or orders of animals and plants do not need to be explained in detail to the students. Teachers should know well the limits content to be delivered to students because not all concepts known by teachers must be taught.

3.4. What are difficulties in teaching these concepts?

Almost all of the teachers said that the limitations of the technology or media used in learning are the main problem in the learning process. Teachers also have difficulty in choosing learning strategy and model. Teachers certainly have its difficulties in learning [12], the difficulties were written by MI is allocation of a limited time, while very much a concept that should be taught. RI and CI says it's hard to teach scholarly language of animals and plants. While EI said it is hard to bring real specimens to class to be introduced to the students.

3.5. What are error of understanding or students misconceptions that is common in the concept?

Teacher's ability to identify errors of understanding or misconceptions of students affected by the teaching experience and knowledge of the content being taught [13]. This is evident from the answers of teachers who teach the kingdom Animalia, which their students often think that animals of the phylum sponges are dying because of similar objects such as rocks, and snakes are invertebrate animals. Teachers expressed misconceptions is the result of a previous learning process and not through individual tests or interviews. Therefore, researchers use the term error of understanding not misconceptions because of misconceptions is a mistake in the learning process of science should be determined through specific testing [12].

3.6. What factors are likely to influence the teaching of the concept to the students?

All the teachers say the same thing about the factors that may affect the pursuit of the concept to students namely strategies, approaches, methods, and models chosen by the teachers. From the interviews conducted, the teacher also said that their understanding of the content being also taught significantly affect the teaching of the concept. MI, RI, and the CI as different teachers say that teaching experience is still less could affect how to teach these concepts. EI as an experienced teacher said that old age had made many of the concepts to be forgotten and should read the material back. EI also said that the conditions of the students and the learning environment is also very influential.

3.7. What is local potential that can be used to help convey the concept?

Exploiting local potential is indispensable in the learning process, the goal is students can more easily understand the concept of an example, which can be seen directly. Exploiting local potential will also be vital for the preservation of local wealth. Moreover, teachers can improve the quality of learning by utilizing and maximizing the potential of locally owned in learning [15]. But in the learning process, especially the concept of kingdom Animalia invertebrate, no one of the teacher can exploit local potential. This is because most of the examples of invertebrates from the sea, while in the district Malangbong not has local potential from the sea. Whereas for example other invertebrates like Insecta widespread in any area, but the teacher does not utilize it. As for the material kingdom Plantae, CI also said that there is no typical Bandung plants that could be introduced to the students. CI only use the plants around the school but are not a common vegetation of the area though there are Garut orange (Malangbong entered district Garut) and Pelung chicken which is typical of plants and animals from West Java.

Some things that may affect the utilization of local potential is teachers' understanding of the natural resources owned by the region and the potential availability of locally owned area is associated with content. Because not all of the content in Biology owned by an area. Therefore, the ability of teachers to identify the content and connect with local potential is needed to create a more meaningful learning process.

3.8. What are the procedures or the order of teaching the concept?

The procedure teaching is tailored to the teacher's lesson plan, but the RI said that the teaching procedures used to focus more on the order of the material in the textbook. The method of teaching utilized by all teachers is the conventional method. Specifically on EI and CI adds the discussion method in learning. But overall learning implementation that teachers use still conventional methods. When asked in an interview, the reason teachers are still using conventional methods because of limited facilities in schools so that there is no other choice than conventional methods. This indicates that the pedagogy of teachers is still less because it can not create meaningful learning on the grounds no technology can be utilized. Whereas the procedures taught by Mitchel and Mitchel is a tactic that selected teachers is a procedure to use, when, how, and why to increase the difference aspects of learning [13]. Although the actual procedure was chosen teaching can not guarantee to learn, but if used in the right way and the right time can influence the thinking of students and may be able to improve their understanding of learning better [13].

3.9. How do you know the students already understand or not the concepts they are studying?

All teachers choose an oral test to determine students who already understand or not the concept have learned at the end of each lesson. The oral test is done through questions and answers to students. However, CI teachers always do a written test in addition to the oral tests were performed. And all the teachers implement a written test end of each chapter of the concepts learned. This is consistent with that expressed by Loughran [13] that the teacher should constantly monitor students' understanding and progress of both formal and informal.

3.10. What are learning resources that you used in this material?

In the learning process carried out, almost all teachers use learning resources in the form of the same textbooks used by students, in addition to using the internet as other supporting material. But CI teachers use the plants in the surrounding environment as a source of learning support. Based on the results of interviews conducted, the teacher said they rarely read journals / articles and textbooks to supplement the information. The reason is that it is easier to understand the content in the textbooks of students while journals and textbooks are tough to understand because of the discussion that is too deep.

Based on the answers to the teacher on the document CoRes, CoRes are then analyzed using a scoring rubric CoRes with coding 1-3 and categorized into a group of upper, middle, or lower. The results of the analysis of the coding of mastery *Content Representation (CoRes)* biology teacher presented in the table below.

Table 2. Summary of analysis results of coding CoRes.

Teachers	Total score coding	Category
MI	16	Lower
RI	16	Lower
CI	22	Middle
EI	19	Middle

Table 2 shows that CoRes mastery of MI and RI are in the lower category while CI and EI are in the middle category. At the lower category, essential concepts that appear in the CoRes can not cover the essential concepts and details. Can not yet determine the important role of the concepts taught to students. Learning procedure is not described in detail and only refers to the textbook. Limitations of school facilities are the reason teachers still prefer the conventional method so that the learning procedure is also not specified in detail. In the middle category, teachers have started to cover the essential concepts and begin to be explained in detail. The determination of the essential concepts for students, not only based on the concept and curriculum structure but also consider the benefits in the daily lives of students. CI learning procedure refers to the RPP have been made and curriculum while EI refers to textbooks and curriculum. But learning methods have remained conventional and discussions. Has not been able to identify misconceptions students in detail and not exploit local potential to the fullest.

Based on the CoRes documents that have been analyzed and the results of interviews have been conducted shows that the control of teachers to the CoRes is still lacking. The lack of supporting facilities such technology in learning make teachers tend to retain the conventional method. In fact, 3 of 4 teachers are young and new teachers to learn about a wide range of strategies, methods, and models of learning, but do not apply them in the learning process to be carried out. All the teachers said that the characteristics of the concept of a determinant in choosing technology and pedagogy to be used. But CoRes results show that the lack of facilities technology makes the teacher choose conventional methods again and ultimately failed to create meaningful learning, or motivate students to learn more about the biology of [16].

4. Conclusion

Conclusion of this research shows that there are no teachers who are in the top category. Two teachers are in the lower category and the other two are in the middle category. This is because the new teachers were first to answer the questions CoRes and have never been required to determine the essential concepts that should be taught to students. Besides the lack of support from the school to the progress of teachers in teaching and lack of evaluation of the performance of teachers often lack the mastery of teachers to CoRes. It is also shown that ability of teacher's TPACK is still lacking.

Advice from researchers to the readers of this article is to conduct research on participants more with the school situation is more varied, ranging from limited to the school with complete facilities. The aim was to see whether the limited means of support also affects the ability to integrate content into technology and pedagogy. Readers can choosing different research areas to see the extent to which culture can influence the learning process. Another suggestion is to compare teachers who taught biology from biology education background and biology nature.

Acknowledgments
Researchers say many thanks to the supervisors Dr. Hj. Siti Sriyati, M. Si. and Dr. H. Riandi, M. Si., and Dr. H. Ari Widodo, M. Ed. which has helped researchers from planning a research proposal to prepare research as well as the writing of this article. Researchers are also very grateful to the participant and schools in Bandung who have been willing to be a part of this study. And thank you very much for my parents who always pray for me, support, and gave funds for this research.

References
[1] Hamalik O 2014 *Psikologi Belajar & Mengajar* (Bandung: Sinar Baru Algesindo)
[2] Sani R F 2014 *Pembelajaran Saintifik untuk Implementasi Kurikulum 2013* (Jakarta: Bumi Aksara)
[3] Koehler M J and Mishra P 2009 *Contemporary Issues in Technology and Teacher Education* **9** 60
[4] Shulman L S 1986 *Educational Researcher* **15** 9
[5] Van Driel, Verloop N and De Vos W 1998 *Journal of Research in Science Teaching* **35** 675
[6] Mulyasa H E 2015 *Guru dalam Implementasi Kurikulum 2013* (Bandung: PT. Remaja Rosdakarya)
[7] Hashweh M Z 1987 *Teaching & Teacher Education* **3**
[8] Sanjaya W 2006 *Strategi Pembelajaran Berorientasi Standar Proses Pendidikan* (Bandung: Prenada Media)
[9] Käpylä M, Heikkinen J P and Asunta T 2009 *international Journal of Science* Education **31** 1399
[10] Williams J, Eames C, Hume A and Lockley J 2012 *Research in Science & Technological Education* **30** 337
[11] Williams J 2012 *Journal of Technology Education* **24** 43
[12] Hadiyanti L N, Widodo A and Rochintaniawati D 2015 *Int. Conf. on Mathematics, Science, and Education* September 2015 Semarang
[13] Loughran J, Berry A and Mulhall P 2012 *Understanding and Developing Science Teacher's Pedagogical Content Knowledge* 2nd Edition, ed J J Loughran *et al* (Rotterdam: Sense Publisher) chapter 3 pp 17-19
[14] Loughran J, Milroy P, Berry A and Gunstone R 2001 *Research in Science Education* **31** 290
[15] Sudiana I M and Surata I K 2010 *Suluh Pendidikan* **8** 44
[16] Berry A and Van Driel J 2012 *Journal of Teacher Education* **64** 125

Peer Assisted Learning Strategy for Improving Students' Physiologic Literacy

S Diana*

Program Studi Pendidikan Biologi, Sekolah Pascasarjana Universitas Pendidikan Indonesia, Bandung, Indonesia

*sariwulan@upi.edu

Abstract. Research about the implementation of the Peer Assisted Learning (PAL) strategy in Plant Physiology lecture has carried out, in which it aims to improve students' physiologic literacy. The PAL strategy began with a briefing by the lecturers to the students tutor about pretest questions, followed by the interaction between student tutors with their peers to discuss response problems, terminated by answering responsiveness questions individually. This study used a quasi-experimental method, one - group pre-test post-test design. This design includes a group of students observed in the pre-test phase (tests carried out before PAL treatment) which is then followed by treatment with PAL and ends with post-test. The other students group (control) was given the pre-test and post-test only. The results showed that the PAL strategy can increase student's physiologic literacy significantly. One of the weaknesses of students' physiologic literacy is that they have not been able to read the graph. The faculties are encouraged to begin introducing and teaching material using a variety of strategies with scientific literacy aspects, for example teaching research-based material. All students respond positively to the PAL strategy.

1. Introduction

Until now plant physiology concepts are considered difficult for students [1; 2] and teachers [3; 4], thus causing misconception [1; 5; 6; 7; 8; 9]. Some ways to overcome this are through the implementation of inquiry based demonstration interactive [10], dual mode teacher training [3], development of teaching materials based on deep dialogue/critical thinking [11], use of conceptual attainment model [8], and conceptual change text-oriented instruction [1; 5].

The scientific literacy must contain context, competencies and knowledge about [12], so that students are not enough to master knowledge only, even if those are still considered difficult. Furthermore, the scientific literacy aspect consists of role of science, scientific thinking and doing, science and society, mathematics and science, science motivation and beliefs [13].

Physiologic literacy aspects in this study lead to scientific thinking and doing, and mathematics and science [13], but the content and context adapted to Plant Physiology course materials as part of Biology. Science literacy especially physiologic literacy skills should be procured in the lecture as a learning innovation. Learning innovations that can be structured in the classroom and monitored directly by the teacher, or even not limited by time and place, is learning involves peer tutoring, namely peer assisted learning (PAL) [14; 15]. The PAL implementation has proven to increase the scientific literacy ability of the pre-service Biology teachers on Plant Morphology lecture [16] and their anatomic literacy ability [17]. PAL program earlier studies indicated that the presence of peer tutors were helpful in the success of Plant Physiology lab works [18], involving students as lab assistant also supports the improvement of lab works quality, such as in Botany Phanerogamae [19], Plant Physiology [20; 21; 22], Plant Anatomy [22], and Plant Morphology course [20; 22; 23]. Application of PAL also can increase students' concepts mastery of Plant Embryology subjects [24]. Scientific studies through research on the application of PAL in the fields of biology, both nationally and internationally, over the years has not revealed its influence on the improvement of physiologic literacy ability. Therefore, based on the successful implementation of PAL in the earlier study, this study will revealed the application of PAL strategy to improve students' physiologic literacy in Plant Physiology course. In addition, this research was particularly carried out to improve the pre-service Biology teachers' physiologic literacy, in which PAL implementation can be done outside of class schedules.

2. Experimental Method

The method used in this study was experimental method, with quasi experimental pre and posttest design. This design includes a group students that observed in the pretest phase (tests carried out before PAL treatment) which is then followed by treatment with PAL and post-test (the post-test items are different with the pretest items, but same in the content). Another group students (control group) were taken the pretest and posttest datas only.

The research subjects were students of Biology Education Study Program of class 2012 and 2013 enrolled in Plant Physiology courses in 2014. The research instruments were questionnaire to solicit tutor, a set of pretest in a form of essay questions about growth, development and plant hormones which contain aspects of scientific literacy, a set of essay questions about growth, development and plant hormones that contains the same aspects as in that of the pretest but contains different discourse. Another research instruments were a set of posttest questions in a form of multiple choice questions to confirm the response results in PAL in which these questions have the same discourse as that of the pretest, and also a questionner to solicit students' response towards PAL program.

PAL program were preceded by discussions between Plant Physiology lecturers with the head of the group (tutor who have previously been selected based on the highest academic achievement and class sociogram results) about pre-test results and its expected answers. Lecturers was not directly informed the tutors about the expected response from the pretest questions, but rather encompass most correct answer from the tutor and indirectly led the tutor to the expected answer. This stage is also a PAL that occurs between the lecturers and tutors.

The next phase was the interaction between the tutors with their tutee, in which they discuss about problem solving test based on lecturer-tutor discussion results, with the smallest lecturers' intervention possible, in which lecturers were merely observing tutors-tutee interaction from a distance. After PAL was executed, all students were assigned to answer a set of test problem solving . Multiple choice-questions postest was then performed to confirm PAL's results. All questions used as an instrument of this research mostly include aspects of the ability of scientific thinking and doing, and mathematics and science [13] and distributed according to various levels of cognitive level of Marzano's new taxonomy [26]. The pretest and posttest datas was tested statistically using the t test [27]. To determine students' physiologic literacy improvement, normalized test-gain (N-gain) was calculated and categorized using Hake's formula [28].

3. Result and Discussion

From Table 1 it appears that prior to the PAL (pre-test), the average students' physiologic literacy in Plant Physiology subjects is still low, which only amounted to 29.3 in experimental group and 33.2 in control group. The pre-test of both group is not different significantly with t value = 0.997< t table = 2.000 (α = 0,005). The highest physiologic literacy ability in experimental group pre-test were in analysing research results, predicting and analysing hypothesis, with 56.8, 54.4 and 51.5 score, respectively. The ability to analysing research results, which solicited through an item in which discussed about the properness of a plant being categorized as long day plant (the plant in question was stored in high temperature treatment followed by long day and short day treatment, in which it could not produce flower). But if this plant was stored in low temperature followed by long day treatment it can produced flower, whereas if it followed by short day treatment it could not produce flower). According to Marzano's new taxonomy, this capability is classified in analysis cognitive level, especially analysing errors. Similarly, another item contain the ability to analyse the hypothesis of results research graph which based on soybean hypocotyl growth research (soybean was administered with auxin then also given kinetin without giving hormones). The question is about the possibility of epicotyl growth that is not too different from the controls when kinetin was given early. It turns out half of all students realize that plants parts hypothesized (epicotyl) is different from the plants parts observed in the graph (hypocotyl). Therefore, based on the questions of the analysis, the ability of the students showed near enough. In contrast to the level of analysis, on other physiologic literacy aspects, the ability of students is still very low, even in terms of giving examples categorized as retrieval according to Marzano's new taxonomy, i.e. only 7.4 from a scale of 100.

The highest ability of physiologic literacy in the control class in pretest are determining independent variable and determining research purposes with each scores 57.8. The ability of determining independent variable is obtained from the item which questioned about independent variable in the research that is if the plant was kept at high temperature and given the treatment of long day or short day, the plant is not flowering. Another case when the plant is kept at low temperatures and given a long day treatment, the plant is flowering, but if it is given short day treatment, the plant is still not flowering. The ability of physiologic literacy in the control class which is the highest among other is determining a research purpose which is collected from item which questioned the purpose of research based on the graphic display of the research result.

From the pretest datas, the level of students' physiologic literacy based on scaffolding teaching theory through the Vygotsky's ZPD theory (the zone of proximal development) [29] only reached 29.3 and 33.2 (Table 1), which is achieved when students learn individualy. These findings are consistent with the results of other scientific literacy studies in senior high school students in Bandung that it were still low [30], even in junior high school students in Sumedang district, also it were very low [31]. As well as the results of other biology literacy studies among students are still low [16; 17].

After PAL implementation, students' physiologic literacy ability, especially in the ability of scientific thinking and doing increased dramatically (Table 1), which reached an average of 82.7 from a scale of 100. This indicates that all students have reached the maximum ZPD (the zone of proximal development) due to scaffolding through PAL from tutor to their peers, amongst tutors and amongst peers, even from tutee to tutor.

The highest aspect of this response was obtained in terms of the ability to determine independent variables (Figure 1) which includes the level of understanding (integration), and compiling the data in the form of a table that also included the level of understanding, but it included as symbolization in the new taxonomy Marzano. In both these aspects all students can master it very well, followed by the ability to analyze the hypothesis, deduce and determine decisions. Response results showed that the ability which was still in the poor category is constructing a hypothesis (58.8), while the ability to overcome problem (66.2) and considers effective variable (73.5) were sufficient. The response was carried out immediately after PAL happens in the classroom and the students answer the response based on an agreement through serious discussion, so that answers tend to be uniform.

In contrast to the physiologic literacy skills obtained from the response after the PAL, physiologic literacy ability obtained from the posttest still relatively low, which reached an average of 40.2 on a scale of 100 in experimental group and 34.9in control group (Table 1), although there are a different significantly between experimental group ang control group with t value = 2.545>t table = 2.000 (α = 0,005). When traced from the item's format, pretest items are about the same with problem solving items test, in which the differences were only in terms of its discourse and graphs. The discourse of posttest items were same with pretest but different only in terms of its items form. The form items in pretest were essay, while the form items in posttest were multiple choice. The most aspects which the students' mastered in the posttest is the ability to analyze hypotheses and to determine the independent variables (Figure 1) that was both at 67.1 in experimental group and in control group was at 65.6 and 56.2 scores respectively (Table 1).

The most interesting things in the posttest, among others, was on aspects closely related to the ability to compile the data in tables/graphs, especially in reading charts contained in item posttest no. 10. On this item, students were asked to count coleoptile growth of wheat sprouts which only given ABA alone, which directly can be seen in the graph of growth hormone-treated with a concentration of 0 um. Of all the students who responded, only about 12% - 16% can answer correctly, most choose another option which involves the administration of hormones. Apparently the ability to create a table is not always in line with the ability to read charts, although both carry the aspect of symbolization.

Table 1. Learning results summary

Physiologic Literacy Aspects	Level of Thinking	Student's Mastery Score				
		PAL			Control	
		Pretest	PST	Posttest	Pretest	Posttest
Determining independent variable	Comprehension (Integrating)	47.1	100	67.1	57.8	56.2
Determining parameter	Comprehension (Integrating)	40.6	89.7	43.6	31.3	28.5
Arranging data into table	Comprehension (Symbolizing)	49.7	100	15.7	43.8	12.5
Constructing concept from research results	Analysis (Generalizing)	40.15	87.5	27.6	32.8	21.2
Analysing research results	Analysis (Analyzing errors)	56.8	-	-	39.1	-
Predicting	Analysis (Specifying)	54.4	75.6	-	48.4	-
Analysing hypothesis	Analysis (Analysing Errors)	51.5	97.1	67.1	42.2	65.6
Drawing conclusion	Analysis(Generalizing)	25.3	97.1	43.4	32.8	40.6
Determining research purposes	Metacognition (Specifying Goals)	14.1	79.4	18.6	57.8	18.7
Composing alternative treatment	Knowledge UtilizationDecision Making)	10.3	77.9	31.4	35.9	31.2
Determining decisions	Knowledge Utilization (Decision Making)	19.4	91.2	25.7	25.0	15.6
Considering effective variables	Knowledge Utilization (Decision Making)	16.2	73.5	44.3	17.2	18.8
Constructing hypothesis	Knowledge Utilization (Experimenting)	21.8	58.8	10.0	18.8	15.6
Determining methods to prove a hypothesis	Knowledge Utilization (Experimenting)	16.2	75.9	39.3	34.4	32.8
Problem Solving	Knowledge Utilization (Problem Solving)	17.6	66.2	38.6	29.7	43.7
Giving example	Retrieval (Recalling)	7.4	82.6	37.3	21.1	40.0
Average		29.3	82.7	40.2	33.2	34.9
N-gain			0.15		0.02	

Another interesting aspect is the decisive decision explored by item posttest showing that the ability of students is very low (25.7) in which this result are very different for the same aspects of the item response results in which showed very high ability (91.2). This item posttest asked students to decide the most appropriate treatment from the two alternatives to obtain optimal α-amylase activity in wheat germ coleoptile, based on some graphs of various phytohormone effect on the growth and activity of wheat germ α-amylase coleoptile. The desired answer is GA, but in posttest answers, the majority of students answer Kinetin - ABA which means Kinetin treatment only.

On the item of problem solving test that ask almost the same thing, students are asked to choose one of the most appropriate treatment alternatives as the act of decision-making in order to get the high number of shoots based on a bud formation graph. The majority of students can answer it correctly. This very noticeable difference is due to students' very low ability to read charts, because there were 6 graphs in one item. Of all the tutors interviewed, sayed that they are very less trained in reading charts and have not even been discovered science literacy problems like this before, so they felt difficult when answering questions. Actually, create the chart at this point can be done by entering data aided by a computer program, and students are used to create the chart. On the other hand the ability to read charts from other data sources is unfamiliar to students. In addition, the student mastery of the course material also affects the views of students on reading the graph.

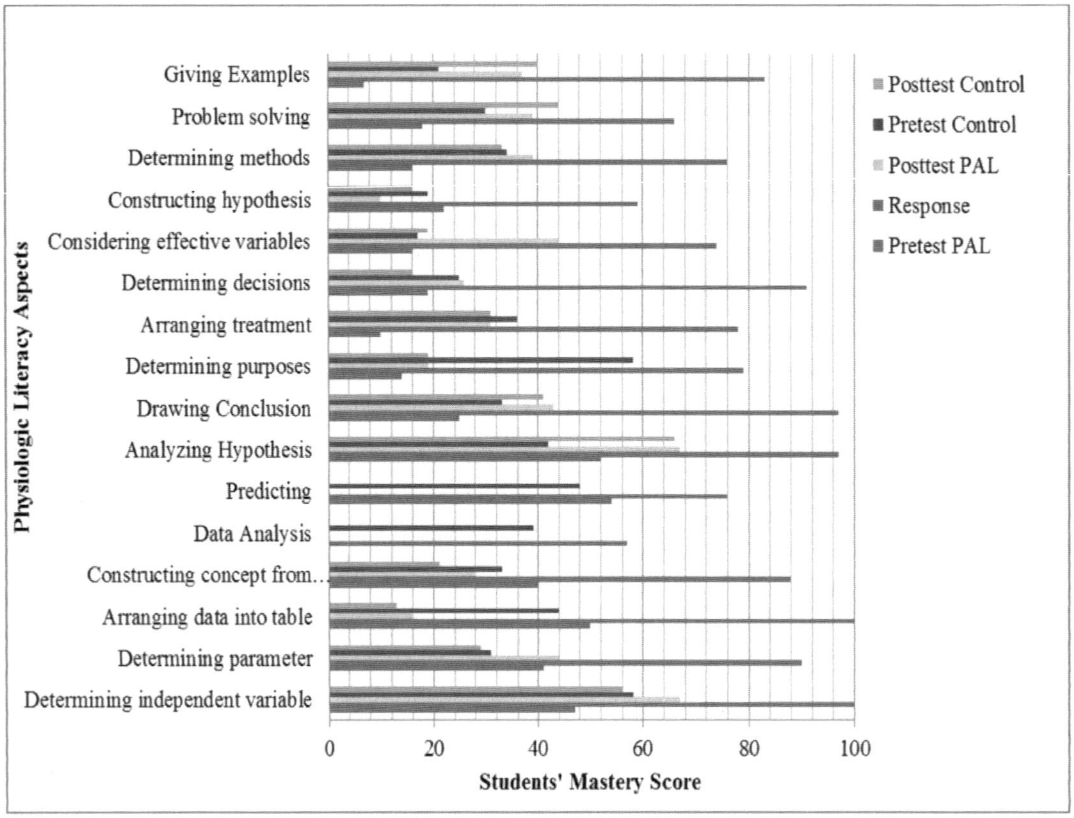

Figure 1. Students' physiologic literacy ability.

The results of test posttest statisticaly show that PAL strategy can improve students' physiologic literacy skills significantly, although the increase is low with N-gain of only 0.15 in experimental group and 0.03 in control group. This finding is somewhat different from the study results that showed moderate score of student's anatomic literacy after PAL implementation [17]. The implementation of PAL also can improve significantly to student's scientific literacy on Plant Morphology lecture with moderate category(N-gain 0.4) [16]. This finding is also somewhat different from the results on Plant Embryology course [24], that the implementation of PAL can improve student mastery of concepts with N-gain of 0.76, although not directly related to science literacy skills. Similarly, with a series of PAL research in the lab, showed that PAL is facilitated by a lab assistant contribute to the success of the lab and improve the students' N-gain for Morphology of Plants and Plant Anatomy lab works from low to medium, both from the test phase to the implementation phase of the program [20; 21; 22; 23]. Despite the many advantages of PAL strategy either directly related to academic achievement, it seems that students' physiologic literacy is still low, but this was not due to PAL itself but rather to the lack of scientific literacy learning frequency in general.

The low ability of physiologic literacy of pre-service biology teacher students corresponds to other scientific literacy abilities at the same level of education and at lower levels. At the university level, the science literacy of prospective teachers in West Java was also low [32], as well as its quantitative literacy skills [33; 34] and microbiological literacy [35; 36]. Therefore it is not surprising that scientific literacy capabilities of students at senior high school and junior high school are extremely poor as evidenced by PISA outcomes [37], and from other studies using the scientific literacy assessment (SLA) [30; 31]. Even according to [38], students in Southern Trinidad who have high academic achievement, their science literacy were relatively low.

It is very likely that learning activity in school and university emphasizes on concepts' mastery, the students less sharpen thinking related to scientific literacy aspects. The students may already good at developing and implementing mini research design which loaded with scientific method, but they are still poorly trained in analyzing the research of others. Therefore, the most appropriate action is to start promote scientific literacy as early as possible. According to questionnaires and classroom observation results, all students like the application of PAL as an appropriate step to deepen the mastery of the course material. The reason the students liked the application of PAL is, among others, the students become more understanding of the material, there was discussion among the students in a more relaxed/ comfortable atmosphere, shared understanding among students, the perception, language better understood by peer tutors, and students feel more intimate with their classmates. Disadvantages experienced by students are they generally less experienced in examining research results and had never answered the questions of research-based course material.

4. Conclusion

The Peer Assisted Learning (PAL) strategy which is carried out in Plant Physiology course to improve physiologic literacy of students significantly, began with a briefing by the lecturers to the students tutor about pretest questions discussion, followed by the interaction between student tutors with their peers to discuss response problems, terminated by answering responsiveness questions individually.

The PAL strategy has a potential to improve students' physiologic literacy. In order to increased students' physiologic literacy, the faculties are encouraged to begin introducing and teaching material using a variety of strategies with scientific literacy aspects, for example teaching research-based material. Learning evaluation tools are also expected to demand aspects of scientific literacy, not just a concept charged.

Because the instrument used in this study mostly only about scientific thinking and doing and mathematics and science, this study needs to be followed up by research on other aspects of scientific literacy, i.e. the role of science, science and society, and affective aspects contained in the components of science motivation and beliefs.

Acknowledgments

We aknowledge the dedicated participation of all students of Biology Education in Departement of Biology Education Universitas Pendidikan Indonesia of class 2012 and 2013 enrolled in Plant Physiology courses in 2014.

References

[1] Cakir OS, Geban O and Yu¨ ru¨ k N 2002 *Biochemistry And Molecular Biology Education* **30** 239
[2] Çimer A 2012 *Educ. Res. Rev.* **7** 61
[3] Widodo A, Riandi and Hana N 2008 Dual Mode Inservice Training As An Alternative Teachers Professional Development Program *Proceeding The Second International Seminar an Science Education* (Bandung: SPs UPI) p 150

[4] Mumuni A A O, Dike J W and Uzoma-Nwogu A 2017 *World Journal of Education* **7** 1
[5] Yenilmez A and Tekkaya C 2006 Journal of Science Education and Technology **15** 1
[6] Köse S 2008 World Applied Sciences Journal **3** 283
[7] Vila F and Sanz A 2012 Biochemistry And Molecular Biology Education **40** 148
[8] Afandi 2014 Remediasi Miskonsepsi Mahasiswa Menggunakan Model Concept Attainment Pada Materi Metabolisme Sel *Prosiding Seminar Nasional 2014* 160
[9] Malińska L, Rybska E, Sobieszczuk-Nowicka E and Adamiec M 2016 *CBE—Life Sciences Education* **15** 1
[10] Martini S 2016 *ejournal.unwir.ac.id* **I** 1
[11] Ningrum DEA, Prihatin J and Pujiastuti 2014 *Pancaran* **3** 155
[12] Bybee R, McCrae B and Laurie R 2009 *Journal of Research in Science Teaching* **46** 865
[13] Fives H, Huebner W, Birnbaum A S and Nicolich M 2014 *Science Education* **98** 549
[14] Tessier J 2007 *Journal of College Science Teaching* January/February 2007 64
[15] Roscoe R D and Chi M T H 2008 *Instr Sci.* **36** 321
[16] Diana S 2016 *JurnalPengajaran MIPA* **21** 82
[17] Diana S 2016 Implementasi Strategi Peer Assisted Learning (PAL) untuk Meningkatkan Literasi Anatomi Mahasiswa Calon Guru Biologi *Proceeding Biology Education Conference* vol 13 p 554
[18] Diana S and Rustaman N 2010 Preliminary Profile of Tutor's Ability in Managing Practical Work on Plant Physiology Through Peer Assisted learning (PAL) Program *Proceeding The 4th International seminar of Science Education* (Bandung: SPs UPI) p B15
[19] Diana S, Rustaman N, Redjeki S and Iriawati 2011 Studi Awal tentang Kemampuan Asisten Mahasiswa dalam Pelaksanaan Program Peer Assisted Learning (PAL) Praktikum Botani Phanerogamae *Prosiding Seminar Nasional Biologi* (Bandung: Jurusan Pendidikan Biologi FPMIPA UPI) p 431
[20] Diana S, Rustaman N, Redjeki S and Iriawati 2011 Peer Assisted Learning (PAL) Program on Plant Morphology and Physiology in Empowering Future Biology Teacher-Students-Assistant *Proceeding 5th International Seminar on Science Education* (Bandung: SPs UPI) p 42.1
[21] Diana S, Rustaman N, Redjeki S and Iriawati 2012 ImplementasiTaksonomi Baru Marzano untuk Pemberdayaan Mahasiswa Asisten Praktikum Fisiologi Tumbuhan dalam Program Peer Assisted Learning (PAL) *Prosiding Seminar Nasional IX Pendidikan Biologi* (Surakarta: Prodi Biologi Jurusan Pendidikan MIPA FKIP USM) p 170
[22] Diana S 2014 Pemberdayaan Asisten Praktikum untuk Pelaksanaan Peer Assisted Learning (PAL) Disertasi SPs UPI Bandung p 151-153
[23] Diana S, Rustaman N, Redjeki S and Iriawati 2012 Program Peer Assisted Learning (PAL) dalam Praktikum Morfologi Tumbuhan untuk Pemberdayaan Asisten Mahasiswa Calon Guru Biologi. *Seminar Program Penelitiandan Pengabdian kepada Masyarakat* (Bandung: LPPM UPI)
[24] Diana S 2014 Penerapan Strategi Peer Assisted Learning (PAL) untuk Meningkatkan Penguasaan Konsep Dalam Perkuliahan Embriologi Tumbuhan *Prosiding Seminar Nasional Mathematics and Sciences Forum* (Semarang: FMIPA Universitas PGRI Semarang) p 417
[25] Creswell J W 2008 Educational Research Planning Conducting and Evaluating Quantitative and Qualitative Research (third ed) (New Jersey: Pearson Merrill Prentice Hall) p 298-354
[26] Marzano R J and Kendall J S 2008 Designing & Assessing Educational Objectives Applying the New Taxonomy (London: Corwin Press) p 1-7
[27] Schefler W C 1987 *StatistikauntukBiologi, Farmasi, KedokterandanIlmu yang Bertautan*Edisi ke-2 (Bandung: Penerbit ITB) p 71-106
[28] Hake RR 1999*The Impact of Concept Inventories On Physics Education and It's Relevance For Engineering Education*http://www.physics.indiana.edu/~hake.
[29] van der Stuyf RR 2002 *Scaffolding as a Teaching Strategy.*

http://books.google.co.id/books?id=L5B9MJjHXQYC&pg=PA363&dq=Scaffolding+as+a+teaching+strategy+by+Van+Der+Stuyf&hl=en&sa=X&ei=hf1sUfOFFczHrQftnIHwBg&sqi=2&ved=0CDEQ6AEwAQ#v=onepage&q=Scaffolding%20as%20a%20teaching%20strategy%20by%20Van%20Der%20Stuyf&f=false

[30] Diana S, Rachmatulloh A and RahmawatiE S2015 Profil Kemampuan Literasi Sains Siswa SMA Berdasarkan Instrumen Scientific Literacy Assesments (SLA) *Prosiding Seminar Nasional XII Pendidikan Biologi* (Surakarta: FKIP UNS) pSP-006-1_285-291

[31] Rachmatulloh A, Diana S and Rustaman N 2016 Profile Of Middle School Students on Scientific Literacy Achievements by Using Scientific Literacy Assessments (SLA) *AIP Conf. Proc.*1708 080008 (2016)

[32] Putra MIS, Widodo W and Jatmiko B 2016 *Jurnal Pendidikan IPA Indonesia* **5** 83

[33] Rheinlander K and Wallace D 2011 *Numeracy* **4** 1

[34] Nuraeni E, Rahmat A, Redjeki S and Riandi 2014 ProfilLiterasi Kuantitatif Mahasiswa Calon Guru Biologi *Prosiding Seminar Nasional Mathematics and Sciences Forum* (Semarang: FMIPA Universitas PGRI Semarang) p BP09

[35] Hamdiyati Y, Sudargo F, Redjeki S and Fitriani A 2017 Biology Education Student's Profile on Microbiology Literacy *Advances in Social Science, Education and Humanities Research (ASSEHR)* **57** 63

[36] Herlianti Y, Rustaman N and Fitriani A 2012 Literasi Mikrobiologi pada Siswa dan Mahasiswa Calon Guru Madrasah Aliah *Proseding Seminar Nasional Pendidikan IPA* (Jakarta: Fakultas Ilmu Tarbiyah dan Keguruan UIN Syarif Hidayatulloh) p 79 -96

[37] OECD 2016 Programme For International Student Assessment (PISA) Results From PISA 2015 Indonesia OECD Publishing

[38] Mahatoo J 2012 *Scientific Literacy and Nature of Science as it Impacts on Students' Achievement in South Trinidad.* http://uwispace.sta.uwi.edu/dspace/bitstream/handle/2139/12709/Judy%20Mahatoo.pdf?sequence=1

IOP Conf. Series: Journal of Physics: Conf. Series **895** (2017) 012127 doi:10.1088/1742-6596/895/1/012127

Students' Cognitive Abilities in Plant Anatomy Practical Work

S Setiono[1,2*], N Y Rustaman[1], A Rahmat[1] and S Anggraeni[1]

[1]Departemen Pendidikan Ilmu Pengetahuan Alam, Universitas Pendidikan Indonesia, Bandung, Indonesia
[2]Departemen Pendidikan Biologi, Universitas Muhammadiyah Sukabumi, Sukabumi, Indonesia

*setiono@student.upi.edu

Abstract. Cognitive abilities is fundamental for the students, as it is closely related to higher thinking skills such as the ability to think critically, creatively, and problem solving. This descriptive study aims to investigate the cognitive abilities of biology prospective teachers in the course of Plant Anatomy Practicum based on the cognitive process dimension and dimensions of knowledge the using the framework of Revision of Bloom taxonomy. A number of biology prospective teachers was involved in this study (n=42). The instrument used to collect data for students' cognitive process mastery in the form of multiple choice with 5 options. Research finding shows that the average student's cognitive ability is 68.10. The acquisition of knowledge mastery of cognitive ability is still under the criterion of mastery in the course of the Plant Anatomy Practicum (75). Validity and reliability of the instrument (0,71) and (0,81). It is necessary to design lecture programs both in the class and laboratory to develop student' cognitive abilities.

1. Introduction

Cognitive abilities are the abilities to carry out brain activity (thinking activity) to gain knowledge. Cognitive abilities based on revised bloom taxonomy [1] include the ability to memorize, understand, apply, analyze, evaluate, and create. These cognitive abilities show the gap ranging from simple cognitive processes to more complex cognitive processes. Cognitive abilities are fundamental to be mastered by students, especially for a pre-service biology teacher. Students' cognitive abilities become an important factor in determining the success of students in their studies [2]; [3]. Cognitive abilities possessed by students become the basis for the formation of other capabilities that must be owned by students. The ability to think critically, creatively, and problem solving requires a complex cognitive process, so that this cognitive ability becomes an important position [3]. The process of cognition encompasses various aspects, such as perception, memory, thought, symbols, reasoning, and problem solving. Based on Piaget's opinion, developing cognitive abilities in the student is important so that students: 1) Are able to develop their perception based on what he sees, hears, and feels so that the child will have a complete and comprehensive understanding; 2) Are able to train his memory of all events and occasions he experiences; 3) Are able to develop his thoughts in order to connect one event with another. 4) Are able to understand the various symbols scattered around the environment; 5) and able to do good reasoning that is occurred through a natural process (spontaneous) or through a scientific process (experiment). Students' success in learning is influenced by mastery of prerequisite materials before accepting new material. The most important factor in influencing the learning process is what the students already know [4]. New concepts or new information must be linked to existing concepts in the student's cognitive structure ". So before students learn a material, relevant concepts must already be exist in the student's cognitive structure, so that students will have no difficulty in accepting new material. A complex concept in science can only be understood if the more fundamental concepts involved in the formation of a new concept have been fully understood [5]. The mastery of prerequisite material for the success of students is important, that is why the teacher always have to direct and improve the concepts that relate to students in order to follow the new lesson.

Teaching-Learning process is one complicated process because students does not merely absorb information from the teacher but also involves various activities and actions that must be done, so that the teaching-learning process should be able to emphasize the various activities and actions by using certain methods and approaches that can develop the active learning, for both teachers and students [6]. Furthermore, Knowledge is built into the mind of the child through assimilation and accommodation [4]. Assimilation is the absorption of new information in the mind. While accommodation is rearranging the structure of the mind because of new information, so that the information has a place in the brain. Another definition of accommodation is that accommodation is the mental process that involves the creation of a new scheme that matches the new stimulation or modifies the existing scheme to fit the stimulus [6]. Constructivism as a cognitive psychological flow argues that knowledge is the meaning of a reality and man must build the meaning to the reality. The implication of the constructivism theory in teaching-learning process is that knowledge can not be transferred entirely from the teacher's mind to the mind of the student. Students themselves who should be mentally active constructing their knowledge, especially related to the physical knowledge and logic.

Plant anatomy is one of the compulsory materials in Biology Education Studies Program. This Plant Anatomy Material is usually packed in a course of Plants Anatomy, in some universities this subject is broken down into two different courses which is Plant Anatomy Course and Plant Anatomy Practicum. In the curriculum structure of Biology Education Program of University of Muhammadiyah Sukabumi (UMMI), the material of plant anatomy is taught on the subject of Plant Anatomy and Plant Anatomy Practicum. This Plant Anatomy Practicum program requires students to: 1. Able to master the concepts, principles, laws and theories about the structure of plant anatomy; 2. Students are able to identify the complexity of network structures that make up plant organs; 3. Students are able to analyze the relationship between the anatomical structure of the plant and its function; 4. Students are able to analyze the relationship between plant anatomy structure and form of adaptation done by plants; 5. Students are able to apply knowledge about plant anatomy structure; 6. Students are able to communicate the results of scientific inquiry in the form of scientific papers; 7. Students are able to do scientific inquiry to develop knowledge about the structure of plant anatomy. Students are also required to be able to master the procedural knowledge in terms of making the preparation, perform observations under the microscope, and able to study microscopic images of observations under a microscope [7]. In the plant anatomy practicum, the students are required to be able to relate the knowledge acquired through the theory from the practical work and have spatial ability to explain the observed images under a microscope[8]. The demands of skills in learning outcome of the course and demands in doing the practical work of Plant Anatomy requires students to think complex, the demands are not only to memorize and understand the material but students are required to be able to apply knowledge, analyze, evaluate, and create.

The purpose of this research is to obtain a picture of the cognitive abilities of students based on the cognitive process dimension from the new bloom taxonomy, and to know the obstacles and problems facing students in the process of acquiring knowledge, especially in Plant Anatomy Practicum. A learning experience that enables students to be physically and mentally involved and facilitates the construction of knowledge processes becomes an important position. Learning experiences organized by teachers or lecturers should provide an opportunity for students to construct knowledge. Learning

IOP Conf. Series: Journal of Physics: Conf. Series **895** (2017) 012127 doi:10.1088/1742-6596/895/1/012127

experience that has to be held in the teaching-learning process of science is inquiry [9]. Inquiry learning can improve the mastery of concepts [10]; [11]; [12]; [13] & [14]. Inquiry-based learning allows students to gain a learning experience like when a scientists construct scientific knowledge. Inquiry-based learning involves students hands on and minds on in the process of constructing knowledge, thus creating meaningful learning. The significance of this process will positively affect the achievement of student competence.

2. Method

This research is a qualitative research that aims to explain the phenomenon of students' cognitive abilities in the context of Plant Anatomy Practicum. This research was conducted in Biology Education Study Program. Its implementation time is in the odd semester of the 2015/2016 school year. The subjects of this research are 42 students who take the course of the Plant Anatomy Practicum. Research data is in the form of cognitive ability test results that are analyzed descriptively and qualitatively.

The implemented activity of plant anatomy practicum in this research is learning experience which is proving or verifying knowledge obtained from theory. Practicum is carried out starting from cell material and its parts, basic tissue system, vascular and organology.

Data collection was performed through cognitive ability tests based on cognitive ability level in revised bloom taxonomy [1] from C2, C3, C4, C5 and C6 levels and interviews with students. The data of the research were then analyzed and compared with predetermined success indicators. Parameters used to measure the achievement of cognitive ability is a score-based parameter of achievement of learning mastery in the course of Plant Anatomy, which is 75.

3. Result and Discussion

The results obtained are quantitative data of students' cognitive ability and qualitative data of the interview result. Data is presented in the form of graphic drawings and descriptions. Figure 1 shows students 'cognitive abilities based on cognitive process dimensions and Figure 2 shows students' cognitive abilities based on the knowledge dimension.

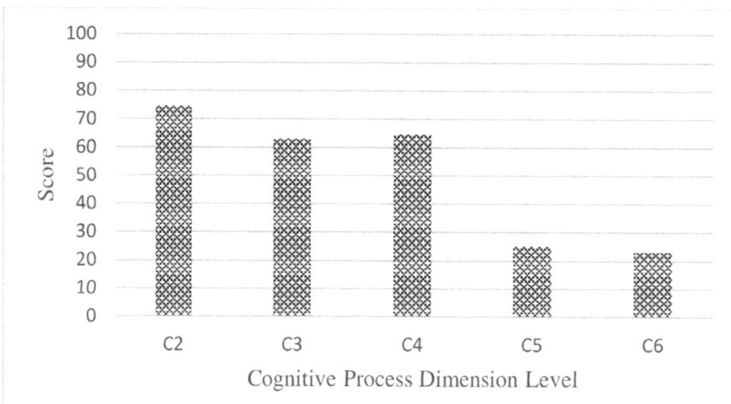

Figure 1. Student cognitive ability based on the cognitive process dimension

Figure 1 above illustrates the mastery of the cognitive abilities of pre-service biology teacher. Of the four levels of cognitive process dimension measured, mastery at the C2 level attainment is the highest, followed by ability at C4, C5 and last level at C6 level. The data in Figure 1 shows that students still have difficulty in mastering the level of cognitive process dimension at C3, C4 , C5 and C6 levels.

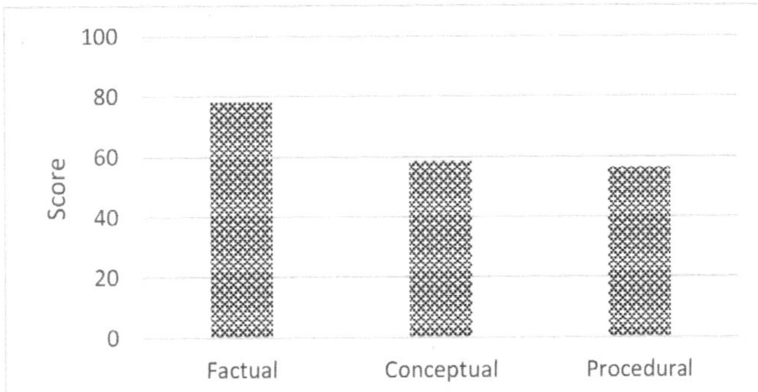

Figure 2. Student cognitive ability based on the knowledge dimension

Figure 2 shows the mastery of cognitive abilities based on the dimensions of factual, conceptual and procedural knowledge. Figure 2 shows that the mastery of the students' procedural knowledge has the lowest attainment compared with the mastery of the conceptual and factual dimensions. The dimension of factual, conceptual, procedural and metacognitive knowledge is a gap that describes the ability of concrete thinking to abstract thinking [1]; [15]. The results of the study illustrate that students have difficulty in abstract thinking. This is due to the mastery of procedural dimensions that are lower than the conceptual and factual dimensions. The achievement of this knowledge dimension shows that the tendency of students to have difficulty in abstract thinking. This will affect the students' complex thinking ability. In addition, the low acquisition of the factual knowledge dimensions is caused by students who does not fully understand several procedures in the plant anatomy practicum. Results of interviews with students show that students does not know some procedures in making the preparation, for example students do not know what reagents should be used when they will observe the berlignin cell wall. Some students also do not know the procedure of making fresh preparation properly.

Achievement on the cognitive process dimension of C2 level is higher than the other level, but the achievement is still below the passing grade which is 75. At the C2 level, students are required to be able to understand the material content of plant anatomy. Students are considered to be understood if they can construct the meaning of the learning messages, whether they are spoken, written or graphically conveyed through the teaching process [1]. Students can understand something well if their conceptual understanding is also good. Conceptual understanding underlies the students in understanding something. The results of the study showed that at level C2, student is facing difficulties to understand the knowledge on conceptual dimensions, student is facing difficulties on reconstructing the meaning of descriptive messages, and student is facing difficulties to describe the characteristics of plant tissue using words.

Achievements at C3 level are lower than C2 and C4 levels. Apply with regard to procedural knowledge [1]. Thus, students' procedural knowledge influences the achievement of the cognitive process dimension at C3 level. The results showed that the students' procedural knowledge is lower than the factual and conceptual knowledge, so it is reasonable that the achievement of C3 level is still low. In the above explanation mentioned that the low achievement of this low level of C3 is the ability of students to think abstract is low and students do not understand some procedures in learning practicum plant anatomy.

Achievements at the C4, C5 and C6 levels show that students still have difficulty mastering the cognitive processes dimensions at the level of analyzing, sythesis and creating. Thinking occurs when someone takes new information to be then is stored in the memory and interconnects or rearranges and extends this information to reach the goal or find possible answers in confusing situations [3]. The process of thinking will be more complex if it requires students to connect information, analyze information, and create inferences from observed phenomena. Cognitive processes at a high level

IOP Conf. Series: Journal of Physics: Conf. Series **895** (2017) 012127 doi:10.1088/1742-6596/895/1/012127

require abstract thinking skills [1], so if the student is facing difficulties in solving the problem of cognitive processes at this high level, it indicates that students' abstract thinking ability is still low.

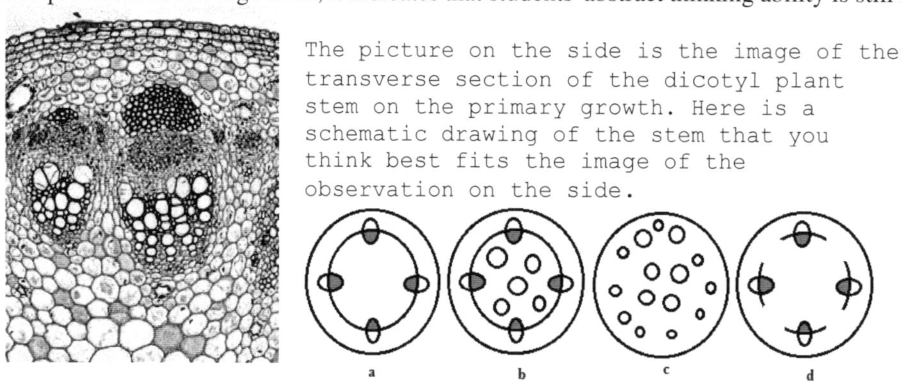

Figure 3. Sample item for cognitive abilities at level C6 (factual)

The process of thinking from C4 to C6 levels is included in complex thinking or better known as "High-Level Thinking", the three aspects are aspects of analysis, evaluation aspects, and aspects of creation [1]. Analyzing includes the ability to parse an issue or object into its elements, cognitive processes in analyzing include: differentiating, organizing, and finding the implied message [15]. Creating is the ability to combine several elements into a unified form and the ability to produce something new by organizing some elements or parts into previously invisible patterns and structures, these capabilities can include: making, planning and producing [15]. The other three aspects of the same realm, namely the aspect of remembering, the aspect of comprehension, and the aspect of the application, fall into the intellectual part of the lower-order thinking. At the C4 level, students are still struggling to organize information. Students struggle when told to determine the relationship between the anatomical structures of the tissue or plants' organ with its function and make the relationship between the anatomical structures observed with the new situation presented. Meanwhile, at the C6 level students are still struggling in formulating hypotheses that meet certain criteria.

The results of interviews with students, shows that students have difficulty in solving problems at high cognitive process level because the students have not been accustomed to work on these questions, so students find it difficult to solve these problems. Learning experience that is experienced by students, especially on practicum learning also affects the ability to think students. The practical learning experience that the students get during this time is still a verificative experiential learning experience (practicum to proof), the students are not familiar with the investigative practicum. This is a factor causing the complex thinking ability of students still low. Verificative learning practicum experience only train the ability of cognitive processes in terms of understanding the contents of material that is constructed through the activities lab.

The plant anatomy practicum requires students to think complexly, students are required to analyze the observed images under a microscope, then analyze the linkage between the structure of the tissue or organ with the function of the tissue or organ [7]; [8]. Ability to analyze not only requires students to combine a lot of information but also requires students to think abstract. If the abstract thinking ability is still low, students will find it difficult to analyze and cognitive processing higher than that, such as to create.

The ability to think complex requires students to connect information, describe the information obtained, make consideration of the obtained information, and produce something new based on the information obtained through the learning process. Incomplete information will make it difficult for students to make connections between information. The results of interviews with some students shows that in the process of learning in the course of Plant Anatomy Practicum students are not required to obtain extensive and in-depth information, students only get information from the explanation of lecturers during the lecture activities carried out. Information that is partially understood will cause students to engage in high-level thinking processes. The practical learning experience that only verificative also does not provide the learning experience to the students to have complex thinking skills.

The acquisition of knowledge is also influenced by the learning experience that is experienced by the students. Principally, the learning experience should enable student involvement in the learning process. Knowledge must be constructed by students themselves through the learning experience experienced by students, so it is very important to involve students in the learning process [8]. Less meaningful learning experiences will make it difficult for students to understand the concept or concept construction. [9] States that one of the standards in teaching science is inquiry. Inquiry is a learning experience that can be held by lecturers in order to build student knowledge. Inquired earning experience will help students construct the knowledge that is required in the learning process. Inquiry learning can strengthen students' abilities in carrying out various cognitive processes [16]. Experience through observing, identifying problems, designing experiments, collecting data, and analyzing data enables students to gain experience and constructed knowledge naturally, like a scientists discover and gain knowledge. Inquiry-based learning can also develop students' ability to develop a number of important basic skills possessed by students such as problem-solving skills, critical thinking skills, and creative thinking skills.

4. Conclusion

The results of this study showed that students' cognitive abilities in Plant Anatomy materials are still low, especially in the high-level cognitive process dimension. Students are still struggling in performing the process of thinking, especially at the level of the ability to think complex such as analyzing, and creating. These three levels of thinking processes require students to connect a lot of information, and it becomes an obstacle for students. The learning experience that is experienced by students during the formation of such knowledge becomes an important factor of student's success in shaping their knowledge. Lecturers must design a learning experience that can assist students in reconstructing their knowledge, especially at the level of high-level cognitive process dimension. It is necessary to design lecture programs both in the class and laboratory to develop student' cognitive abilities

Acknowledgments

This research was suported by Kementrian Riset, Teknologi dan Pendidikan Tinggi Indonesia for financial support and Departemen Pendidikan Biologi Universitas Muhammadiyah Sukabumi.

References

[1] Anderson LW and Krathwohl D R 2001 *A Taxonomy for Learning, Teaching, and Assesing: A Revision of Bloom's Taxonom y of Educatioanl Objectives* (New York: Addison Wesley Longman, Inc)
[2] Kim et al. 2015 *Journal of International Students* **5** pp 526-540
[3] Rosén M, Yang Hansen K and Wolff, U 2017 *Cognitive Abilities and Educational Outcomes* (Springer: New York, USA)
[4] Foster J L et al. 2017 *Journal of Experimental Psychology: Learning, Memory, and Cognition* **10** pp 1-13
[5] King P M 2009 *Journal of College Student Development* **50** pp 597-620
[6] Demetriou A and Spanoudis G 2017 *Cognitive Abilities and Educational Outcomes* (Springer: New York, USA)
[7] Nuraeni E et al. 2015 *Jurnal Ilmu Pendidikan* **21**(2) pp 98-105
[8] Ermayanti E, Rustaman N Y and Rahmat A 2016 *Jurnal Pengajaran MIPA* **21**(2) pp 78-84
[9] National Research Council 2000 *National science education standards* (Washington DC:

National Academy Press)
[10] Kudish P et al. 2015 *Bioscene* **41** pp 10-15
[11] Gray C et al. 2015 *Biochemistry and Molecular Biology Education* **52** pp 521-527
[12] Putra R A, Sudargo, F Redjeki S and Adianto 2014 *IJSR* **3** (3) pp 7-12
[13] Bugarcic et al. 2012 *Biochemistry and Molecular Biology Education* **40** pp 174-180
[14] Siritunga D et al. 2011 *CBE-Life Sciences Education* **10** pp 287-297
[15] Rinaldi L and Smith A K 2017 *Journal of Intelligence* **5** pp 1-26
[16] Lazonder AW and Harmsen R 2016 *Review of Educational Research* **86** pp 681–718

IOP Conf. Series: Journal of Physics: Conf. Series **895** (2017) 012128 doi:10.1088/1742-6596/895/1/012128

The Learning of Compost Practice in University

T W Agustina [1,2*], N Y Rustaman[1], Riandi[1] and W Purwianingsih[1]

[1]Program Studi Pendidikan Ilmu Pengetahuan Alam, Sekolah Pasca Sarjana Universitas Pendidikan Indonesia, Bandung, Indonesia
[2]Program Studi Pendidikan Biologi, Jurusan Pendidikan Matematika dan Ilmu Pengetahuan Alam, Fakultas Tarbiyah dan Keguruan, Universitas Islam Negeri Sunan Gunung Djati, Bandung, Indonesia

*tri_agustina@student.upi.edu

Abstract. The compost as one of the topics of the Urban Farming Movement in Bandung city, Indonesia. The preliminary study aims to obtain a description of the performance capabilities and compost products made by students with STREAM (Science-Technology-Religion-Art-Mathematics) approach. The method was explanatory sequential mixed method. The study was conducted on one class of Biology Education students at the one of the universities in Bandung, Indonesia. The sample was chosen purposively with the number of students as many as 44 people. The instruments were making Student Worksheets, Observation Sheets of Performance and Product Assessment, Rubric of Performance and Product, and Field Notes. The indicators of performance assessment rubrics include Stirring of Compost Materials and Composting Technology in accordance with the design. The product assessment rubric are a Good Composting Criteria and Compost Packaging. The result of can be stated most students have good performance. However, the ability to design of compost technology, compost products and the ability to pack compost are still lacking. The implication of study is students of Biology Education require habituation in the ability of designing technology.

1. Introduction

Currently, urban areas are experiencing the problem of reducing agricultural land [1]. Urban Farming Movement as an alternative to solving the narrowing of urban land. Garbage in the city can be a potential for Urban Farming. Garbage in urban areas is mostly biological waste (organic waste). However, the urban community involvement in waste management is still lacking [2]. Urban waste can be used for composting. The compost is used as a medium for growing of plant in urban farming [3].

One of the classic problems in Indonesia is the increase of the graduation rate each year that is always followed by a rise of unemployment [4]. In Indonesia, to synergize the University graduates with the needs of the workforce, Presidential Decree number 08 of 2012, regarding KKNI (Indonesia National Qualifications Framework) was released [5]. The condition is suitable to KKNI in term learning outcomes that should be done by graduates according to the under graduate profiles. The under graduate profiles on Biology Education Studies Program of one State University on Bandung are Biology Educators of secondary school, Educational Researcher and Edupreneur [6].

One approach to learning that can be done to prepare the under graduate profile through STREAM (Science-Technology-Religioun-Art-Matematics). STREAM is the development of STEM that have been developed in other countries, example United States [7], Japan, Australia [8], Turkey [9], Malaysia [10]. STREAM is following the steps in the form of cycle design engineering process that include: Ask, Imagine, Plan, Create, Improve back to ask and so on. Ask: starting with step questions about the problems being faced. Imagine: to think and use their knowledge and creative ideas to solve the problems. Plan: think of the ingredients/materials/resources for making technology. Technology is used to solve the problem. Create: create a design and follow a plan that has been created in the previous steps. Improve: analyzing technology made whether to work or not, change back design to produce better technologies [8]. Religion indicator associated with religious attitude. Religious attitude are to respect, appreciate and practice the teachings of the religion professed [11]. Religion adopted in this research is Islam because in the Islamic University. Art is meant in this study that creativity, innovation, technology and design in making the resulting product [12] [13]. The Next Generation Science Standards issued a framework for students ranging from kindergarten to grade XII. This framework helps students to build three-dimensional learning. Three dimensions include scientific and engineering skills; the concept of cross-cutting (cross-cutting concepts) in studying science and engineering; and understanding of the core material in the study of science [7].

STREAM suitable learning approach using performance assessment and product. In the realm of skills by observing and evaluating student demonstration. In the realm of product can assess skills with steps to creating quality products and products that can be evaluated [14]. In the assessment of performance and product use guidelines on the assessment rubric. The one form is an analytic rubric with specific criteria [15]. The creating of performance rubrics that represents all stages of ability that needs to be assessed [16].

The study has been conducted in Biology Education has a biology garden facility and close to market. There are many organic waste in market. Based on this background made formulation of the study problem are description of the performance capabilities and compost products made by students with STREAM approach. Further the study questions were made, among other: the student's ability in stirring of compost materials, the student's ability in composting technology, the student's ability to produce of compost, and the student's ability in packaging of compost. The preliminary study aims to obtain a description of the performance capabilities and compost products made by students with STREAM approach.

2. Method

The method of explanatory sequential mixed methods design. The first, collecting quantitative data and then collecting qualitative data [17]. The quantitative data derives assessment of performance and product assessment by making a percentage. Purposive sampling is a Biology Education student of 6th semester of class A. Purposive sampling because class A has more male students than class B. The sample number of students as many as 44 people. It consist of 10 male students and 34 female students.

Students have been divided into composting groups. The one group consists of 3 people to facilitate observation and enable student performance in composting. The qualitative data comes from field notes during the descriptive practice.

The instruments has been validated by 2 fellow lecturers, 1 expert of Biology Education and 3 experts as advisory commission. The indicators of performance assessment rubrics include Stirring of Compost Materials and Composting Technology in accordance with the design. The product assessment rubric are a Good Composting Criteria and Compost Packaging.

3. Result and Discussion

3.1. The student's ability in stirring of compost materials (performance assessment)

Figure 1 shows the performance results in stirring of compost materials indicator. Figure 1 shows that the student performance of 40.91% is categorized very good: 52.27% is good categorized: and 6,82% is enough catogorized. The results show that most of the students have good performed in compost materials stirring. Based on field notes, students have been seen actively doing materials compost stirring. The materials used come from the remaining vegetables from market. Students also have utilized the tools and materials that exist around in the Biology Garden.

Figure 1 shows the performance ability of most students is good. The results are in line with the Septiani study that the performance assessment applied in the STEM approach to plant media preparation materials has been able to reveal the skills of the science process [18].

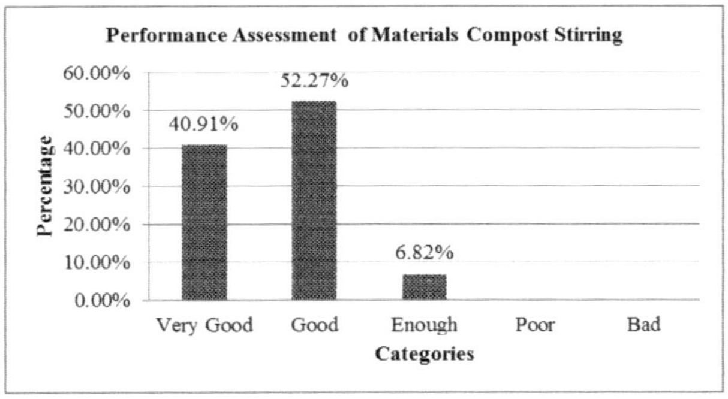

Figure 1. Figure of performance assessment of materials compost stirring

3.2. The student's ability in composting technology (performance assessment)

Figure 2 shows the performance results on composting technology that match the design. Before of the compost practice, students have been assigned to make composting mini-proposal according to worksheet. The mini-proposal contains the problem of composting, how to solve compost problems, compost theory, determining tools, materials, work methods, and composting design. Students have been assigned to draft compost cost budget and composting relationship with religion aspect. Figure 2 shows all under categorized students in designing of composting technology. Students have difficulties in designing of composting technology. Based on field notes and observations that the tools, materials, and workings of composting that have been written in mini-proposal were inconsistent with the compost practice. Students were not familiar with STREAM worksheet. Students felt limited time to made mini-proposal.

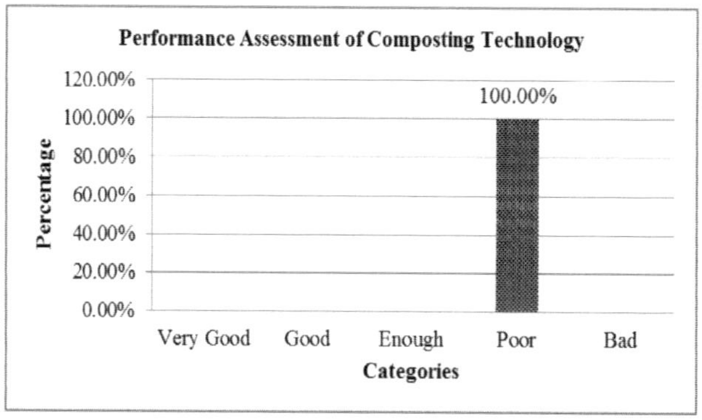

Figure 2. Figure of performance assesment of composting technology

3.3. The student's ability to produce of compost (product assessment)

Figure 3 shows compost product assesment produced by students. The results showed that the compost product among others 14.00% is very good categorized: 12.00% is enough categorized: and 74.00% is poor categorized. These results have shown that most of the compost that students have made is not good enough and has done repetition of composting.

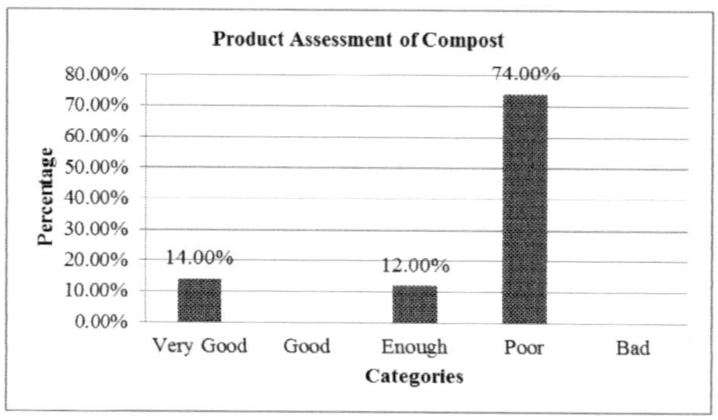

Figure 3. Figure of product assessment of compost

Based on figure 3 most of the students have done repetition of composting and until this report has not produced compost. It is in accordance with the STEM stages of the Improve Stage by analyzing the technology made whether to work or not, changing the design back better [8].

3.4. The student's ability in packaging of compost (product assessment)

Figure 4 shows the assessment in packing of compost. The compost packaging deals with the Art in STREAM approach. Figure 4 shows that students' ability to pack of compost among others 14.00% is very good categorized: 14.00% is enough categorized: and 72.00% is poor categorized. These results have shown that most of the compost packing is still the same as other groups. Based on field notes, students did not give reasons for selection of packaging and less attractive packaging. The compost package chooses was a plastic that has no pores.

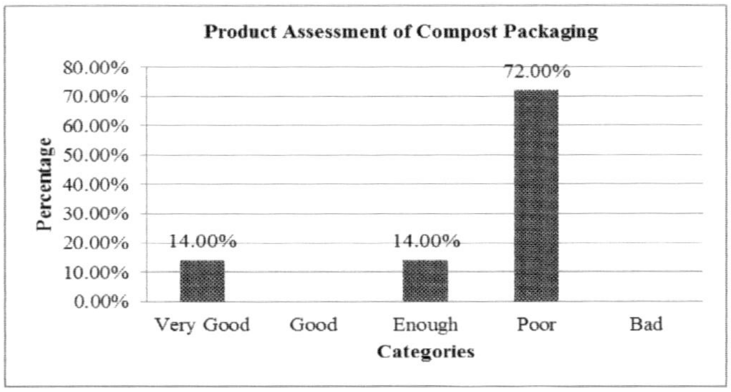

Figure 4. Figure of Product Assessment of Compost Packaging

STREAM perspectives used in this study is a perspective originated from STEM ie technology, engineering, and mathematics that are intercepts are in the science [19]. STEM added aspects of religion and art become of STREAM. It is chosen because the technology, religioun, engineering, art and mathematics involved are interrelated and supportive of science. Crosscutting Analysis Concept of STREAM which refers to STEM. Cross-cutting concept facilitates students to make interdisciplinary connections: Science, Technology, Religion, Engineering, Art and Mathematics. Crosscutting concept will make students better understand science and engineering in theory and practice [7]. Cross-cutting Concepts Analysis of Composting is presented in Table 1. In general, based on the results of study shows that students' ability in designing technology is still lacking. This condition causes the compost and packing ability to be less good.

Table 1. Cross-cutting concepts analysis of composting

Concepts that Surround of the Subject Matter (Science)	
The Compost	Process: Ask-Imagine-Plan-Create-Improve
1. The type and composition of compost materials 2. The success factors of composting 3. The good compost traits	Ask : The identifiying of problem and made of problem formulas Imagine : The looking for solution to the problem of composting Plan : The determining tools, materials, and designing compost technology Create : The making compost according to the design that has been made Improve : The analyzing and testing of composting result according to the function. The compost product has been produced according to the criteria good compost. If the compost product is not suitable then done redesign to get better results
Cross-Cutting Concept	Explanation
1. Cause and effect: mechanism and explanation 2. Scale, proportion and quantity 3. Energy and matter: Flows, Cycles and Conservation	1. Cause and effect: Students were guided to analyzed the causal effects of the good compost can affected plant growth. In this study, the application of compost in seeding of kale plants for the aquaponic system. The next learning was aquaponic system. 2. Scale, proportion and quantity: A bridge between science and mathematics in determined the composition of compost materials and making of EM_4 (effective microorganism 4) solution. 3. Energy and matter: Flows, Cycles and Conservation: The composting requires energy from the environment for the fermentation process. The fermentation process of aerobic produces humus, carbon dioxide, and heat [3].
Technology	**Religioun**
1. The using simple tools in composting 2. The using container to accommodate the result of mixing of	1. The appreciate the teachings of Islam: taking care of God's natural resources of taking care of compost by placing in a shady place. The observing every day to avoid drought on compost. When the dried compost is added with water and stirred until evenly distributed. The observation has been done until the compost into humus is marked by the color change to black (colour soil), crumbled (crumb / dry

compost materials (Composter)	structure) and odorless. The teachings live of Islam: thank God for the available natural resources. The rest of the organic waste and the presence of microorganisms can be utilized for plant life and fertilize the soil. 2. Implement the teachings of Islam (doing, fulfilling, delivering): the composting to give goodness to the environment as a form of worship that will earn reward from God.
Engineering	**Art**
1. The design composition of compost materials and tools already available in the surrounding environment **2.** The made of materials composting container design that have been stirred (composter)	1. The creating innovative and creative of composting technology 2. The interesting packaging of compost. The compost packaging still maintained the compost quality.
Mathematics	
1. The calculating on the tools and materials needed for composting 2. The determining on the composition of materials compost that have been used 3. The calculating of the EM_4 (effective microorganism 4) solution that has been used as starter composting **4.** The calculating on the required cost budget for composting	

(The Cross-cutting Analyses adopts Bybee's , 2013 and developed by Agustina, et. al, 2017)

4. Conclusion
The conclusion of preliminary study is students' performance ability in composting of compost materials is mostly good. However, performance ability in designing technology is still lacking. Most of the students are still not good in compost and compost packaging. Thus, students of Biology Education require habituation in the ability of designing technology for other learning especially STREAM learning.

Acknowledgments
The authors would like to thank to the Directorate General of Islamic Higher Education who has funded this preliminary study.

References
[1] Dinas Pertanian dan Ketahanan 2012 *Rencana Strategis Dinas Pertanian dan Ketahanan Pangan Kota Bandung* (Online) available at https://portal.bandung.go.id/storage/konten-lama/download/Transparansi/Perencanaan/Rencana_Strategis_Dinas_Pertanian_dan_Ketahanan_Pangan_2009-2013.pdf

[2] Irawati I, Salahudin and Selvianti 2009 Konsep Industri Sampah (KIS) sebagai Konsep Pengelolaan Sampah Perkotaan di Kota Bandung. *Prosiding Seminar Nasional Pendidikan Lingkungan Hidup* (Malang: Univlersitas Negeri Malang) p 214-227

[3] Nisa K 2016 *Memproduksi Kompos dan Mikroorganisme Lokal (MOL)* (Jakarta: Bibit Publisher)

[4] Badan Pusat Statistik 2015 *Februari 2015, Tingkat Pengangguran Terbuka (TPT) sebesar 5,81 Persen* (Online) available at https://www.bps.go.id/Brs/view/id/1139

[5] Direktorat Pembelajaran dan Kemahasiswaan 2014 *Buku Kurikulum Pendidikan Tinggi* (Jakarta: Direktorat Jenderal Kementrian Pendidikan Tinggi Kementrian Pendidikan dan Kebudayaan)

[6] Pendidikan Biologi 2015 *Kerangka Kualifikasi Nasional Indonesia dan Profil Lulusan Pendidikan Biologi* (Bandung: Fakultas Tarbiyah dan Keguruan. Unpublished)

[7] Bybee 2013 *The Next Generation Science Standards and The Life Sciences* (NSTA's K-12 Journals)

[8] Rosicka, C 2016 *Translating STEM Education Research into Practice* (Victoria: Australian Council for Educational Research)

[9] Ceylan and Z Ozdilek 2014 Improving a Sample Lessons Plan for Secondary Sciences Courses within the STEM Education *Procedia-Social and Behavioral Sciences* **177** p 223-228

[10] Bahri N M , Suryawati E and Osman K 2014 Students Biotechnology Literacy: The Pillars of Stem Education in Malaysia *International Society of Educational Research* **10** (3): 195-207

[11] Kementrian Pendidikan dan Kebudayaan 2014 *Panduan Penilaian Pencapaian Kompetensi Peserta Didik Sekolah Menengah Pertama* (Jakarta: Direktorat Pembinaan Sekolah Pertama)

[12] Wijaya A D, Karmila N, and Amalia M R 2015 Implementasi Pembelajaran Berbasis STEAM (Science, Technology, Engineering, Art, Mathematics) pada Kurikulum Indonesia *Seminar Nasional Fisika dan Aplikasinya* (Bandung: Universitas Padjadjaran)

[13] Oner A T, Nite S B, Capraro R M and Capraro M M 2016 From STEM to STEAM: Students' Beliefs About The Use of Their Creativity *The STEAM Journal* **2** (2): 1-14.

[14] Stiggins R J 1994 *Student-Centered Classroom Assesment* (New York: Macmillan College Publishing Company)

[15] Zainul A 2001 *Alternative Assesment* (Universitas Terbuka: Jakarta)

[16] Wulan A R 2008 *Skenario Baru Bagi Implementasi Asesmen Kinerja Pada Pembelajaran Sains di Indonesia* (Online) available at http:www.file.upi.edu.

[17] Creswell J W 2012 *Educational Research: Planning, Conducting, and Evaluating Quantitative and Qualitative Research* (Boston: Pearson)

[18] Septiani A 2016 Penerapan Asesmen Kinerja dalam Pendekatan Stem (Sains Teknologi Enjineering Matematika) untuk Mengungkap Keterampilan Proses Sains *Prosiding Seminar Nasional Pendidikan dan Saintek* (ISSN: 2557-533X)

[19] Hsu 2014 *Perspective of STEM.* (Online) available at http://perspectivestemeducation.com/services.htm.

IOP Conf. Series: Journal of Physics: Conf. Series **895** (2017) 012129 doi:10.1088/1742-6596/895/1/012129

Students Mental Representation of Biology Diagrams/Pictures Conventions Based on Formation of Causal Network

A W Sampurno[1], A Rahmat[2*] and S Diana[2]

[1]Program Studi Pendidikan Biologi, Sekolah Pasca Sarjana, Universitas Pendidikan Indonesia, Bandung, Indonesia
[2]Departemen Pendidikan Biologi, Fakultas Matematika Dan Ilmu Pengetahuan Alam. Universitas Pendidikan Indonesia, Bandung, Indonesia

*adirahmat@upi.edu

Abstract. Diagrams/pictures conventions is one form of visual media that often used to assist students in understanding the biological concepts. The effectiveness of use diagrams/pictures in biology learning at school level has also been mostly reported. This study examines the ability of high school students in reading diagrams/pictures biological convention which is described by Mental Representation based on formation of causal networks. The study involved 30 students 11th grade MIA senior high school Banten Indonesia who are studying the excretory system. MR data obtained by Instrument worksheet, developed based on CNET-protocol, in which there are diagrams/drawings of nephron structure and urinary mechanism. Three patterns formed MR, namely Markov chain, feedback control with a single measurement, and repeated feedback control with multiple measurement. The third pattern is the most dominating pattern, differences in the pattern of MR reveal the difference in how and from which point the students begin to uncover important information contained in the diagram to establish a causal networks. Further analysis shows that a difference in the pattern of MR relate to how complex the students process the information contained in the diagrams/pictures.

1. Introduction

Representations can be a mental representation (MR), which varies depending on the abstractions, for example, is a representation may be similarities, photos, portraits, descriptions, symbols, picture/diagram, art, etc. [1]. The level of mental representation that is formed is obtained from the relevance of the detailed information of the Causal network [2]. Intelligence development of learners must be a concern in the learning process, it is useful for learners to be able to think well. In each of the learning process, the ultimate goal is to establish a pattern of thinking, because thinking activities involving high mental activity [3]. Difficulties students in understanding abstract concepts of biology is because the human mind was not able to give the best response to representations that are abstract [4]. One form of information presentation visual media through diagrams/pictures, ie in the form of diagrams/pictures conventions, diagrams/pictures conventions using color, arrows, and notation describing the structure and relationships between functions or processes to represent any of the information contained in the picture, it will describe the non-structural features, such as picture/diagram of a process or behavior [5]. Picture has a role in understanding science concepts including learning system concepts in biology, the ability to represent, understand and read the picture becomes very important in education, especially learning activities [6].

Previous research by revealed, when students read images of convention representations, MR of students shows different patterns, the formation of a mental representation and the student's ability to interpret the component information in the picture, is an important aspect in order to fully understand the picture [7] however in this study the pattern of representation refers to the pattern developed by Ito. Data obtained by the causal network is the result of a process of observation and analysis, a model of causal network will be shaped like the mechanism of the system [8]. To represent a variety of visual media would require a model of representation, CNET Protocol involves activities of working memory to create a cognitive schema in active form, cognitive schema contains information that is interconnected, and its formation is very dependent on working memory involved in receiving and processing information [9]. Processing information in working memory is fundamental to represent the mental, so that the information obtained necessarily refer to the process of acquiring information, information integration, coding of information to be stored in the form of a scheme, and coding the information to be represented back [2]. Working memory will vary between each individual, so that each individual will produce a mental representation that is different in reading a variety of instructional media [10]. Therefore the mental representation forms is determining the formation of cognitive schemas and comprehension levels of students during the learning [11].

The main objective of this study was to analyze the pattern of students' mental representations with topological order of causal network when reading picture representation convention on the concept of the excretory system. Knowing the pattern of student mental representation when reading diagrams/pictures, teachers can select and define diagrams/pictures that can be used appropriately as a media of learning biology, so it will affect the control of the cognitive load that may arise against material abstract that exist in learning biology. This study is one that can serve as a bridge to help teachers in planning and seeing the students' ability in shaping knowledge through learning media as a whole by promoting the cognitive processes to achieve affective and psychomotor better. By knowing the pattern of mental representation of students in reading conventions diagrams/pictures, the teacher can knows that conventions diagram/picture can help students in the form of their knowledge, and may assist teachers in selecting appropriate learning media in accordance with the material being studied. Also helpful for students in developing their ability to read and fully understand the diagrams/pictures.

2. Research Method

This research uses descriptive method, aims to give an overview of how different patterns of mental representations of the students when reading the picture as measured by the establishment of a network of causal instruments worksheet-based CNET-protocol. Worksheet contains a number of conventions diagrams/pictures, statements and/or questions, done after students get the teaching materials. Formation of students mental representation is the result from the representation of the information elements contained in Figure 1. The results of the analysis is qualitative data in the form described by variations of the formation pattern of causal networks. The pattern formations of causal networks analyzed with topological order that can be show the different pattern with a level of causal network [12].

Participants are 30 students of class XI MIA, determined by convenience sampling. Research conducted at one of senior high school at Banten area. Implemented in the first semester of the academic year 2016/2017, exactly in the March-April 2017.

3. Result and Discussion

The findings showed a tendency RM pattern formed on the picture representation convention. The picture presented is a picture on the Relationship of Structure Formation Process nephron and urine (Figure 1).

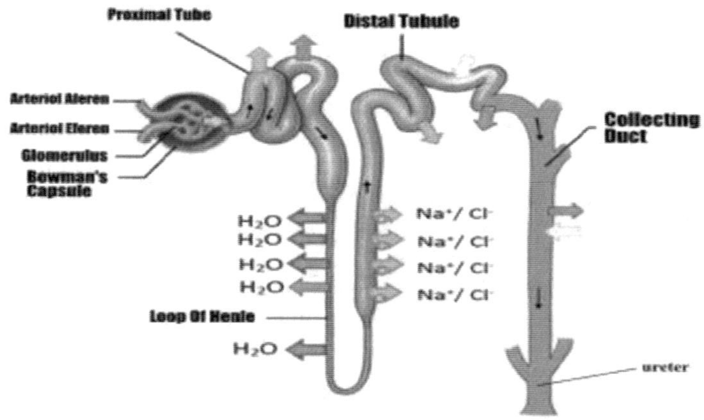

Figure 1. The relationship nephron structure and process of the formation of urine [13]

Patterns formed mental representations adapted to the basic pattern of the development of Ito [12]. Research findings showed a variation of the pattern of causal networks formed by students, for more details, here is a breakdown of the data on the MR students to the conventions picture. Representation of mental of Figure 1 is formed by determining the variable part of the information elements and sort the elements of the information correctly, and can build a causal network relationships among elements information. The findings showed three variations of patterns formed, the pattern of the first refers to the pattern of Markov Chain, second pattern refers to the pattern of feedback control with a single measurement, and third pattern refers to the pattern repeated feedback control with multiple measurement. Percentage of pattern formation tendency of students' mental representations can be seen in Figure 2.

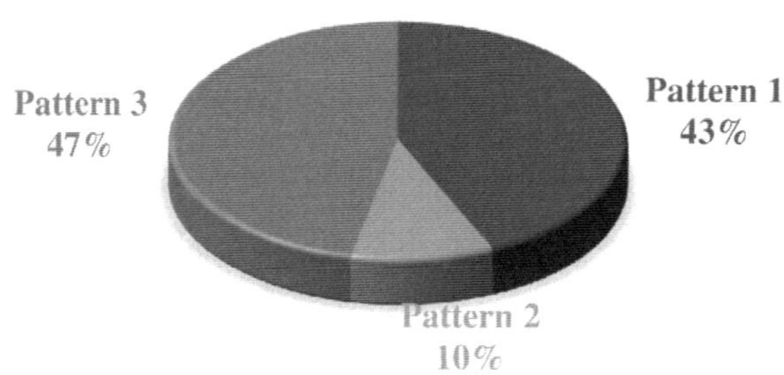

Figure 2. Percentage of pattern formation of students mental representations

This study found three variations of the pattern of mental representations found. The first pattern consists of respondents that can form a pattern like a work flow or process involving the whole or some of the elements in the picture information related to the structure and process. The second pattern consist of respondents can form different patterns flows between the elements of the process and structure information but still in one relationship, which involves an element of information and the overall process or some elements of the structure information. The third pattern consist of respondents can form different patterns flows same as the second pattern between structural elements and process information but still in one relationship, but it involves more than one element of the overall process and the information or some element of structure information. Of the three kinds of patterns such representations, third representation patterns are patterns that are expected for the representation of Figure1. Charta shown in Figure 2 can be seen that the pattern of the third most dominating with a percentage of 47%, higher than pattern 1 by 43% and pattern 2 by 10%.

IOP Conf. Series: Journal of Physics: Conf. Series **895** (2017) 012129 doi:10.1088/1742-6596/895/1/012129

Three patterns variations of representation formed based on completeness and sequence of information elements, and relationships between information elements. In the picture of the relationship nephron structure and process of formation of urine contains several elements of information that is divided into two, that is information element of structure and process. Elements of structure information that is glomerulus, Bowman's capsule, proximal tube, loop of henle, distal tubule, collecting duct. As for the information elements associated with that process, that is filtration, reabsorption, and augmentation. The pattern of this representation if it is linked with the picture relationship nephron structure and process of the formation of urine, then the two parts separate information elements, element structure and process information, but the second part of the information elements to be interconnected to perform activities of the excretory system. The pattern of mental representations are obtained by an worksheet instrument, can be seen in figure 3.

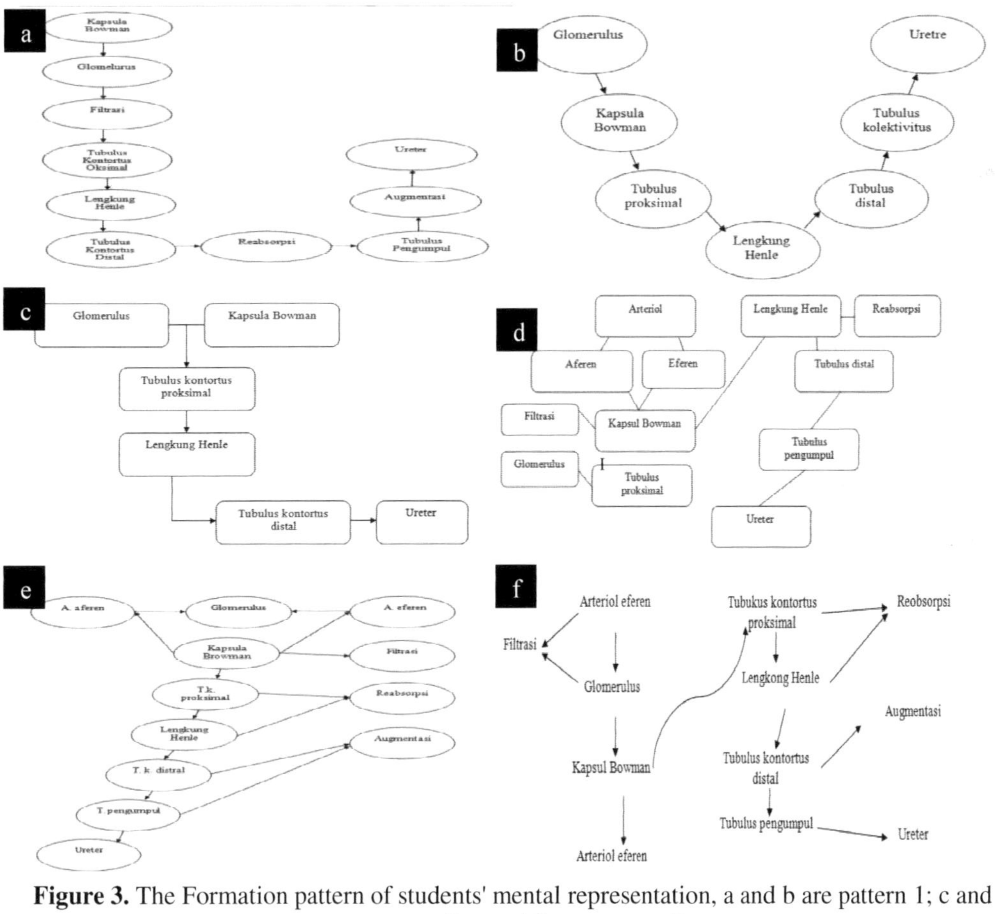

Figure 3. The Formation pattern of students' mental representation, a and b are pattern 1; c and d are pattern 2; e and f are patterns 3

In figure 3 shows some examples of the variation formation pattern of MR, the pattern proves that the MR pattern formation is formed based on the completeness of the information elements and the order of the information elements and relationships between information elements. On the a and b (Figure 3) visible difference from the amount of information and the order of elements of different information and the relationship between process elements and different structures, and both are shaped like flows of process mechanism, on the part of a contained element structure information and processes into one, while part b only structural information elements are formed flows. On the c and d

(Figure 3) shows the pattern of MR which has one separate elements into an element called element attached, seen from the difference networking causal formed between the two have differences of completeness of information, sequences, and the relationships among elements of information, part c visible presence of the element bound ie glomerulus/Bowman's capsule subsequently forming a flow, and on d seen meandering but seen only form a flow with reabsorption as elements bound, while the filtration should be elements of other dependent seen getting into the same flow. In part e and f seen students are able to distinguish between information elements and processes but interconnected between one and the other becomes a mechanism that looks more complex. The third variation of patterns formed describe that each pattern MR on different students, starting from which students form their networks to form a causal networks that showed comprehension and complexity thinking of students.

Patterns of representation will relate to the establishment of causal networks, the study showed differences in causal networks are different for each student. Third pattern (RepeatedFeedback Control With Multiple Measurement) shows a more complex representation compared with other patterns. Variety of patterns formation will vary depending on the students not only depend on the completeness of the information elements in the picture, but also the students' understanding of the picture reading and the knowledge they had before. More information students have correlated with the greater network of causal formed, the establishment of causal networks is determined by processing information (information processing) in working memory, because the MR formation is influenced by the activity of working memory [2]. Thus the amount of information does not necessarily prove that the students can and understand the intent of the picture which, prior knowledge related to the speed of processing information in working memory [10]. Thus the formation of MR depending on processing information in working memory that influenced by prior knowledge of students, students' understanding of the picture, and the complexity of the information in the picture. The amount of information that students will understand correlated with students prior knowledge and form a unity, so that the working memory will form a representation of the picture derived from the prior knowledge and the amount of information that students understand.

Proven with varied formation patterns of representation, it indicates that not all students are able to understand the information on the picture with a high degree of complexity. The difference from the pattern formed representations show the complexity of the picture represents the students in the form of causal networks. The information that has a high level of complexity that can limit the cognitive process and reduce the level of comprehension of students [14]. When given a picture with a high complexity early take on the role as a knowledge base for understanding the picture, understanding the picture portrayed through the representation of the pattern formed by the students. Prior knowledge lack of proper knowledge about a concept or information held by students, can lead to decreased ability students to interpret the information element in a diagram/picture [15]. Forming a good mental representations needed a good prior knowledge, understanding of the picture to identify each element of the information properly, so that it will form a good representation of the picture, although the level of complexity of the picture is high. So that different patterns of mental representations that reveal how and where the information element representation of picture which begins to form a causal network that shows the difference in the complexity of thinking of each student.

4. Conclusion

Difference from the Formation patterns of MR show the complexity of the picture represents the students in the form of causal networks. The formation of a MR is influenced by three factors, that is the prior knowledge of students, students understanding of the picture, and the complexity of the pictures presented. Mental representations needed a good prior knowledge, understanding of the picture to identify each element of the information properly. MR reveal how and where the information element representation of picture which begins to form a causal network. The ability of student when reading conventions diagram/picture describe the performance of working memory each student represents the diagram/picture becoming a pattern of causal network. Causal network reveal complexity of thinking each student, than shows different mental representation of students towards the diagram/picture.

By knowing the pattern of mental representation of students when reading conventions diagrams/pictures, the teacher can knows that conventions diagram/picture can help students understand concepts of relationship between structure and process of the body, and may assist teachers in selecting appropriate learning media that convenient with the concepts, also helpful for students in developing their ability to read diagrams/pictures to understand the concepts better.

This research is still far from perfect, lacking in some things like the lack of measurement of the prior knowledge of students, so it will be correlated with the pattern of representation is formed, and therefore in further research needs to be some things in order to strengthen and support research into better.

Acknowledgments
The authors would like to thank the parents who have given support for the continuity of this research, Mr. Asep Joko and Mrs. Yaya who already provided place and assist in this research.

References

[1] Paivio A 1990 *Mental Representation: A Dual Coding Approach* [ninth edition] (New York: Oxford University Press, Inc)
[2] Arentze T, Dellaert B G C and Timmermans H J P 2008 Modeling and Measuring Individuals' Mental Representations of Complex Spatio-Temporal Decision Problems *Environment and Behavior* **40** (6): 843-869
[3] Moseley D, Baumfield V, Elliott J, Gregson M, Higgins S, Miller J and Newton D P 2005 *Frameworks for Thinking: a handbook for teaching and learning* (Cambridge: Cambridge University Press)
[4] Treagust D V and Tsui C Y 2013 Multiple Representations in Biological Education, Models and Modeling in Science Education (Springer Science+Business Media B.V)
[5] Cheng M and Gilbert J 2015 Students' Visualization of Diagrams Representing the Human Circulatory System: The Use of Spatial isomorphism and representational conventions *International Journal of Science Education* **37** (1): 136-161
[6] Frek V, Vrtacknik M, Blejec A and Gril A 200. Students Understanding of Mulecular Structure Representation *International Journal of Science Education* **25** (10): 1227-1245
[7] Rahmat A, Soesilawati S A and Nuraeni E 2016 *Studi beban kognitif siswa SMA dalam pembelajaran biologi: representasi mental siswa ketika dihadapkan pada diagram isomorfisme spasial dan representasi konvensi* (Laporan Peneltian Penguatan Kompetensi. Universitas Pendidikan Indonesia)
[8] Yazdani Az, Yazdani Ak and Boerwinkle E 2016 Conceptual Aspects of Causal Network in Applied Context *J Data Mining Genomics Proteomics*
[9] Rahmat dkk 2016 Studi Beban Kognitif Siswa SMA Dalam Pembelajaran Biologi: Representasi Mental Siswa Ketika Dihadapkan Pada Diagram Isomorfisme-Spasial.
[10] Gathercole S and Alloway T 2009 *Memori Kerja dan Proses Belajar [Working Memory and Learning* (Jakarta: PT. INDEKS)
[11] Plass J L, Moreno R and Brunken R 2010 *Cognitive Load Theory* (Cambridge: Cambridge University Press)
[12] Ito S 2016 *Bayesian Networks and Causal Networks, Information Thermodynamics on Causal Networks and its Application to Biochemical Signal Transduction, Springer Theses* (Springer Science+Business Media Singapore)
[13] Campbell N A, Race J B and Mitchell L G 2008 *Biologi Edisi Kelima Jilid III* (Jakarta: Erlangga)
[14] Eseryel D, Ifenthaler D, and Ge X 2013 Validation study of a method for assessing complex ill-structured problem solving by using causal representations *Education Tech Research Dev* **61**

443–463
[15] Andres H and Petersen C 2002 Presentation Media, Information Complexity, and Learning Outcomes *Journal Educational Technology Systems* **30** (3): 225-246
[16] Novick L and Catley K 2014 When Relationships Depicted Diagrammatically Conflict With Prior Knowledge: An Investigation of Students' Interpretations of Evolutionary Trees *Science Education* **98** (2): 269-304

IOP Conf. Series: Journal of Physics: Conf. Series **895** (2017) 012130 doi:10.1088/1742-6596/895/1/012130

Junior High School Students' Perception about Simple Environmental Problem as an Impact of Problem based Learning

M C Tapilouw[1,2*], H Firman[1], S Redjeki[1] and D T Chandra[1]

[1]Universitas Pendidikan Indonesia, Bandung, Indonesia
[2]Yayasan Salib Suci, Bandung, Indonesia

*marisatapilouw@student.upi.edu

Abstract. Environmental problem is a real problem that occur in student's daily life. Junior high school students' perception about environmental problem is interesting to be investigated. The major aim of this study is to explore junior high school students' perception about environmental problems around them and ways to solve the problem. The subject of this study is 69 Junior High School Students from two Junior High School in Bandung. This study use two open ended question. The core of first question is environmental problem around them (near school or house). The core of second question is the way to prevent or to solve the problem. These two question are as an impact of problem based learning in science learning. There are two major findings in this study. The first finding, based on most students' perception, plastic waste cause an environmental problem. The second finding, environmental awareness can be a solution to prevent environmental pollution. The third finding, most student can classify environmental pollution into land, water and air pollution. We can conclude that Junior High School Students see the environmental problem as a phenomenon and teacher can explore environmental problem to guide the way of preventing and resolving environmental problem.

1. Introduction

Nowadays, many environmental problem happen in the earth. We should see this environmental problem as a complex problem that occur in student's daily life. Natural environment is important part of our lives and science education because we must teach Junior High School students how to appreciate natural environment. There are many environmental problem around us, for example air pollution caused by vehicles and industries emission. Junior high school students can see these problems everyday whether it is near school or near their house.

Problem based learning can be used in science learning, especially in environmental concepts. Through problem based learning, students can solve a complex, ill-structured problem that discipline-based content [1]. In problem based learning, students are introduced to real world problem (environmental problem) so students will place environmental problem as their needs and will gather information that match to environmental problem in their daily life. Science learning must be connected to natural environment as an integral part of it. Problems are introduced to act as triggers for science learning [2]. Throughout environmental problems, students can learn to work collaboratively in order to improve problem solving skill and critical thinking. Science teacher play important role as a facilitator in science learning and responsible to guide students in making science learning as a meaningful and joyful learning.

Junior high school students' perception about environmental problem is interesting to be investigated. This statement as a foundation of this study. The reason is we can see student's perception about environmental problem around them as an impact of problem based learning. In this article, there will be an exposure of student's perception about environmental problem around them.

2. Experimental Method

This study is a descriptive study. The procedure of this study is described in Figure 1. Environmental concepts in science learning as an initiation of this study. Science teacher use problem-based learning approach within environmental concepts, so that students see environmental problem around them as real world problem. The student of Class VII in two Junior High School located in Bandung as a subject of this study. About 69 junior high students (36 students from Junior High School A and 33 students from Junior High School B) are the subject of this study.

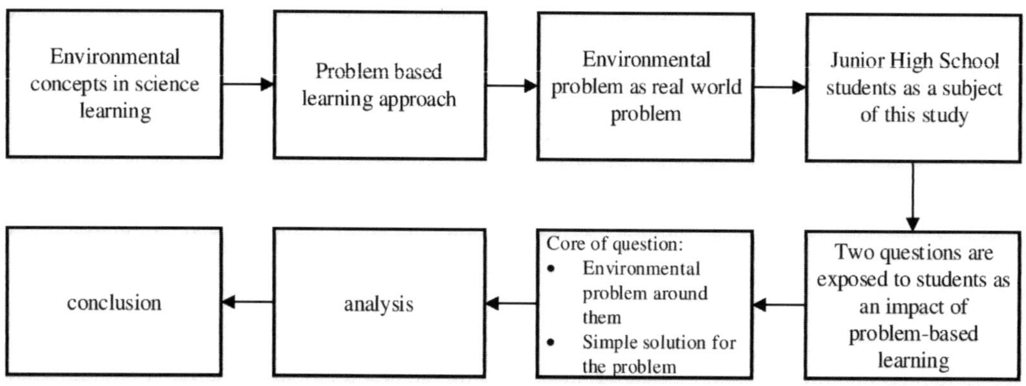

Figure 1. Study procedure

Two questions are exposed to Junior High School students as an impact of problem-based learning in science learning especially in environmental concepts. The questions are:

1. Mention two environmental problems around you (near school or near house)!
2. Mention two simple solutions for preventing or resolving the problem!

These question need short answer to see spontaneous answer. All answer are collected to be analysed.

3. Result and Discussion

3.1. Environmental problem

Each student gives two example of environmental problem, so that there are 72 answer for Junior High School A. Based on student's answer about environmental problem around them, there are 14 kinds of answer. As we can see in Figure 2, the answer are less water infiltration (2.78%), less field (1.39%), burning garbage (6.94%), flood (22.22%), cigar smoke (1.39%), smell of garbage (6.94%), vehicle emission (8.34%), polluted river (6.94%), garbage strewn (30.56%), industrial emission (2.78%), dirty toilet (2.78%), plastic garbage (1.39%), Clean environment* (2.78%), No Garbage* (2.78). We can see two interesting answers which are clean environment and no garbage around students.

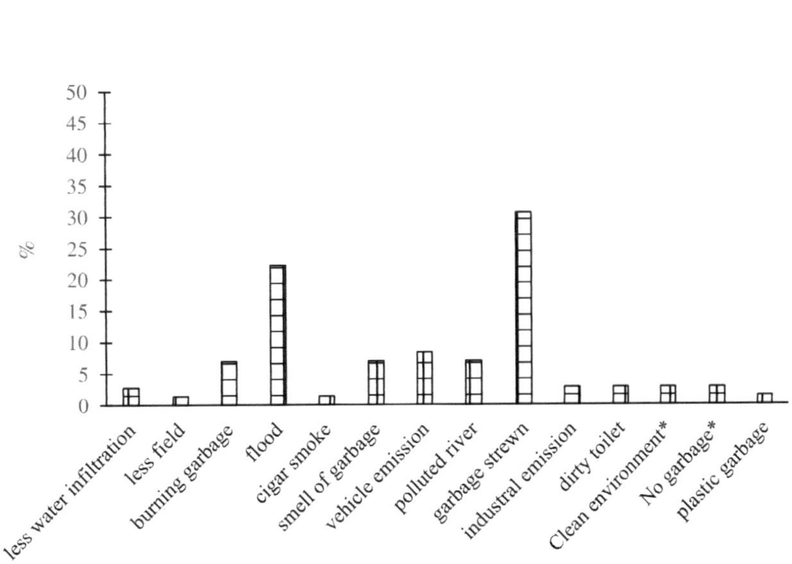

Figure 2. Environmental problem (School A)

On the other hand, there are 66 answer for Junior High School B because each student gives two answers. Based on student's answer about environmental problem around them, there are 12 kinds of answer. As we can see in Figure 2, the answer are vehicle emission (18.18%), CFC gas (3.03%), garbage strewn (27.27%), dirty sewage (9.09%), burning garbage (12.12%), polluted land (7.57%), plastic garbage (3.03%), polluted river (12.12%), pesticide uses (1.52%), flood (3.03%), industrial emission (1.52%), cigar smoke (1.52%).

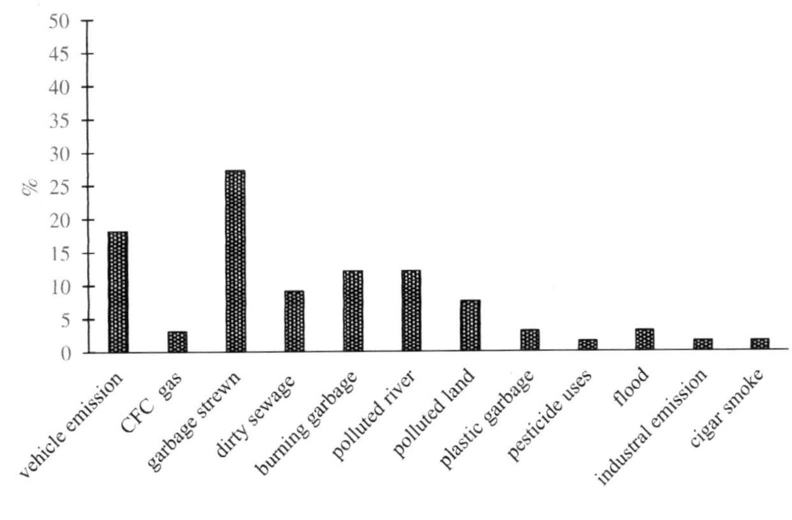

Figure 3. Environmental problem (School B)

From Figure 3 and Figure 4, we can see seven same answer about environmental problem. They are flood, cigar smoke, burning garbage, vehicle emission, polluted river, garbage strewn and industrial emission. Junior high school students see environmental problem as part of their daily life. Beside of that, most students mention garbage strewn as an environmental problem. Science teacher needs to teach using environmental problem in order to educate students about sustainable development in the future [2]. Environment concepts as one part of science learning. Vehicle emission consist of xenobiotic compounds that are foreign to the biosphere [4]. This kind of emission cause environmental problem especially air pollution and also acid rain [5]. The problem that exist is dirty toilet at School A, include sometimes students forget to turn off the tap. Turning off the tap is like a domino effect. By turning off the tap, we can save the energy [6]. Based on Figure 3, students mention that CFC gas as an environmental problem, and ozone layer become thinner each year [7].

According to student's answer, pesticide can cause land pollution. So, earthworms can be a bio-indicator of polluted land [8]. Overusing of pesticide and fertilizer can cause land pollution [9]. From Figure 3 and Figure 4, we can see polluted river and dirty sewage become problem for our natural environment because these two problem can decrease water quality. Bad water quality in river/ sewage can cause bad human health [10] and even affect water biota/ ecosystem [11]. As an example, there are water crisis in China [12] and also bad air quality [13] and the China's government tried to overcome these environment problem. The plastic also become an environmental problem around the Students. Even in Canada, plastic harm water ecosystem and become a serious problem [14]. This case happened also in China [15] where there are environmental problem cause of plastic waste. All explanation about environment problem can eager science teacher to be creative in teaching environment concepts by choosing problem as a foundation science learning.

3.2. Simple solution

From the second question, each student gives two simple solutions so that there are 72 answer (Junior High School A). There are 15 kinds of answer, wise in land uses (1.39%), put trash into trash can (23.61%), Bio-pore (1.38%), Garbage Bank (26.39), Recycle (9.72%), cleaning environment together (4.17%), picking up trash (5.56%), sorting out (5.56%), 4R (4.17%), Water filter (1.39%), environmental awareness (5.56%), keep toilet clean (4.17%), plant trees (2.78%), use bicycle (2.78%), organic fertilizer (1.39%).

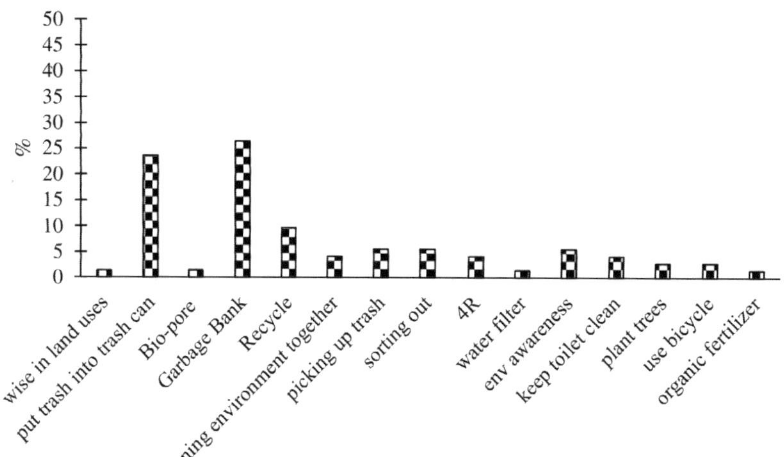

Figure 4. Simple solution (School A)

There are 66 answers from Junior High School B Students about simple solution for environmental problems. From the answer, there are 11 kinds of simple solution for environmental problem, use bicycle (19.69%), plant trees (9.09%), picking up trash (1.52%), put trash into trash can (33.33%), treating waste (10.61%), environment awareness (10.61%), Bio-pore (1.52%), less garbage burning (4.55%), water filter (3.03%), less air conditioner (4.55%), 4R (1.52%).

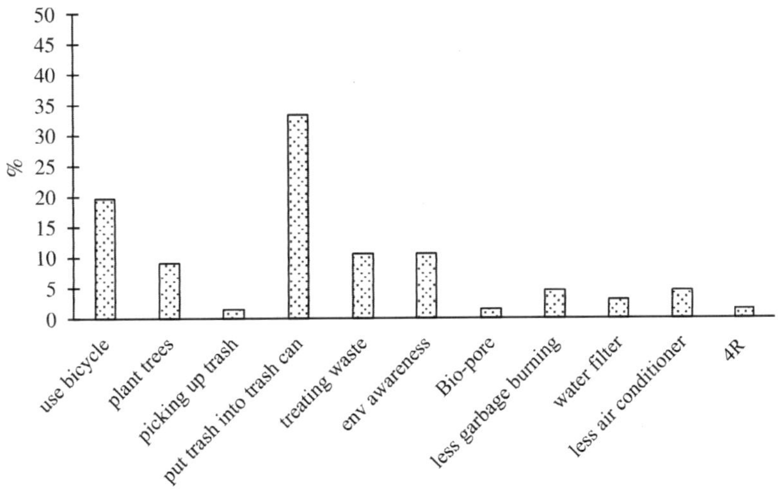

Figure 5. Simple solution (School B)

Based on Figure 4 and Figure 5, there are six same answers about simple solution to prevent or resolve environmental problem. They are put trash into trash can, 4R (Reduce, Reuse, Recycle, Replace), water filter, environment awareness, plant trees and use bicycle. There are many solution for preventing and resolving environment problem. Base on Figure 4, recycling is a simple solution. A research shows that recycling can overcome environment problem especially land pollution [16]. Heavy traffic can cause terrible air pollution [17] and we can prevent it by using bicycle to school [18]. In every industry, they should have waste water treatment [19] in order to resolve water/ river pollution. The way student give simple solution for preventing or resolving environment will help them to think critically. The science teacher become facilitator for students themselves.

4. Conclusion

From the exposure and analysis above, there are two conclusion. First, Junior High School Students see the environmental problem as a phenomenon, for example environmental pollution around them. Second, science teacher can explore environmental problem to guide the way of preventing and resolving environmental problem by exposing example of real environment problem around them.

Acknowledgments

This study was supported by science teacher and Junior High School students from two Catholic Junior High School located in Bandung. I dedicate this study for Science learning improvement.

References

[1] Ertmer P A 2006 *Interdisciplinary Journal of Problem-Based Learning* **1(1)** 40-54
[2] Nariman N and Chrispeels J 2015 *Interdisciplinary Journal of Problem-Based Learning* **10(1)** 1-16
[3] Hoang T T P and Kato T 2016 *Sustainable Environment Research* **26** 274-286
[4] Godheja J, Shekhar SK, Siddiqui SA and Modi DR 2016 J *Environ Anal Toxicol* **6(5)** 1-18
[5] Berger T W, Turtscher S, Berger P and Lindepner L 2016 *Environmental Pollution* **216** 624-635
[6] Thi N B D, Lin C-Y and Kumar G *Sustainable Environment Research* **26** 197-202

[7] Sharma S, Sharma P, Khare M, Kwarta S 2016 *Sustainable Environment Research* **26** 142-148
[8] Carter L J, Ryan J J and Boxall A B A 2016 *Environmental Pollution* **213** 922-931
[9] Bech T B, Rosenborn A R, Sorensen S R and Jacobsen C S 2017 *Environmental Pollution* **222** 404-411
[10] Tong Z, Yang B, Hopke P K, Zhang K M 2017 *Environmental Pollution* **220** 1112-1120
[11] Kay P, Hughes S R, Ault J R, Ashcroft A E and Brown L E 2017 *Environmental Pollution* **220** 1447-1455
[12] Han D, Currell M J and Cao C 2017 *Environmental Pollution* **220** 1222-1233
[13] Chen J, Li C, Ristovski Z, Milic A, Gu Y, Islam M S, Wang S, Hao J, Zhang H, He C, Guo H, Fu H, Miljevic B, Morawska L, Thai P, Lam Y F, Pereira F, Ding A, Huang X and Dumka U C 2017 *Science Total Environment* **579** 1000-1034
[14] Anderson J C, Park B J and Palace V P 2016 *Environmental Pollution* **218** 269-280 Kay P, Hughes S R, Ault J R, Ashcroft A E and Brown L E 2017 *Environmental Pollution* **220** 1447-1455
[15] Li W C, F H, TSE and FOK I 2016 *Science Total Environment* **566** 333-349
[16] Guo L and Wu D-Q 2017 *Sustainable Environment Research* **27** 1-6
[17] Markiewicz A, Bjõrklund K, Eriksson E, Kalmykova Y, Störmval A-M and Siopi A 2017 *Science Total Environment* **580** 1162-1174
[18] Johansson C, Lövenheim B, Schantz P, Whalgren L, Almstörm P, Markstedt A, Störmgren M, Forsberg B and Sommar J N 2017 *Science Total Environment* **584-585** 55-63
[19] Klomjek P *Sustainable Environment Research* **26** 274-286

Profile of High School Students' Propositional Network Representation when Interpreting Convention Diagrams

M Fatiha[1], **A Rahmat**[2*] **and R Solihat**[2]

[1]Program Studi Magister Pendidikan Biologi, Sekolah Pascasarjana, Universitas Pendidikan Indonesia, Bandung, Indonesia
[2]Departemen Pendidikan Biologi, Universitas Pendidikan Indonesia, Bandung, Indonesia

*adirahmat@upi.edu

Abstract. The delivery of concepts in studying Biology often represented through a diagram to easily makes student understand about Biology material. One way to knowing the students' understanding about diagram can be seen from causal relationship that is constructed by student in the propositional network representation form. This research reveal the trend of students' propositional network representation patterns when confronted with convention diagram. This descriptive research involved 32 students at one of senior high school in Bandung. The research data was acquired by worksheet that was filled by diagram and it was developed according on information processing standards. The result of this research revealed three propositional network representation patterns are linear relationship, simple reciprocal relationship, and complex reciprocal relationship. The dominating pattern is linear form that is simply connect some information components in diagram by 59,4% students, the reciprocal relationship form with medium level by 28,1% students while the complex reciprocal relationship by only 3,1% and the rest was students who failed to connect information components by 9,4%. Based on results, most of student only able to connect information components on the picture in linear form and a few student constructing reciprocal relationship between information components on convention diagram.

1. Introduction

The delivery process of concepts in studying Biology often represented through a diagram. The teacher very dependent to diagram and graphic for teaching Biology material with well [1]. The diagram can be represented in two ways, spatial isomorphs and convention representation [2]. The spatial isomorphs diagram is a diagram which is resembles the form of actual source while the convention representation illustrates the form that is represented through symbolic features, color and icon. Someone will interpret the diagram with well if he/she able to identify the elements in the diagram by spatial isomorphism or convention representation [3]. One way to knowing the students' understanding about the diagram can be seen from causal relationship which is constructed by student in form of propositional network representation.

Studying a meaningfulness, cognitive psychology develop the propositional representation. Proposition as representation concept play a role in information processing model (illustrating every kind of information) and often interpreted connect with picture [4]. Information processing involve working memory to get experience (knowledge) which is stored in long-term memory. Empirical component that is stored the information according to experience can be explained in form of propositional network [4]. According to that opinion, propositional network representation can be a way to knowing student's understanding about the diagram through causal relationship that is constructed by student. Proposition which is known by subject can be helping to choose information and factual that is relevant in combining cognitive units with appropriate way to solve the problem [5].

Symbolic knowledge representation is a rather neglected problem, especially studying network that can represent and study the infinite propositional rules [6]. Diagram interpretation can form the visual representation of person's mental. The formation may be guided by previous knowledge that involve associative connection between mental visual representation and information that is triggered by diagram [3]. Propositional network representation explain declarative/conceptual knowledge which is represented in propositional network similar with semantic network representation [7]. Research about propositional network representation may never have been done. But the student's mental visualization picture when they reads diagram was already done a lot [3]. In a section other than education, propositional network representation has been widely applied in Human Factors and Ergonomics Society section to illustrate knowledge object in a work system scenario. Propositional network consists of one set of nodes which are represented information source, agent, and objects that are connected in certain causal paths [8].

2. Methods

This research is a descriptive research which aims to give a student's mental representation description and the causal relationship is represented in form of propositional network representation when interpreting convention diagram. This research also reveal the factors which are influence propositional network representation pattern that is constructed by student. The data was collected qualitatively to see the trend of propositional network representation pattern then it's described in order to explain student's mental representation description when interpreting diagram. The subject in this research was 32 students from one of senior high school in Bandung that was determined according to convenience sampling technic or willingness to involve in this research. Data collection is acquired by using written instrument in form of worksheet which aims to identify and analyze the trend of propositional network representation which is made in propositional network according to diagram that have been available.

Worksheet is developed according to 4 standards information processing [9] and adapted to be three standards that is information identification, information integration (interpretation and information synthesis) and information relevance analysis. Identification and information integration is made in form of question that is made in description form while information relevance analysis is explained in form of propositional network representation that is constructed by student. Worksheet instrument development is executed through judgment preparation step for validate the instrument by expert lecturer. Judgment implementation by expert lecturer about worksheet instrument aims to know the content of validity, conformity with framework indicator that is adopted, question conformity with answer key and language structure.

3. Result and Discussion

The research data is analyzed according to results of respondents answer in the worksheet and observation result according to questionnaire and findings obtained while in the field. The given of worksheet instrument have held in the moment after studying sense system material. Previous knowledge (prior knowledge) will very guides student to connect the connection of associative in mental visual representation with information that is triggered by diagram [3]. This also aligned with previous research that first knowledge influence interpretation diagram/picture ability [10]. Research finding that is presented in form of student's propositional network representation pattern description when interpreting convention diagram according to student's answer analysis result in the worksheet that have been given. The developed diagram in this research has gone through judgment step by expert lecturer. Something that is considered according to conformity analysis with curriculum that's used, conformity analysis of material concept according to information component indicator and completeness which are show the process mechanism. Diagram that is used in the research to get student's representation information is a diagram about eyeball structure and mechanism of seeing process below.

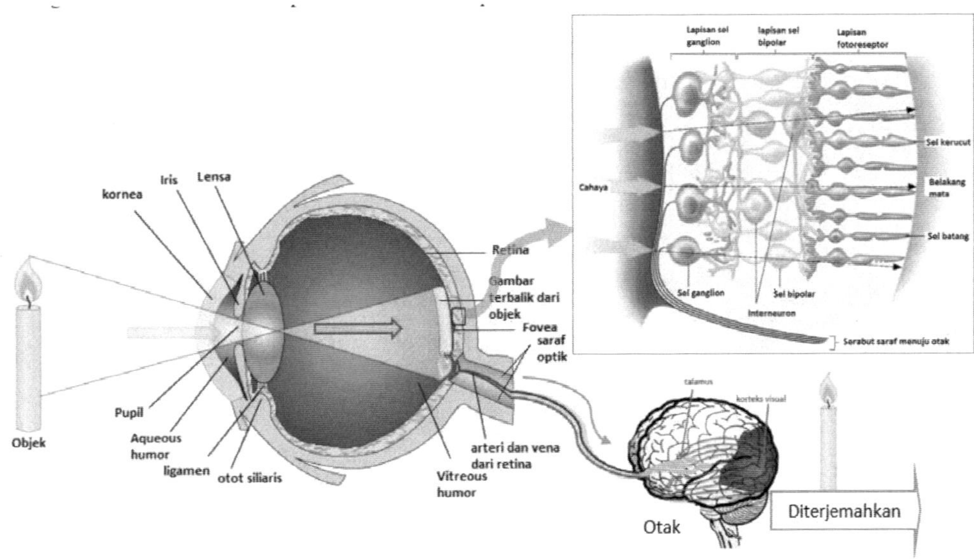

Figure 1. Diagrams of the eyeball structure and the mechanism of seeing process

Figure 1 describing information component of structure which is compile the eyeball and seeing process mechanism from the beginning, light that is came in to the eye as an object which is seen till the object's shadow is translated by brain to be a real object. Some question in worksheet guide student to connect between information component concepts in form of propositional network pattern. The question require respondent to identify information component on the diagram that is related with seeing process mechanism only (light as object, cornea, aqueous humor, pupil, iris, lens, vitreous humor, retina, fovea, photoreceptors that is consists of stem cells, cone cells, bipolar cells, interneuron, ganglion cells, inverted shadow from object, optical nerve, brain, objects real picture) and describing the function of every information component with a truly concept and connecting between information component concept to build propositional network representation.

3.1. The trend of student's proppositional network representation pattern when interpreting convention diagram

Based on research finding, from 32 respondent which are filling the worksheet instrument, it is found three student's propositional network representation big pattern on the eyeball structure diagram and seeing proses mechanism according to complexity level of the relationship between information component concepts that is used. The three patterns are linear form (simple), reciprocal relationship form (causal) with medium level and complex causal relationship form. Propositional network representation pattern is related with causal network formation that is constructed by student. Causal network illustrate all of the student's knowledge to object/diagram that is useful to rate the student's way taking decision to the information [11]. So, can be seen the variation of relationship pattern between information components in propositional network because the student's comprehension to the information about diagram that is interpreted also different.

The linear pattern is a pattern that is very dominating in this research and constructed by students by 59.4% from all of respondent. Based on analysis result from linear pattern, student construct the propositional network with a different information component number and also the relationship number between the information component concepts. In the second pattern, 28.1% student already able to construct reciprocal relationship between the concept from information component that is found on picture, although a few student only. Meanwhile, on the third pattern, 3.1% student already able to connect the information element concepts to be more complex. The complex form which is formed is interpreted as a relationship which is involved information component number that is connected one to another so much compared with previous pattern. Besides of the three pattern, the rest is student that is failed to connect between information component by 9.4% and in detail, the student percentage in every pattern can be seen on figure 2.

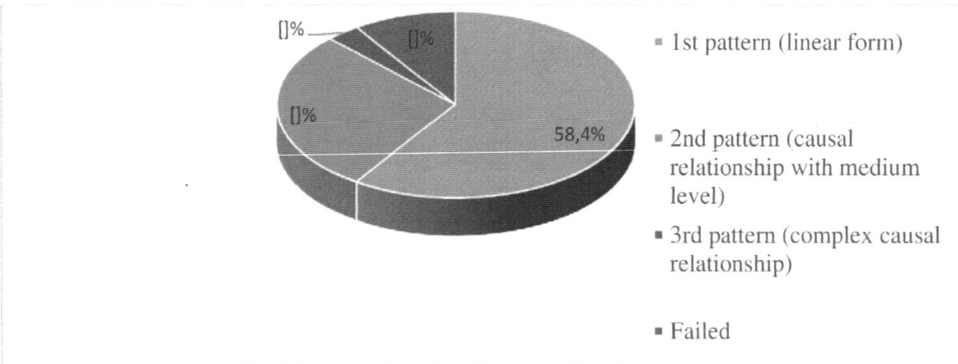

Figure 2. Student percentage on every trend of propositional network representation

Three pattern which are represented student propositional network representation description on figure 2 above also have an ability to processing the different information in every pattern. This information processing represent from visual form and verbal become verbal system. On this worksheet, student not only asked to find and synthesize important information component on the convention picture, but also able to describe with causal relationship form from information component. Information processing center in working memory is different between visual and verbal information [4]. When the information component is interpreted, information processing in working memory every student to do the activity will also different. Working memory that is recorded in form of mental representation can give implication to studying activity if it able to choose a better method that will be used in studying process [12]. This is known from student's answer when identifying information, describing information on the eyeball structure diagram and seeing process mechanism and then connecting it in form of propositional network. The following is representation pattern that is elaborated according to information interpretation on picture 1.

3.1.1. 1st pattern: propositional network in linear form (simple). On propositional network pattern in linear form show the relationship between concepts with one-way traffic. Average respondent in this pattern, in minority (31%) already able to identify the information component that is expected (light as object, cornea, aqueous humor, pupil, iris, lens, vitreous humor, retina, fovea, photoreceptors that is consists of stem cells, cone cells, bipolar cells, interneuron, ganglion cells, inverted shadow from object, optical nerve, brain, objects real picture) while by 42.1% of respondent also writing the information component that is not too necessary in seeing process mechanism (ciliary muscles, artery and vena and also ligament). Meanwhile about 26.3% of respondent writing some important information component, but less than complete. This is caused by the student firstly knowledge is very react in student to know a concept and new information on diagram, so they do a wrong interpreting

[10]. Aligned with previous research, someone cannot interpret and represent information component in diagram without knowing and understanding that information [3].

Table 1. Ability to process student information on the 1st pattern (linear)

1st Pattern (Linear)		
No	Information Identification	Respondent Percentage
1	Writing almost all of the information components on diagram	42,1%
2	Writing important information component about seeing process mechanism	31,6%
3	Less complete in writing information component on diagram	26,3%
No	Information Integration	Respondent Percentage
1	Less complete in describing information component but the concept is already correct	94,7%
2	Describing information component completely and the concept is correct	5,3%
No	Information Relevance Analysis	Respondent Percentage
1	Connecting information component properly and correctly	73,7%
2	Connecting most of information component with incorrect concept	26,3%

Based on table 1, student already able to identify all of important information component and able to describe the role in seeing process mechanism in the first pattern. But still, linear pattern that was described by student unable to connect from every information component that is already identified and described previously, so the trend that has been formed be in the same direction. The assumption that can be reported is because student assume a process that happen in mechanism, only shows the process from coming of the light till it's translated by brain. Information component that can be connected also limited light lane through eyeball structure to brain, without thinking about function linkage from every information component on the convention diagram. Information component that is described and connected in propositional network representation by student are still found the incorrect concept in explaining seeing process mechanism. Even though, most of students already correct and properly in interpreting the information component and explaining seeing process mechanism on figure 1. Propositional network pattern is linear form (simple) that is constructed by student can be an example to figure 3.

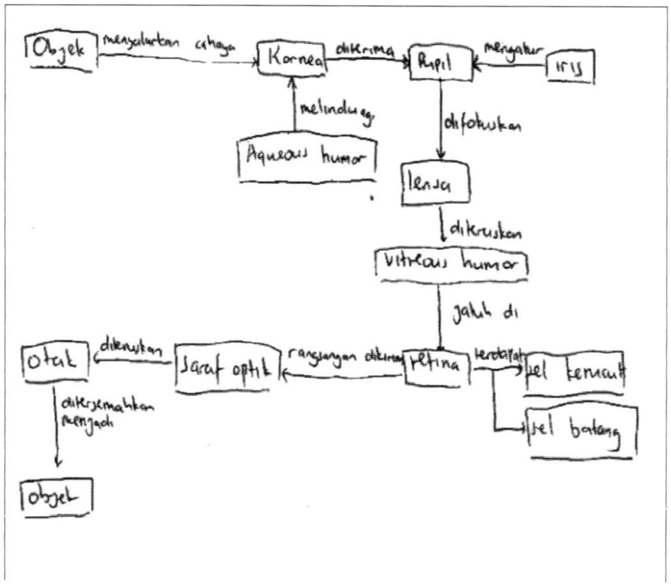

(a)

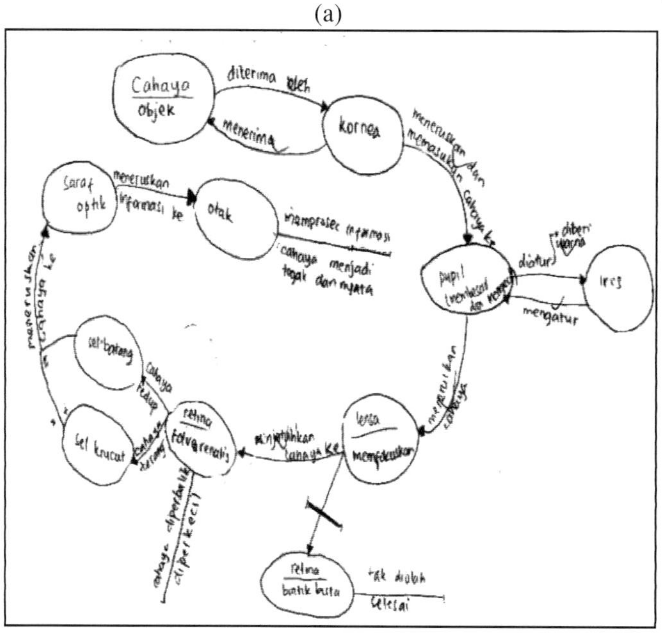

(b)

Figure 3. Propositional network pattern is linear form with the correct concept (a) and incorrect concept (b)

3.1.2. 2nd Pattern: Propositional network forming some reciprocal relationship (medium). Propositional network pattern which is forming some reciprocal relationship of information component already connecting the correct concept, in accordance with when identifying constructed information component. One of example from propositional network pattern that show some reciprocal relationship described to figure 4.

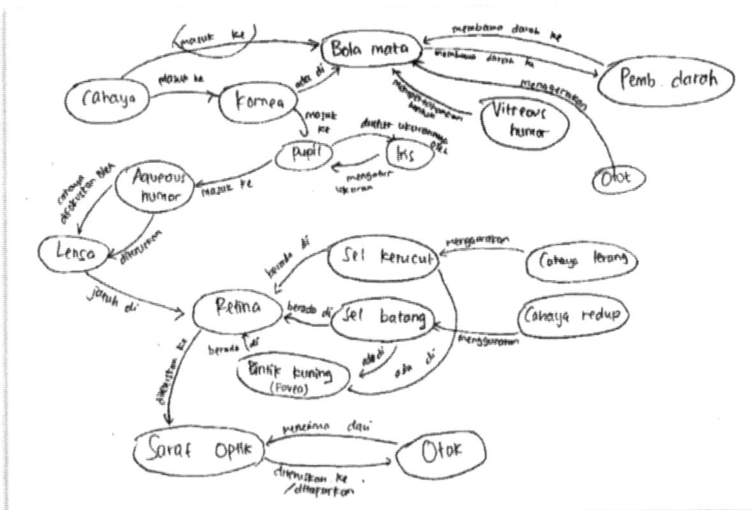

Figure 4. Propositional network pattern that show some reciprocal relationship (medium)

The reciprocal relationship happened because information component number that is obtained more information. More information that is owned by student, then the causal relationship that is made is increase [11]. This is caused of the student beginning knowledge is related with information processing speed in working memory so causal network can be formed [13]. Based on table 2, most of it (77,8%) identifying all of the important information component expected (light as object, cornea, aqueous humor, pupil, iris, lens, vitreous humor, retina, fovea, photoreceptors that is consists of stem cells, cone cells, bipolar cells, interneuron, ganglion cells, inverted shadow from object, optical nerve, brain, objects real picture) and also writing the information component that is not too necessary in seeing process mechanism (ciliary muscles, artery and vena and also ligament). Meanwhile about 22.2% of respondent are writing some important information component, but less than complete.

Table 2. Ability to processing student information on the 2nd pattern (reciprocal relationship with medium level)

2nd Pattern pattern (Reciprocal relationship with medium level)		
No	Information Identification	Respondent Percentage
1	Writing almost all of the information components on diagram	77,8%
2	Less complete in writing information component on diagram.	22,2%
No	Information Integration	Respondent Percentage
1	Less complete in describing information component but the concept is already correct	94,7%
2	Describing information component completely and the concept is correct	5,3%
No	Information Relevance Analysis	Respondent Percentage
1	Connecting most of information component properly and correctly	100%

3.1.3. 3rd Pattern: Propositional network forming more reciprocal relationship (complex). On the pattern that was constructed with connecting information component in a complex way aligned with the respondent's way when identifying almost all of the information component on the diagram. The reciprocal relationship complexity is happened because the information component that is obtained relevance with the purpose of diagram that is wanted to be delivered. Student who can choose and interpret the diagram so they can connect information component correctly. This is also related with information processing speed and student's beginning knowledge [13], [3]. Important information component that is identified almost all (light as object, cornea, aqueous humor, pupil, iris, lens, vitreous humor, retina, fovea, photoreceptors that is consists of stem cells, cone cells, bipolar cells, interneuron, ganglion cells, inverted shadow from object, optical nerve, brain, objects real picture) but also writing information component that is not too important in seeing process mechanism (ciliary muscles, artery and vena and also ligament). Beside, information component that is identified and described also completely and correctly concept. This is proven with reciprocal relationship that can consists of many information component and also the concept is correct. Propositional network pattern with complex relationship that is constructed by student can be an example to figure 5.

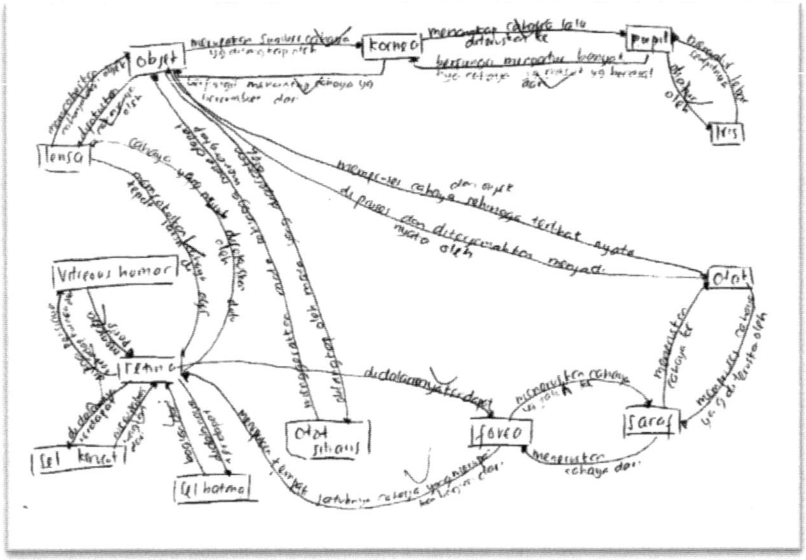

Figure 5. Propositional network pattern that shows complex reciprocal relationship

3.2. *Factors that influence student's propositional network representation*

Various internal or external factor that influence student to build propositional network representation when are faced to a convention picture may need to know. On the questionnaire, consists of 15 statement and can be grouped to be two parts about student's problem in understanding the diagram and constructing propositional network. According to questionnaire's answer analysis, most of student agree that teacher have used a diagram to explain material in studying activity. The explanation of material using a diagram from teacher more preferable by most of student and also easy to understand compared with verbally explanation. Beside via teacher's explanation, student are also prosecuted to be able to find and explain the diagram in front of the class as one of biology assignment in groups. Student are given a freedom to use a diagram from various source like text book or another relevance source like internet. Most of the student feel that diagram on the text book is very complex and hard to understood. When explaining the diagram, most of student feeling easy to explain the diagram that is used because student have many information from diagram so it can help them understand the material that have been studied. But if it compared with verbally explanation, student prefer to explain the diagram through verbally explanation compared with using diagram.

4. Conclusion

Student's propositional network representation when interpreting convention diagram reveal three causal network pattern which are linear, some reciprocal relationship and complex reciprocal relationship. The pattern variation is influenced by the different of information processing that is happened in student's working memory. Linear pattern is more dominating when student connecting the concept between information component that consists in convention diagram and guided by prior knowledge that is obtained by student in studying process moment. The revealed propositional network representation quality in this research cannot guarantee that using a diagram can push the cognitive load. The implication that is could reported to biology learning process is the need of other strategy that can strengthen the student beginning knowledge first.

Acknowledgments

Researchers say many thanks to the supervisors Dr. rer. nat Adi Rahmat, M. Si. and Dr. Rini Solihat, M. Si., which has helped researcher from designing proposal till preparing the research and also helping the writing of this article. Researcher also say many thanks to the respondents and biology teachers who have been ready to be a part of this research. And many thanks to Nana Sunarya, Fitra Ramadan, Rifki Ramdhan, and Ari Wiguna Sampurno who has become the incredible partner.

References

[1] Cook M 2011 Teachers' use visual representation ini the science classroom *Science Education Internasional* **22** (2) pp 175-184
[2] Hegarty M 2011 The cognitive science of visual-spatial displays: implications for design *Topics in Cognitive Science* **30** (3) pp 446-474
[3] Cheng M and Gilbert J 2015 Students' visualization of diagrams representing the human circulatory system: The use of spatial isomorphism and representational conventions *International Journal of Science Education* **37** (1) pp 136-161
[4] Paivio A 1990 *Mental representation: a dual coding approach 9th edition* (New York: Oxford University Press Inc)
[5] Broers N J 2002 Selection and Use Of Propositional Knowledge In Statistical Problem Solving *Learning and Instruction* **12** (2002) pp 323-344
[6] Pinkas G 1995 Reasoning nonmonotonicity and learning in connectionist networks that capture propositional knowledge *Artificial Intelligence* **77** (1995) pp 203-247
[7] Anderson J R and Kline P J 1979 A general Learning Theory and Its Application To Schema Abstraction *The Psychology of Learning and Motivation* vol **13**
[8] Walker G H, Stanton N A, Stewart R., Jenkins D, Wells L, Salmon P and Baber C 2009 Using an integrated methods approach to analyse the emergent properties of military command and control. *Applied Ergonomics* **40** pp 636–647
[9] Marzano R J, Pickering D and McTigheJ 1993 *Assessing Student Outcomes, Performance Assessment Using The Dimension of Learning* (Alexandria: Association for Supervision and Curriculum Development)
[10] Novick L dan Catley K 2014 When relationships depicted diagrammatically conflict with prior knowledge: an investigation of students' interpretations of evolutionary trees *Science Education* **98** (2) pp 269-304
[11] Arentze T, Dellaert B G C and Timmermans H J P 2008 Modeling and measuring individuals' mental representations of complex spatio-temporal decision problems *Environment and Behavior* **40** (6) pp 843-869
[12] Kuswana W S 2011 *Taksonomi berpikir* (Bandung: PT Remaja Rosdakarya)
[13] Plass J L, Moreno R and Brunken R 2010 *Cognitive Load Theory* (Cambridge: ambridge University Press)

IOP Conf. Series: Journal of Physics: Conf. Series **895** (2017) 012132 doi:10.1088/1742-6596/895/1/012132

Perceptions of Prospective Biology Teachers on Scientific Argumentation in Microbiology Inquiry Lab Activities

E Roviati[1,2*], A Widodo[1], W Purwianingsih[1] and R Riandi[1]

[1]Pendidikan IPA Sekolah Pascasarjana, Universitas Pendidikan Indonesia, Bandung, Indonesia
[2]Pendidikan Biologi, Institut Agama Islam Negeri Syekh Nurjati Cirebon, Cirebon, Indonesia

*evi1roviati@gmail.com

Abstract. Inquiry laboratory activity and scientific argumentation in science education should be promoted and explicitly experienced by prospective biology teacher students in classes, including in microbiology courses. The goal of this study is to get information about perceptions of prospective biology teachers on scientific argumentation in microbiology inquiry lab activities. This study reported the result of a survey research to prospective biology teachers about how their perception about microbiology lab classes and their perception about inquiry and argumentation in microbiology lab activities should be. The participants of this study were 100 students of biology education department from an institute in Cirebon, West Java taking microbiology lecture during the fifth semester. The data were collected using questionnaire to explore the perceptions and knowledge of prospective biology teachers about microbiology, inquiry lab activities and argumentation. The result showed that students thought that the difficulties of microbiology as a subject were the lack of references and the way lecturer teaching. The students' perception was that argumentation and inquiry should be implemented in microbiology courses and lab activities. Based on the data from questionnaire, It showed that prospective biology teacher students had very little knowledge about scientific argumentation and its implementation in science education. When the participants made arguments based on the problems given, they showed low quality of arguments.

1. Introduction

Microbiology is regarded as the basic science of biology, because microbes are ideal systems in investigating basic questions in Biology, such as the question of sex origins, specialization, adaptation, cellular function, genetics, biochemistry, and the physical properties of all other living organisms [1]. Besides, microbes also determine the quality of human and other living organisms' life, such as in the fields of health, industry, ecology, biotechnology and genetically engineering. Microbes have important roles in the future, because the technology has more and more attractions in developing new techniques investigating the utilization of microbes to produce food, to overcome environmental damage, to cure diseases and even to create biological weapons [2]. Microbiology is one of the compulsory courses that should be taken by students majoring in biology education as prospective biology teachers.

Learning microbiology as science for prospective biology teacher is not only learning theory in lectures but also learning to experience experimentation skills through laboratory activities.

Therefore, lab activity is an essential activity and an integral part of microbiology learning. Lab activities are activities that involve students in observation, manipulation of objects and real material. They interact with materials and equipment in order to understand how the universe works so that they are able to link between the domain of knowledge and thinking skills with the real life [3]. Microbiology laboratory activities in this study provide students with basic skills to isolate microbes from their sources, to grow microbes in the medium, to examine the presence of microbes in some environments, to examine antimicrobial substances and to examine the role of microbes in some fields.

Science learning through inquiry activities, including microbiology lab activities that integrate hands-on and minds-on activities, is in accordance with science education objectives according to National Science Education Standards [4]. It is to create well-educated students who are able to experiment properly and have the urge to find out and understand the universe, using scientific processes and appropriate principles in making decisions.

Argumentation is one of important skills that should be taught in science education. Argumentation is defined as the connection between claims and data through justification and evaluation of knowledge [5]. The term argumentation is not always referred to debates, although it is one of many ways of argumentation. Argumentation is a process of thinking and social interaction where students can construct and criticize arguments [6]. In the other hand, argument is defined as a series of proposition. Students' involvement in scientific argumentations can lead to a better understanding of the concepts and the processes of science [5].

Kuhn [7] argues that argumentation and inquiry skills are two important skills in science learning that need to be possessed by prospective science teacher students. Prospective science teacher students who have the opportunity to design inquiry-oriented laboratory activities that are enriched with critical discussion, have the opportunity to support argumentation, so they can be encouraged to promote argumentation in their classroom when they teach [8]. In order for argumentation to take place in the classroom in science learning, students must play an active role in the discussions [9]. Unfortunately, science learning has so far been conducted with an emphasis on question and answer interaction and teacher-dominated lab activities that does not involve students in activities that support discussion, argumentation or social construction of knowledge [10]. Students' opportunities to participate in authentic arguments in science classes are still very rare.

Students' perceptions of science affect their learning and achievements in learning science [11]. Research shows that there is a relationship between students' attitudes toward science learning in school with their learning and achievement in learning science [12]. Research on the perception of prospective biology teacher students on arguments in lectures and microbiology lab activities can be useful for understanding students' difficulties in applying arguments in science classes. There is limited understanding of how prospective science teacher students' perceptions on argumentative discourse in scientific argumentation, their knowledge and ability to participate in a scientific arguments. Therefore, the main purpose of this study is to investigate perceptions of prospective biology teachers on scientific argumentation in microbiology inquiry lab activities.

2. Method

This field study research was conducted in one of the Education Institute for Educational Staff (LPTK) in Cirebon West Java. This study was conducted by involving 100 students as respondents who are in the 5th semester taking the course of microbiology. The students' age varied from twenty to twenty three.

Respondents were asked to provide information using a set of questionnaire accordance with their knowledge and views on the course of microbiology, particularly on lectures and laboratory activities as well as their knowledge and abilities of scientific argumentation. The questionnaire consisted of 5 parts based on the group of question asked. The first part was about microbiology course, while the second part was about microbiology lab activities in general. The third part included questions in perceptions about inquiry and argumentation in microbiology courses and lab activities. The forth part was the questions about students' knowledge on scientific argumentation and the last part was the questions to test the students' skills of scientific argumentation used to give opinions and arguments to some microbiological problems.

This field study research used descriptive qualitative approach. Data obtained through a set of questionnaire. Questionnaires are used to obtain student background data, students' perceptions of lectures and microbiology practice, and to determine the level of knowledge and ability of students about scientific argumentation. The questionnaire data was analyzed descriptively as a description of what happened in the field. While the results of the level of scientific argumentation were analyzed by the modified Toulmin's Argumentation Pattern (TAP) Scheme, by Toulmin (1958) [13].

3. Result and Discussion

3.1. Students perception about microbiology as a course

The prospective biology teacher students in this study did not think that microbiology is the most difficult course. They believed that statistics was the most difficult (96%) subject in this semester followed by genetics and biochemistry, no one of the students answered microbiology as the most difficult course. While, microbiology included courses with moderate difficulty (79%). According to the students, they did not regard microbiology as a difficult course on the grounds that despite using scientific terms, the discussion is related to daily life. So, for them it was more triggering curiosity, easier to remember and easier to be learned.

Factors that determined the difficulty in studying microbiology was the lack of reference books (52%) and the course material itself (32%). Students realized the importance of reference books to study the courses, so they find it difficult to learn course material if reference books are not available. While, 40% of students considered that the way lecturers teach was enough to affect the degree of difficulty of microbiology courses.

Then students were asked to choose the subjects considered to be the most difficult in the course of microbiology. Most of them (80%) answered microbial genetics, followed by microbial nutrition and metabolism (7%) and microbial classification (7%). While the cause of the difficulty of the subject matter was that it was a matter of process (45%) and scientific terms used (36%). As many as 57% of students are interested in studying further microbiology and 24% of them were very interested.

3.2. Students perception about laboratory deactivities in microbiology

The students' perception about the most important benefit of lab activities in microbiology was to learn investigation and research (49%) and to help them understand the concept (33%) followed by develop scientific attitudes (14%). Almost all of the students (97%) also assumed that lab activities can help them understand the microbiological concepts being studied. As a prospective science teacher, students understood very well the benefits and importance of lab activities in science learning, including in learning microbiology. As revealed by Millar [3], that laboratory practice is an integral activity in learning science and is a process of learning through experience. Lab activities aim to motivate students, develop experimental techniques and skills, learn scientific approaches and enhance theoretical understanding of subjects.

The most difficult laboratory activities were isolation and transfer of microbes (57%), followed by antimicrobial effectiveness test (22%) and microbial dyeing technique (15%). While the cause of the difficulty is due to the complexity of the method (63%) and the unsuspecting participant condition (17%). In composing the lab report, Students thought the most difficult part was in writing the discussion (80%) and followed by determining the tools, materials and methods (7%). While the main source of reference in the discussion is from the internet in the form of blogs and articles (74%) followed by reference books (14%).

3.3. Students perception on inquiry and argumentation in microbiology courses and lab activities

Table 1 shows the result of prospective biology teacher student's perception on inquiry and argumentation in microbiology courses and lab activities.

Table 1. Student's perception on inquiry and argumentation

QUESTION	YES	NO
Is microbiology lab activity already oriented inquiry?	68	32
Should microbiology lab activities be inquiry-oriented?	94	6
Is there a discussion session and argumentation in a microbiology lecture?	81	19
Is it necessary to have discussion sessions and arguments in microbiology lectures?	95	5
Is there a discussion session and argumentation in a microbiology lab activities?	84	16
Is it necessary for discussion and argumentation sessions in microbiology lab activities?	90	10

Based on the questionnaire result as shown on table 1, students considered that the lab activities conducted in microbiology courses had been inquiry oriented (89%) and microbiology lab activities need to be inquiry oriented (94%). Although lab activities that have been done by students in the research site were considered to be inquiry-oriented, but it still dominated by low level inquiry, such as discovery learning, interactive demonstration or inquiry lesson, and actually the lab activities done have not fully applied inquiry lab [14]. The activities were mostly still being guided by the manual books.

As many as 81% of students considered there were discussion sessions and arguments in the lecture of microbiology while in practice as much as 84%. They also argue that there should be more discussion sessions and arguments in lectures (95%) and in lab activities (90%). According to students so far, more discussion activities conducted in lectures than lab activities. However, during the lab activities, the instructor always actively provided questions related to step by step and the reason for a method in the experiment. Therefore, students assumed that there are already discussion sessions and arguments in lectures and lab activities.

3.4. Student knowledge about scientific argumentation

Table 2 shows the result of the forth part questions in the questionnaire asked about Student knowledge about scientific argumentation.

Table 2. Student knowledge about scientific argumentation

Question	True	Half	False
Understanding on the term of scientific argumentation	28	42	40
The importance of argumentation in science learning	12	36	52
Parts of an argument	0	22	78

Data about students' knowledge of the argument showed on table 2, that most students did not know much about scientific argumentation in science learning. Students did not know much about the meaning of scientific argumentation, the importance of scientific argument in science learning and components in argumentation. But, some students had better answer when describing their perception on scientific argumentation, although they had never specifically studied about argumentation. Students emphasized the need for evidence in expressing scientific opinion. But they had not paid any attention to the word or phrase linking the claim or opinion with the evidence presented.

3.5. Students' ability to make arguments

Table 3 shows the result of the questions that indicate students' ability to provide arguments.

Table 3. Ability to make arguments

No.	Topics of problem	Level			
		1	2	3	4
1.	Antibiotics	63	37	0	0
2.	Sterilization	46	54	0	0
3.	Tempe Fermentation	34	54	12	0

While the ability of students in drawing arguments for a problem, as shown on table 3, was still relatively low, by looking at the answers to the questions solely on most of claims without data supported or accompanied by simple data. Even the claims given were not all valid.

4. Conclusion

Based on the results of the study, it can be concluded that the prospective biology teacher considered the course of microbiology as a course with medium difficulty level. The difficulties faced by students are the lack of reference books and microbial genetics topics as the most difficult subject. Students understand the importance of lab activities in microbiology and the need for inquiry-oriented lab activities and scientific arguments. Students had low knowledge and argumentation skills.

Acknowledgments

The authors are grateful to those who have provided financial and moral support for this research. This study was supported by the Ministry of Religious Affairs, Republic of Indonesia (Kementerian Agama RI). My special thanks to the research assistants and students who have been very supportive to this study.

References

[1] Tortora G J, Funke B R and Case C L 2010 *Microbiology an Introduction* (San Fransisco: Pearson Education Inc.)

[2] Schaechter M, Kolter R and Buckley M 2004 *Microbiology in the 21st Century: Where Are We and Where Are We Going?* (Washington DC: American Academy of Microbiology; American Society for Microbiology)

[3] Millar R 2004 *The Role of Practical Work in the Teaching and Learning of Science* (Washington DC: University of York)

[4] National Research Council 1996 *National Science Education Standards* (Washington: National Academy Press)

[5] Erduran S and Jiménez-Aleixandre M P 2007 Argumentation in Science Education: An Overview. In S. Erduran & M.P. Jimenez-Aleixandre (Eds.) *Argumentation in Science Education: Perspectives from Classroom-Based Research* (Dordrecht: Springer)

[6] Golanics J D and Nussbaum E M 2008 Enhancing online Collaborative Argumentation through Question Elaboration and Goal Instructions *Journal of Computer Assisted Learning* **24** 167-180.

[7] Shaughnessy M E 2004 An Interview With Deanna Kuhn *Educational Psychology Review* **16** (3) 267-282

[8] Ozdem Y, H Ertepinar, J Cakiroglu and S Erduran 2013 The Nature of Pre-Service Science Teachers' Argumentation in Inquiry-Oriented Laboratory Context *International Journal of Science Education* **35** (15) 2559-2586

[9] Gray R and N H Kang 2014 The Structure of Scientific Arguments by Secondary Science Teachers: Comparison of Experimental and Historical Science Topics *International Journal of Science Education* **36** (1) 46-65
[10] Newton P, Driver R and Osborne J 1999 The place of argumentation in the pedagogy of school science *International Journal of Science Education* **21** 553–576
[11] Koballa Jr. T R, Crawley F E and Shrigley R L 1990 A Summary of Science Education-1998 *Science Education* **74** (3) 369-381
[12] Osborne J, Erduran S and Collins S 2003 Attitudes toward Science: A Review of The Literature and its Implications *International Journal of Science Education* **25** (9) 1049-1079
[13] Toulmin, S E 1958 *The Uses of Argument* (Cambridge University Press, Cambridge)
[14] Wenning C J 2005 Levels of Inquiry: Hierarchies of Pedagogical Practices and Inquiry Processes *Journal of Physics Teacher Education* **2** (3) 3-11

IOP Conf. Series: Journal of Physics: Conf. Series **895** (2017) 012133 doi:10.1088/1742-6596/895/1/012133

The Analysis of *Sustainable Development* Content in the Syllabus of Environmental Knowledge and Plants Ecology Lecture

A Putra[1*], A Rahmat[1] and S Redjeki[1]

[1]Science Education Study Program, Post-Graduate School, Universitas Pendidikan Indonesia, Bandung, Indonesia
[2]Biology Education Study Program, Universitas Samawa Sumbawa Besar, Sumbawa, Indonesia

*armansyah@student.upi.edu

Abstract. This research aims to find out how much the content of sustainable development exist in the content of environmental knowledge and plant ecology courses. The focus indicators of sustainable development indicators is the environment. This research is a qualitative research type with qualitative descriptive approach. The analyzed variables are only 2 courses, which are environmental knowledge and plants ecology. The results showed that the syllabus contents analysis of environmental knowledge and plants ecology courses in private Lembaga Pendidikan Tenaga Kependidikan (LPTK) in the province of Nusa Tenggara Barat is already good enough and the sustainable development contents is very large, almost all syllabus contents has already prioritize the sustainable development load of both the subject of environmental knowledge and plants ecology, although there are still some syllabus contents that was not includes sustainable development load, but the percentage is quite small, especially in the course of Plant Ecology.

1. Introduction

The environmentally sustainable development we use in this research is a very popular translation of "sustainable development" used in Western countries. [1]. Environmentally Sustainable Development is used in UU no. 23 of 1997 on Environmental Management. [2] Sustainable development and biodiversity has to be a key components in a nature-oriented curriculum. [3] Sustainable development describes a commitment to advance peoples' welfare. [4] Sustainable development has a wide appeal and a small specificity, but several joint development and environment as well as equality are found in many attempts to describe it. [5] Sustainable development can also be done through formal education, not just informally. [6] Looking at the structure, scope, and application, the sustainable development is closely related to natural science and social science. [7] There are several countries in Asia applying sustainable development and successful in the implementation, those countries are Korea, Japan. India, Thailand and the Philippines. [8]

The explanation above shows that the implementation of environmentally sustainable development is important to be implemented because it is closely related to the utilization of natural resources as an asset to realize people's welfare. In accordance with the rules mandated by the United Nations and the

European Union in 2007. [9] Given Sustainable development can not only be done formally and informally, the researchers tried to apply in the lecture system.

The lectures planned by a lecturer so that students can actively participate in the teaching-learning process that requires steps that are clear, systematic, and targeted. Unstructured lecture planning can be fatal, because the lecture does not reach the right target. Therefore it is very important for a lecturer in preparing the lecture also to re-analyze the weaknesses in the syllabus. To find out how much contents of sustainable development in the syllabus in several private LPTKs in the province of Nusa Tenggara Barat, researchers tried to analyze the examples of syllabus used, both in Lembaga Pendidikan Tenaga Kependidikan (LPTK) A, B and C.

Based on the description of the background above, the basis of researchers in conducting research and assessing, the extent to which contents of sustainable development in the syllabus in private LPTK in Nusa Tenggara Barat.

2. Experimental Method

2.1. Research method

The method used in this study is a qualitative method consisting of three private LPTKs namely A, B and C.

2.2. Research design

This research uses qualitative or naturalistic method because it is done in a natural condition. According to Creswell, qualitative research is a process of research and understanding that is based on the methodology of investigation of social phenomena and human problems. [10] This research begins by collecting the syllabus of several LPTKs in Nusa Tenggara Barat, then analyze how much of the contents of the syllabus that are about sustainable development.

2.3. Research subject and location

The research was conducted in three private LPTKs namely A, B and C located in Nusa Tenggara Barat.

3. Result and Discussion

From several syllabus formats used as an example of analysis, the following findings are outlined as (see table):

Table 1. Environmental knowledge's syllabus contents comparison about *sustainable development*

a) LPTK A

Syllabus Content	***Sustainable Development Contents***					
	1	**2**	**3**	**4**	**5**	**6**
Courses Contract and Review						
Environmental Characteristics			√	√	√	√
Environmental Problems			√	√	√	√
Population and its' Issues:						
• Dynamics Population			√	√		√
• Population and Density Issues						
Ecology Concepts as the Basis of Environmental Knowledge:			√	√		√

Syllabus Content	*Sustainable Development Contents*					
	1	**2**	**3**	**4**	**5**	**6**
• Populations, Community, ecosystem, habitat, nichia						
• Biotic Tolerance Range (Alive)						
Environmental Materials and Energy:						
• Food Chains						
• Food Webs		√				
• Food Pyramids						
Humans and Environments:						
• Reciprocal relationship between humans with biotic and abiotic environments				√		√
• The impact of human activities on the physical, chemical, biological environments						
Mid-Semester Test						
Natural Resources and Environments:						
• Types and Characteristics of Natural Resources						
• Soil and Land						
• The Exploitation of natural resources					√	
• Mineral and energy						
• The Management and Conservation of natural resources						
• Damage to natural resources and environments						
Health and Environment:						
• Epidemiology of the disease						
• Food contamination				√		
• Environmental and health effects						
Water and Soil Pollution					√	
Air and Sound Pollution				√		
Environmental Management Strategy				√		√
Final Exam						

Note: 1=Climate Change, 2=Energy, 3=Biodiversity, 4=Health and Environment, 5=Waste and Natural Resources, 6=Economy and Environment.

Table 1 a): Shows that the syllabus contents on the course of environmental knowledge has a very good sustainable development load. As can be seen from the contents of each meeting syllabus. But of all of the contents, there are still some contents existed that are not in accordance with indicators of sustainable development, which is the indicator of "climate change".

b) LPTK B

Syllabus Content	*Sustainable Development Contents*					
	1	**2**	**3**	**4**	**5**	**6**
Courses Contract and Review						
History of Environmental Science and Units in Ecosystems: Individuals, Population, Communities, ecosystem, and habitats			√	√		√
Components and Functions of Ecosystems, and Concepts of Energy in Ecological System		√				

Syllabus Content	*Sustainable Development Contents*					
	1	2	3	4	5	6
Succession						
Definition and Types of Environments			√	√		√
Types of environmental pollution and Source of pollutants				√	√	√
The Impacts of environmental pollution and environmental pollution mitigation solutions				√	√	√
Mid-Semester Test						
Global environmental issues: acid rain, ozone holes, greenhouse effects, and global warming effects	√			√	√	
Classification and Variety of Natural Resources and the Utilization					√	
Natural Resources Bio-conservation					√	
Food Chain, Food Web, and Food Pyramids		√				
Population Dynamics, Population and Density Issues, Effects of Population Density on Health				√		
AMDAL				√	√	√
Environmental Management Strategy				√		√
Final Exam						

Note: 1=Climate Change, 2=Energy, 3=Biodiversity, 4=Health and Environment, 5=Waste and Natural Resources, 6=Economy and Environment.

Table 1 b): Show that the syllabus contents of environmental knowledge course in LPTK B has excellent sustainable development load level. As can be seen from the syllabus contents of each meeting, all syllabus contents is in accordance with indicators of sustainable development.

c) LPTK C

Syllabus Content	*Sustainable Development Contents*					
	1	2	3	4	5	6
Courses Contract and Review						
History of the development of environmental science, History of emergence of environmental problems, Regulations on the environment				√		√
Relationship Between Living Beings in Ecosystems, Symbiosis, Mutualism, Komensalism, Parasitism			√			
Environmental Materials and Energy: • Food Chains • Food Webs • Food Pyramids		√				
Energy Flow and Biogeochemistry Cycle		√				
Human and Environment: • Reciprocal relationship between humans and biotic and abiotic environments • The impact of human activities on the physical, chemical, biological environment				√	√	√
Mid-Semester Test						
Natural Resources and Environments:			√		√	
Mineral and energy • The Management and Conservation of natural resources		√		√	√	

Syllabus Content	*Sustainable Development Contents*					
	1	**2**	**3**	**4**	**5**	**6**
• Damage to natural resources and environments						
Health and Environment:						
• Epidemiologi of diseases				√		
• Food Contamination						
AMDAL				√	√	√
Environmental Management				√		√
The Impacts of Water and Soil Pollution environment and health				√		
Air and Sound Pollution				√		
Environmental Issues				√	√	√
Final Exam						

Note: 1=Climate Change, 2=Energy, 3=Biodiversity, 4=Health and Environment, 5=Waste and Natural Resources, 6=Economy and Environment.

Table 1 c): Similar with the table A, the syllabus contents of environmental knowledge course on LPTK C has a very high sustainable development load level, but from all the content, there is still a syllabus contents that is not in accordance with the indicator of sustainable development, which is the indicator of "climate change".

Table 2. Plants ecology's syllabus contents comparison about *sustainable development*

a) LPTK A

Syllabus Content	*Sustainable Development Contents*					
	1	**2**	**3**	**4**	**5**	**6**
Courses Contract and Review						
Relation between plants ecology with another study subjects			√	√		√
Plants Society			√			
Vegetation Classification						
Plants Community			√			
Relation between Vegetation with Environment			√	√		√
Mid-Semester Test						
Terrestrial Biomes	√					
Vegetation Biomes in Indonesia	√					
Vegetation Techniques and Analysis						
Final Exam						

Note: 1=Climate Change, 2=Energy, 3=Biodiversity, 4=Health and Environment, 5=Waste and Natural Resources, 6=Economy and Environment.

Table 2 a): Show that the syllabus contents of Plants Ecology course has excellent sustainable development load level. As can be seen from the syllabus contents of each meeting, although there is still a syllabus content that is not in accordance with indicators of sustainable development, which is the indicator of "Waste and Natural Resources".

b) LPTK B

Syllabus Content	*Sustainable Development Contents*					
	1	2	3	4	5	6
Courses Contract and Review						
Definition, ecological modelling, and ecological development of plants and their benefits and basic understanding in plant ecology			√			
Plants in the environment (macro and micro), limiting factors and adaptation strategies		√	√			
Abiotic environmental factors: Light and its effects on plants			√	√		√
Abiotic environmental factors: light and atmosphere and its role for plants					√	
Abiotic Environmental factors: Temperature and effect on plants and adaptation strategies to extreme temperatures				√	√	√
Abiotic Environmental Factors: Topography and soil and its role for plants, distribution and adaptation based on topography and soil conditions			√			
Mid-Semester Test			√			
Biotic Factor and the Interactions			√			
Population and Distribution of Plants			√			
Vegetation and its characteristics; Method and Analysis of Vegetation						
Ecosystem			√			
Ecosystem Development (Succession)			√			
Terrestrial Ecosystem			√			
Beach Ecosystem			√			
Final Exam						

Note: 1=Climate Change, 2=Energy, 3=Biodiversity, 4=Health and Environment, 5=Waste and Natural Resources, 6=Economy and Environment.

Table 2 b): Show that the syllabus contents of Plants Ecology course has excellent sustainable development load level. As can be seen from the syllabus contents of each meeting, although there is still a syllabus content that is not in accordance with indicators of sustainable development, which is the indicator of "Climate Change".

c) LPTK C

Syllabus Content	*Sustainable Development Contents*					
	1	2	3	4	5	6
Courses Contract and Review						
Definition, historical development of plant ecology, Approaches of plant ecology			√	√		√
Environmental factors that play a role in plant ecology	√		√	√		√
Water and Fire as a limiting factor	√					
Organic and anorganic material in the soil as a limiting factor		√				
Plants, Animals, Microorganisms, and Human as a limiting factor				√	√	√
Mid-Semester Test						
Classification of the community, the concept of observing			√			

Syllabus Content	*Sustainable Development Contents*					
	1	2	3	4	5	6
the pattern of plant communities						
Types of endemic and cosmopolite, Distribution with and without external help, Steps and Types of Succession			√			
Vegetation of tropical rainforests, temperate deep forests, boreal forests, grasslands, tundra, dry arid zones and deserts, coastal forests, mangrove, freshwater plants, brackish water plants, estuariam, coral reefs			√			
Types of Vegetation Maps and Data Collection Technique			√			
Make the minimum area curve. And point and line methods						
Quadratic and Quarter Methods						
Density, frequency, dominance, importance, and ordination techniques						
Application in agriculture, forestry and urban area development						
Final Exam						

Note: 1=Climate Change, 2=Energy, 3=Biodiversity, 4=Health and Environment, 5=Waste and Natural Resources, 6=Economy and Environment.

Table 2 c): Show that the syllabus contents of Plants Ecology course in LPTK C has excellent sustainable development load level. As can be seen from the syllabus contents of each meeting, all syllabus contents is in accordance with indicators of sustainable development.

4. Conclusion

Based on the results of the analysis that has been done, the conclusion obtained was that the analysis of syllabus contents of environmental knowledge and plants ecology in private LPTKs in Nusa Tenggara Barat is good enough and sustainable development load that is very large, almost all syllabus contents has already prioritize sustainable development load in the course of environmental knowledge as well as plant ecology, although there are still some syllabus contents that was not includes any sustainable development load, but the percentage is quite small especially in the course of plant ecology.

Acknowledgments

Thus the author can convey about the results of research contained in this article. On this occasion the authors would like to express the appreciation and gratitude to BUDI-DN and LPDP as sponsors and all the parties that has big impacts to the success and completion of this research. The author realizes that this article is far from perfect, therefore constructive criticism and suggestions are desirable for the perfection of this article. Hopefully this research will be useful for the readers.

References

[1] Kates et al. 2005 *Science and Policy for Sustainable Development* **47** (3) 8–21
[2] UU No. 23 Tahun 1997 tentang "Pengelolaan Lingkungan Hidup". Jakarta. Indonesia
[3] Ruzickova et al. 2015, Environmental education for sustainable development and biodiversity conservation in selected study programmes at the Faculty of Natural Sciences of Comenius University in Bratislava (Shortened version in English). E-Journal for environmental education
[4] Adelle et al. 2009 *Sustainable Development Indicators* European Communities
[5] Rajalaksmi S 2016 *Sustainable development through environmental ethics* International Journal of Applied Research

[6] Mathar, Reiner, 2015 *Chapter 2 in Schooling for Sustainable Development in Europe*. Springer: Heidelberg, New York, Dordrecht, London

[7] Sunal, Szymanski 1993, Social Studies and the elementary/middle shcool student, (USA). Japan Council on the UN Decade of Education for Sustainable Development,. Buku Pedoman Pendidikan untuk Pembangunan Berkelanjutan di Asia PBB and Uni eropa. 2007, Suistainable Development indicator.

[8] Creswell J W and Clark V L P 2007 *Designing and Conducting Mixed Methods Research* (California: Sage Publication)

IOP Conf. Series: Journal of Physics: Conf. Series **895** (2017) 012134 doi:10.1088/1742-6596/895/1/012134

How can Smartphone-Based Internet Data Support Animal Ecology Fieldtrip?

I S Kurniawan[1,2], F S Tapilow[1,3] and T Hidayat[1,3*]

[1]School of Postgraduate Studies, UPI Bandung
[2]Department of Biology Education, Universitas Pasundan Bandung
[3]Department of Biology Education, UPI Bandung

*topikhidayat@upi.edu

Abstract. Identification and classification skills must be owned by the students. In animal ecology course, the identification and classification skills are necessary to study animals. This experimental study aims to describe the identification and classification skills of students on animal ecology field trip to studying various bird species using smartphone-based internet data. Using Involving 63 students divided into 7 groups for each observation station. Data of birds were sampled using line transect method (5000 meters/station). The results showed the identification and classification skills of students are in sufficient categories. Most students have difficulties because of the limitations of data on the internet about birds. In general, students support the use of smartphones in field trip activities. The results of this study can be used as a reference for the development of learning using smartphones in the future by developing application about birds. The outline, smartphones can be used as a method of alternative learning but needs to be developed for some special purposes.

1. Introduction

Animal Ecology field trip is directly related to nature. Fieldtrip provide positive benefits for students because students can directly observe the actual objects they can find in the field [1,2]. Field trip activities should be developed with new methods or approaches in line with the development of science and technology. The advancement of digital technology, one of which mobile technology can be developed to assist students in conducting experiments based on the environment, facilitate student learning so as to improve experimental group learning performance and develop positive attitude in the investigation and provide more meaningful learning opportunities [3,4]. The affordability of mobile phone technology can improve acceleration in learning, facilitate data collection, analysis and presentation so as to encourage collaboration and enhance interaction among students to conduct discussions and will create new meanings and understandings in learning [5,6,7]. The progress of some smartphone applications with increasingly sophisticated features is not maximally utilized in learning. Smartphone applications are mostly used as a means of social interaction, games (games) and entertainment. It would be better if smartphone applications can be used in learning and field activities to support more effective learning [8,9].

Smartphones and other digital devices can be a potential source of support for learning. Mobile devices such as laptops, smartphones, digital assistants have great potential in supporting learning inside and outside the classroom [10,11,12]. Several studies have confirmed the use of smartphone technology is effective to increase the ability of the student identification. This identification skill is important in supporting conservation efforts. Improving the identification skills can be done collaboratively with the help of the internet, video recording and Facebook applications as a simple new method [13,14,15]. The identification skills can be extracted by the real evidence available in nature freely, learning is not always limited to the classrooms, nature has provided an authentic learning resource to be learned in revealing the facts contained in the environment [16,17]. In addition to the identification skills, the essential classification skills are possessed by students in studying the species. Classification skills can assist in categorizing the species based on its specific characteristics. An important step in the classification process is the segmentation of data that is the assignment of boundaries between different classes of behavior within a certain timeframe [18] classification capability is indispensable to promote ecological progress [19].

Birds are an interesting species to be studied with various characteristics. Learning birds can be done with various methods, one with the help of smartphone-based internet data. Research on birds at this time is necessary because there has been a decline in some species of birds due to hunting or business purposes. Thus, the decline of bird populations indirectly affects ecology and conservation efforts are needed with various approaches [20]. The Internet provides a variety of data that we need, but not all data can be presented completely, it is a flame [21,22] for that it needs to be developed special features [23,24,25]. This study aims to describe the ability of identification and classification of students in studying the various species of birds found around the location of observations by using smartphone-based internet data. In the long term, students' identification and classification skills can provide benefits, especially in ecological conservation.

2. Experimental Method

This research used the experimental method. Data was analyzed using the quantitative descriptive technique. This research involved 63 students divided into 7 groups during field activities in animal ecology. This research was conducted in Binuangeun area of Banten province from 23 to 25 October 2016.

2.1. Data collection

The identification and classification data was obtained from the field. Data was collected in the morning from 07.00-11.00 and in the afternoon from 14.00-17.00. Each sampling point is only given for 20 minutes. Bird data collected based on observations by the naked eye without the aid of tools such as telescopes. The results of field observations are further identified with the help of smartphones using internet data to measure student identification and classification skills tailored to each indicator. Data collection is also done through a questionnaire of student responses to the use of smartphones in field activities.

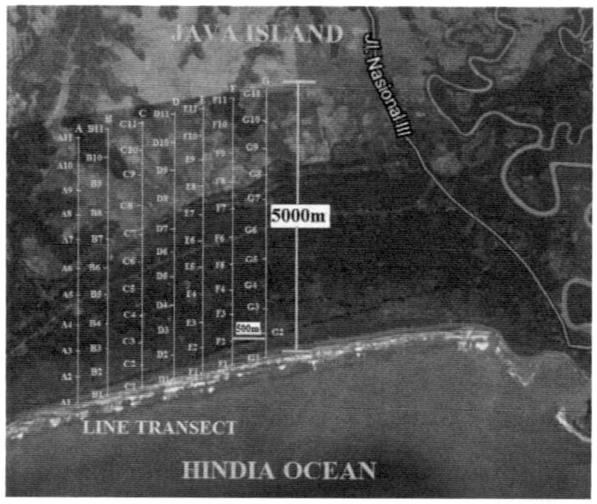

Figure 1. Line transects of 7 groups (A-G) each group has 11 bird observation points. Observations were made using transect line method divided into 7 stations (from A-G), each observation station was divided into 11 observation points. The observation is done as far as 5000m, the distance between station is 500m

2.2. Data analysis

The identification and classification skills were analyzed to determine its category. References to the criterion is based on the average score ideal (*Mi*) and the score of the standard deviation of the ideal (*SDi*) is achieved by a sheet instruments. This study used a questionnaire scale of 5 with the conversion value and scores, this study using a formula developed by [26]. Correlation test using *r test* (Pearson correlation test). Questionnaires captured on student responses to the activities of field study and use of smartphones-based internet data.

Table 1. Conversion score on a scale of 5

Numb	Score	Criteria
1	$x > (Mi + 1{,}8\ SDi)$	Excellent
2	$(Mi + 0{,}6\ SDi) < x < (Mi + 1{,}8\ SDi)$	Good
3	$(Mi - 0{,}6\ SDi) < x \leq (Mi + 0{,}6\ SDi)$	Sufficient
4	$(Mi - 1{,}8\ SDi) < x \leq (Mi - 0{,}6\ SDi)$	Low
5	$x \leq (Mi - 1{,}8\ SDi)$	Lowest

3. Result

The average score of identification and classification skills were 2.77 and 2.69 respectively from a maximum score of 4 (Fig. 2), which is categorized sufficient (Table 2).

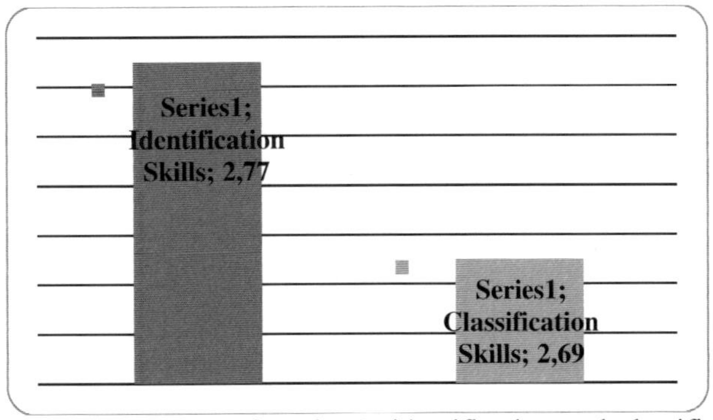

Figure 2. Mean score of student's identification and classification skills to birds on fieldtrip activities

Table 2. Conversion scale 5 of identification skills

Identification Skills		Classification Skills		Category
Numb	Score	Numb	Score	
1	$x > 3.01$	1	$x > 3.01$	Excellent
2	$2.84 < x < 3.01$	2	$2.81 < x < 3.01$	Good
3	$2.66 < x \leq 2.84$	3	$2,59 < x \leq 2,81$	Sufficient
4	$2.49 < x \leq 2.66$	4	$2,39 < x \leq 2,59$	Low
5	$x \leq 2.49$	5	$x \leq 2.39$	Lowest

Identification skills are essential in classifying. These two abilities are closely related to each other. Thus it can be assumed that if students have good identification skills then they will also have good classification capabilities as well. There is a significant correlation between the identification and the classification skills of *r*-Pearson correlation (.973), these correlation test results show a strong relationship between the two variables. The results of data analysis in this study showed that the identification and classification skills of students included in the category sufficient. To dig deeper into this finding, the researchers distributed a questionnaire of student responses related to these findings (Table 3).

Table 3. Data analysis of student questionnaire

No.	Questions	Frequency	
		Yes	No
1	Is smartphone use in field activities helpful?	54	9
2	Does the identification of birds using internet-based smartphones provide convenience?	12	51
3	Does the bird classification process use Internet-based smartphones ease?	13	50
4	Are there any constraints in accessing internet data?	59	4
5	Is the data available complete enough on the internet?	3	60
6	Is the internet network adequate at the observation location?	29	34
7	Does it need a special application in studying birds using a smartphone?	61	2
8	Do you agree that field activities using smartphones are more effective?	33	30
9	Are you interested in reusing smartphones in learning?	46	17

IOP Conf. Series: Journal of Physics: Conf. Series **895** (2017) 012134 doi:10.1088/1742-6596/895/1/012134

4. Discussion

Student identification skills have declined as they are not implemented in schools so they are not used to identifying species when this is essential to an effort to conserve biodiversity and support sustainable development [16]. Students should be equipped with the ability to identify biodiversity so they understand the importance of species sustainability. Identification capabilities can be extracted with the real evidence available freely in nature, teaching and learning methods for the identification and knowledge of species for biodiversity education and sustainable development should always include project-based experiences and methods in authentic environments [1,17]. The identification process using smartphone-based internet data has limited data, making it difficult for students to identify birds in research locations.

The use of smartphones has contextual limitations that result in difficulty in obtaining complete data [21,24] Network services are a very important factor in using smartphones to obtain data from the internet. Inadequate network services will have an effect on data retrieval access in search of broader topics. The use of mobile learning can assist graduate students in accessing resources and research communities on a broader topic so as to generate ideas and build new [22]. The classification skills begin with the identification skills, if the student has good identification skills should the classification skills are also good. In this research, category of student classification ability is enough. As well as classification capabilities, some of the obstacles of using internet-based data are the lack of availability of the complete content needed to study birds. Content on the internet is so limited that we will have difficulty in obtaining complete data for specific needs [23].

Based on questionnaire data can be explained that most students are greatly helped by using smartphones in learning. But there are some obstacles in identifying and classifying birds among others. Difficulty in obtaining relevant data on the internet, data available on the internet is incomplete. Most of the students agree that smartphone use in learning needs to be developed and used in learning. We need to think about the implications of the use of smartphones in learning in the future. App development may be necessary to learn specific content. To overcome the limitations of the Internet network can be developed smartphone applications with offline data so that activities are not disrupted due to the bad network. In general, the results of this study explain that the smartphone can be used to support the field trip activities [1,2].

5. Conclusion

Animal ecology course can not be separated from field activities, indirectly will always be related to nature. The smartphone is a technological product that is growing rapidly with increasingly sophisticated features. However, smartphones are used very little in supporting learning, generally more used for personal communication, social media, games, music, and video players. This study tried to optimize the use of smartphones in learning, especially on field activities. The results of this study can be used as a reference in developing learning by using a smartphone. Smartphones can be used as a method of alternative learning, although they must be modified in such a way for a particular purpose. How smartphones are designed and developed to support innovative learning, this should be something to think about.

Acknowledgments

Thanks to our colleagues for their contribution, also to the Department of Biology Education and Faculty of Mathematics and Natural Sciences Education (Universitas Pendidikan Indonesia) for special support. Thanks a lot to students as participants in Plant Physiology course at 2016 and for the data contribution.

References

[1] Thomas R L and Fellowes M D E 2016 Effectiveness of mobile apps in teaching field-based identification skills *Journal of Biological Education* **9266** 1–8

[2] Nelson A R, Cormier R L, Humple D L, Scullen J C, Sehgal R and Seavy N E 2016 Migration patterns of San Francisco Bay Area Hermit Thrushes differ across a fine spatial scale *Animal Migration* **3** 1–13

[3] Ahmed S and Parsons D 2013 Abductive science inquiry using mobile devices in the classroom *Computers and Education* **63** 62–72

[4] O'Bannon B W and Thomas K 2014 Teacher perceptions of using mobile phones in the classroom: Age matters! *Computers and Education* **74** 15–25

[5] France D, Powell V, Mauchline A L, Welsh K, Park J, Whalley W B and Rewhorn S 2016 Ability of students to recognize the relationship between using mobile apps for learning during fieldwork and the development of graduate attributes *Journal of Geography in Higher Education* **8265** 1–11

[6] Gikas J and Grant M M 2013 Mobile computing devices in higher education: Student perspectives on learning with cellphones, smartphones & social media *Internet and Higher* Education **19** 18–26

[7] Vázquez-Cano E 2014 Mobile Distance Learning with Smartphones and Apps in Higher Education *Educational Sciences: Theory & Practice* **14** (4) 1505–1520

[8] Beri P S 2016 Migratory Birds Tracking System using RF as Communication Medium and Android Mobile Base Station *International Journal on Recent and Innovation Trends in Computing and Communication* **4** (5) 543–545

[9] Bouten W, Baaij E W, Shamoun-Baranes J and Camphuysen K C J 2013 A flexible GPS tracking system for studying bird behaviour at multiple scales *Journal of Ornithology* **154** (2) 571–580

[10] Sung Y.-T., Chang K.-E. and Liu T.-C 2015 The Effects of Integrating Mobile Devices with Teaching and Learning on Students' Learning Performance: A Meta-Analysis and Research Synthesis *Computers & Education* **94** 252–275

[11] Shih B.-Y., Chen C.-Y. and Chen Z.-S 2006 An Empirical Study of an Internet Marketing Strategy for Search Engine Optimization *Human Factors and Ergonomics in Manufacturing* **16** (1) 61–81

[12] Alqaryan S, Alkhalifa M, Alharbi M, Alabaishi S and Aldrees T 2016 Smartphones and professionalism : A cross-sectional study on interns and final-year medical students (5) 198–202.

[13] Kontkanen J, Kärkkäinen S, Dillon P, Hartikainen-Ahia A and Åhlberg M 2016 Collaborative processes in species identification using an internet-based taxonomic resource *International Journal of Science Education* **38** (1) 96–115

[14] Mortimer J and Greene T 2017 Investigating bird call identification uncertainty using data from processed audio recordings *New Zealand Journal of Ecology* **41** (1) 1–8

[15] Stagg B C and Donkin M 2013 Teaching botanical identification to adults: experiences of the UK participatory science project "Open Air Laboratories" *Journal of Biological Education* **47** (2) 104–110.

[16] Palmberg I, Berg I, Jeronen E, Kärkkäinen S, Norrgård-Sillanpää, P, Persson C and Yli-Panula E 2015 Nordic–Baltic student teachers' identification of and interest in plant and animal species: The importance of species identification and biodiversity for sustainable development *Journal of Science Teacher Education* **26** 6 549-571

[17] Robinson B S, Inger R and Gaston K J 2016 A rose by any other name: Plant identification knowledge & socio-demographics *PLoS ONE* **11** (5) 1–13

[18] Schuld C, Franz S, Hedel H J A Van, Moosburger J, Maier D, Abel R, … Rupp R 2015 International standards for neurological classification of spinal cord injury: classification skills of clinicians versus computational algorithms *Spinal Cord.* **53** 324–331

[19] Resheff Y S, Rotics S, Harel R, Spiegel O and Nathan R 2014 AcceleRater: a web application for supervised learning of behavioral modes from acceleration measurements *Movement Ecology* **2** 27

[20] Zuckerberg B, Fink D, Sorte F A La, Hochachka W M and Kelling S 2016 Novel seasonal land cover associations for eastern North American forest birds identified through dynamic species distribution modeling *Diversity and Distributions* **22** 717–730.

[21] Saltan F 2016 Blended Learning Experience of Students Participating Pedagogical Formation Program: Advantages and Limitation of Blended Education *International Journal of Higher Education* **6** (1) 63

[22] Cheng-lin H and Jian-wei C 2016 A Target Design of a Mobile App Providing Supportive Service for Flipped Classroom *Journal of Educational and Social Research* **6** (1) 27–32

[23] Jackson E A, Sc, B, Gardens A and Da K 2016 M-Learning devices and thier impact on postgraduate researchers scope for improved intergration in the research community *The Online Journal of New Horizons in Education* **6** (1) 104–113

[24] Pempek T A and McDaniel B T 2016 Young children's tablet use and associations with maternal well-being. *Journal of Child and Family Studies* **25** 8 2636-2647

[25] Hopkins N, Tate M, Sylvester A and Johnstone D 2016 Motivations for 21st century school children to bring their own device to school *Information Systems Frontiers* 1–13.

[26] Jumadi 2012 *Pemetaan Kompetensi Pedagogik, Profesional, Kepribadian dan Sosial Guru Fisika SMA/MA di Daerah Istimewa Yogyakarta*. (Makalah). LPPKM UNY. Unpublished.

How is the Inquiry Skills of Biology Preservice Teachers in Biotechnology Lecture?

M S Hayat[1,2*] and N Y Rustaman[1]

[1]Program Studi Pendidikan IPA, Sekolah Pascasarjana Universitas Pendidikan Indonesia, Bandung, Indonesia
[2]Program Studi Pendidikan Biologi FPMIPATI Universitas PGRI Semarang, Semarang, Indonesia

*m.syaipulhayat@upgris.ac.id

Abstract. This study was to investigate the inquiry skills of biology pre-service teachers in one teachers college in Central Java in biotechnology lecture. The method used is a case study of 29 biology preservice teacher. Data were collected using observation sheets, questionnaires, and interview guidelines. Research findings collected through questionnaires show that most students are accustomed to asking questions and formulating biotechnology issues; Skilled in conducting experiments; Skilled in obtaining relevant information from various sources; As well as skilled at processing, analyzing and interpreting data. Based on observation: lectures are not dominated by lecturers, students are able to solve problems encountered and conduct investigations. Based on the interview towards lecturers: students are always actively involved in questioning, investigation, inquiry, problem solving and experimenting in lectures. Why do most students show good inquiry skills? Because students are accustomed to invited inquiry in biology lectures. The impact, the students become more ready to be invited to do more advanced inquiry, such as real-world application inquiry, because the skill of inquiry is essentially trained.

1. Introduction

Biology is part of a science that continues to develop, especially entering the second decade in the 21st. Biological science content is very promising to develop, because it is influenced by the rapid advances in technology. The integration of biological and technological advancements breeds innovations that can be useful for increasing the welfare and quality of life of the people, for example the discovery of sophisticated tools for detecting health, telemedicine, artificial intelligence, bioengineer, and synthetic biology [1]. To face the development of science, especially in biology, reliable resources needs to be prepared. Education has a central role in preparing resources according to need, which has competency in facing 21st century challenges. There are ten skills that must be owned by 21st century society and categorized into four groups, namely: ways of thinking (creative and innovation; critical thinking, problem solving and decision making; learning to learn, metacognition); ways of working (communication; collaboration); tolls for working (information literacy, ICT literacy); and living in the world (citizenship-local and global, life and career, personal and social responsibility-including cultural awareness and competence [2].

In the midst of high demand for reliable Human Resource (HR) and literate on the progress and challenges of the 21st century, there seems to be a lack of awareness among Indonesia. These indicators include many graduates of universities who are not directly absorbed by the workforce because of lack of competition, lack of special skills or creativity owned by the community in empowering and utilizing local potentials into economically valuable products, on the other hand many untapped natural potentials become waste and pollute the environment. Science lessons that emphasize mastery of concepts without learner lifelong learning also exacerbate the conditions that are not promising despite having a certificate on various levels (primary, secondary, high education), thus challenging educators prospective teachers and education practitioners to find a solution [3].

To meet these challenges, learners should be more invited to solve problems with the context that is often faced in everyday life. Real-world application inquiry can be one of the strategies offered to address the challenges faced by Indonesian society today. Real-world application inquiry guides students to be able to solve problems that occur in the real world. Real-world application inquiry can build students to solve problems that occur in everyday life [4]. Students apply experiences they have learned into new situations, so they find authentic answers related to problems solved by individual or cooperative and collaborative ways. The learning experience that students do can be problem-driven & project-based approaches like those done by scientists and engineers [5], for example energy, global warming, healthy food, and others.

Before the implementation of real-world application inquiry strategy, it is necessary to first analyze the readiness of biology pre-service teacher in the course. Are the students already accustomed to inquiry? What type of inquiry are introduced?. The things that must be prepared in advance to the students, so they are skilled in doing questioning, solving problems, experimenting and hands-on activity. Inquiry is a strategy to teach students about the process of researching and explaining foreign phenomena on the conceptions of scientific methods [6]. Inquiry as a process of defining and investigating problems, formulating hypotheses, designing experiments, finding data, and illustrating the conclusions of those problems [7]. The National Science Education Standards [8] emphasizes that inquiry is an activity that involves observation; asking question; Check books and information from other sources to see what is known; Investigative planning; Review what is already known from the experimental evidence; using tools to collect, analyze, and interpret data; Proposing answers, explaining, predicting; and communicate results.

The success of students in achieving these indicators is certainly supported by various factors, such as the teaching strategies applied by the lecturers, the significance of the challenges faced by the students, as well as the supporting facilities in the continuity of the lectures. The teachers college has full responsibility in preparing prospective teachers to become professional teachers and meet the standards of 21st century competence. Therefore, based on the above background, the problem formulation in this study is "How is the inquiry skills of biology preservice teachers in one teachers college in Central Java on biotechnology lecture? ".

2. Method

This article is the result of qualitative descriptive analysis of a case study. The research was conducted in Biology Education Department in one teachers college in Central Java with the subjects involved were students as many as 29 people selected by purposive method from six semesters in academic year 2016-2017 and four lecturers. The sample selection of students from the sixth semester is based on the experience of taking enough courses, so their responses are enough to represent the readiness of the biology students as a whole in the course of real-world application inquiry.

The focus of the research is on the readiness of biology pre-service teacher to follow lectures oriented towards real-world application inquiry. The matters revealed in this study were adapted from NRC [8] and Wenning [4] on inquiry aspects, there are: 1) students' habits in asking questions and formulating problems; 2) the skill of experimenting; and 3); The skills of finding relevant information and analyzing and interpreting data. The data were collected using the main instrument of questionnaire, consisting of 11 statements. In addition, supported by data collected through observation, which is done in lecturing process in four subjects that serve as the subject of research;

And interviews conducted on four lecturers to confirm the implementation of the lectures that had been his face.

3. Result and Discussion

How ready the students to follow the real-world application inquiry based on the lectures collected using various instruments is described in two major sections. The things that can be revealed from the results of this study include the general description of how ready the students in following the lectures based on real-world application inquiry, and the specific analysis of how ready the students based on the aspect of inquiry.

3.1. General description readiness of students in following lectures based on inquiry

In general, how ready the students in dealing with lectures designed with Inquiry is represented from three data retrieval techniques, namely student questionnaire, lecture observation, and interview with lecturer.

3.1.1. Student questionnaire data. Based on the questionnaire which contains 11 statements about how ready the students in dealing with biology lectures based on Inquiry, the overall data analysis results are represented as in Figure 1. Figure 1 showed that 31.42% strongly agreed to the statement on the questionnaire; 46.93% agreed; 18.39% stated disagree; and 3.26% stated strongly disagree. This data indicates that overall the average student states are ready to follow lectures designed based on Real-world Application Inquiry. That is, so far most of the students have been accustomed to do inquiry process in the lectures and actually feeling the impact.

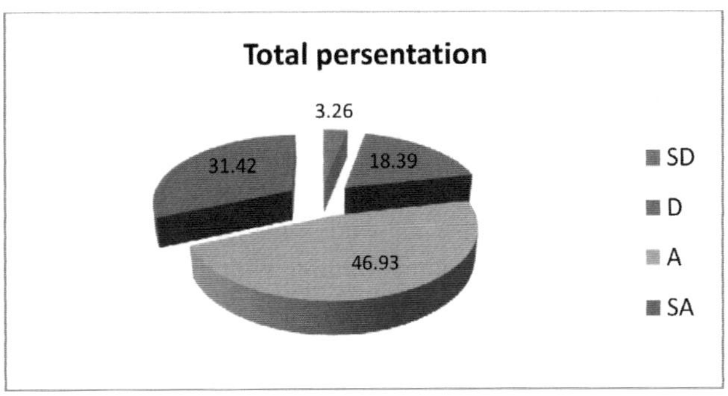

Figure 1. Total presentation of student questionnaires

3.1.2. Data results of lectures observations. This section represents some of the findings of field observations in one of the courses, which are.

a. Lecturing process are not dominated by lectures done by the lecturers, but are done by varied methods. This means that lectures are based on student centered learning with various methods or strategies applied by lecturers;
b. Students are guided to solve problems by first identifying the problems to be solved in a structured way;
c. Students are directed to investigate problems by searching for various sources / literatures, then the data obtained are categorized by group.

3.1.3. Data of interview with the lecturers. Interviews were conducted on lecturers who were a subject of the research, the aim was to cross-check and compare the data that have been collected from the questionnaire and observation. From several questions asked, represented by some important questions that represent the purpose of the study. Interview data are shown in Table 1.

Table 1. Data results of interview with the lecturers

No	Questions	Lecturers' Answers
1	What methods / approaches / learning models have been used in lectures?	Each course and even each material use different learning strategies modified to fit the characteristics of the material. Ordinary students are invited to discuss, question, answer, argue, inquiry, practicum, problem-solving, even on a certain subject we invite to experiment produce a product.
2	Are students already used to being actively involved in asking and answering questions? If yes, What percentage of students who are active in a class?	Students are accustomed to be active in the lecturing process, especially doing questions and answers. Their involvement between 40-70% actively ask.
3	Are students accustomed to being invited to formulate and solve problems?	In the lecturing process, students are usually invited to formulate and solve problems

3.2. Specific analysis of how ready the student based on inquiry aspect

The student questionnaire consisting of 11 statements is based on five aspects of inquiry. Determination of the five aspects of the objective is to represent the readiness of students in following the lecture real-world application inquiry, as well as see the habits built in lectures that have been done. The data of the research on the readiness of students specifically based on inquiry aspect is illustrated in Table 2-4.

Table 2 represents Aspect I data, which is the student's habit of asking questions and formulating problems. On average 0.86% of students strongly disagree with the four statements; 13.79% disagree; 58.62% agreed; and 26.72% strongly agree. This data shows that most students are used to asking questions and formulating problems in lectures. Even if analyzed more specifically from Table 3, there is a very interesting fact to be discussed: 1) statement-2, shows the most strongly agreeing percentage of the other options, which is 44.83%. This means that in the lecture, students are very involved to develop scientifically oriented questions; 2) the percentage agree on statement-3 approaching at absolute percentage, that is 82.76%. This fact indicates that the lecturer is really training students to ask questions and answer questions, by giving further questions until students understand the concepts/ materials learned.

Table 2. Aspect data I: the students' habits in asking questions and formulating problems

Aspect I	**SD**	**D**	**A**	**SA**
Statement 1	0	17,24	55,17	27,59
Statement 2	3,45	10,34	41,38	44,83
Statement 3	0	3,45	82,76	13,79
Statement 4	0	24,14	55,17	20,69
Average	0,86	13,79	58,62	26,72

IOP Conf. Series: Journal of Physics: Conf. Series **895** (2017) 012135 doi:10.1088/1742-6596/895/1/012135

Aspect data II, that is experimental skill is shown in Table 3. On average 4.60% of students stated strongly disagree with the statement; 24.14% disagree; 39.08% agreed; And 32.18% strongly agree. The data represent the majority of students used to solve problems by experimenting. However, the percentage of disapproval is also quite large on this aspect. Moreover, if you look more specific on the 5 statement, the percentage does not agree greater than the agreed percentage, which is 34.48%> 27.59%. This means that the students are not yet fully used in following up the problems into scientific investigation, observation, practicum and experiment.

Table 3. Aspect data II: the skill of experimentation

Aspect I	SD	D	A	SA
Statement 5	3,45	34,48	27,59	34,48
Statement 6	3,45	13,79	48,28	34,48
Statement 7	6,90	24,14	41,38	27,59
Average	4,60	24,14	39,08	32,18

Aspect III, on the skills of finding relevant information and analyzing and interpreting data, is illustrated in Table 4 that an average of 3.45% of students responded strongly disagreeing with the statement; 18.97% disagree; 43.97% agreed; And 33.62% strongly agree. Specific data on each statement also represent the same with the average data. This fact indicates that most of the students are accustomed and skilled at obtaining relevant information from various sources and able to analyze and interpret the data.

Table 4. The skills of finding relevant information and analyzing and interpreting data

Aspect III	SD	D	A	SA
Statement 8	10,34	20,69	44,83	24,14
Statement 9	0	13,79	44,83	41,38
Statement 10	0	20,69	41,38	37,93
Statement 11	3,45	20,69	44,83	31,03
Average	3,45	18,97	43,97	33,62

The data that has been shown previously generally has represented the readiness of students in facing lectures based on real-world application inquiry. One indicator that can be used as a reference is the student's habits in doing inkuiri. The findings are in line with the results of the research by Sen & Vekli, that in inquiry learning students are invited to undergo a research process such as a scientist, determining research questions, hypotheses, obtaining results by designing experiments and developing science process skills [9]. Different studies confirm that, students who are more often invited to do inquiry by their teachers are more active in doing scientific practice and more interactive in the learning process [10]. In addition, the habit of doing inquiry trains students to develop problem-solving skills, especially those that are often faced by people in everyday life. Fulfillment of these indicators can be used as a recommendation that students have demonstrated readiness to follow lectures designed with the development of real-world application inquiry.

4. Conclusion

Based on the data collected, students have shown how ready they are in following the lectures based on inquiry. The facts shown by most of the percentage of students responding agree and strongly agree on each statement on the questionnaire. In addition, based on the results of field observations and interviews with lecturers, students have been actively engaged in question and answer, investigation, inquiry, solving problems and experimenting in lectures.

Acknowledgments
Acknowledgments are conveyed to all parties who assist in the collection of research data. Secondly, to the Ministry of Research and Technology and the Ministry of Finance of the Republic of Indonesia who gave scholarships to the authors during the study.

References

[1] Tjandrawinata R R 2016 *Industri 4.0: revolusi industri abad ini dan pengaruhnya pada bidang kesehatan dan bioteknologi* (Jakarta: Working Paper from Dexa Medica Group)

[2] Binkley M, Erstad O, Herman J, Raizen S, Ripley M, Ricci M M and Rumble M 2012 *Handbook on Assessment and teaching of 21st century skills* ed P Griffin, B McGaw and E Care (Australia: springer) chapter 2 pp 17–66

[3] Rustaman N Y 2016 Pemberdayaan *Entrepreneurship*: Implementasi Teori-U dalam Bioteknologi Praktis Berbasis STEM *Pros. Seminar Nasional Sains dan Entrepreneurship III* (Program Studi Pendidikan Biologi FPMIPATI Universitas PGRI Semarang)

[4] Wenning C J 2005 Levels of Inquiri: Hierarchies of Pedagogical Practices and Inquiri Processes *J. of Physics Teacher Education* **2** 3

[5] Wenning C J 2010 The Levels of Inquiry Model of Science Teaching *J. of Physics Teacher Education* **6** 2

[6] Joyce B, Weil M and Calhoun E 2000 *Models of Teaching* (Boston-London: Allyn and Bacon)

[7] Trowbridge W and Bybee W 1990 *Becoming a Secondary School Science Teacher* Fifth Edition. (Ohio: Merrill Publishing Company)

[8] NRC (*National Research Council*) 1996 *The National Science Education Standards* (Washington DC: National Academic Press)

[9] Sen C and Vekli G S 2016 The Impact Of Inquiry Based Instruction On Science Process Skills And Self-Efficacy Perceptions Of Pre-Service Science Teachers At A University Level Biology Laboratory *Uni. J. of Edu. Research* **4** 3

[10] Lindsay B W, Randy L B, Brooke A W, and Jennifer L M 2014 The Science Elf: Assessing The Enquiry Levels Framework As A Heuristic For Professional Development *International Journal Of Science Education*

IOP Conf. Series: Journal of Physics: Conf. Series **895** (2017) 012136 doi:10.1088/1742-6596/895/1/012136

Enhancing Scientific Inquiry Literacy of Prospective Biology Teachers through Inquiry Lab Project in Microbiology

K Kusnadi[1,2*], N Y Rustaman[1,2], S Redjeki[1] and I N P Aryantha[3]

[1]Sekolah Pascasarjana,Universitas Pendidikan Indonesia, Bandung, Indonesia
[2]Departemen Pendidikan Biologi, Universitas Pendidikan Indonesia, Bandung, Indonesia
[3]Sekolah Ilmu Teknologi Hayati, Institut Teknologi Bandung, Bandung, Indonesia

*kusnadi@upi.edu

Abstract. The implementation of the inquiry laboratory based project to enhance scientific inquiry literacy of prospective biology teachers in Microbiology course has been done. The inquiry lab based project was designed by three stages were debriefing of basic microbiology lab skills, guided inquiry and free inquiry respectively. The Study was quasi experimental with control group pretest-posttest design. The subjects were prospective biology teachers consists of 80 students. The scientific inquiry literacy instrument refers to ScInqLiT by Wenning. The results showed that there was significant difference of scientific inquiry literacy posttest scores between experiment and control (α 0,05) and was obtained N-gain score was 0.49 (medium) to experiment and 0.24 (low) to control. Based on formative assessment showed that development of student's scientific attitude, research and microbiology lab skills during conducting project were increased. Student's research skills especially in identification of variables, constructing a hypothesis, communicating and concluding were increased. During implementation of inquiry project also showed that they carried out mind and hands-on and so collaborative group investigation lab activities. Our findings may aid in reforming higher-education, particularly in microbiology laboratory activities to better promote scientific inquiry literacy, scientific attitude, research and laboratory skills.

1. Introduction

Inquiry is the essence of science education. Inquiry is the process of finding out by searching for knowledge and understanding. Inquiry also refers to the activities of students in which they develop knowledge and understanding of scientific ideas, as well as an understanding of how scientists study the natural world [1]. Scientific inquiry is the element of science literacy. Scientific inquiry refers to the diverse ways in which scientists study the natural world and propose explanations based on the evidence derived from their work [2]. Scientific inquiry is a powerful way of understanding science content. Students learn how to ask questions and use evidence to answer them. In the process of learning the strategies of scientific inquiry, students learn to conduct an investigation and collect evidence from a variety of sources, develop an explanation from the data, and communicate and defend their conclusions [3]. Scientific inquiry is an extension of the scientific method, that applied in action in learning science and life [4]. Scientific inquiry literacy is the person ability of understanding and applying science concepts in inquiry process or scientific investigation [5]. Scientific inquiry literacy is the component of science literacy which has to be possessed of every citizens.

IOP Conf. Series: Journal of Physics: Conf. Series **895** (2017) 012136 doi:10.1088/1742-6596/895/1/012136

The quality of meaningful teaching and learning process in higher education should have the following characteristics: 1) invite the curiosity of students; 2) challenge students to learn; 3) activate mental, physical and psychological; 4) facilitate lecturers in managing the class; 5) develop student creativity; and 6) developing students' understanding of the content learned. To accommodate the principles of learning at higher education, particularly for the adult education, the experiental learning is often used as learning model, that leads to autonomy and self-directed learning [6].

Project-based learning is an innovative learning model, which emphasizes contextual learning through complex activities [7,8,9,10]. The focus of learning lies in the core concepts and principles of a discipline, involving learners in problem-solving investigations and the activities of other meaningful tasks, giving learners the opportunity to work autonomously for constructing their own knowledge, and reaching its peak by producing tangible products [11]. Syntax of project based leraning model has been developed by The George Lucas Educational Foundation [12]. Project based learning is one type of inquiry based learning model. The main goals of inquiry-based learning are: 1) developing students' curiosity and motivation to learn the principles and concepts of science; 2) developing students' scientific skills so they can work like a scientist; 3) familiarize students to work hard to gain knowledge. In principle, inquiry instruction helps students to formulate questions, seek answers or solve them to satisfy their curiosity and to help their theories and ideas about the world. Inquiry based learning also aims to develop thinking skills [2]. Intellectual abilities that is expected in the inqury-laboratory based learning is integrated skills as follows: 1) measure metric (standard unit), 2) build empirical law based on proof and logic, 3) designing and conducting scientific investigations, and 4) using technology and mathematics during investigations [13]. The American Association for the Advancement of Science recommended that science educators have to promote independent learning, engage students in experimental design and execution, emphasize science as it is practiced, and promote scientific communication skills [14, 15]. The science education researcher also emphasizes the importance of application of inquiry-based learning in biology at university level [16].

There have been many research reported on project-based learning and inquiry in science, but there are still few reports of scientific inquiry particular in microbiology lab activities. This study is expected to provide meaningful information for debriefing of scientific inquiry for biology teacher candidates in Institute of Teachers Education (LPTK).

2. Experimental Method

This study used a quasi-experimental design, with the control group pretest-posttest design [17]. The subjects are prospective biology teachers who enrolled in Microbiology. Study was carried out at one of public Institute of Teacher's Education (LPTK) in Bandung. The sum of subjects were 80 students (experimental was 42 and the control was 38 students). The study was conducted in three stages as follow: 1) debriefing of basic microbiology lab skills, such as making of sterilized media, bacterial staining technique, isolation and enumeration of microbes, and bacterial identification; 2) guided inquiry for debriefing of scientific inquiry skills; and 3) designing and implementing of microbiology project (free inquiry); called as "Mini-research" project. The scientific inquiry literacy of prospective biology teachers was measured by a well-chosen multiple choice test refers to the ScInqLiT instrument [5]. The student's research and laboratory skills, and scientific attitude development were assessed by formative assessment during the learning process. Data Analysis used empirical statistical approach and Normalized gain (N-gain) score category [18].

3. Result and Discussion

Based on the research results in measuring of scientific inquiry literacy of prospective biology teachers was obtained recapitulation of pretest-posttest data (Experiments and control class) as shown in Table 1.

Table 1. Recapitulation of Pretest-Posttest scores of scientific inquiry literacy

Parameters	Control		Experiment	
	Pretest	*Posttest*	*Pretest*	*Posttest*
Sum of students	38	38	42	42
Score averages	32,84	47	35,29	61,81
Deviation standard	5,69	10,47	8,51	10,44
Minimun scores	18,00	23,00	16,00	37,00
Maximum scores	47,00	68,00	42,00	82,00
N-gain	0.24		0.49	
Normality test	0,43 Normal	0,97 Normal	0,50 Normal	0,59 Normal
Homogenity test	0,87 Homogen	0,25 Homogen	0,002 Not Homogen	0,12 Homogen
Two difference t' Test of scientific inquiry literacy posttest scores between experiment and control at α 0.05			**0,00**	

Based on Table 1 shows that there is a significant difference between experiment and control posttest averages score at level significance α 0.05 (t' test 0,00 < 0,05). It indicates that inquiry lab project provides meaningful experience for prospective biology teachers for debriefing of scientific inquiry. Table 1 also shows that N-gain score is different between experiment and control class. N-gain of experiment class is 0.49 (medium category), while control is 0.24 (low category). It means that there is an increase in N-gain score in experimental higher than the control class. Furthermore, the comparison of N-gain score of scientific inquiry literacy for each indicators between experiment and control class is shown at Figure 1.

Based on Fig 1 indicates that the N-gain score between experiment and control class of scientific inquiry literacy of prospective biology teachers is different, particular in indicators of designing experiment procedures, identification and control variables and construct the hypothesis. Thereby the inquiry project-based learning has contributed to debriefing of prospective biology teachers to understand principles and procedures of scientific inquiry. During conducting microbiology project they were doing sciences as wells as scientists were doing research. This indicates that inquiry based projects provide meaningful learning experiences for prospective biology teachers, so suggest that they can design inquiry project-based learning at schools in the future.

Furthermore development of research skills of prospective biology teachers through formative assessment during implementation of microbiology project is shown at Table 2.

Table 2. Development of averages research skills scores of prospective biology teachers

No	**Indicators**	**Guided inquiry lab 1**	**Guided inquiry lab 2**	**Free inquiry Lab**
1	Formulate the problems	77,38	80,00	80,71
2	Ask questions	76,90	77,86	79,52
3	Identification of variables	76,90	81,67	82,14
4	Control of variables	75,48	79,76	80,36
5	Construct hypothesis	70,60	79,05	80,24
6	Design procedures of investigation	79,29	80,24	80,83
7	Communicate results of investigation	80,71	83,57	84,29
8	Analysis and interpreted data	77,02	80,24	80,12
9	Draw conclusion	74,88	81,43	82,26
	Averages	**77,98**	**81,37**	**81,88**

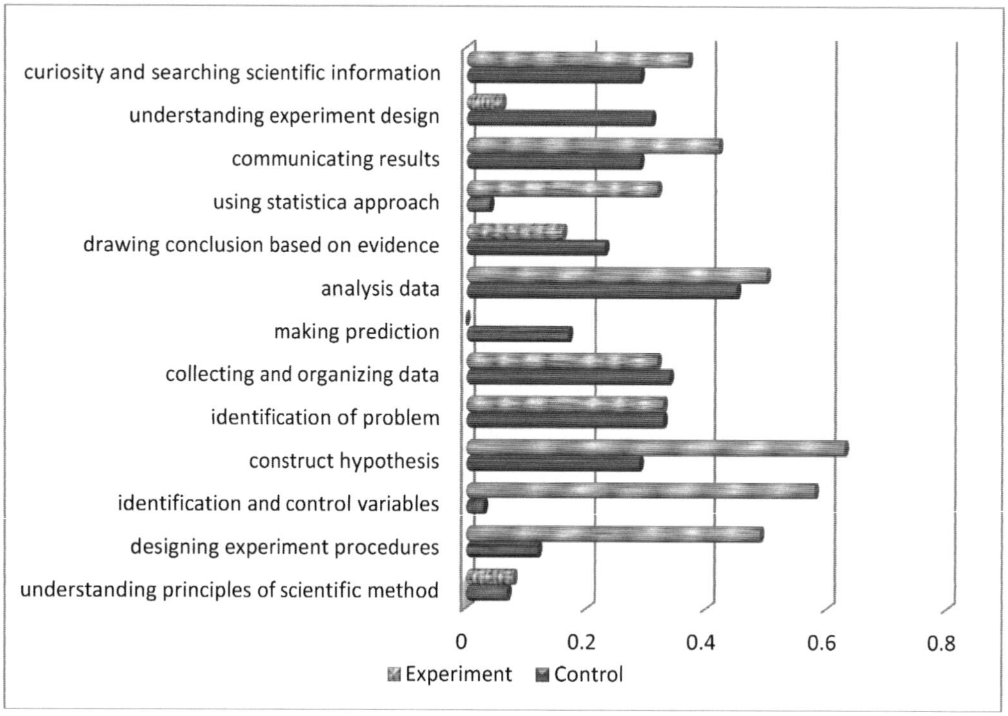

Figure 1. The comparison of N-gain scores scientific inquiry literacy between experiment and control class of each indicators

Based on data in Table 2 indicates that there is development of student's research skills on each inquiry stages from guided inquiry to free inquiry. Thereby the inquiry-project based learning has contributed to debriefing of prospective biology teacher's research skills. They engaged fully in designing and implementation of investigation. They learn content as well as practice for applying microbiological concepts to solve problems in everyday life. They learn to cultivate their own authentic research from individual work and then work collaboratively in small groups. Each student learns how to design scientific research, to produce meaningful products. This study is also carried out formative assessment through performance assessment of microbiology-laboratory skills of prospective biology teachers during both practicum and inquiry project activities. The data are shown at Fig 2.

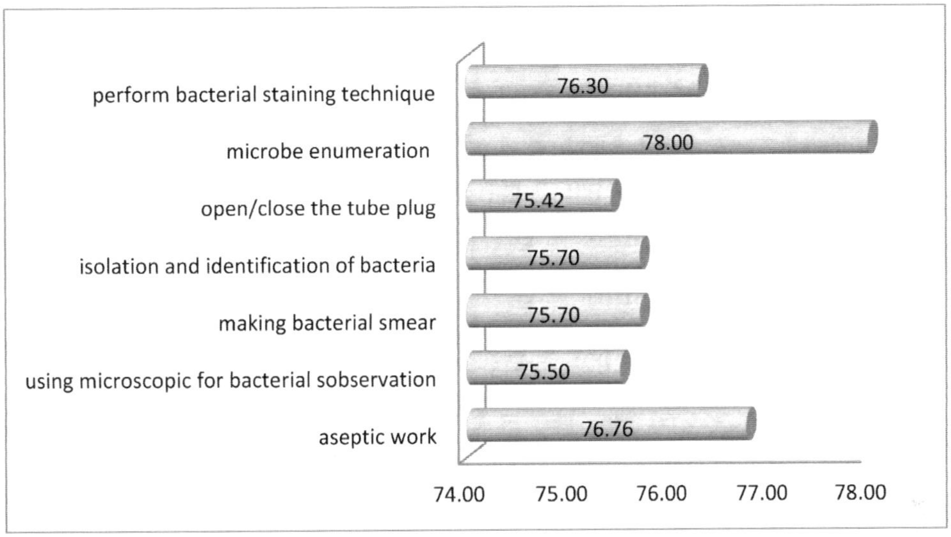

Figure 2. Averages of student's microbiology lab skills scores of each indicators

Based on Fig 2 shows that the averages of microbiology lab skills scores during the project implementation of prospective biology teachers is high enough. It means that the inquiry project - based learning has contributed to the debriefing of student's microbiology lab skills especially in microbe enumeration, aseptic work and perform bacterial staining technique. It's appropriate with research results of appearance of specific student's microbiology laboratory skills through inquiry project-based learning in Microbiology course [19]. Thereby the inquiry project based learning has to develop for coaching laboratory skills. Furthermore through formative assessment is obtained data of student's scientific attitude during microbiology project implementation. The data are shown at Figure 3.

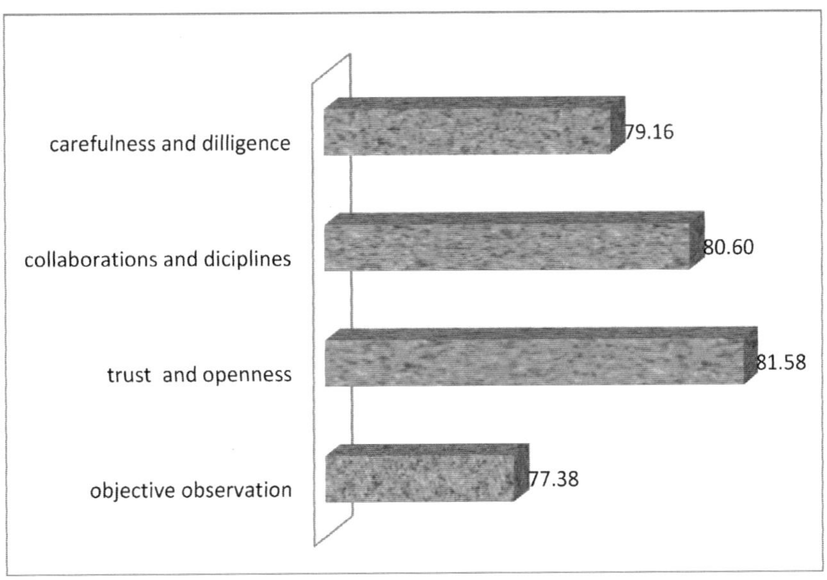

Figure 3. Averages of student's scientific attitude scores during project implementation

Based on Fig 3 shows that averages of student's scientific attitude scores during the project implementation is high enough, particular in trust and openness as well as collaboration and disciplines. It means that the inquiry project-based learning in microbiology has contributed to build student's scientific attitude that be needed for working and getting career in the future. And scientific attitudes also are an important belongs to every citizens, who are essential to work and to social life in the future. Furthermore it is necessary to convey the title of the group "Mini-research" inquiry-project were designed and implemented as shown in Table 3. Based on Table 3 indicates that the group project titles cover representing all areas of applied microbiology consists of health microbiology (4 groups), food microbiology (4 groups), environmental microbiology (3 groups) and industrial microbiology (3 groups). Group inquiry projects also produce various representative products such as: pure cultures of bacteria or fungi, packaging products such as skin ointment and fermented foods or beverages, and so posters. This product is very meaningful for both of scientific and general public community.

Table 3. The Group Inquiry-Lab Project Titles

Group	Inquiry Projects Titles	Area of Microbiology	Representative products
1	Utilization of guava leaf extract (*Psidium guajava*) to inhibit the growth of acne bacteria (*S. aureus*)	Health Microbiology	Pure culture, skin ointment, poster
2	Effect addition of *Beta vulgaris* tuber extract to organoleptic value of goat milk yoghurt using inoculum of *Lactobacillus bulgaricus* and *Streptococcus thermophillus*	Food Microbiology	Goat milk yoghurt, poster
3	Effect of *Rhizobium sp* inoculum concentration to greenpeal (*Phaseoulus radiatus*) growth	Environmental Microbiology	Pure culture, liquid Bio-fertilizer, poster
4	Effect of Mengkudu juice (*Morinda citrifolia*) concentration to inhibit fungi growth of *Candica albicans*	Health Microbiology	Pure culture, mengkudu Juice, poster
5	Effect of lactic acid bacteria (LAB) concentration to organoleptic and save value of bread fermentation	Food Microbiology	Bread product, poster
6	Effect of Incubation Temperature on Bioethanol Production of Banana peel Wastes with Acid Hydrolysis and Fermentation By *Saccharomyces cereviceae*	Industrial Microbiology	Pure culture, bioethanol, poster
7	Effect of biofertilzer *Azotobacter sp.* inoculum to Chillies Growth	Environmental Microbiology	Pure culture, liquid fertilizer, poster
8	Fermentation of Guava, Mango and Tomatoes Juices Using *Lactobacillus acidophilus* as Probiotic Drink	Food Microbiology	Probiotic drink, poster
9	The Effects of *Saccharomyces cerevisiae* and *Aspergillus niger* Inoculum on Bioethanol Production from Pineapple Waste (*Ananas comosus*)	Industrial Microbiology	Pure culture, bioethanol, poster
10	Production of Anti bacterial Ointment of *Chromolaena odorata* Leaf Extract to Inhibit *S. aureus* Growth	Health Microbiology	Pure culture, skin ointment,poster
11	Effect of Addition of *Azotobacter sp* to Straw Compost Mixture on Chilli Growth (*Capsicum Frutescens*)	Environmental Microbiology	Pure culture, liquid bio-fertilizer,poster
12	Fermentation of Pomegranate Juice using Probiotic Lactic Acid Bacteria (*Lactobacillus delbruccki* sub sp bulgaricus)	Food Microbiology	Yoghurt pomegranate juice, poster
13	Effect of Pinus Stem Bark Extract to inhibit *Staphylococus aureus* growth	Health Microbiology	Pure culture, skin ointment,poster
14	Effect of *Saccharomyces cerevisiae* Inoculum Concentration on Bioethanol Production of Fermented Rotten Fruit	Industrial Microbiology	Pure culture, bioethanol,poster

4. Conclusion

Inquiry project-based learning in Microbiology has improved scientific inquiry literacy of prospective biology teachers. The Normalized gain score increased in medium category (N-gain = 0.49). Based on formative assessment showed that student's scientific attitude, research and microbiology lab skills were increased. The student's response to the programme reported that they gained greater experience and confidence in the scientific inquiry and laboratory skills activities. Our findings may aid in reforming of microbiology lab activities to better promote scientific inquiry literacy, scientific attitude, research and laboratory skills in the College. Based on research finding it may develop and implement the inquiry-project based learning for debreifing of prospecrive biology teachers in Institute of Teacher's Education (LPTK).

References

[1] Chiappeta E L 1997 Inquiry-based Science;Strategies and Techniques for Encouraging inquiry in the Clasroom. *The Science teacher*
[2] National Research Council 1996 *The National Science Education Standards* (National Academy Press. Washington DC)
[3] National Science Teacher Association (NSTA) 2004 *Standards For Science Teacher Preparation* (Washington, DC: NSTA and AETS)
[4] Rustaman N Y 2010 Pengembangan Pembelajaran Sains Berbasis Kemampuan Dasar Bekerja Ilmiah. Teori, Paradigma, Prinsip dan Pendekatan Pembelajaran MIPA dalam Konteks Indonesia. FPMIPA UPI
[5] Wenning C J 2007 Assessing inquiry skills as a component of scientific literacy *J. Phys. Tchr. Educ.* **4** (2)
[6] Fry H, Ketteridge S and Marshall S 2009 *A Handbook for Teaching and Learning in Higher Education Enhancing Academic Practice* Third edition Routledge Taylor and Fancis Group, New York
[7] Krajcik J, Blumenfeld P C, Marx R W, Bass K M and Fredricks J 1998 Inquiry In Project-Based Science Classroom: Initial Attemps By Midle School Students *The journal of the leraning science* **7** (3) 313-350
[8] Krajcik J, McNeill K L and Reiser B J 2007 Learning Goal Driven Design Model: Developing Curriculum Materials That Align With National Standards and Incorporate Peoject-Based Pedagogy *Journal Science education* **92** 1-32
[9] Thomas J W, Mergendoller J R and Michaelson A 1999 *Project-Based Learning: A Handbook For Middle And High School Teachers* (Novato, CA: The Buck Institute for Education) [online.] http://www.bgsu.edu/
[10] Mergendoller J R and Thomas J W 2009 Managing Project Based Learning: Principles From The Field
[11] Thomas J W 2000 *A Review Of Research On Project-Based Learning. California: The Autodesk Foundation Brown MP and Austin K, The New Physique* , pp. 25–30.
[12] The George Lucas Educational Foundation. 2005. *Instructional Module Project Based Learning.* [online.] http://www.edutopia.org/project-based-learning
[13] Wenning CJ 2011 The Levels of Inquiry Model of Science Teaching. *J. Phys. Tchr. Educ.* Online, **6** (2)
[14] American Association for the Advancement of Science. 1993. *Benchmarks for Science Literacy.* New York, NY: Oxford University Press
[15] National Research Council 2000 *Inquiry and the National Science Education Standards: A Guide for Teaching and Learning* (Washington DC: National Academy Press)
[16] Caccavo F Jr 2011 An Open-Ended, Inquiry-Based Approach to Environmental Microbiology *The American Biology Teacher* **73** (9) 521-525
[17] Gall M, Gall J P and Borg 2003 *Educational Research an Introduction* (New York: Pearson Eduation inc)

[18] Meltzer D E 2002 The relationship between mathematics preparation and conceptual learning gains in physics: A possible ''hidden variable'' in diagnostic pretest scores *American Association of Physics Teachers. Am. J. Phys.* **70** (12) 1258-1268

[19] Kusnadi, Rustaman N Y, Redjeki S and Aryantha I N P 2012 Analisis Kemunculan Keterampilan Spesifik Lab Mikrobiologi Melalui Pembelajaran Mikrobiologi Berbais Proyek Inkuiri "Mini Riset" Mahasiswa Biologi *Jurnal Pengajaran MIPA* **17** (1)

International Conference on Mathematics and Science Education (ICMScE) IOP Publishing
IOP Conf. Series: Journal of Physics: Conf. Series **895** (2017) 012137 doi:10.1088/1742-6596/895/1/012137

How to Improve the Mastery of Students'Concept on Photosynthesis Topic?

K Ulfa*, S Anggraeni and B Supriatno

Program Studi Pendidikan Biologi, Sekolah Pascasarjana Universitas Pendidikan Indonesia, Bandung, Indonesia

*khalida_ulfa@student.upi.edu

Abstract. The PPDP learning strategy in this research is the acronym of Practicum method, Presentation-discussion method, Demonstration method, and Presentation-discussion method. This study aims to describe the effect of applying PPDP learning strategies to mastery of high school students' concepts on photosynthesis topic. The research method is a weak experiment, with the research design "The One-Group Pretest-Postetst Design". The implementation of the study involved 35 students in one of the high schools in the city of Palembang. The research instrument used is in the form of test equipment, assessment rubric and questionnaire. Data were analyzed using Microsoft Excel and SPSS 24 Program. The statistical result showed that PPDP learning strategy had an effect on improving conceptual and Effective on the achievement of the value on the minimum criteria set by the school. This is due to the transformation of knowledge from hands-on to minds-on through the discovery of facts about the concept of photosynthesis. This fact leads to the construction of further understanding through cognitive sharing when the activities of the discussions formed similarity and consolidation of the concept of photosynthesis intact. In addition, demonstration activities also cause students' logic of thinking to develop through observation of factors that may affect the rate of photosynthesis. This PPDP learning strategy can be utilized by teachers in explaining photosynthetic topic.

1. Introduction

Biology as one of the disciplines of natural science is perceived and not easy to be understood or mastered by learners [1-2]. This can be seen from the acquisition of National Biology Test Scores of high school level in 2015-2016 is still relatively low [3]. In essence, matter in Biology learning consists of a number of facts, concepts and principles. The fact leads the student to the observation of an object / phenomenon so as to produce concrete knowledge. The principle shows how the object / phenomenon occurs while the concept shows a mental abstraction of an object or phenomenon. One of the material that contains the concept felt difficult to be understood by teachers and students is the material of photosynthesis. Basically photosynthetic material invites students to understand an abstract energy transformation process, capturing and using light energy, then stored in the form of chemical energy that occurs in the chloroplast, through various processes to form starch and release oxygen[2]. According to some scientist, the concept of photosynthesis is an important concept in biology [4-7]. The principle of a large concept of abstract and complex photosynthesis processes must be able to direct the knowledge that students have from conceptual (abstract) to be transformed into factual (concrete).

Some research on the concept of photosynthesis proves that, still found students who are confused and misconception [8-11]. The confusion and misconceptions on the concept of photosynthesis were

IOP Conf. Series: Journal of Physics: Conf. Series **895** (2017) 012137 doi:10.1088/1742-6596/895/1/012137

found in junior high school students [12-13], high school students [9, 14-16], Student level [17-19] and to prospective teachers [20-23].

One of the learning strategies considered to assist students in developing conceptual mastery is the PPDP learning strategy. The PPDP learning strategy in question is the acronym of the learning method consisting of the Practicum method, the Presentation-discussion method, the Demonstration method, and the Presentation-shortened methodology into PPDP. The PPDP learning strategy consists of four goals. First, integrated lab work aims to generate many facts related to the concept of photosynthesis. Second, in the presentation-discussion activities conducted cognitive sharing of the facts found that aims to equate and consolidate the concept of photosynthesis. Third, demonstration activities using photosynthesis kits aims to develop students' thinking and logic skills to factors that influence the rate of photosynthesis. Fourth, the presentation-aims of discussion aimed to equate and consolidate the concept of photosynthesis in a holistic way. The PPDP learning strategy is expected to reduce the level of confusion and misconception among students.

2. Experimental Method

The research method used is the weak experiment method with the research design "The One-group Pretest-Postetst Design. Subjects in the study were 35 students of class XI at one school in the city of Palembang. The PPDP learning strategy is divided into three stages. The first stage is the application of classroom practicum method completed with Student Worksheet (LKS). It aims at developing conceptual mastery by practicing Ingenhousz, Sachs, Pristley, and leaf disks. In the second stage, two methods of learning are applied, namely presentation-discussion method and demonstration method. Method of presentation-discussion conducted aims to discuss the findings or facts of practical activities that have been done previous students, resulting in similarity and stability in the mastery of the concept. After completion of the presentation-discussion activity, continued the demonstration method. The demonstration method aims to develop students' high-level thinking skills. During the demonstration activities, teachers used photosynthesis kits as a demonstration medium. Through the use of photosynthesis kits, students are trained in analyzing and evaluating activities undertaken. The third stage, applied presentation-discussion method. This method aims to discuss the results of observations made by students during the demonstration activities. This method is selected by researchers to guide students to connect all information that has been obtained into a whole concept of photosynthesis. The research instrument used is in the form of concept mastery test, assessment rubric and questionnaire. Data were analyzed using Microsoft Excel and SPSS Program 24.

3. Result and Discussion

Statistical data show pretest, posttest, and mean value of n-gain mastery of student concept after applying PPDP learning strategy can be seen in Table 1.

Table 1. Data statistics mastery of concepts

Statistics	value	
	Pretest	Posttest
average	38,12	81,30
variance	21,5	83,2
at least	27,8	60,8
maximum	46,8	97,5

Based on Table 1, shows an increase in the average mastery of the concept of 43.18 after applied PPDP learning strategy. Once analyzed, the average N-gain shows a value of 0.70 in the range of $0.61 \leq$ N-gain ≤ 0.80 with high category interpretation. Achieving the improvement of conceptual mastery caused during the learning process takes place, the teacher strives to create an active and enjoyable learning environment. The discovery of many facts about the concept of photosynthesis, the existence of cognitive sharing and the development of mind logic led to the formation of constructions of understanding in students. The concept of a person is formed by experience, and each student has the potential to process the information received [24-25]. Understanding the concept of photosynthesis can also be seen from the improvement based on the criteria of understanding according to Morgil and Yoruk [26] in Figure 1.

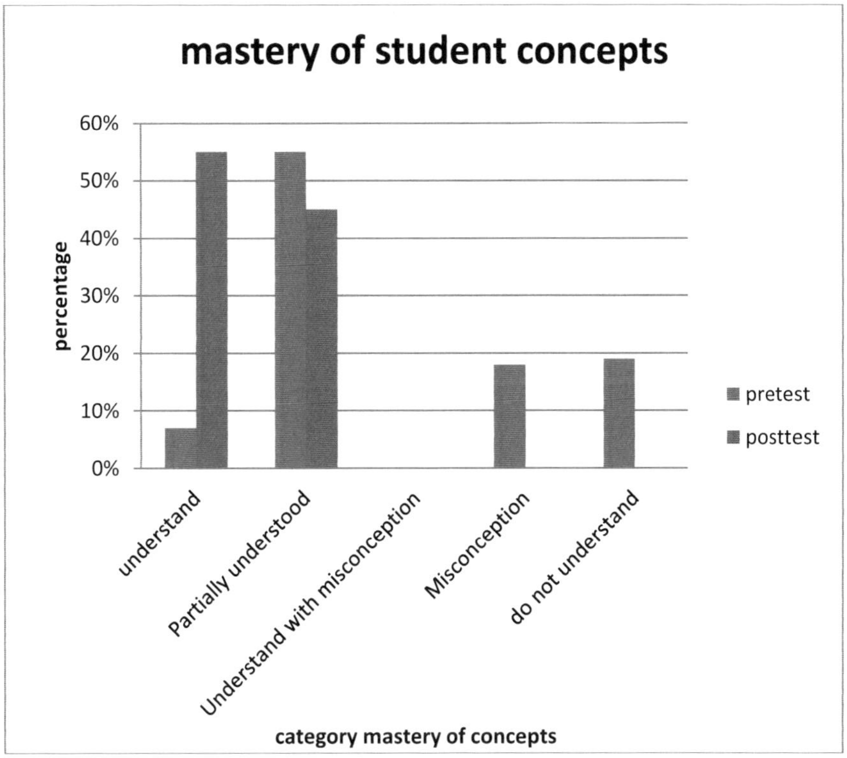

Figure 1. Increased mastery of student concepts

Based on Figure 1, shows significant differences in mastery of student concepts before and after the applied PPDP learning strategy on photosynthesis material. Direct experience in the discovery of facts leads to the transformation of knowledge from hands-on to mind-on. This affects the students' understanding of photosynthesis material. If studied in the "discovery" format, then the observed facts form the basis of concept or principle formation. If learning is "verification", then the observed facts become clear evidence of the correctness of the learned concepts or principles [27]. Based on statistical analysis, the effect of applying PPDP learning strategy can be seen in table 2.

Table 2. The influence of PPDP learning strategies on mastery of student concepts

One-Sample Test						
	Test Value = 0					
					95% Confidence Interval of the Difference	
	t	df	Sig. (2-tailed)	Mean Difference	Lower	Upper
pretest	48,672	34	,000	30,114	28,86	31,37
posttest	52,721	34	,000	64,229	61,75	66,70

Table 2 above shows the significant effect of PPDP learning strategy on mastery of student concept which is seen from p-value < 0,05 significance. The influence of the application of PPDP learning strategy can be seen from the increase in mastery of concepts in students.

4. Conclusion

Based on the above explanation, it can be concluded that the improvement of conceptual mastery is caused by the implementation of appropriate learning strategy. Through the application of PPDP learning strategies, students can understand the concept of photosynthesis through the discovery of many facts. This strategy also stimulates students to be active in explaining and arguing while learning takes place. In addition, logical thinking and reasoning skills can develop through activities that can stimulate the ability to analyze, predict, interpret and conclude. Implementation of learning strategies requires good time management to be accomplished.

Acknowledgments

The authors would like to thank all those who have provided financial and moral support for this research.

References

[1] Cimer A 2012 What Makes Biology Learning Difficult and Effective: Students' Views. *Educational research and review* **7** (3) 61-71

[2] Supriatno B 2013 Development of lecture program of ancorbary school-based biology education development to develop the ability to design and develop the design of laboratory activities. *Disertation*. Bandung: Sekolah Pascasarjana Universitas pendidikan Indonesia

[3] Kementrian Pendidikan dan Kebudayaan (Kemendikbud). 2016. In Press Conference in Jakarta. Koran Solopos.com. access in 9 Mei 2016. [online] at : http://www.solopos.com/2016/05/09/hasil-ujian-nasional-mendikbud-nilai-rerata-un-2016-sma-turun-ini-penyebabnya-717741.

[4] Finley F N, Stewart J and Yarroch W L 1992 Teachers perception of important and difficult science content *Science education* **66** (4) 531-538

[5] Bahar M, Johnstone A H and Hansell M H 1999 Revisiting Learning Difficulties in Biology *Journal of Biological Education* **33** (2) 84-86

[6] Lawson A E and Thompson L D 1988 Formal Reasoning Ability and is conceptions Concerning Genetice and Natural Selection *Journal of Research in Science Teaching* **25** (9) 733-746

[7] Storey R D 1989 "Textbook Errors & Misconceptions in Biology: Photosynthesis" *The American Biology Teacher* **51** (5) 271-274

[8] Wanderse J H 1983 Students' Misconceptions About Photosynthesis: A Cross age Study *Proceedings of the International Seminar on Misconceptions in Science and Mathematics* (Ithaka, NY: Cornell University) p 441-466.

[9] Haslam F and Treagust D F 1987 Diagnosing secondary students misconception of photosynthesis and respiration in plants using a two-tier multiple choice instrument *Journal of biological education* **21** (3) 203-210

[10] Eisen Y and Stavy R 1988 Students' understanding of photosynthesis *The American Biology Teacher* **50** (4) 208-212

[11] Anderson C W et al. 1990 The effect of instruction on college non majors' conception of respiration and photosynthesis *Journal of research in science teaching* **27** (8) 761-776

[12] Bell B 1985 Students' ideas about plant nutrition: What are they? *Journal of Biological Education* **19** (3) 213-218

[13] Eisen Y, Stavy R and Yaakobi D 1987 How students ages 13–15 understand photosynthesis *International Journal of Science Education* **9** (1) 105-115

[14] Amir R and Tamir P 1994 In-depth analysis of misconceptions as abasis for developing researeh-based remedial instruction: The ease of photosynthesis *The American Biology Teacher* **56** 94-100

[15] Özay E and Öztaş H 2003 Secondary students' interpretations of photosynthesis and plant nutrition *Journal of Biological Education* **37** 68-70

[16] Marmaroti P and Galanopoulou D 2006 Pupils' understanding of photosynthesis: A questionnaire for the simultaneous assessment of all aspects *International Journal of Science Educuation* **28** (4) 383-403

[17] Hazel E and Prosser M 1994 First-Year University Students' Understanding of Photosynthesis, Their Study Strategies & Learning Context *The American Biology Teacher* **56** (5) 274-279

[18] Carlsson B 2002 Ecological understanding: Ways of experiencing photosynthesis *Journal of Science Education* **24** (7) 681-699

[19] Köse S 2008 Diagnosing student misconceptions: Using drawings as a research method *World Applied Sciences Journal* **3** (2) 283-293

[20] Köse S and Uşak M 2006 Determination of prospective science teachers' misconceptions: Photosynthesis and respiration in plants *International Journal of Environmental and Science Education* **1** (1) 25 – 52

[21] Orbanić N D and Battelli C 2011 Misconceptions in science. In: M. Cotič, V. Medved Udovič, & S. Starc (Eds.*) Developing various literacy* (Koper: University of Primorska, Annales) pp. 275-282

[22] Ahopelto I, *et al.* 2011 Future elementary school teachers' conceptual change concerning photosynthesis *Scandinavian Journal of Educational Research* **55** (5) 505-515

[23] Ameyaw Y 2016 Evaluating Students' Misconceptions Of Photosynthesis And Respiration In A Ghanaian Senior High School *International journal of advanced biological research* **6** (2) 202-209

[24] Dahar R W 1996 *Learning theoriest* (Jakarta : Erlangga)

[25] Semiawan C 1987 *Approach to process skills* (Jakarta: Gramedia)

[26] Morgil I and Yoruk N 2006 Cross-Age Study of The Understanding Some Concepts In Chemistry Subjects In Science Curriculum *Journal of Turkish Science Education* **3** (1) 15 – 27

[27] Rustaman, N.Y. (1995). "The role of practicum in biology learning". *Paper.* Presented on training for technicians and college labs. Cooperation of FPMIPA IKIP Bandung with Directorate General of Higher Education. Bandung : FPMIPA IKIP.

Profile of High School Students' Understanding of Scientific Inquiry

N Anggraeni[*], **Y H Adisendjaja and A Amprasto**

Program Studi Pendidikan Biologi, Universitas Pendidikan Indonesia, Bandung, Indonesia

[*]anggraeniniki@gmail.com

Abstract. Scientific inquiry is the main objective of education. The purpose of this study is to investigate the understanding of scientific inquiry from high school students in grade 11. This study has involved 32 high school students. The scientific inquiry understanding is measure by using VASI (Views About Scientific Inquiry) questionnaire. Then, all responses are categorized as informed, partially informed, and naïve. The result showed group of respondents have a comprehension in informed category at the relationship among procedures and research questions. In the other side, there are some group of respondents that still have lack comprehension in the aspect of characteristic scientific research and the relation between questions. There's group of respondents still didn't comprehend about the differences between scientific evidence and scientific data's, the reason to select procedures, and also the relation between conclusion and data's collection.

1. Introduction

Science education has important role to establish the students with some meaningful skills such as critical thinking, creative, logic and insiative in response to the society issues. The science educators in the whole world emphasize the importance of the nature of knowledge and scientific study as basic component and essential from science literate [1]. Furthermore, explained that scientific inquiry is become main curriculum standard in some countries. Scientific inquiry referred to various ways about how scientist study about the nature and formulate the explanations based on the evidence acquired form they research. Inquiry also referred to the student's activity about how they can develop their knowledge and their point of view from scientific ideas, and also how scientist study about the nature [2].

This is very important to engage the students in each activity that involve thinking skill and intervened as a scientist [3]. Inquiry also includes students activities to evolved knowledge and comprehend some scientific ideas, also for understand about how scientist work in study about the nature [4;5]. Scientific inquiry described as a process to do work and produce the knowledge from the scientist [6]. Scientific inquiry includes the methodology of knowledge and integrate with scientific knowledge, scientific reasoning and critical thinking [7]. Biology as one area of science subject has four aims, such as: teaching some Biology facts, evolving the skills, teaching the skills and encourage the tangible attitude [8].

Scientific inquiry refers to the ways which scientist do some research about nature and proposed explanation based on the evidence form what they did [9]. The study about point of view from the students as Biology teacher prospective and Science teachers regarding scientific inquiry has indicates understanding of the students as Biology teacher prospective are better than Science teachers in some scientific inquiry aspect and group of respondents still have less understanding or included in partial informed category at some scientific inquiry aspects [9]. Besides that, in the other study about the nature of science and scientific inquiry has indicates teachers and students have in view which not including in informed category [10;11]. According to the study result before and because of still slightly the research about scientific inquiry, it's needed the research in purpose for describe the scientific inquiry from high school students.

2. Method

This study is using descriptive research which explore the understanding of high school students about the scientific inquiry. The subjects in this study are 32 high school students in science program grade 11 at SMA Negeri 1 Singaparna. The instrument use in this study for explore understanding about scientific inquiry is Views About Scientific Inquiry (VASI) developed by Lederman, *et al* (2014). Some important things about this instruments as the following: 1) question number 1, in the research instrument will give a scientific research case as descriptive research; 2) question number 5, respondent will give a task for choosing design that explained in discourse according on the question. The answer from this question is design A much better (>) than design B; 3) question number 6, presented the data from the research with a deviant data. A task for respondent is for create a conclusion and the reason why they choose that conclusion; 4) question number 7, respondent has a task to choose the result reconstruction fossil and try to explain the reason also asked the information required to explained their choose. All the answers from the respondents will analyse and represented by three categories, which is: informed, partially informed, and naïve [6] and then making percentage. The categories will analyse from the reason given after respondents find the answers or choose the answers. The informed category if the answers relate with the view science constructivist, the partially informed if the answers relate with the view science constructivist about science but there are less for some stuff, the naïve category if the answers doesn't relate with the view of science constructivist.

3. Result and Discussion

The results are summarized at table 1. Table 1 shows the percentage of results from the student respondents about scientific inquiry based on each category. Data is presented as a percentage of each category to make it easier to see comparisons between the three categories.

Table 1. Percentage students' understanding of scientific inquiry

No	*Aspect of scientific Inquiry*	*Category (%)*		
		Informed	Partially informed	Naïve
1	*Is that investigation scientific?*	43,75	53,13	3,12
	Is that investigation an experiment?	31,25	9,38	59,38
	Is scientific investigation can follow more than one method?	18,75	62,5	18,75
2	Is scientific investigation always begins with scientific question?	25	68,75	6,25
3	If question and procedure are the same, are scientists reaching the same conclusion?	59,38	12,5	28,12
	If questions are the same but procedures different, whether scientists reach the same conclusions?	59,38	31,25	9,38
4	Are scientific data same with scientific evidence?	3,12	21,88	75
5	Questions guide procedure. A better than B	15,62	37,5	48,88
6	Conclusions must consistent with data collected	34,38	28,12	37,5
7	Best construction	53,13	34,38	12,5
	Information needed	21,88	18,75	59,38

3.1. Scientific investigation

For the first question are consists of three questions, is that investigation scientific? is that investigation an experiment? Is scientific investigation can follow more than one method?. Based on table 1. For the question is that investigation is scientific? Categorization of respondents pertained partially informed that is equal to 53.13%. Most students answer "yes" that the investigation (discourse in question) is scientific, but gives an unfavourable reason. This shows that respondents do not yet know the full characteristics and characteristics of an immoral activity. Although they have experience in conducting scientific investigations but little increase the understanding of the nature and process of science [12] The reason given by the students is that the person is conducting the research because of researching the bird based on the food, it is seen that the students' understanding is not yet intact. The next question relates to the student experimental research included in the naïve category, based on table 1. The percentage of 59.38% of the reasons presented is that he experimented because he could find the causal relationship of the research and the hypothesis. This shows that students have not understood the kinds of research. In general, research investigation has three types: descriptive, experimental and correlational [13]. For questions relating to various methods in the investigation of students included in the category of partially informed that is equal to 62, 5% and a balanced percentage between students who include informed and naïve category that is equal to 18.75%.

3.2. Scientific investigation begins with question

To the next question whether scientific inquiry always begin with a scientific question? Categorization of respondents is partially informed that is equal to 68.75%. These results indicate that students are less aware of the function of the question in an investigation. An inquiry begins with a question and a question will direct the investigation procedure [6]. With the question in the research it will show the purpose of research and can guide the research procedure so as to generate a conclusion. Given the importance of the questions in a study, the questions in the research must be formulated clearly so that research can be directed and generate data in accordance with the questions we formulate.

3.3. Procedures and conclusion

To the question whether scientists will get a same conclusion if they have same questions and same procedure? 59.38% of students respond according to the informed category. The reasons they have for the most part have understood that there are differences of conclusions even though the procedures and questions are the same or different. Among the reasons that they say most answer that everyone must be different thinking then the conclusion will be different. In other words the procedure will affect the results [6].

3.4. Procedure inquiry guided by question

To evaluate respondents' understanding about relation between procedure and research question, respondents asked to choose one of two design A or B that agree with research question and they asked to explain their reasons. A total of 48.88% of students answered less precisely and fall into the naïve category. Of the appropriate answers were only about 15.62% of students who are able to provide reasons including informed categories, and about 37.5% of students who still have an understanding that belongs to the category of partially informed. To answer the same question, the scientist may design different procedures [6].

3.5. Scientific data and scientific evidence

For the question whether the data and scientific evidence are different? Respondents provided answers that belonged to the naïve category. The answer they put forward still confuses the difference between the evidence and the scientific data, although student answers state mostly that the evidence with the scientific data is different. Based on table 1. The percentage of naïve reached 75% almost as much as the number of students. In fact, the data is the result of observations collected by scientists during the investigation. The evidence is the result of data analysis that is interpreted and directly related to the specific question [6].

3.6. Conclusion must consistent with data collected

The respondent understanding of the relationship between the conclusions and the data collected, respondents were given a chassis about the duration of lighting at the rate of plant growth. Based on table 1 between the three categories yielding balanced respondents' answers. The naïve category ranged from 37.5% while the informed respondents were 28.12% and the informed category was 34.38%. Judging from the answers presented by students most of the students answered in accordance with the data presented they make the appropriate conclusions based on the table. Each research conclusion should be supported by evidence derived from data collected [6].

3.7. Explanation developed of combination data collected and previous information

For the last question about relationship between investigation results and information were needed to support their finding, all respondent exactly chose conclusion and the best structure of possible Dinosaurus fossil. Most respondents chose the correct arrangement and referred to the drawings so that the informed category of 53.13%, but for reasons of information used by scientists to provide explanation only slightly touched the previous information so that included in the category of naïve with a percentage of 59, 38%.

4. Conclusion

In general group of respondents have the understanding in informed category on the relationship between procedure and study questions aspect. But there are still a lot group of respondents have less understanding on aspect of the characteristics scientific investigations and relation between questions. Group of respondents also still don't understand about the differences between scientific evidence and scientific data, the reason choose procedure and the relation between conclusion and collection data. According to the finding from this study, the suggestion given to investigate another study towards difference subject for example to physic or chemistry subject.

References

[1] Senler B 2015 Middle School Students' View of Science Inquiry: An International Comparative Study *Science Education International.* **26** 166-179

[2] NRC 1996 *Inquiry and the National Science Education Standards Washington* DC: National Academy Press

[3] Ozgelen, S., Yilmaz-Turzun, O., & Hanuscin, D. L 2012 Exploring the development of preservice science teacher' view on the nature of science in inquiry-based laboratory instruction. *Researchin Science Education.* **43** 1551-1570

[4] Anderson, R. D 2002 Reforming Science Teaching: What Research Says About Inquiry? *Journal of Science Teacher Education.* **13** 1-12

[5] Colburn, A. 2000 An Inquiry Primer. *Science Scope.* (March 2000), 42-44.

[6] Lederman, J.S., Lederman, N.G., Bartos, S.A., Bartels, S.L. Meyer, A. A., & Schwartz, R. S 2014 Meaningful Assessment of Learners' Understanding about Scientific Inquiry-The Views About Scientific Inquiry (VASI) Questionaire. *Journal of Research in Science Teaching* **51** 65-83

[7] Lederman, N.G. 2006. "Research on Nature of Science: Reflections on the Past, Anticipations of the Future". *Asia-Pasific Forum Science Learning and Teaching.* **7** (1)

[8] Rustaman, N. 2003 *Kemampuan Dasar Bekerja Ilmiah dalam Sains.* Makalah Seminar, Bandung: Universitas Pasundan

[9] Adisendjaja, Y. H., Rustaman, N., Redjeki, S., Satori, D. 2016. Gambaran Pandangan Mahasiswa Calon Guru Biologi dan Guru IPA tentang Inkuiri Ilmiah. [Online] Tidak diterbitkan

[10] Lederman, J. S. & Lederman, N. G, 2004 Early Elementary Students' and Teachers' Understanding of Nature of Science and Scientific Inquiry: Lesson Learned From Project ICAN. Paper Presented at the Annual Meeting of the National Association for Research in Science Teaching. Vancouver, British Columbia. April, 2004.

[11] Schwartz, R.S., Lederman, N., Khishfe, R., Lederman, J.S., Mathews, L., & Liu,S. 2002 Explicit/Reflective Instructional Attention to Nature of Science and Scientific Inquiry: Impact on Student Learning. *Paper Presented at The 2002 Annual International Conference of The Association for The Education of Teachers in Science*

[12] Schwartz, R.S., Lederman, N.G. & Crawford, B.A., 2004. Developing Views of Nature of Science in an Authentic Context: An Explicit Approach to Bridging the Gap between Nature of Science and Scientific Inquiry. Science Teacher Education **88** 610-645

[13] Lederman, J. S., 2009. Teaching Scientific Inquiry: Exploration, Directed, Guided and Open Ended Levels. In National geographic science: Best Practice and Research Base 8-20. Hapton Brown Publisher

IOP Conf. Series: Journal of Physics: Conf. Series **895** (2017) 012140 doi:10.1088/1742-6596/895/1/012140

Learning Experience on Transformer Using HOT Lab for Pre-service Physics Teacher's

A Malik[1,2*], A Setiawan[1], A Suhandi[1] and A Permanasari[1]

[1]Sekolah Pascasarjana, Universitas Pendidikan Indonesia, Bandung, Indonesia
[2]Program Studi Pendidikan Fisika, UIN Sunan Gunung Djati Bandung, Bandung, Indonesia

*adammalik@uinsgd.ac.id

Abstract. This study aimed at investigating pre-service teacher's critical thinking skills improvement through Higher Order Thinking (HOT) Lab on transformer learning. This research used mix method with the embedded experimental model. Research subjects are 60 students of Physics Education in UIN Sunan Gunung Djati Bandung. The results showed that based on the results of the analysis of practical reports and observation sheet shows students in the experimental group was better in carrying out the practicum and can solve the real problem while the control group was going on the opposite. The critical thinking skills of students applying the HOT Lab were higher than the verification lab. Critical thinking skills could increase due to HOT Lab based problems solving that can develop higher order thinking skills through laboratory activities. Therefore, it was concluded that the application of HOT Lab was more effective than verification lab on improving students' thinking skills on transformer topic learning. Finally, HOT Lab can be implemented in other subject learning and could be used to improve another higher order thinking skills.

1. Introduction

Laboratory practice is a set of activities in the laboratory including demonstrations, computer simulations complementing experiment works for hand on, offering students a rich learning experience, gaining a conceptual understanding of disciplines and developing practical skills [1, 2]. The demands of laboratory practice in schools to face the 21st century have been widely suggested by science educators since many benefits are derived from these laboratory activities [3, 4]. According to Carin [6] through laboratory practice, learners will acquire various skills such as manipulating materials, observing, grouping, measuring, using number skills, recording data, replicating, identifying variables, interpreting data, predicting, formulating hypotheses, inferring, generalizing, creating models, and making decisions.

Laboratory activities conducted in schools have not been managed effectively; the type of experiment practiced is too simple and less worthy. Students are only required to report the observations, but are rarely asked to analyze the interrelationship between the observed variables, test predictions, or selecting some possible explanations of the research results; the activities carried out are still structured, and less arouse students' curiosity [5, 7]. The existence of the needs of providing laboratory practice to improve critical thinking skills stimulates the development of Higher Order

Thinking (HOT) Lab. Steps in the HOT lab are the results of merging the creative problem solving and problem-solving lab. From the merging of these models, the HOT Lab design steps are then compiled, and they consist of five general processes: 1) understanding the challenges, 2) producing ideas, 3) preparing laboratory practice activities, 4) conducting laboratory practice, and 5) communicating & evaluating the results of activities [8]. Every activity in the HOT Lab promotes both convergent thinking skills (critical thinking) and divergent thinking skills (creative thinking).

The topic of transformers is significant to learn, considering the application of the concept of transformer is widely used in electronic devices in everyday life. A transformer is a device for introducing mutual coupling between two or more electric circuits. The term iron-core transformer identifies the coupled coils, which are wound on a magnetic core of laminated specialty steel to confine the flux and maximize the coupling. Air-core transformers are found in electronic and communications applications. A third group consists of coils wound over one another on a nonmetallic form, with a movable slug of magnetic material within the center for varying the coupling [9]. In its simplest form, the AC transformer consists of two coils of wire wound around a core of iron. Transformer consists of two types of step up transformer and step down transformer. The relationship between the voltage and the coil in the transformer is written in the equation (1).

$$V_2 = -\frac{N_2}{N_1}V_1 \qquad (1)$$

When $N_2 > N_1$, the output voltage V_2 exceeds the input voltage V_1. This setup is referred to as a step-up transformer. When $N_2 < N_1$, the output voltage is less than the input voltage, and we have a step-down transformer [10]. In an ideal transformer, where there are no losses, the power $I_1 V_1$ supplied by the source is equal to the power I_2V_2 in the secondary circuit, in equation 2 [10].

$$I_1V_1 = I_2V_2 \qquad (2)$$

The studies on improving critical thinking skills have used a variety of strategies and learning models such as problem-based approach [11], the application of writing strategy [12], collaborative approach work [13], and guided inquiry learning model [14]. However, studies on improving critical thinking skills through laboratory activities, conducted in colleges, are considered as innovation. This research used mix method with the embedded experimental model. The results showed that the critical thinking skills of students in the experimental group were higher than the control group. Finally, HOT Lab can be implemented in other topic learning.

2. Experimental Method

The research used was mix method with the embedded experimental model. This type is defined by having qualitative data embedded within an experimental design. The priority of this model is established by the quantitative, experimental methodology, and the qualitative data set is subservient within that methodology. This model can either be used as a one-phase or a two-phase approach, in which the timing reflects the purpose for including the qualitative data [15]

The population of this study was all students of Physics Education of UIN Sunan Gunung Djati Bandung at semester V academic year 2015/2016. The samples used in this study were 60 students consisting of 30 students (12 male and 18 female) of the experimental group and 30 students (9 male and 21 female) of group control. The sampling technique used was simple random sampling.

The experimental group, in practicum on the transformer used HOT Lab design, consisting of 11 phases of activities including real-world problems, determining and evaluating ideas, asking experimental questions, providing materials and equipment, predicting, applying question method; exploring, measuring, analyzing, concluding, presenting. While the control group used a design verification lab consisting of 9 stages of activities including giving purposes, delivering basic theory, providing tools and materials, giving a preliminary assignment, giving trial procedures measuring, analyzing, concluding, and end task.

The instruments used in this research were observation sheet, interview protocol and critical thinking skill test in the form of essay. Observation sheets are used to view the activity of all phases in practicum guide HOT Lab. Assessment of HOT Lab implementation used scale 1-3. The developed Critical Thinking Skills (CTS) indicator refers to the Binkley et al. framework, which includes aspects of CTS 1 Explain. State results, justify procedures, and present arguments; CTS 2 Analyze how parts of a whole interact with each other to produce overall outcomes in complex systems; CTS 3 Interpret information and draw conclusions based on the best analysis; CTS 4 Synthesize and make connections between information and arguments; CTS 5 Inference. Query evidence, conjecture alternatives, and conclude, and CTS 6 Evaluate. Assess claims and arguments [16].

The improvement of critical thinking skills determines with the gain score <g> from Hake [17]. The difference enhancement of critical thinking skills between the experimental and control group, the calculation of t-test score was carried out, previously examined for the normality test using One-Sample Kolmogorov Smirnov Test and the homogeneity test used Test of Homogeneity of Variances. Besides to determine the impact of the implementation of the HOT Lab design on improving students' critical thinking skills relative to the application of verification lab, the effect size calculation was performed. The effect size of this study was determined by computing the standardized standard difference (d) with Cohen et al. [18].

3. Result and Discussion

Based on the result of student practicum report and observation sheet in the experimental group show the students understand the concept of a transformer, able to solve real problem, not fixated on the mathematical equation applicable to the topic of a transformer. Also, understand the relationship between varieble, understand the function of each tool component, experimental tools in accordance with method questions, able to explore and analyze experimental data and present experimental results clearly and correctly. The opposite result occurs in the student practicum report and the observation sheet in the control group.

Based on the results of interviews with students in the experimental group, they feel the benefits of the application of HOT Lab, among others: to be more critical in understanding the problems, improving conceptual understanding, practicing creative thinking skills in practicum with the limitations of existing tools and materials, assisting in practicum. Besides, the student can identify relationships among variables, improving skills in using measuring tools, practicing analytical skills in comparing predictions and experimental results. The students' constraints both in the experimental and control groups show that they find it difficult to make the appropriate coil with the problem making it difficult to make the appropriate comparisons.

Transformers allow for easy changes in alternating voltage. Because energy (and therefore power) are conserved $I_1 V_1 = I_2 V_2$ to relate the currents and voltages in the primary and secondary windings of a transformer. Transformers are useful for transmitting power over long distances. Because the generator voltage is stepped up, the current in the transmission line is reduced, and hence I^2R losses are reduced. In practice, the voltage is stepped up to around 230 000 V at the generating station, stepped down to around 20 000 V at a distributing station, then to 4 000 V for delivery to residential areas, and finally to 120–240 V at the customer's site [10].

Based on the results of the research that, the improvement of students' critical thinking skills was shown by calculating the individual N-gain <g>, and then it was averaged and interpreted into the N-gain category. The data required to determine the amount of the improvement of the critical thinking skills were the pretest and posttest results data for both the experimental group and the control group. N-gain critical thinking skills for experimental and control groups can be seen in Table 1.

Table 1. N-gain scores of critical thinking skills of experimental and control groups

Description	Experimental Group		Control Group	
	Pretest	Posttest	Pretest	Posttest
Maximum score (%)	44.44	77.77	30.55	52.78
Minimum score (%)	30.55	91.67	19.44	36.11
Average score (%)	38.25	83.24	25.19	46.57
Standard deviation	1.43	1.25	1.05	1.28
Average Gain (%)	45.00		21.39	
N-gain (%)	72.94		28.63	

Table 1 shows that critical thinking skills scores for both groups were improving. The average improvement score (gain) of students' critical thinking skills of the experimental group was 45.00% while the improvement score (gain) of the control group was 21.39%. Based on the calculations, the average score <g> of students' critical thinking skills of the experimental group at 0.73 was categorized high. It was higher than the control group score 0.29 classified into low.

To strengthen the analysis, then the calculation of <g> on each aspect of critical thinking skills of both groups was performed. The calculation results were shown in Figure 1.

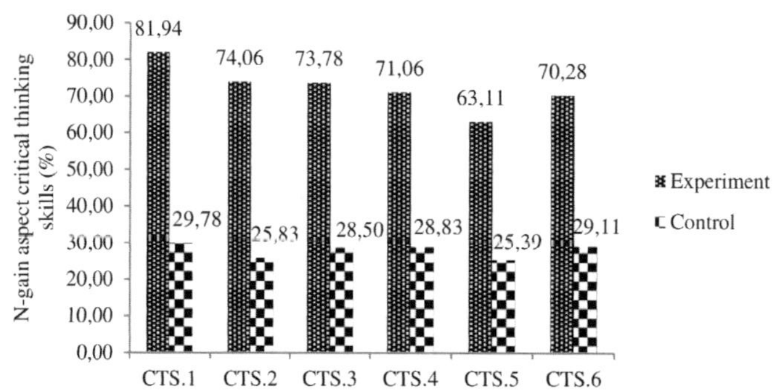

Respectively:

1. CTS 1. Explain. State results, justify procedures, and present arguments
2. CTS 2. Analyze how parts of a whole interact with each other to produce overall outcomes in complex systems
3. CTS 3. Interpret information and draw conclusions based on the best analysis
4. CTS 4. Synthesize and make connections between information and arguments
5. CTS 5. Inference. Query evidence, conjecture alternatives, and conclude
6. CTS 6. Evaluate. Assess claims and arguments

Figure 1. Mean N-gain scores of each aspect of critical thinking skills

Figure 1 shows that the critical thinking skills of students in all aspects in the experimental group ware higher than the control group. The five aspects of critical thinking skills (CTS 1, CTS 2, CTS 3, CTS 4 and CTS 6) in experiment group were categorized into the high category unless the CTS 5 aspect is included in the moderate category. All aspects of critical thinking skills in the control group fall into the low category.

The N-gain score of the aspect of CTS 1 was the biggest improvement in both the experimental group and control group. Students applying HOT Lab design were able to express and present the results, procedures, or arguments. These were developed at the stage of exploration and presentation in the HOT Lab design. At those stages, students designed and examined circuits or working system of the determined choice of answers to solve problems and present experimental results obtained in the form of verbal and non-verbal. Students applying the verification lab design were told what to measure, execute the established procedures, and not to present the obtained results of the experiments.

The aspect of CTS 2 was the smallest improvement in N-gain occurred in both the experimental and control groups. The increase aspect of CTS 2 students in the experimental group was higher than in the control group. The stages of analyzing activities in the HOT lab design provide a better opportunity for students to think on analyzing the results to examine the compliance with the predictions and analyst produced as the basis for concluding. An aspect of CTS 2 in the control group was less developed since the students analyzed the results obtained to be adjusted with the reference data and analyzed the factors causing the differences. It supports the results of previous studies showing that physics and engineering studies encountered difficulties in analyzing the results of the experiment [19, 20].

The aspect of CTS 3 in the experimental group was greater improved than the control group. Students applying HOT Lab design were able to categorize information, ideas, arguments and claims. It was developed in the prediction, exploration, and measurement stages in the HOT Lab design, therefore students were able to predict the relationship between the variables to be measured in the test and interpret the result data of experiment into tables and graphics. Meanwhile, students applying the verification lab design must follow the determined procedures. Students' interpreting skills should be developed, since analyzing and interpreting data are necessary for the practice of physics and science in general [21].

The N-gain score of aspect CTS 4 was improved in the high category in the experimental group and the low category in control group. The stages of questioning dealing with experiment and method in the HOT Lab design provided a better opportunity for students to connect the given information by delivering arguments. Students applying verification lab design merely decided details of the analysis and the emphases on the concept (quantitative). It is following the developed focus of competencies on the verification lab design explaining and supporting what students learn, teaching experimental techniques, and less training reasoning skills [22].

The N-gain score of students' critical thinking skills on the aspect of CTS 5 in the experimental group was improved greater than those in the control group. The improvement was categorized into the medium category in experimental group while in control group include the low category. Students applying HOT Lab were able to collect evidence to solve problems and to draw conclusions based on the analysis. It was developed in the measuring and concluding stages, the students measured the variables related to the problems to be solved and completed the experimental results based on the data analysis. Students applying the verification lab design measured established variables and summed up the results of experiments by the purpose of the predetermined analysis.

The N-gain scores of the aspect of CTS 6 in both experimental and control groups were improved. Students in the experimental group were higher increase than those in the control group. The stages of determining and evaluating ideas and conclusions in the HOT Lab design provided a better opportunity for students to evaluate claims and arguments, and to reevaluate the presented conflicting evidence. Students applying the verification lab design should follow the established procedures. The previous research that the implementation of the cookbook experiments rarely demonstrates the comprehension of what has been carried out [23].

The result of statistical calculation of students' critical thinking skills in experimental class using HOT Lab and control class using verification is shown in Table 2.

Table 2. Statistic calculation of students' critical thinking skills

Data type	Normality ($\alpha = 0.05$)				Homogeinity ($\alpha = 0.05$)	Uji t ($\alpha = 0.05$)	
	Level Significance (2-tailed)		Information			Level Significance (2-tailed)	Information
	Experimental	Control	Experimental	Control			
Critical thinking skills	0.953	0.209	Normal	Normal	0.419 Homogeneous	0.000	Significant

Based on Table 2 it can be concluded that there is a significant difference in the critical thinking skills of students who were applying the laboratory activities with HOT Lab compared to the verification.

The results of calculation show that the effect size of students' critical thinking was 7.85. According to Cohen [18], this acquisition was considered as big effects. Therefore, it was concluded that the application of HOT Lab design was effective in improving the critical thinking skills of students compared to the verification lab design.

The HOT Lab design has characteristics that enable students to improve their critical thinking skills. The HOT Lab design is the development of creative problem solving and problem-solving laboratory models, with the following characteristics: containing context-rich issues that must be solved through lab practice, applying physics concepts to solve problems, having constraints in problem-solving, demanding creative thinking in problem-solving. Various alternative options are not trivial to solve the problems, and the results of the problem solving are presented.

Lab practice offers students rich learning experiences, gains the conceptual understanding of disciplines and develops practical skills [1, 2]. The contribution of lab practice in instructional science should be able to foster abilities to develop conceptual thinking, generating imagination, stimulating desires and sharpening methodological acumen as a part of an experimental experience. Lab practice is important in learning science in schools, and it determines students' traits toward science in schools and science outside the group room [24].

Based on the results of the above research, it is clear that lab practice implementation using lab verification design does not develop many critical thinking skills of students because it only trains low-level thinking skills. The implementation of cookbook lab model does not benefit the students, particularly the debriefing of science skills, hands-on skills, and even minds-on skills [25].

4. Conclusion

The implementation of HOT Lab design is more effective in improving students' critical thinking skills compared to the implementation of verification lab design on the electrical circuit materials. Therefore the HOT Lab design implementation seems worth taking into consideration for use in teaching physics in other materials and learning physics in more formal education.

Acknowledgments

The researcher would like to thank the students who participated in this research. Special thanks are also to the chairperson of the Program Studi Pendidikan Fisika UIN Sunan Gunung Djati Bandung, Lab operator and laboratory assistant of the Program Studi Pendidikan Fisika UIN Sunan Gunung Djati Bandung who gave permission and helped conduct the research.

References

[1] Byers W 2002 Promoting active learning through small group laboratory classes *University Chemistry Education,* **6** p 28–34
[2] Kirschner P A and Meester M A M 1988 The laboratory in higher science education: problems, premises and objectives *Higher Education* **17** p 81–98
[3] Wenning C J 2006 A framework for teaching the nature of science *Journal of Physics Teacher Education Online.* **3** (3) p 3-10
[4] Hofstein A and Lunetta V N 2004 The laboratory in science education: Foundation for the 21st century *Science Education* **88** (1) p 28–54
[5] Carin A and Sund R B 1997 *Teaching Science Thourgh Discovery* 6th ed (Ohio: Meril Publishing Company)
[6] Lamanna J R and Eason P K 2011 Building creative scientists in the classroom laboratory: applications for animal behavior experiments *The American Biology Teacher* **73** (4)

[7] Malik A and Setiawan A 2016 The Development of Higher Order Thinking Laboratory to Improve Transferable Skills of Students *Procedings 2015 International Conference on Innovation in Engineering and Vocational Education* (Bandung), Vol 56 (Amsterdam: Atlantis Press) p 36–40

[8] Nahvi M and Edminister J A 2003 *Theory and Problems of Electric Circuits 4th ed* (New York:The McGraw-Hill Companies)

[9] Serwey R A and Jewett J W 2004 *Physics for Scientists and Engineers* 6th ed Thomson Brooks/Cole

[10] Trilling B and Fadel C 2009 *21st Century Skills: Learning for Life in Our Times* (San Francisco: John Wiley & Sons, Inc)

[11] Alismail H A and McGuire P 2015 21st Century Standards and Curriculum: Current Research and Practice *Journal of Education and Practice* **6** (6) p 150–155

[12] Jackson D 2014 Business graduate performance in oral communication skills and strategies for improvement', *The International Journal of Management Education*, **12** (1) p 22-34

[13] Boyles T 2012 21st Century Knowledge, Skills, And Abilitiesand Entrepreneurial Competencies: A Modelfor Undergraduate Entrepreneurship education *Journal of Entrepreneurship Education* **15** p 41-55

[14] Grifin P, Care E and McGaw B 2012 *The Changing Role of Education and Schools* (*Assessment and Teaching of 21st Century Skills*) eds Grifin P McGaw B and Care E (New York: Springer) chapter 1 p 1-15

[15] Darling-Hammond L 2012 *Policy Frameworks for New Assessments* (*Assessment and Teaching of 21st Century Skills*) eds Grifin P McGaw B and Care E (New York: Springer) chapter 6 p 301-339

[16] Cowden C D and Santiago M F 2016 Interdisciplinary explorations: promoting critical thinking via problem-based learning in an advanced biochemistry class. *Journal of Chemical Education*, **93** (3) p 464–469

[17] Stephenson N S and Sadler-Mcknight N P 2016 Developing critical thinking skills using the Science Writing Heuristic in the chemistry laboratory *Chem. Educ. Res. Pract* **17** 72–79

[18] Fung D C To H and Leung K 2016 The influence of collaborative group work on students' development of critical thinking: the teacher's role in facilitating group discussions. *Pedagogies: An International Journal*, **4818** (May) p 146–166

[19] Malik A 2010 Inquiry learning model using virtual laboratory and real laboratory to improve mastery of concept and critical thinking skill of high school student on dynamic electric topic. *Tesis* (Bandung: Universitas Pendidikan Indonesia)

[20] CreswellJ W 2008 *Educational Research. Planning, Conducting, and Evaluating Quantitative and Qualitative Research, Third Edition* (USA: Pearson Prentice Hall)

[21] Binkley M, Erstad O, Herman J, Raizen S, Ripley M, Miller-Ricci M and Rumble M 2012 *Defining Twenty-First Century Skills* (*Assessment and Teaching of 21st Century Skills*) eds Grifin P McGaw B and Care E (New York: Springer) chapter 2 p 17–66

[22] Hake R R 1999 Analyzing change/gain scores, American Educational Research Association, [online] di:http://www.physics.indiana.edu/~sdi/AnalyzingChange-Gain, pdf

[23] Cohen L, Manion L, Lecturer P, Morrison K and Lecturer S 2007 *Research Methods in Education* 6th ed (Oxon: Routledge)

[24] Eshach H and Kukliansky I 2007 Statistics in the physics laboratory. in: 12th Biennial Conference of the European Association of Research on Learning and Instruction (Budapest, Hungary)

[25] Kukliansky I 2009 The characterization of data analysis and understanding of related concepts in the physics laboratory among science and engineering students. *Unpublished doctoral dissertation*, Ben-Gurion University of the Negev

[26] Bowen G M, and Roth W M 2005 Data and graph interpretation practices among preservice science teachers *J Res Sci Teach* **42** (10) p 1063–1068
[27] Wenning C J 2011 Experimental inquiry in introductory physics courses *Journal of Physics Teacher Education* **6** (2).
[28] Von Aufschnaiter C.and Von Aufschnaiter S 2007 University students' activities, thinking and learning during laboratory work *European Journal of Physics* **28** (3) S51–S60
[29] Sharpe M R 2012 Secondary school student's attitudes to practical work in school science *Thesis PhD* University of York Education.
[30] Heuvelen A V 2000 The workplace, student minds, and physics learning systems *Journal American of Physics* **69** (11)

IOP Conf. Series: Journal of Physics: Conf. Series **895** (2017) 012141 doi:10.1088/1742-6596/895/1/012141

Assessment in Science Education

N Y Rustaman*

Departemen Pendidikan Biologi, Universitas Pendidikan Indonesia, Bandung, Indonesia

*nuryanirustaman@upi.edu

Abstract. An analyses study focusing on scientific reasoning literacy was conducted to strengthen the stressing on assessment in science by combining the important of the nature of science and assessment as references, higher order thinking and scientific skills in assessing science learning as well. Having background in developing science process skills test items, inquiry in its many form, scientific and STEM literacy, it is believed that inquiry based learning should first be implemented among science educators and science learners before STEM education can successfully be developed among science teachers, prospective teachers, and students at all levels. After studying thoroughly a number of science researchers through their works, a model of scientific reasoning was proposed, and also simple rubrics and some examples of the test items were introduced in this article. As it is only the beginning, further studies will still be needed in the future with the involvement of prospective science teachers who have interests in assessment, either on authentic assessment or in test items development. In balance usage of alternative assessment rubrics, as well as valid and reliable test items (standard) will be needed in accelerating STEM education in Indonesia.

1. Introduction

Entering year 2020, career of science graduates will be focused in competencies needed in the 21st century such as critical thinking and creativity as ways of knowing, communication and collaboration as ways of working [1], which is most familiar as Four C competencies. Stressing too much on science content mastery will be time and energy consuming. It is mostly expected to develop higher order thinking skills in the form of valid test items as well as combination of self- and peer-assessment on inquiry based instruction and meaningful student learning experience.

Inquiry is actually the nature of science itself. Inquiry based instruction has four to five forms: as a method, as an approach, as a learning model, as tools to develop personality, and as competency to be develop and assessed [2]. Meanwhile levels of inquiry had been introduced as discovery learning, interactive demonstration, inquiry lesson, inquiry laboratory, real world application, and hypothetical explanation [3].Each level of inquiry has pedagogical purpose and intellectual scientific skills. Therefore the "competencies" developed in inquiry based learning is very useful to be used as the scientific skills as what are mostly expected in 2013 curriculum in Indonesia, In spite of its scientific approach through all subject matter, scientific skill and reasoning skill is very important for science in junior high school as well as for science major (Biology, Chemistry, Physics) in senior high school.

There is closed relationship between science instruction and science assessment. The nature of assessment is needed to be developed in the nature of science as inquiry, active assessment for active science. Assessment can be differentiated into 'assessment of learning', 'assessment for learning', and 'assessment as learning' based on the goal, whereas based on its function it can be differentiated into formative assessment and summative assessment [4]. Formative assessment directs the learners to learn further, while summative assessment tends to direct the learner to compete each other.

Science process skills had been studied before, whether as learning or as assessment [5, 6, 7, 8, 9, 10]. Their studies showed that there are some characteristics in their science process skill assessment, especially in test items, among others are as follows. They used the concepts as context, measured each science process indicators specifically, and put some information before asking something in their stem of test items to be interpreted by the respondents. Therefore the characteristics of test items on science process skills are quite different from test items for measuring concept mastery, as they should be able to differentiate those who cannot answer the test because they do not have science process skills from those who do not understand the concepts to be mastered.

Inquiry have been defined in its many form [2]. Inquiry had also been differentiated based on its level of approaches [3]. There are many more definition of inquiry in science instruction, as it can be found in National Science Education Standard [11] and in Science Inquiry [12]. Unfortunately assessment for inquiry learning is still not well defined, and also the studies about inquiry assessment during the instruction. The leveling on science inquiry will be a good start for preparing assessment for scientific inquiry, because they are spectrum of intellectual sophistication from low to high and range of locus of control from teacher to student.

This article is trying to define how to assess inquiry, propose a model, and introduce simple rubrics as assessment for learning and some types of test items to assess the scientific reasoning skills (intellectual scientific skills) as assessment of learning. The results are still in progress. Actually the title of this article is proposed as "A concept model about scientific reasoning literacy: Case study on science learning assessment".

2. Method

a) Meta-analyses study from S2 Science students who have been teaching science in Junior high school for more than 6 years (2015, 2016, 2017) who try to implement performance assessment on science learning, especially in STEM based learning.
b) Observation and discussion with participants during biology teachers workshop in determining the cognitive process dimension and during biology learning in lab school (March-May 2017)who still have difficulties in planning & constructing instructional objectives as well as test items on cognitive process dimension, especially in determining from the simple one (C1) to more and most complex (C6). They still confused on the key words for each level of taxonomy (C1-C6).
c) Observing science teacher's activities and response during inquiry based science teaching and learning in workshop (May 3rd.- May 13th. 2017) who still have difficulties in responding scientific reasoning [13], and tend to have dual perspective scoring rubric. The rubric was prepared from the discussion among all workshop participants.
d) Studying documents from Wenning's works by his permission.

Table 1. Scientific practices and intellectual skills [13]

Inquiry Levels	Intellectual skills			#
Discovery Learning	**Rudimentary intellectual skills**	• Classifying; **Conceptualizing**; Concluding • Contextualizing; **Ordering**; Generalizing • Problematizing	inductive reasoning inductive reasoning inductive reasoning	**7 skills**
Interactive Demonstration	**Basic intellectual skills**	o Estimating; **Predicting,** Using conditional thinking o Explaining	deductive reasoning inductive reasoning	**4 skills**
Inquiry Lesson	**Intermediate intellectual skills**	• Applying information • Describing relationships; using combinatorial Reasoning; using correlational reasoning; • Making simple sense of quantitative data	deductive reasoning inductive reasoning inductive reasoning	**5 skills**
Inquiry Labs	**Integrated intellectual skills**	o Defining precisely a problem to be studied; o Interpreting quantifiable data to establish laws using logic o Defining precisely the system to be studied o Designing& conducting controlled scientific	inductive reasoning inductive reasoning deductive reasoning deductive reasoning	**4 skills**

Inquiry Levels	Intellectual skills			#
		investigations;		
Real World Application	**Culminating intellectual skills**	• Using causal reasoning to distinguish co-incidence from cause and effect	inductive reasoning	**6 skills**
		• Using causal reasoning to distinguish correlation from cause and effect	deductive reasoning	
		• Using data & math in the solution of real-world problems	deductive reasoning	
		• Summarizing for the purpose of logically justifying a conclusion on the basis of empirical evidence.	deductive reasoning	
		• Using proportional reasoning to make prediction	deductive reasoning	
		• Determining if an answer to a problem or question is reasonable including size and/or units	deductive reasoning	
Hypothetical Explanation	**Advanced intellectual skills***	o Generating predictions through the process of deduction; **Thinking deliberately**	deductive reasoning	**6 skills**
		o Generating & evaluating analogies	inductive reasoning	
		o Thinking analogically; **Thinking reflectively**	inductive reasoning	
		o Thinking to assimilate concepts	inductive reasoning	

*not yet developed [14]

Data analysis were carried out qualitatively through triangulation and assertion of questions. Some results of observation, documentation and framework on science researcher were used as tools in this study. These analysis resulted in a model on scientific reasoning plus rubrics for inquiry based teacher competencies and student's characteristic in science learning through inquiry as a draft.

3. Result and Discussion

a) The result of meta-analyses study shows that high school student still have difficulties in scientific processes conducted by S2 science students especially in reasoning or intellectual ability in the last three years (2014, 2015, 2016). Nevertheless when they learned science through well planned STEM based science instruction, they did improve their scientific processes as well as their engineering design practice [14, 15]. In STEM education there are three partite as integration (cross cutting concept, core ideas of discipline, and engineering design practice. Students learn through inquiry based learning and integrated with project based learning. When the process was planned systematically through D-D-O (define-develop-optimize) in the engineering design practice in STEM education [16] the students will be able to improve their scientific ability as well as their concept mastery according to Cognitive Bloom's taxonomy [17]. Therefore to improve student ability in scientific process-intellectual (=reasoning) the teacher can start from science or scientific literacy through inquiry based learning, and then gradually develop their ability or literacy through gradual type inquiry based learning. Later the students can improve their technology and engineering literacy.

Table 2. Scientific Practices and Cross Cutting Concepts of the NGSS Framework [18]

Scientific Practice	Engineering &Technology	Cross-cutting Concept
Asking questions	Defining problems	Patterns
Developing models	Using models	Scale, proportion, quantity
Planning investigation	Carrying out investigation	Systems and Systems Models
Using mathematics	Using computational thinking	Stability and Change
Constructing explanation	Designing solution	Energy and Matter
Engaging argument from evidence	Engaging argument from evidence	Cause and Effect
		Structure and Function
Obtaining & com-municating information	Evaluating and communicating information	Connection to engineering, techno-logy & application of science

The cross-cutting concepts enable teachers and students to focus the science teaching learning process toward the important things of the connection interdisciplinary concept. Therefore stressing on concept mastery will not be needed any longer, instead its application which is called as literacy (applying concepts meaningfully).

b) Based on observation resulting from science teacher's activities and their response during inquiry based science learning in workshop held from May 3rd. toMay 13th. 2017, it was found that science teachers and science trainers ("Widyaiswara") still are not used to scientific reasoning type test items [19]. There was a tendency that they even have lack experience in preparing such type of questions in constructing scientific processes and higher order thinking skills. Type of questions used by science teachers will determine the quality of test items and imaginative context whether lower order or higher order thinking skills [20].

c) Based on the interactive discussion results during the workshop on inquiry based science learning held on May 3rd to May 13th 2017, it was found that the participants could think merely from teacher's perspective, not from student perspective. So to direct them first they have to think from their own perspectives, then coincident they think from student's perspective. Later based on this result, task and rubrics were developed and then were used to monitor the improvement of science teachers and their students in inquiry based learning. This instrument was prepared as dual perspective scoring rubric. The rubric was prepared through discussion, and judged through monitoring and evaluation process during trial-out implementation by model teachers at partner schools.

d) Based on the thorough study on Wenning's documentation about his works and his students work by his permission (May 6th.,2017), it was found that some efforts had been done by other scientists or science educatorsabout assessment on test item construction. Most of their works related to certain framework. Related to scientific processes and reasoning ability, some frameworks are used as references in developing sets of test items, such as scientific literacy in PISA [21], Nature of Science Literacy Test [22],Scientific Inquiry Literacy Test[23]; cognitive process according to Cognitive Bloom's Taxonomy [17], and Classroom Test of Formal Reasoning [24]-[25]. Scientific Practices and Intellectual Skills was created as framework for Science Reasoning Literacy Test (SciReLiT) based on Levels of Inquiry Model of Science Teaching by Carl J. Wenning [13]. Some example of test items were studied to get the main idea of framework dealing with how to assess scientific or intellectual skill as the results of inquiry based science learning.

Refer to the 21st. century skills, and considering the whole of the study resulted in from the triangular analysis, at last it is decided to introduce "A concept model about scientific reasoning literacy: Case study on science learning assessment". The model proposed has characteristics as follows. Firstly, it assesses science process skills focusing on intellectual or reasoning aspects in the form of paper and pencil test set, and in the form of performance assessment (task and rubrics) of est items instruments and using alternative assessment in balance at the same time. All these show that the model uses the nature of science, the nature of science literacy, and the nature of assessment.

By definition, 21st. century learning tasks can be open-ended, involve unbounded sets of information, and may involve on-going redefinitions of the goal of task. It is important that students develop skills to establish and adapt goals according to available information, seek out relevant and valid information for the task, and continually monitor their own progress. The teacher's role is to set highly motivating tasks with achievable goals and to provide sufficient structure and scaffolding based on a thorough understanding of the students' interests and needs. The students also sets goals and targets for their own learning, and move forward with a clear understanding of the usefulness and application of the new skills and understanding they are developing [26].

Another approach that fits well with the teaching of 21st. century skills is tailored and differentiated instruction. It was stated that teachers in differentiated classroom draw upon strategies such as a small-group instruction, materials presented at a variety of reading levels, personalized rubrics, learning contracts, a variety of product and task options with common learning goals, and independent studies. Small-group instruction may be particularly helpful in targeting learning tasks, allowing students to shape their own learning goals and to seek out and select materials and information of relevance to the task. The task of teacher is to provide the most effective structure through the establishment of smaller groups based on similar abilities or to provide opportunities for peer mentoring.

A developmental approach to assessment and learning is directed to the aim of the approach to using assessment for teaching 21st century skills is to move a student's learning forward along a path or progression of increasingly complex knowledge and capabilities [26]. The focus is on recognition of a student's readiness to learn and the process of building upon the current stage of learning. By contrast, a deficit approach to assessment and teaching focuses on discovering, the things that a student cannot do, and teaching is then designed to address those deficits.

Variation for students at different stages on cognitive or social skill progression can be differentiated into four stages id est. stage A to D [26]. In Levels of Inquiry, instead of simultaneous development of scientific reasoning, it was propose to plan learning sequences. Based on the learning sequences, certain skills which are still missing can be directed and then be assessed [3]. So both approaches enable the students to improve their learning by the help of more knowledgeable Others (MKO), and Zone of proximal development (ZPD) through scaffolding [27]. The concept of the zone of proximal development (ZPD), which can be thought of as an ideal space in which people can learn most effectively. People learn effectively within their ZPD because they have enough prior knowledge to scaffold their learning of more complex skills or information, but not so much as to lead disengagement because they are being taught concepts or material that are too simple.

4. Conclusion

A combination of paper and pencil test focusing on multiple choice test item with reason, and performance or alternative assessment is proposed as a model of scientific reasoning/intellectual literacy as science learning outcome of high school students. The alterative assessment is in the form of task and rubrics that reflects 21$^{st.}$ century skills which consider as part of four competencies (4Cs). The test on scientific reasoning skills literacy and rubrics for assessing the essence of each level of inquiry can be developed gradually with the involvement of graduate students through their investigation in their thesis writing. In the perspective of the nature of science, the nature of science literacy and the nature of assessment, it is focused to empower the assessment for learning and as learning to complete the assessment of learning.

Acknowledgments

The writer would like to extend their deepest gratitude to Carl J Wenning from Illinois State University for the permission to overview Hanson's thesis for the workshop in general and especially for this study with the support of World Bank and PPPPTK IPA.

References

[1] Binkley M, Erstad O., Herman J., & Raizen S., Ripley, M., Miller-Ricci, M., & Rumble, M. 2012. "Defining twenty-first century skills". In Griffin, P., McGaw, B., & Care, E. (Eds.) *Assessment and Teaching of 21st Century Skills*. New York: Springer Science-Business Media B.V.DOI 10.1007/978-94-007-2324-5.

[2] Rustaman, N.Y. 2007. "Basic Scientific Inquiry in Science Education and Its Assessment". Paper presented as Keynote speech in *First International Seminar in Science Education*. Held by Science Education Study Program, Graduate School, Indonesia University of Education.

[3] Wenning C J 2010 Levels of Inquiry: Using inquiry spectrum learning sequences to teach science *J. of Physics Teacher Education Online* **5** (3) 11-20

[4] Popham W J 2011 *Classroom Assessment: What Teachers Need to Know* Fourth edition (Boston: Allyn and Bacon)

[5] Bell J, Black P, Johnson S, Murphy P, Quatter A and Russell T 1989 *Assessment Performance Unit: Science at age 15* (London: Her Majesty's Stationery Office)

[6] Schilling M, Hargneaves L, Harlen W, and Russell T 1990 *Written Tasks* (London: Paul Chapman Publishing Ltd.)

[7] Harlen W, Macro C., SchilingM, Malvern D, Reed K 1990 *Progress in Primary Science: Workshop Materials for Teacher Education* (London: Routledge)
[8] Russell T, and Harlen W 1990 *Assessing Science in the Primary Classroom: Practical Tasks* (London: Paul Chapman Publishing Ltd.)
[9] Rustaman N Y, Firman H. and Rustaman A. 1992. *Pengembangan dan Validasi Butir Soal Keterampilan Proses Sains sebagai Persiapan Wajib Belajar Pendidikan Dasar Sembilan Tahun.* Laporan Penelitian yang didanai DP2M Dirjen Dikti.
[10] Rezba R J, Sprague C and Fiel R L 2006 *Learning and Assessing Science Process Skills* (Dubuque, Iowa 52002: Kendall/Hunt Publishing Company)
[11] The National Research Council 1996 "Assessment in Science Education" In *National Science Education Standard* (Washington D.C.: National Academy Press) pp. 75-101
[12] National Research Council. 2000. Inquiry.
[13] Hanson S T, and Wenning C J. (2017). Framework for Science Reasoning Literacy Test (SciReLiT) version 8, February 9.
[14] Afianti E, Rustaman, N and Suwarma, IR 2017 Performance Assessment Implementation in STEM-Based Learning to Investigate Students' Creativity on the Cell Topic. Article draft
[15] Septiani A, and Rustaman N Y, 2016. Implementation of performance assessment in STEM education to detect science process skills. [online] 2016 [cited 2017 May 2] Available from: URL: http://publikasiilmiah.ums.ac.id
[16] Bybee R W 2013 *The case for STEM education: Challenges and opportunity* (Arlington, VI: National Science Teachers Association (NSTA) Press)
[17] Anderson LW and Krathwohl D R Eds. 2001 *A Taxonomy for Learning, Teaching, and Assessing: A Revision of Bloom Taxonomy of Educational Objectives* (New York: Addison-Wiley Longman)
[18] National Research Council 2013 *Next Generation Science Standards Framework* (Washington DC: National Academies Press)
[19] Hanson S T. 2016. *The Assessment of Scientific Reasoning Skills of High School Students: A Standardized Assessment Instrument.* Thesis, Submitted in Partial Fulfillment of the Requirements for the Degree of Master of Science, School of Teaching and Learning, Illinois State University
[20] Lazear D 2004 *Higher-Order Thinking: The Multiple Intelligences Way* (Chicago: Zephyr)
[21] OECD 2007 *PISA 2006 Science Competencies for Tomorrow World: Volume I Analysis* (OECD Publishing)
[22] Wenning C J 2006 Assessing nature-of-science literacy as one component of scientific literacy *J. of Physics Teacher education Online* **3** (4) 3-14
[23] Wenning C J 2007 Assessing inquiry skills as a component of scientific literacy *J. of Physics Teacher Education Online* **4** (2) 21-24
[24] Lawson A E, Adi H and Karplus R 1979 Development of correlational reasoning in secondary schools: do biology courses make a difference? *The American Biology Teacher* **41** 420-425
[25] Lawson A E 2005 What Is the Role of Induction and Deduction in Reasoning and Scientific Inquiry? *Journal of Research in Science Teaching* **42** (6)716-740
[26] Woods K., Mountain R., and Griffin P. 2015. Linking developmental progressions to teaching. In Griffin, P., McGaw, B., & Care E. (Eds.). *Assessment and Teaching of 21st Century Skills: Methods and Approach. Educational Assessment in an Information Age: Series Editor* (New York: Springer Science-Business Media Dordrecht)
[27] Vygotsky L S 1978 *Mind and society: The development of higher mental processes* (Cambridge, MA: Harvard University Press)

IOP Conf. Series: Journal of Physics: Conf. Series **895** (2017) 012142 doi:10.1088/1742-6596/895/1/012142

VLP Simulation: An Interactive Simple Virtual Model to Encourage Geoscience Skill about Volcano

E Hariyono[1,2], Liliasari[1*], B Tjasyono[3] and D Rosdiana[4]

[1] Program Studi Pendidikan IPA, Sekolah Pascasarjana Universitas Pendidikan Indonesia, Jl. Dr. Setiabudi No. 229 Bandung 40154, Indonesia
[2]Jurusan Fisika FMIPA Universitas Negeri Surabaya, Kampus Ketintang Surabaya, Indonesia
[3] Program Sains Kebumian, Institut Teknologi Bandung, Labtek IX, Jl. Ganesa 10, Bandung 40132, Indonesia
[4] Departemen Fisika, Universitas Pendidikan Indonesia, Jl. Dr. Setiabudi No. 229, Bandung 40154, Indonesia

*liliasari@upi.edu

Abstract. The purpose of this study was to describe physics students predicting skills after following the geoscience learning using VLP (Volcano Learning Project) simulation. This research was conducted to 24 physics students at one of the state university in East Java-Indonesia. The method used is the descriptive analysis based on students' answers related to predicting skills about volcanic activity. The results showed that the learning by using VLP simulation was very potential to develop physics students predicting skills. Students were able to explain logically about volcanic activity and they have been able to predict the potential eruption that will occur based on the real data visualization. It can be concluded that the VLP simulation is very suitable for physics student requirements in developing geosciences skill and recommended as an alternative media to educate the society in an understanding of volcanic phenomena.

1. Introduction

Volcanoes phenomenon was an interesting issue in geoscience learning [1, 2], however, the lack of learning interactive media has not been able to facilitate the optimal activity in the classroom. Therefore, it was necessary designing a simulation media that was able to accommodate the students' requirement in learning about geosciences concept interactively as an effort to encourage geoscience skills [3]. Many volcanic simulation programs have been developed by geoscience educators and researcher at the international level, including Computer Aided Instruction or CAI [3], Interactive Virtual Earth Science Teaching or InVest [4], V*hub* [5], MIAVITA [6]. These programs proved able to support the community to understand the phenomenon of the volcano but it must be modified and adapted according to the physics students need in order to improve geoscience knowledge and predicting skills based on volcanic activity in Indonesia.

VLP (Volcano Learning Project) is a simple virtual simulation model that has developed to accommodate futuristic geoscience learning. This software provides an opportunity for students to take advantage of technology facilities [5], interaction with a wider and access more diverse range of authentic information sources based on the real data of volcanic activity in Indonesia which less underutilized as a learning resource. This model becomes a virtual media to provide knowledge about geoscience and can be used to exercise the geosciences skills to equip students to be more capable to solve the problems based on the real of geological data [7], to describe basic geosciences phenomena in the various contexts [8], to get better understanding the science behind phenomena [9], and help students to understand volcanic activity that cannot be directly observed [10]. Become a hope, the software capable of supporting the physics student has a high-level order thinking [11].

The fundamental problem to be answered through this research is how the volcano simulation in optimizing geoscience learning. The research focuses are describing characteristics of the volcano simulation software that was developed, improving predicting skill for physics student, and describing student's perception about VLP in the geoscience learning context. Collecting data from the student learning result and questioner about student perception and conducting analysis through qualitative and quantitative.

2. Experimental Method

This research is an implementation of the development of virtual simulation software about volcano. That was implemented in the Department of Physics in one of the state university in East Java. The design used is quasi experiment with one sample pre-test post-test design. This research was conducted to 24 physics student in the class of 2014 in February 2017 who was taking the course of Earth and Space Sciences in academic years of 2016/2017. The measured ability is the students' predicting skill based on VLP simulation and student perception analysis to software that was developed for getting to know its suitability to the requirement of student in geoscience learning.

3. Result and Discussion

3.1. VLP simulation

Figure 1 shows as a design of simple simulation software was developed by researchers. This software were inspired by the "Eruption software! A volcanic crisis simulation" developed by Carl E. Renshaw that was accessed from www.dorthmouth.edu. The software is about the volcanic disaster crisis and is designed interactively to develop decision-making skills [3]. As a novelty of the VLP simulation is the eruption parameters equipped with the more of volcanic earthquake types (low frequency, hybrid, deep earthquakes, shallow earthquakes, long distance tectonic and local tectonic) in accordance with the data obtained from PVMBG and coupled with environmental temperatures. These parameters are not founded from the developed previous software. In addition, the software is simpler and very easy to be operated by all user levels. It is very likely to help in educating the community in the understanding of volcano activity.

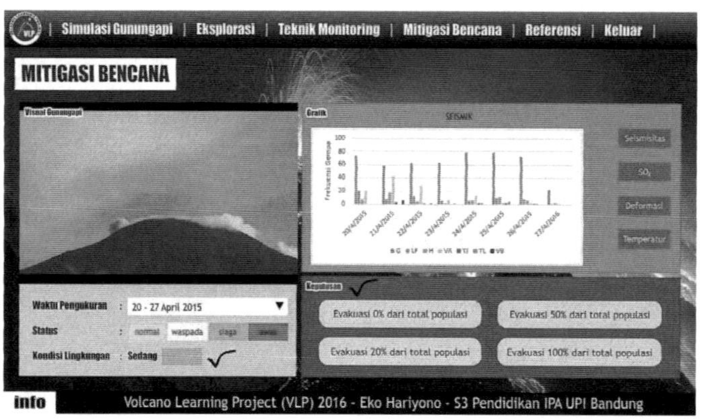

Figure 1. Design of VLP interface with *Mitigasi Bencana* (Disaster Mitigation) feature

The data used by researchers in software development is the real data of volcanic activity. That is accessed from the Centre of Volcanology and Geological Hazard Mitigation or *Pusat Volkanologi dan Mitigasi Bencana Geologi* (PVMBG). The data used include volcano-seismic data, SO_2 flux when the volcano is active, volcano deformation, and changes in environmental temperature due to volcanic activity. This context as adapted from the New Zealand's geoscience emergency and management [12]. Students are tasked to analyse eruption activity based on visualization and various eruption parameter data.

Table 1. The results of student analysis of volcanic activity based on physics parameters

Activity status	Activity indicators based on physical parameters
Normal	Dominated by low frequency earthquakes with an average frequency of 15 Hz, the average flux rate of 120 -160 tons/day, unstable deformation (up and down) and environmental temperature average is 22°C.
Watchful	Dominated by a long distance tectonic with an average frequency of 65 Hz, the number of SO_2 fluxes increases compared to normal status (230-260) tons/day, the deformation is stable and tends to rise and unstable temperatures in the ranges of 20-26°C and tend to rise.
Standby	Dominated by deep earthquakes with frequency> 70 Hz, the number of SO_2 flux is increased up to 350 tons/day and temperature tends to rise 23-28°C.
Watch out	Dominated by shallow earthquakes with frequencies between 70-100 Hz, the number of SO_2 fluxes reaches 420 tons/day, soil deformation tends to rise, and the average environmental temperature reaches on 25-30°C.

Table 1 shows a description of volcanic activity identified by visualization and various eruption parameter data. The results of this analysis were used as a basis for predicting volcanic activity based on the cases and presented in the form of group activities.

3.2. Students description of volcanic activity based on VLP

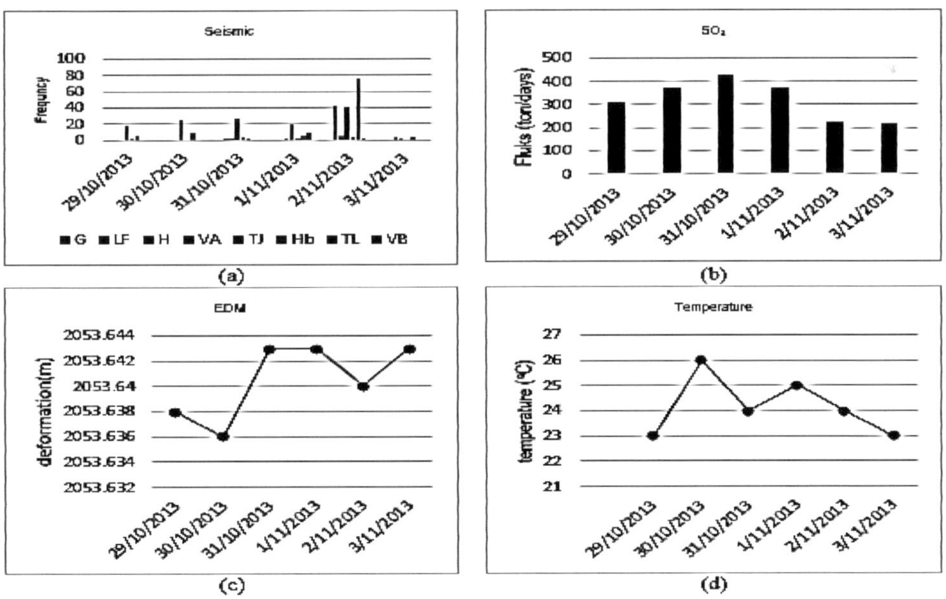

Figure 2. One example of representation of the volcanic activity based on eruption parameter from the VLP (a) seismic (types and frequency), (b) SO_2 flux released (ton/days), (c) deformation (m), and (d) Environmental temperatures (°C)

Figure 2 shows the volcanic activity based on eruption parameters, such as the seismic, flux, deformation, and environmental temperature. The real data accessed from the recording of volcanic activity at 29 October to 31 November 2013. All data show up in graphics. Using interpretation in graphical form can help students to make easy to understand numerical data through images [13].

Based on student's analysis of volcanic activity in VLP simulation, students understanding about eruption parameter is good enough. Students can explain the types and characteristics of volcano earthquakes and find the dominant type of earthquake as a marker of eruption based on their frequency. In addition, the students also managed to identify the amount of SO_2 flux released into the atmosphere at the time of volcanoes will be an eruption. The deformation changes and environmental temperature fluctuation are well explained.

Table 2. The ability of each group in analysing volcanic activity data based on VLP software

Groups	Volcanic status	Logical reasoning
I	Watchful	Dominated by type A earthquake, SO_2 flux decreased but still more than 160 tons/day, deformation experienced inflation, and environmental temperature between 23-26°C.
II	Standby	The amount of SO_2 flux between 250-400 tons/day, the environment temperature between 20-27°C.
III	Watchful	Dominated by deep earthquakes, the number of SO_2 fluxes, deformation increases, and temperature decreases.
IV	Watchful	Seismicity is dominated by an earthquake in (VB), the amount of SO_2 flux, deformation and environment temperature is decreases.
V	Watch out	Increased SO_2 flux (>350 tons), deformation tends to increase, shallow earthquakes (VB) is high.
VI	Standby	Volcanic activity is dominated by shallow earthquakes, the amount of SO_2 flux between 250-400 tons/day and environment temperatures between 20-27°C.

Based on table 2, students can determine the status of volcanoes and present of them about logical reasoning although the volcano status was determined differently. This skill was strongly influenced by the ability to analyse data and the ability to make conclusions according to the benchmarks made in the previous phase. Based on these results, groups II and group VI can determine the status of volcano correctly, but for other groups is incorrect.

3.3. Assessing students predicting skills and several findings

After following the learning activities using VLP, students perform the final test using predicting skill test was developed by the researcher.

Table 3. Description of physics students predicting skill by using paired sample test

	Paired Differences					t	df	Sig. (2-tailed)
				95 % Confedence Interval of the Difference				
	Mean	Std. Deviation	Std. Error Mean	Lower	Upper			
Pair 1 Pretest-Posttest	-3.229E1	21.39988	4.36823	-41.32804	-23.25529	-7.392	23	.000

Based on the results of the statistical analysis using paired sample test (table 3), we can be seen that the learning by using VLP simulation significantly influence the predicting skills of physics students. This can be seen from the significance value (2-tailed) of 0.000 <0.050. It can be concluded that learning by using VLP simulation media can improve predicting skill of physics students.

Nevertheless, there are several findings related to the ability of students to develop predicting skills consists of the ability to processing information, presenting in the graphs, the ability to make conclusions based on graphics, and predicting skill based on data.

Table 4. Profile of student ability in developing predicting skills.

Aspects	Average ability (%)
Ability to processing information	89.16
The ability to present data in graphical form	59.72
Ability to formulate conclusions	66.67
Predicting skill based on data	66.67

Table 4 illustrates the students' ability to develop predicting skills. The gain of student's geoscience concept through learning with VLP interactive media can help students in predicting volcanic activity. The ability of students in data processing is the best aspect (89.16%) and the presenting data in the graphical form is the lowest ability (59.72%), while the ability to formulate a conclusion and predicting skill based on data is good enough (66.67 %). Based on the data it can be seen that understanding the graphic is easier than making a graphical representation. Drawing a graph as a visual representation requires an understanding of the content, context, and construction [14]. The success of drawing graphs is closely related to the science process skill [15], especially in making the conclusions and prediction.

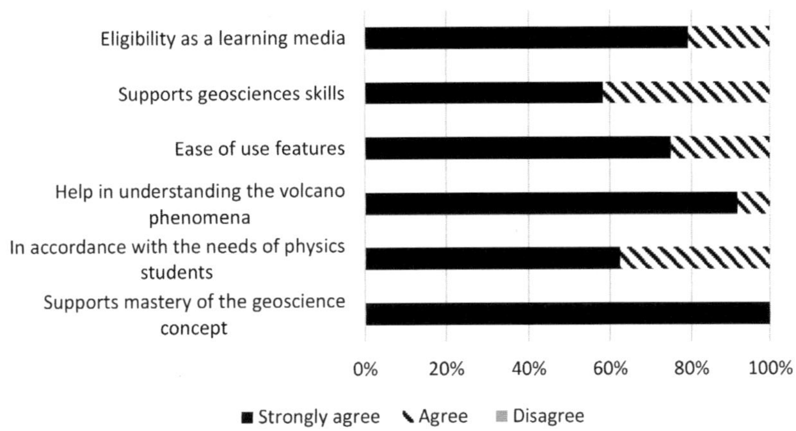

Figure 3. Student's perception about VLP software

To ensure that the VLP software is in line with the student's needs, conducted an analysis of the questionnaire results is distributed to the students. The results of the analysis on student perceptions about VLP software can be seen in figure 3. Most of the physics students stated that the VLP software is a great fit applied in geoscience learning. The software is very helpful in understanding the geoscience concept and volcanic phenomenon, more than 50% of students strongly agree if the software is able to support geosciences skills (predicting skills), and almost 80% of students said the VLP is very appropriate as a media in geoscience learning. Based on these results can be said that the VLP as a computer-based simulation is feasible to be used as a media to improve student's skills in geoscience learning.

Some suggestions are regarding the refinement of VLP software is the amount of observation data needs to be added so that the more options can be analysed by users. In order to be widely used by the public, it is strongly recommended that the software is applied to the smartphones, making it more accessible in helping people to understand about volcano.

4. Conclusion

VLP interactive simulation is an effective media in developing geoscience skills for physics students. Through this software, students are easier to understanding volcanic phenomena. This conclusion is supported by the results of the statistical analysis. The VLP simulation was significant to influence the physics students predicting skills. This can be seen from the significance value (2-tailed) of 0.000<0.050. Furthermore, students can explain the types and characteristics of volcano seismic and find the dominant seismic type as a marker of an eruption. In addition, the students also managed to identify the amount of SO_2 flux released into the atmosphere at the time of volcanoes will erupt, deformation changes and environmental temperature can well be explained. As a note, the student's skills are still low in making a graphical representation based on data. Students should have a good graphical representation to get the perfect of predicting skill. Based on the results of the analysis of student perceptions can be seen that the VLP software is a suitable simulation for physics students in studying the volcanoes phenomenon and recommended as a media to educate the community about volcanoes.

Acknowledgments
Our thanks to Mr. Hendra Gunawan from the office of the Pusat Vulkanologi dan Mitigasi Bencana Geologi Bandung-Indonesia which has provided guidance in the development of VLP simulation media and provides important data about volcanic activity in Indonesia. Also, all academic staff at FMIPA Universitas Negeri Surabaya who has facilitated this research.

References
[1] Harpp K S and Sweeney W J 2002 *Journal of Geoscience Education.* **50** 410
[2] Nunn J A and Braud J 2013 *Journal of Geoscience Education.* **61** 28
[3] Barclay E J, Renshaw C E, Taylor H A and Bilgie A R 2011 *Journal of Geoscience Education* **59** 85
[4] Parham Jr T L, Cervato C, Gallus Jr W A, Larsen M, Hobbs J, Stelling P, Greenbowe T, Gupta T, Knox J A, and Gill T E 2010 *Journal of Geoscience Education.* **58** 177
[5] Palma J L, Courtland L, Charbonnier S, Tortini R and Valentine G A 2014 *Journal of Applied Volcanology.* **3**:2
[6] Texier-Teixeira P, Chouraqui F, Perrillat Coulomb A, Lavigne F, Cadag J R and Grancher D 2014 *Nat. Hazards Earth Syst. Sci.* **14** 2374
[7] Wysession M E 2014 *Earth Future.* **2** 299
[8] Fermeli G, Steininger F, Dermitzakis M, Melendez G and Page K 2014 *Geophysical Research Abstracts.* **16** EGU2014-8678.
[9] Luo W, Pelletier J, Duffin K, Ormand C, Hung W, Shernoff D J, Zhai X, Iverson E, Whalley K, Gallaher C and Furness W 2016 *Journal of Geoscience Education.* **64** 60
[10] Harpp K S, Koleszar A M and Geist D J 2005 *Journal of Geoscience Education.* **53** 173
[11] Chang C Y 2010 *Per Sci Educ.* **40** 103
[12] Dohaney J, Brogt E, Kennedy B, Wilson T M and Lindsay J M 2015 *Journal of Applied Volcanology.* **4**:12
[13] Guidezzoli A, Diamanti T, Ponti F D, Neri A, Gori R, Imboden S, Bisson M, Pareschi M T, Cavazzoni C, Ongaro T E, Calori L, Erbacci G and Menconi G 2006. *An interactive virtual environment to communicate Vesuvius eruptions numerical simulations and Pompeii history.* Proceeding Siggraph'06 ACM SIGGRAPH 2006 Educator's program, Article No. 55. Boston, Massachusetts- July 30-August 03, 2006.
[14] Francis K, Jacobsen M and Friesen S 2014 *Journal of Information Technology Education Research.* **13** 233
[15] Evagorou M, Erduran S and Mäntylä T 2015 *International Journal of STEM Education.* **2**:11

IOP Conf. Series: Journal of Physics: Conf. Series **895** (2017) 012143 doi:10.1088/1742-6596/895/1/012143

Using Performance Assessment Model in Physics Laboratory to Increase Students' Critical Thinking Disposition

E Emiliannur[1*], I Hamidah[2], A Zainul[3] and A R Wulan[4]

[1]Sekolah Pascasarjana, Universitas Pendidikan Indonesia, Bandung, Indonesia
[2]Departemen Pendidikan Teknik Mesin, Universitas Pendidikan Indonesia, Bandung, Indonesia
[3]Profesor Pendidikan Ilmu Pengetahuan Sosial, Universitas Pendidikan Indonesia, Bandung, Indonesia
[4]Departemen Pendidikan Biologi, Universitas Pendidikan Indonesia, Bandung, Indonesia

*emiliannur_sy@student.upi.edu

Abstract: Performance Assessment Model (PAM) has been developed to represent the physics concepts which able to be devided into five experiments: 1) acceleration due to gravity; 2) Hooke's law; 3) simple harmonic motion; 4) work-energy concepts; and 5) the law of momentum conservation. The aim of this study was to determine the contribution of PAM in physics laboratory to increase students' Critical Thinking Disposition (CTD) at senior high school. Subject of the study were 11th grade consist 32 students of a senior high school in Lubuk Sikaping, West Sumatera. The research used one group pretest-postest design. Data was collected through essay test and questionnaire about CTD. Data was analyzed using quantitative way with N-gain value. This study concluded that performance assessmet model effectively increases the N-gain at medium category. It means students' critical thinking disposition significant increase after implementation of performance assessment model in physics laboratory.

1. Introduction

While a great deal of research has examined students' critical thinking skills, less is known about students' tendencies to use these skills [1]. Critical thinking has always been a central goal of education, but having critical thinking skills does not necessary meant that the person will use these skills even when the situation requires the application of such skills [2]. Critical thinking disposition is also believed to be essential for students, primarily to ensure they use critical thinking skills in the classroom and later when they enter the workforce, because people must also be disposed to use what they have learned [3, 4]. Some researchers extend the definition of critical thinking include both abilities and dispositions [5, 6]. Critical thinking dispositions define as the tendencies to think critically when faced with problems to solve, ideas to evaluate, or decisions to make [7, 8].

Educators need to measure critical thinking dispositions, it will help to decide on the appropriate intervention to implement. There are instruments to measure critical thinking dispositions such as: 1) California Critical Thinking Disposition Inventory/CCTDI by Facione; 2) Critical Thinking Disposition

Questionnaire/CTDQ by Ricketts; and 3) Critical Thinking Disposition Scale/CTDS by Akbiyik. On the other hand, pre-service physics teachers' critical thinking disposition can improved through virtual lab modern physics model [9]. Based on the results of analysis journals, there is no instrument used to measure the critical thinking disposition in physics subject at the senior high school level. Field study shows the assessment used in senior high school is more oriented towards the end result, not the process. The facts reflected in local and national assessments using traditional assessment tools such as paper and pencil test that have not been able to measure students' higher order thinking. Assessment expert Lorrie Shepard and others have found that, when educators teach directly to the content and format of specific high-stakes tests, students are frequently unable to transfer their knowledge to items that test it in different ways [10]. Furthermore, students' ability to answer multiple-choice questions does not mean they have the ability to answer the same questions in open-ended form. Indeed, their scores often drop precipitously when answers are not provided for them, and they do not have the option to guess. Thus, a focus on multiple-choice testing gives false assurances about what students know and are able to do [11]. This is why a growing number of educators and policymakers have argued that new assessments are needed.

The increase in popularity of performance assessments during the late 1980s and 1990s came about in part because of dissatisfaction with traditional, multiple-choice tests [12]. In a performance assessment, rather than choosing among pre-determined options, students must construct an answer, produce a product, or perform an activity. Performance assessment encompasses a very wide range of activities, from completing a sentence with a few words (short-answer), to writing a thorough analysis (essay), to conducting and analyzing a laboratory investigation (hands-on) [13]. Because they allow students to construct or perform an original response rather than just recognizing a potentially right answer out of a list provided, performance assessments can measure students' cognitive thinking and reasoning skills and their ability to apply knowledge to solve realistic, meaningful problems. Literature survey on the assessment of critical thinking dispositions and habits of mind indicates that critical thinking dispositions has been assessed using approaches such as direct observation, rating scores, essays [14, 15]. Some of these methods are used in combination. Therefore, this study was planned to determine the contribution of performance assessment model in Physics Laboratory to increase students' Critical Thinking Disposition (CTD) at senior high school.

2. Experimental Method

This study was conducted on a senior high school in Lubuk Sikaping, West Sumatera, during first semester of 2016-2017 academic years. Participants of the study were 11th grade consist of 32 students. Sampling was carried out by purposive sampling. The research used one group pretest-postest design..

The instrument in this study was developed from seven subscales of CTD by Facione [15]. Data was collected through essay test and questionnaire as a part of PAM. Each instrument in this study was integrated physics practicum material. PAM has been developed to represent the physics concepts which able to be devided into five experiments: 1) acceleration due to gravity; 2) Hooke's law; 3) simple harmonic motion; 4) work-energy concepts; 5) the law of momentum conservation. CTD essay test and rubric consists of 18 item. Subscale of CTD essay test are inquisitiveness (3 items), open-mindedness (3 items), systematicity (3 items), truth-seeking (3 items), analyticity (2 items), self-confidence (2 items), maturity (2 items). The reliability of CTD essay test is 0.802. CTD questionnaire consists of 50 item Likert type response. Subscale of CTD questionnaire are inquisitiveness (9 items), open-mindedness (8 items), systematicity (7 items), truth-seeking (9 items), analyticity (6 items), self-confidence (6 items), maturity (5 items). The reliability of CTD questionnaire is 0.836. Data was analyze using quantitative way with N-gain value. In this study only used one class by comparing the N-gain value between pretest and posttest.

3. Result and Discussion

3.1. Pretest, posttest and n-gain analysis of CTD essay test

Mean score pretest, posttest, and gain of essay test were analyzed descriptively. There is an increase in students' critical thinking disposition score from 56.59 to 84.11. Based on the calculation of the mean value <g> is 0.63 which is categorized as medium [16].

Table 1. Mean scores of essay test for each subscales of the critical thinking disposition

Subscales	Pretest	Posttest	N-gain (%)
Inquisitiveness	46.87	74.47	53.23
Open-mindedness	55.20	85.41	65.10
Systematicity	61.97	88.02	66.40
Truth-seeking	65.10	86.45	66.14
Analyticity	48.43	88.28	77.60
Self-confident	64.84	87.50	56.77
Maturity	52.34	79.68	44.79

Table 2. Normally on pretest and posttest score of essay

	Kolmogorov-Smirnov[a]			Shapiro-Wilk		
	Statistic	df	Sig.	Statistic	df	Sig.
pretest	0.109	32	.200*	0.967	32	0.422
posttest	0.104	32	.200*	0.976	32	0.681

Table 3. Independent sample test of essay

		Levene's Test for Equality of Variances		t-test for Equality of Means						
		F	Sig.	t	df	Sig. (2-tailed)	Mean Difference	Std. Error Difference	95% Confidence Interval of the Difference Lower	Upper
pretest_posttest	Equal variances assumed	1.538	.220	-14.01	62	.000	-27.5	1.96	-31.44	-23.59
	Equal variances not assumed			-14.1	57.958	.000	-27.5	1.96	-31.44	-23.59

Table 1 shows students' pretest, posttest, and gain score of essay test for each subscale of critical thinking disposition. As sees from Table 1, all the critical thinking disposition indicators have increased, marked a positive value on each % N-gain. The highest pretest main score was found on the truth-seeking indicator (65.1) and the lowest on the inquisiveness indicator (46.87). The highest mean posttest was found on the analyticity indicator (88.28) and the lowest on the inquisiveness indicator (74.48). The largest % <g> is obtained on the analytical indicator (77.6) and the lowest on the maturity indicator (44.79).

Kolmogorov-Smirnov and Shapiro-Wilk test on whether the data is distributed normally or not were applied. As sees from Table 2, the score distribution of essay pretest dan posttest displays a normal distribution (sig.>0.05). The result one-way analysis of variance on the significance difference between pretest and posttest essay test mean score are given in Table 3. The equality of the variances was checked through Levene F test and variance were found out to be homogeneous (sig.>0.05). As seen from Table 3, the different between mean score of pretest and posttest is significant (sig.(2-tailed)<0.05).

3.2. Pretest, posttest and n-gain analysis of CTD questionnaire

Mean score students' pretest, posttest, and gain of questionnaire were analyzed descriptively. There is an increase in students' critical thinking disposition score from 76,82 to 84.8. Based on the calculation of the mean value <g> is 0.316 which is categorized as medium.

Table 4 shows students' pretest, posttest, and gain score of questionnaire for each subscale of critical thinking disposition. As sees from Table 4, all the critical thinking disposition indicators have increased, marked a positive value on each % N-gain. The highest pretest main score was found on the maturity indicator (77.03) and the lowest on the open-mindedness indicator (67.48). The highest mean posttest was found on the maturity indicator (92.66) and the lowest on the analiticity indicator (77.60). The highest % <g> is obtained on the maturity indicator (64.60) and the lowest on the analiticity indicator (25.53).

Table 4. Mean scores of questionnaire for each subscales of the critical thinking disposition

Subscales	Pretest	Posttest	N-gain (%)
Inquisitiveness	76.82	84.81	31.61
Open-mindedness	67.48	80.57	36.52
Truth-seeking	72.40	80.47	25.62
Systematicity	75.11	85.04	36.55
Analyticity	68.23	77.60	25.53
Self-confident	72.66	83.20	32.52
Maturity	77.03	92.66	64.60

Table 5. Normally on pretest and posttest score of questionnaire

	Kolmogorov-Smirnov[a]			Shapiro-Wilk		
	Statistic	df	Sig.	Statistic	df	Sig.
pretest	0,128	32	.200*	0.967	32	0.422
posttest	0,131	32	.200*	0.976	32	0.681

As sees from Table 5, the score distribution of questionnaire pretest dan posttest displays a normal distribution (sig.>0.05). The result one-way analysis of variance on the significance difference between pretest and posttest questionnaire mean score are given in Table 6. The equality of the variances was checked through Levene F test and variance were found out to be homogeneous (sig.>0.05). As seen from Table 6, the different between mean score of pretest and posttest is significant (sig.(2-tailed)<0.05).

Table 6. Independent sample test of questionnaire

		Levene's Test for Equality of Variances		t-test for Equality of Means						
		F	Sig.	t	df	Sig. (2-tailed)	Mean Difference	Std. Error Difference	95% Confidence Interval of the Difference Lower	Upper
	Equal variances assumed	0.957	0.33	-8.959	62	.000	-20.1	2.24	-24.61	-15.63
pretest_posttest	Equal variances not assumed			-8.959	60.983	.000	-20.1	2.24	-24.61	-15.63

4. Conclusion

Performance assessment refers to variety of task and situations in which students are given opportunities to demonstrate their understanding and to thoughfully apply knowledge, skill, and habits of mind in a variety of contexts [13]. One reason that performance assessments embedded in classroom instruction may help support stronger learning for students is that they ensure that students are undertaking intellectually challenging tasks [12]. In this study, we can conclude that implementation a performance assessmet model in physics subject effectively increases students' critical thinking disposition. It means students' critical thinking disposition increases because of variety and challenging task of performance assessment.

In the literature, we didn't encounter a study about assessment performance model effect upon individual's critical thinking disposition. But in experimental studies, it can be seen that individuals' critical thinking disposition increased with different application [17-20]. This research is limited to physics subject in senior high school level. Similar studies should be conducted in different subject such as biology and chemistry, and different workgroup.

Acknowledgments
Thanks to Drs. Riska Khaira, M. Si as head master SMA N 1 Lubuk Sikaping, Erna Dewita, M. Si and Elfajridel, S. Pd as physics teacher who has provided many inputs and improvements as long as my research. Big thanks also for all participants who helps this research.

References

[1] Stupnisky R H, Renaud R D, Daniels L M, Haynes T L and Perry R P 2008 The Interrelation of First Year College Students' Critical Thinking Disposition, Perceived Academic Control, and Academic Achievement *Springer Science+Bunsiness Media. Res High Educ* **49** 513–530
[2] Connie S L 2006 Approaches to Evaluate Critical Thinking Dispositions *APERA Conference* 2006
[3] Pithers R T and Soden R 2000 Critical thinking in education: A review *Educational Research* **42** (3) 237–24
[4] Facione P A, Facione N C and Giancarlo C A 2000 The disposition toward critical thinking: Its character, measurement, and relationship to critical thinking skill *Informal Logic* **20** (1) 61–84

[5] Ennis R H 1987 A taxonomy of critical thinking dispositions and abilities In Joan Baron & Robert Sternberg (Eds.) *Teaching thinking skills: Theory and practice* (New York: W.H. Freeman) pp. 9-26

[6] Facione P A 1990 *Critical Thinking: A Statement of Expert Consensus for Purposes of Educational Assessment and Instruction. Executive Summary* "The Delphi Report," The California Academic Press [Online]. Available at: http://www.insightassessment.com/dex.html. Retrieved May 13, 20013 from source.

[7] Ennis R H 1989 Critical thinking and subject specificity: clarification and needed research *Educational Researcher* **13** 13-16

[8] Facione, P. A., Facione, N. C., & Giancarlo, C. A. (1997). The motivation to think in working and learning. In E. A. Jones (Ed.), *Preparing competent college graduates: Setting new and higher expectations for student learning—New directions for higher education* (Vol. 96, pp. 67-79). San Francisco: Jossey-Bass.

[9] Gunawan. (2011). Developing Virtual Lab Physics Model to Increases Pre-Service Physics Teachers' Generic Skill and Critical thinking Disposition. *Dissertation*. UPI. Unpublished.

[10] Shepard L A 2008 Commentary on the National Mathematics Advisory Panel recommendations on assessment *Educational Researcher* **37** (9) 602-609

[11] Kahl S 2008 The Assessment of 21st Century Skills: Something Old, Something New, Something Borrowed. Paper presented at the *Council of Chief State School Officers 38th National Conference on Student Assessment* (Orlando, FL)

[12] Darling H L and Adamson F 2010 *Beyond basic skills: The role of performance assessment in achieving 21st century standards of learning* (Stanford, CA: Stanford University, Stanford Center for Opportunity Policy in Education)

[13] Marzano R J, Pickering D, and McTighe J 1993 *Assessing student outcomes: Performance assessment using the Dimensions of Learning Model* (Alexandria, VA: ASCD)

[14] Norris S P 1992 Testing for the disposition to think critically *Informal Logic* **2** 157-164

[15] Facione P A, Facione N C and Giancarlo C A 1995 The Disposition Toward Critical Thinking *Journal of General Education* **44** (1) 1-25

[16] Hake, R.R. (1999). Reconstrcuting Theory from Practical Experience. In Woolnough, B.E. (ed). *Practical Science, 66-77.* Milton Keynes: Open University Press.

[17] Evren A, Bati K, and Yilmaz S 2012 The Effect of Using V-Diagram in Science and Technology Laboratory Teaching on Preservice Teacher's Critical Thinking Disposition *Procedia – Sosial and Behaviorial Sciences* **46** 2267-2272

[18] Yenice N 2011 Investigating Pre-service Science Teachers' Critical Thinking Dispositions and Problem Solving Skills in Terms of Different Variables *Educational Research and Reviews* **6** (6) 497-508

[19] Neyestani M R, Heydari M H and Ghamarani A 2012 Relationship between Critical Thinking Dispositions and Dialogue-based Education among teachers *Journal of American Science* **8** (8)

[20] Karami M, Pakmehr H and Aghili A 2012 Another View to Importance of Teaching Methods in Curriculum: Collaborative Learning and Students' Critical Thinking Disposition. *Procedia - Social and Behavioral Sciences* **46** 3266 – 3270.

Science Teachers' Pedagogical Content Knowledge and Integrated Approach

M J Adi Putra[1,2*], A Widodo[3] and W Sopandi[4]

[1] Jurusan Pendidikan Dasar, Sekolah Pasca Sarjana Universitas Pendidikan Indonesia, Bandung, Indonesia
[2] Jurusan Ilmu Pendidikan, Fakultas Keguruan dan Ilmu Pendidikan, Universitas Riau, Pekanbaru, Indonesia
[3] Departemen Biologi, Universitas Pendidikan Indonesia, Bandung, Indonesia
[4] Departemen Kimia, Universitas Pendidikan Indonesia, Bandung, Indonesia

*jaya.adiputra@lecturer.unri.id

Abstract. The integrated approach refers to the stages of pupils' psychological development. Unfortunately, the competences which are designed into the curriculum is not appropriate with the child development. This Manuscript presents PCK (pedagogical content knowledge) of teachers who teach science content utilizing an integrated approach. The data has been collected by using CoRe, PaP-eR, and interviews from six elementary teachers who teach science. The paper informs that high and stable teacher PCKs have an impact on how teachers present integrated teaching. Because it is influenced by the selection of important content that must be submitted to the students, the depth of the content, the reasons for choosing the teaching procedures and some other things. So for teachers to be able to integrate teaching, they should have a balanced PCK.

1. Introduction

Pedagogical content knowledge is a conceptualization of specialized content area knowledge required for effective teaching [1]. Shulman defines and describes his knowledge of PCK as' a unique amalgam of unique teachers' content and pedagogy, a form of professional understanding of their own. Shulman argues that PCK includes the most useful representational forms, analogies, illustrations, examples, explanations, demonstrations, the strongest pedagogical techniques, and knowledge of what makes the concept. However, the early conceptualization of Shulman PCK can lead to the blending of knowledge types with the quality of knowledge, suggesting that there is no weak PCK. Professional teachers prefer quality in teaching, paying great attention to the quality of learning and the quality of cognitive outcomes received by their students. So needed a good planning in teaching science to students in class. Pedagogical content knowledge is a comprehensive knowledge that brings together two pieces of knowledge: content knowledge and pedagogic knowledge in one unity that leads to better knowledge for students [2] [3].

In general, elementary teachers often have little or no training in school subjects like science and technology. Not surprisingly, some studies show that they often have difficulty while teaching these subjects [4]. With regard to the ability of teachers to teach in the classroom, developing relevant knowledge is more effective for measuring using PCK [5]. But in science learning in elementary schools has a specificity in teaching because science teaching uses the integrated approach. Integrated teaching can be a powerful tool for reintegrating the curriculum and illuminating the isolated, reductional nature of teaching around the discipline rather than experience. It requires a lot of hard work, initial design, plus substantial restructuring of teacher relationships and class schedules [6].

In this paper we presented teachers' PCK with and their ability to integrate science lessons with other subjects in a series of integrated approaches. Subjects that accompany the subject of this science are selected Civics that have a specificity in learning, while the theme chosen is the theme is often taught in the class of the theme of "entertainment".

2. Experimental Method

This PaP-eR is based on an investigation involving six elementary teachers who teach in third grade. Teachers teach in elementary school for 9 to 12 years in Pekanbaru. All teachers are female. The theme of integrated teaching in this research is entertainment with subject science and civics. Instruments used are CoRe and PaP-eR [7], and open interviews relate to teachers' ability to integrate science and civics. Teacher representation content (CoRe) is collected as part of a learning plan while pedagogical experiences (PaP-eR) are gathered after teaching activities. The interview was done after the teacher finished teaching. The data obtained are analyzed with the rubrics already available, presented with graphs and tables.

3. Result and Discussion

Changes in the nature of objects, because of burning, warming, and in touch with the open environment are topics related to the object and its nature. In a large scope the teacher's understanding of this content is the nature of the object and its nature and the constituent elements. It takes a teacher's ability to represent the content appropriately so that learning activities on this material do not occur misconceptions to students [8], so that there is an element of pedagogy.

Ideally integrated teaching can be taught by teachers starting from designing curriculum, teaching methods, and assessments around selected themes. The representation of a teacher's content demonstrates a teacher's collective ability to teach in classroom [9]. Its can be numerated by appraisal using the CoRe instrument [7] [10]. The following is a representation of teacher content teaching science by using integrated approach.

Figure 1. Presented PCK teachers who plan integrated learning on the "entertainment theme". Four teachers demonstrate the dominance of the ability to manage learning in the form of teaching skills (pedagogy) rather than content representation. Teaching in the lower classes requires skills in managing learning (pedagogy) higher than the content. Because, the content is superficial and still easy to observe by students. In addition, low-grade science lessons are more emphasized in phenomena that appear to be students (i.e. hand function etc.) [6]. Therefore, even though teachers do not teach deeply about science content, it is important to understand science appropriately. Some studies indicate that the quality of teaching is very important for student learning, for example, classroom management and classroom interaction, teaching characteristics have greater influence and the school environment has a similar effect on students' cognitive competence [11] [12]. The understanding of pedagogy in classroom management as described above is more widely gained from the experience of teachers in teaching [13] [14] [15], than experienced teachers certainly have better teaching skills and generate better understanding of students [11] [12].

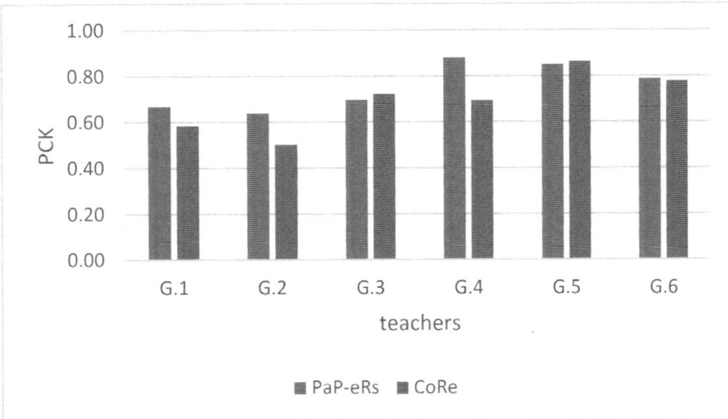

Figure 1. PCK (CoRe & PaP-eR) science teachers with integrated approach

Teaching skills appear when teachers respond to PaP-eR, and answers to teacher interviews. Teachers with high CoRe, have a stronger reason when asked about how they connect the material to the theme (Table 1). Because it is influenced by the selection of important content that must be submitted to the students, the depth of the content, the reasons for choosing the teaching procedures and some other things. Integrated teaching is based on the idea that people acquire knowledge when learning in a coherent "whole" context and when they can relate what they learn to the real world. Integrated teaching seeks to place the teaching of cognitive skills such as reading, math, science, and writing in the context of real-life experience subjects that are specific enough to be practical, and broad enough to enable creative exploration.

From the interviews conducted there is one teacher who still understands the subject of science as a learning that is separate from other subjec civic (Level 1). The teacher presents a divided thinking about science and other learning without considering the theme. Besides this the teacher is more inclined to teach the content of science in a separate way just caught on the subject that is put on the theme being taught **(No Integration)** [16]. This teacher is more inclined to teach the content of science in a separate way just caught on the subject that is put on the theme being taught and teachers' PCK is low.

Furthermore, we find teachers who can generally provide an overview of integrate science and civics. The basic understanding that the integration of two subjects is very likely to occur and integration is described as sequential and the domain may not have in-depth content (Level 2). Integration is described as a thematic teaching where subject areas are organized around topics or themes **(Beginning Integration)** [16]**.** In the initial integration, teachers have recognized the integration of science with other lessons that may reasonably be difficult to realize. However, because the teacher understands the content of science well, it allows teachers to connect science with another subject. Teachers' PCK at this level are low and unbalanced between content and pedagogy.

Next, we found teacher responses that reflect an understanding of the integration of the Civic-science as a one-way process. There is an explicit focus on a single domain, with the second domain being used to support or facilitate the primary domain (Level 3). The introduction of relationships between domains signifies the understanding of integration that arises, but the relationship is tentative (**Emerging Integration**) [16]. Teachers at this level begin to build confidence in the integration of learning practices between the two subjects that are difficult to unite. They can discuss the shared attributes (e.g., civics function more on the value contained not in the content learning process) this indicates that the teacher already understands the subject matter that is not taught only one direction, but two ways. At this level the teacher's PCK is medium and balanced.

Finally, we find teachers who implicitly understand integration as a reciprocal relationship. There is a relationship between two domains although their contents in both domains may not be discussed in depth but are strongly interconnected (Level 4). At this level, teacher's responds indicate that teachers have developed strategies based on their personal experience in applying science and civic teaching by integrated teaching (**fundamental integration**) [16]. Teachers at this level, have the belief that teaching science and civic with integrated teaching. This belief is shown by a high and balanced teacher PCK.

Table 1. Level teachers to integrate science subjects in a integrated approach with CoRe and PaP-eRs science teachers

	CoRe	PaP-eRs	Teachers	Teachers respond
Level 1	0,50	0,64	G.2	Novice: rule-based and inflexible, Separate content domains **(no integration)**
Level 2	0,69	0,88	G.4	Advanced beginner: global, Basic understanding;
	0,58	0,67	G.1	'knowing that' **(Beginning Integration)**
Level 3	0,78	0,79	G.6	Competent performer: organized plan
	0,72	0,70	G.3	Unidirectional **(Emerging Integration)**
Level 4	0,86	0,85	G.5	Proficient performer: analytic decision making Reciprocal processes; 'knowing how' **(fundamental integration)**

From Table 1. Describes the ability of teachers to integrate science and Civics learning in the theme of "entertainment". CoRe and PaPer values indicate conformity with the level of science-science integration capability. This indicates that the balanced teacher PCK both from the side of the content representation as well as their repertoire Experience is more stable in relating the subject of science and Civics in this theme.

4. Conclusion

From the description above it appears that teachers with PCK teachers give shape to the ability to teach science with thematic approach. The effect is due to the teacher's valuable and stable teacher pattern. PCK teachers as a means to connect domains, so they are very easy to choose teaching with thematic approach.

Acknowledgments

We special say thank to Education Office of Pekanbaru as responden and all participants who participated in this research.

References

[1] L S Shulman 1986 Those Who Understand: Knowledge grawth in Teaching *Educational Research* **15** (2) 1-22

[2] A Widodo 2017 Teacher Pedagogical Content Knowledge (PCK) and Students' Reasoning and Wellbeing *Journal of Physics: Conference Series* 1-7

[3] S Park and J S Oliver 2008 Revisiting the Conceptualisation of Pedagogical Content Knowledge (PCK): PCK as a Conceptual Tool to Understand Teachers as Professionals *Research in Science Education* **38** (3) 261–284

[4] M Hultén and E Björkholm 2016 Epistemic Habits: Primary School Teachers' Development of Pedagogical Content Knowledge (PCK) In a Design-Based Research Project *International Journal of Technology and Design Education* **26** (3) 335-351

[5] K Appleton 2006 science Pedagogical Content Knowledge and Elementary School Teacher in *Elementary Science Teacher Education: International Perspectives on Contemporary Issues and Practice* (New Jersey, Lawrence Erlbaum Associates) pp. 31-53

[6] Badan Standar Nasional Pendidikan, "Standar Isi untuk Satuan Pendidikan Dasar dan Menengah: Standar Kompetensi dan Kompetensi Dasar SD/MI," BSNP, Jakarta, 2006.

[7] J Loughran, A Berry and P Mulhall *Understanding and Developing Science Teachers' Pedagogical Content Knowledge* (Rotterdam: Sense Publishers, 2012)

[8] M V Juhler 2016 "The Use of Lesson Study Combined with Content Representation in the Planning of Physics Lessons During Field Practice to Develop Pedagogical Content Knowledge," *Journal of Science Teacher Education,* vol. **27** (5) 533–553

[9] A Bertram 2014 CoRes and PaP-eRs as a strategy for helping beginning primary teachers develop their pedagogical content knowledge *Educación Química* **25** (3) 292-303

[10] W R Veal and J G MaKinster 1999 Pedagogical Content Knowledge Taxonomies *Electronic Journal Of science Education* **3** (4)

[11] M Steffensky, B Gold, M Holdynski and K Möller 2016 Erratum to: Professional Vision of Classroom Management and Learning Support in Science Classrooms—Does Professional Vision Differ Across General and Content-Specific Classroom Interactions? *International Journal of Science and Mathematics Education* **14** (5) 1001–1002

[12] R Tshewang, V Chandra and A. Yeh 2016 Students' and teachers' perceptions of classroom learning environment in Bhutanese eighth-grade mathematics classes *Learning Environments Research* **12** (2) 1-20

[13] Y Li and R Huang 2008 Chinese elementary mathematics teachers' knowledge in mathematics and pedagogy for teaching: the case of fraction division *ZDM Mathematics Education* **40** (5) 845–859

[14] P Tamir 1988 Subject Matter and Related Pedagogical Knowledge in Teacher Education *Teaching and Teacher Educatio* **4** (2) 99-110

[15] J K Lannin, K Chval, M Webb, F Arbaugh, S Hicks, C Taylor and R Bruton 2013 The Development of Beginning Mathematics Teacher Pedagogical Content Knowledge *Journal of Mathematics Teacher Education* **16** (6) 403–426

[16] T Stoddart, A Pinal, M Latzke and D Canaday 2002 Integrating Inquiry Science and Language Development for English Language Learners *Journal of Research In Science Teaching* **39** (8) 664–687

IOP Conf. Series: Journal of Physics: Conf. Series **895** (2017) 012145 doi:10.1088/1742-6596/895/1/012145

Integration of Various Technologies in Biology Learning

M Safitri*, R Riandi, A Widodo and W R Nasution

Biology Education Studies Program, School of Postgraduate Universitas Pendidikan Indonesia, Bandung, Indonesia

*mega.safitri@student.upi.edu

Abstract. Current technological developments require teachers to be able to create effective and efficient learning by integrating technology into teaching. The purpose of this study was to identify the type of technology used by teachers in biology teaching. This research uses descriptive method through observation and a semi-structured interview. Participants involved in this research are three teachers of X class biology from different Senior High School in Indonesia. The findings of the study were analyzed using the rubric of technology integration in the implementation of the use of technology in learning and teaching developed by Britten & Casady. The results of the study revealed that the various technologies used in the biological material of Animalia biology included in the technology that supports learning by using conventional methods or lectures on the material Kingdom Animalia. Teacher's consideration in using technology on biology learning is limited facilities, the limited ability of teachers, complex content, and learning method used. The number of teachers' considerations in integrating technology suggests that teachers in Indonesia need the development of the ability to integrate different types of technology in the lessons learned.

1. Introduction

The development of information technology today requires teachers to have various competencies in implementing effective and efficient learning. Through the various technologies developed today, teacher and students must be able to use technology in learning [1]. In order for teachers to plan and implement effective and efficient learning, teachers should have good knowledge of pedagogical technology and knowledge in addition to having good conceptual mastery [2]. Integration of technological knowledge, pedagogy, and content is conceived within the TPACK conceptual framework developed from the conceptual framework of PCK Lee Shulman [3]. Technology, Pedagogical, content knowledge (TPACK) is a framework built on the interaction of three basic knowledge of technology, pedagogy, and content to produce effective learning by integrating technology [4].

TPACK concept has seven components consisting of 1) technological knowledge, 2) pedagogical knowledge, 3) content knowledge, 4) knowledge of technology content, 5) pedagogical content knowledge, 6) pedagogical knowledge of technology, 7) pedagogical technology Knowledge of content. However, the results show that teachers who have followed the TPACK development program consider technology integration and discipline an important aspect in TPACK development [5]. This research is supported by the results of other studies which say that the components that most influence TPACK teachers are elements of technology knowledge (TK). Kindergarten is the ability of teacher's knowledge and technique in using technology that significantly contributes to TPACK [6].

Effective learning requires the use of efficient technology [7]. In this situation, teachers are expected to keep up with the latest technological developments, so that teachers can learn by using technology that suits the needs of students to become effective and efficient learning. Obstacles in Indonesia, unequal technological facilities require teachers to utilize technology following conditions around the school or residence. Local technology is the equipment used to achieve individual goals in specific situations or environments. The material in question is the equipment used to support the learning process of Biology in Senior High School in Indonesia. Thus the purpose of this study was to identify the type of technology used by teachers in biology learning.

2. Experimental Method

This research is a descriptive study that describes the kind of TPACK technology teachers used in biology learning on the concept of Kingdom Animalia. Participants recruited for the study were high school / MA biology teachers in rural (rural) schools are communities that behave in the same way as the environment and the livelihoods of farmers and traders. The selected research subjects were biology teacher of X class of science as many as three teachers from three different schools in Malangbong district. Subjects have been chosen based on the location of the school and the willingness of teachers to become participants.

The research data was collected through an observation sheet of instructional learning adapted from Harris, grandgenett, and Hofer research [8] and followed by semi-structured interviews. Data collection was done three times in a row. Also, data were also collected from semi-structured interviews with each biology teacher. The collected data were then analyzed using a Rating Technologies (TIAI) analysis developed by Britten & Casady [9].

3. Result and Discussion

The findings of the field described in Table 1. as a whole in district Malangbong biology teacher already using the technology in the form of a whiteboard and the Internet via mobile phones (HP). As for the other technologies are used, adapted to the conditions of the existing facilities and infrastructure in each school teachers. Use of the Internet through HP utilized teachers as the source and means of learning still using the traditional lecture method. Selection of HP as a learning support tool because almost every student's HP that can be used to search for information on the internet. However, the use of HP by teachers themselves in learning is not used but only assigned to students. Results of interviews about the use of HP by the teacher mentioned that HP used by teachers only for personal communication purposes if related to the search and preparation of teaching materials teachers prefer to use a computer-based technology owned by each teacher. Observations indicate the use of HP's learning by students is used for Resourcing and Searching. Resourcing the aim of learning as a teaching resource or information repository for information and data relating to the learning materials, while Searching is a search process to complete the learning resource materials that will be delivered to students. In the course of searching the information sought is information that is not yet known, so the search for information using search engines, which is one of the facilities provided in an application to search for the information we want. Application search engines of the popular used today by people of Indonesia is Google, but many other applications are offered specific sites such as Yahoo, Alta Vista, and others [10]. According to teachers, the use of the internet via mobile phones intended to allow students more actively seek various sources of material for learning activities. In line with the opinion of the research results Sistriyani [11] states that the Internet as a medium on the material kingdom Animalia may help to understand the material and increase the activity and motivation of students so that students do not feel bored in learning activities.

Table 1. Various technologies used by teachers in learning as much as three times meeting on the concept of Kingdom Animalia

Participants	Technology Used
Participant 1	Mobile phone (HP) Whiteboard & marker Textbooks
Participant 2	Whiteboard and markers HP / internet HVS Paper coloring apparatus
Participant 3	Whiteboard and markers Projector Infocus Internet / HP Textbooks

The second technology is used by a biology teacher at the school was a whiteboard. Whiteboard is a device (hardware) called media or messaging/materials to be presented together with a software [12]. Other equipment used by the teacher is projector and laptops. Just like a whiteboard, the projector used as a vehicle for channeling messages or information learned. Another benefit of media or means of directing messages in learning can provide a thorough experience of something concrete to abstract and can arouse students' motivation to learn.

Based on the results of the analysis of the use of various learning technologies using the rubric of technology integration assessment (TIAI) in the implementation of the use of technology in learning and teaching developed by Britten & Cassady [9] showed that the use of learning technology in the district of Malangbong enter the category of technology components to support the implementation of learning. Technology facilitates biology learning, but learning objectives can be achieved without the use of technology. Observations show that the technology used by teachers in the form of HP, projector, and whiteboard used for presenting the material using lecture method. The lecture method is a conventional method of teaching, the teacher just tells just by the book or slide shown through projector infocus [13]. According to the teachers of the interview to mention the limited facilities available in the school made the teacher cannot use technology at all times. Another obstacle is found in the use of technology is the lack of teachers' skills in using technologies such as application usage search in the search for of materials and installation of technological devices such as projector infocus to spend a lot of time in learning that requires a lot of time to prepare lessons.

Lack of knowledge and skills of teachers in the use of computer-based technology or HP is one of the cause's maximal integration of technology in learning. Results 3 of 4 teacher interview mentions that skills in using computer-based technology can only use Microsoft Word, Excel, PowerPoint, and use search google apps, but it is used for personal purposes. While the HP-based technologies used by teachers for communication via messenger or social media that does not include the sake of learning. Lack of knowledge because teachers have not been trained on technology in learning at the school and district levels.

The Role of instructional media is the role of technology in learning that can simplify and accelerate the work of the students, as well as fun students for interacting with colors, images, sound, video, and instant something. The use of technology becomes a very important factor for effective learning because it can evoke positive emotions during the learning process [5][14]. The role of a variety of technologies depends on the content material that will be taught by the teacher. The material consists of a kingdom Animalia phyla classified in invertebrates and vertebrates. According to teachers important concept that should be known the student of the material kingdom Animalia is the common characteristics of each phylum, basic grouping of Animalia, animal examples, as well as the role of animals in the world. By using technology in the form of the projector, whiteboard, and HP in Animalia concept, teachers feel that the lesson has not been effective yet using technology, but the teacher has not been able to develop media based on tools/technology. The variety of media technology as developed by several researchers in materials such as kingdom Animalia handouts illustrated with a map of the concept [15]; E-book interactive multimedia flash-based [16]; Cue Identification Card of Animalia[17]; and media books pop-up[18]. The result of the development of the media can help students understand the concept of Kingdom Animalia, increase motivation and student activity during learning, and improve students' verbal-linguistic intelligence after being given media books *pop-up* on the concept of crustaceans.

4. Conclusion

Various technologies used by biology teachers in Malangbong Regency Garut Regency including technology that supports learning. The type of technology used consists of Whiteboard, HP, projector & laptop, as well as technology / other tools tailored to the learning needs of the implemented and facilities that can be used. However, the integration of the technology still encounters many obstacles such as access to the use of limited technology, complicated material delivered, difficult to apply learning strategies, and lack of knowledge and skills of teachers in using and developing technology as media and biology learning resources. Therefore, the researchers conclude the need for the development of teacher competence in Malangbong district to develop technology as media and learning resources tailored to the existing technology in the school environment.

Acknowledgments

The researchers would like to thank Mrs. Dr. H. Siti Sriyati, M.Si who has helped researchers from the planning of the research study. Researchers also like to thank Dedi Pribadi, S.Pd, Imas Surya Hartini, S.Pd., Andri Suryadi, M.Kom., Gemi Febrian, S.ST., who has provided direction for the conduct of research , And thanks also presented by researchers to participants already willing to help this research.

References

[1] Suryawati E, Firdaus L N, and Hernandez Y 2014 *Jurnal Biogenesis* **11**
[2] Harris J and Hofer M 2015 Journal Of Research On Technology In Education **43** p 211-229
[3] Mishra P and Koehler M J 2006 *Teachers College Record* **108** p 1017-1054
[4] Koehler M J and Mishra P 2009 Contemporary Issues in Technology and Teacher Education **9** p 60-70
[5] Niess M L 2005 *Teaching and Teacher Education.* **21** p 509-523
[6] Sheffield R, Dobozy E, Gibson D, Mullaney J, and Campbell C 2015 *Educational Media International*
[7] Ertmer P A and Ottenbreit-Leftwich A T 2010 *Jurnal of Reasearch on Technology in Education* **42**
[8] Harris J, Grandgenett N, and Hofer M 2010 Testing a TPACK-based technology integration assessment instrument. In C. D. Maddux, D. Britten J S and Cassady J C 2005 *Computers in the Schools*, **22** 49
[9] Adri M 2003 Pemanfaatan Internet sebagai sumber pembelajaran. *IlmuKomputer. com.*
[10] Sistriyani D 2012 *Journal of Innovative Science Education* **1** 1
[11] Riyana C 2008 Peranan Teknologi Dalam Pembelajaran. *Universitas Indonesia, Jakarta.*
[12] Harsono B 2009 *Jurnal Pendidikan Teknik Mesin* **9** p 54-60
[13] Suryadi A 2007 *Jurnal Pendidikan Terbuka dan Jarak Jauh* **8** hlmn 83-98
[14] Nerita S 2017 *BioCONCETTA*, **2**
[15] Ashshidiqqi H 2017 *Pengembangan E-book Multimedia Interaktif Berbasis Flash pada Konsep Kingdom Animalia* (Bachelor's thesis, Perpustakaan Ilmu Tarbiyah dan Keguruan).
[16] Maesaroh T 2016 Pengembangan Perangkat Pembelajaran Animalia Menggunakan Identification Cue Card Dalam Kaitannya Dengan Kemampuan Retensi dan Keterampilan Komunikasi Siswa SMA (Studi Pendahuluan).
[17] Lismayanti M, Hamidah A, and Anggereini E 2016 *Jurnal Penelitian Universitas Jambi: Seri Humaniora* **18** p 44-48

Mapping of Students' Learning Progression Based on Mental Model in Magnetic Induction Concepts

R Hamid* **and D B Pabunga**

FKIP Universitas Halu Oleo, Jl. H.E.A. Mokodompit Kendari 93232, Indonesia

*rimbahamid@gmail.com

Abstract. The progress of student learning in a learning process has not been fully optimally observed by the teacher. The concept being taught is judged only at the end of learning as a product of thinking, and does not assess the mental processes that occur in students' thinking. Facilitating students' thinking through new phenomena can reveal students' variation in thinking as a mental model of a concept, so that students who are assimilative and or accommodative can be identified in achieving their equilibrium of thought as well as an indicator of progressiveness in the students' thinking stages. This research data is obtained from the written documents and interviews of students who were learned about the concept of magnetic induction through *Constructivist Teaching Sequences* (CTS) models. The results of this study indicate that facilitating the students' thinking processes on the concept of magnetic induction contributes to increasing the number of students thinking within the "progressive change" category, and it can be said that the progress of student learning is more progressive after their mental models were facilitated through a new phenomena by teacher.

1. Introduction

Science as one of the important lesson materials in schools, especially in elementary schools that present contextual concepts that should appeal to students if teachers can relate it to the social context of each student, but the fact that there is so many misconceptions found in learning science [1], [2], [3], [4]. This is highly likely because teachers have not optimally incorporated prior student experience as the basis of new knowledge builders. Teachers also need to have the ability to catalyse [5] to trigger a process of developing a structured and growing understanding of concepts leading to a more scientific concept as well as having the ability to assess the progress of each student relating to the concepts taught, therefore conceptual changes at each stage of learning can be identified.

As a part of science learning, science-physics learning is closely related to nature that has various phenomena around students' lives, and this is a part that should not be separated from the set of learning when the teacher teaches the concept. The research results from [6] reveal that physics knowledge is a necessary condition for the development of creativity, which means that basic science teachers are not very impressed with the use of creative questions as part of classroom teaching practice, or teachers have not applied the creativity elements of the mainstream science curriculum effectively. In line with this, [7] emphasize the necessity to build relationships between theory and phenomena or objects. In understanding of science learning today's, consideration of students' prior knowledge is important to improve [8] .

The fact that in studying complex material, like the concept learned in a science class, can occur at least three conditions different from the original knowledge. First, a student may not have the prior knowledge of the to be learned concepts, although they may have some related knowledge. In this case, prior knowledge is missing, and learning is consist of adding new knowledge. Second, students may have some correct prior knowledge about the to be learned concepts, but that knowledge is incomplete. In this incompleteness knowledge case, learning can be conceived as a *gap filling*. In the both missing and incomplete knowledge condition, knowledge acquisition is of enriching kind. In the third condition, a student may have ideas that he has gained, either in school or from everyday experience, that are "in conflict with" the to be learned concepts [9].

This condition should be the basis for teachers to intervene for a science learning process that prepares sufficient space to facilitate and explore and assess the development of concepts that students have understood, ranging from what they think about the concept before learning, how the process of conceptual change at each level of a particular concept, by conducting a formative assessment of changes in the progress of students' conceptual understanding based on learning progression (LP).
LP gives some perspective in science learning, in addition can be used as a cognitive reflection, can also provide information for the emergence of various ways and patterns of development of students' knowledge in a learning process, whether the student's knowledge actually grows in a linear way or progressive or more random. LP on the other hand also provides a true and practical way to describe the students' cognitive pathways. This will certainly contribute to the development of curriculum, classroom learning and assessment [10]. Therefore, determining the appropriate action in a learning that allows students to more easily understand the concept, is closely related to the ability of teachers to know the variation of the characteristics of students' thinking on the concept. The pattern of student concept movement should be the basis for teachers in choosing approaches, methods or techniques in learning, so that learning can be more effective and efficient..

Referring to one of the principles of science education [11], which asserts that assessment has a key role in science education. Formative assessment of the learning done by the students and the summative assessment of its learning progress should be applied to all learning objectives. For an assessment of learning, there is another principle in science learning that the advancement of great ideas must result from studies on topics of interest to students and relevant in their lives.
This is possible because students have basically had various alternative conceptions or natural (early) conceptions, which can negatively affect the results of their judgments and their ability to argue about natural phenomena [12]. Still related to the assessment, The National Research Council, 2001 [13], [14] asserts that each assessment is based on three interrelated elements or foundations: aspects of achievement to be assessed (cognition), The tasks used to collect evidence of student achievement (observation), and the methods used to analyze the evidence resulting from the task (interpretation).

2. Experimental Method

This research uses mixed methods research with convergent parallel design that collect quantitative and qualitative data simultaneously and sequentially [15]. Based on the design of this study, the stages of the study include: (1) quantitative and qualitative data collection, (2) analyzing the two sets of data separately, (3) comparing the results of the analysis of the two data sets, and making interpretations, centering (converging) or more having a varied (divergent) spread. If there are any irregularities, more explanation of the data is required, re-analyzing its database, or checking the quality of the two data forms more carefully. The direct comparison of both sets of data is convergent. The subjects of this study were selected through non-probability sampling technique by purposive sampling. This method is based on the consideration of the researcher according to the particular characteristics being sought [16], namely the fourth grade of elementary school students as many as 25 people who learn about magnetic induction concept.

Learning progression is a series of changes in the students' frame of mind on a concepts learned whose development occurs gradually in the learning phases, through a facilitation process by the teacher. The mapping of the student's frame of mind refers to the category of mental model it has, and the opportunity for the movement / change of ideas has several alternatives: 1) consistent with intuition; 2) consistently analogous to everyday experience; 3) change randomly (ultimately not yet scientific); Or 4) change more progressively toward a scientific final concept.

The techniques used in data collection are: 1) Interviews with representatives of groups of mental model frameworks for students to know the basic or background of the mental model framework related to the concepts they have in interpreting the phenomena observed, 2) video-based observations in the classroom Documentation, especially on the implementation of CTS-based learning by teachers that allows valid data to be obtained in relation to aspects of application activity at each stage of learning. Interaction data between teacher and student and between student and fellow student, teacher facilitation process toward idea emergence from student group. This observation is a reflection for the next learning, and 3) Documentation of students' written responses or responses from observations of the observed phenomena and responses after facilitation and questions given by the teacher at the end of the learning stages to map learning progression and conceptual level patterns change that refers to the student's thinking construction over the time span during the lesson.

3. Result and Discussion

The study of the concept of magnetic induction is the last part of the magnetic concept. Prior knowledge of the magnetism of previously concepts learned will contribute in part to the formation of new knowledge of students about magnetic induction. One indicator of this is the emergence of more mental model group variations than other subspects. In addition, using activities with abstract phenomena, there is less time to facilitate students' thinking through appropriate experience. If an elementary school teacher wishes to succeed in assisting students with misconceptions, learning activities should be limited to observable concepts that can be effectively facilitated, and in accordance with the stages of development for students at the basic levels. This is done to help students understand the magnet in a more meaningful way [17].

When students observe a paper clip that is sticky to the magnet and then the paper clip is held close to one other paper clip, and it turns out to be interested, not a magnet, but a magnet. Against this phenomenon there are two major groups of ideas, the first being that "there is a magnetic flow of energy or magnetic energy" that is understood by 40% of students, and the second is that. "The strongest attraction in the first direct clamp on the magnet" by 44% of the students. In addition, there are students who consider that this phenomenon occurs because of the pressure factor as a link between magnet and paper clip and temperature similarity factor between one clip with another clamp, as their paperwork as seen in Figure 1.

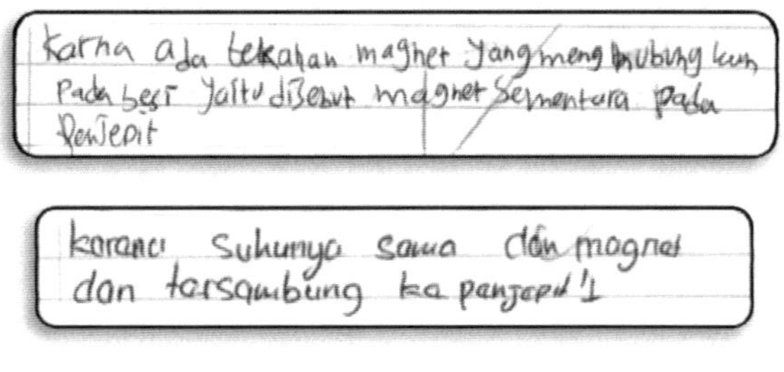

Figure 1. The student's mental model of magnetic induction

Overall how the portrayal of student idea movement related to the concept of magnetic induction is shown in Figure 2.

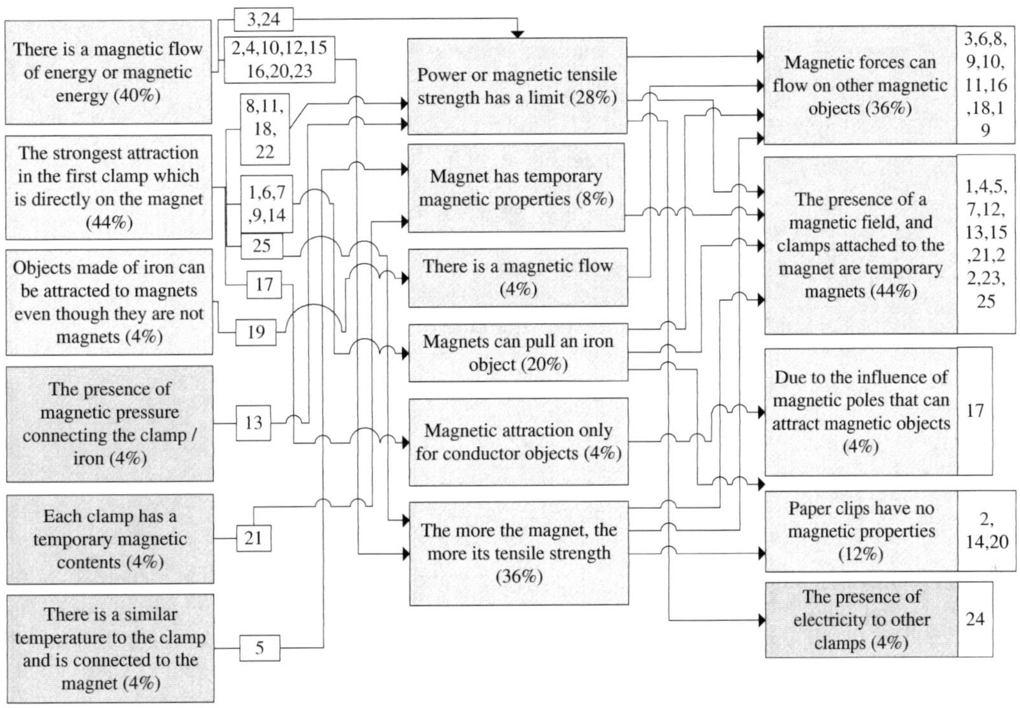

Figure 2. Flowchart of students' thinking movement on the magnetic induction concepts

Caption:

- : Ideas of students with more scientific concepts
- : The idea of students who tend to be analogous to their experience
- : Students' ideas are more intuitive

Based on Figure 2 it is known that in addition to the two dominant ideas groups as described earlier, there are three variants of the intuition-based idea expressed by the students, although each comes from a student, but the idea is unexpected. For example, student 13 reveals that the phenomenon is due to "the presence of magnetic presses linking the iron brace", and the student 5 with a more intuitive comment that "there is a temperature equation in the clamp and is connected to the magnet." Nevertheless, the two students after having facilitated their thinking by the teacher, then both can change the idea and move in a scientific direction. Student 13 has a new idea that "power / magnet tensile strength has a limit" and student 5 turns his idea into "magnet has a temporary magnet.

The facilitation process is done in a similar way to the initial exploration of knowledge, only in the facilitation stage one by one the paper clip is clamped onto a paper clamp that has been sticky before until it is no longer sticky bias. Based on the facilitation, a new framework of ideas emerged: "magnetic attraction only for conductor objects" embraced by one student and "more magnetism more and more appeal" is a new idea of 40% of students who generally derive from the idea that more scientific.

The path of student idea movement on learning of the magnetic induction concept is very dynamic and tends to be random, although at the end of learning there are 80% of students have a more scientific concept, 16% of students are still arguing with an anaolgi basis on their experiences, and 4% With an intuitive idea stating that there is a flow of electricity to the clamp, whereas the student is at the beginning of the lesson and in the facilitation phase has a more scientific idea. This fact indicates that the level of students' confidence in their understanding is still low and volatile, regardless of the mental model it has.

IOP Conf. Series: Journal of Physics: Conf. Series **895** (2017) 012147 doi:10.1088/1742-6596/895/1/012147

An analysis of the student's idea movement profile during magnetic induction learning to determine the distribution of students based on the learning progress categories they have shown in Figure 3 below.

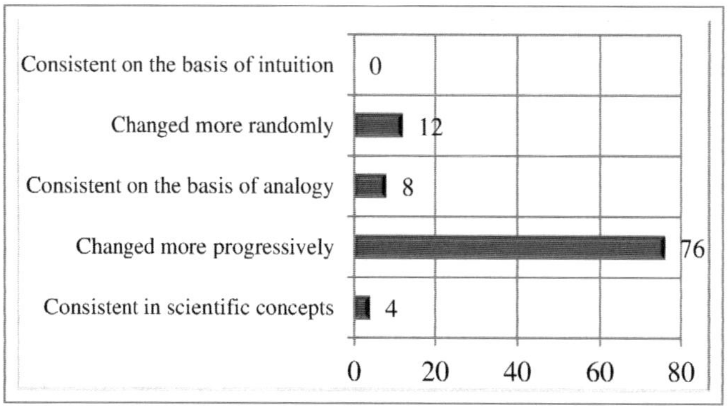

Figure 3. Distribution of the percentage of students based on learning progress on the concept of magnetic induction

The student profile is based on the learning progress of the concept of magnetic induction as shown in Figure 2, that type II (more progressive change) is a category with the highest proportion of students, and from all the concepts of electric and magnetism learned, the concept of magnetic induction is also the highest on (76% of students). This fact can be seen from the point of view of the student's travel process in constructing his knowledge during the study of electrical and magnetic concepts, and indirectly a part that perfects the student's understanding.

The data in Figure 3 also shows something important in the analysis of student learning progress on the concept of magnetic induction, where only 4% of students are consistent with more scientific ideas, while the number of students who have a more scientific mental model is quite large, ie 40%. This also indicates that the movement of student ideas in a learning is very dynamic. The phenomenon is reinforced by several research results that show that there are variations of student ideas on a science phenomenon that observed [18], [19], [20], [21].

The varied pattern of student movements in the learning of the concept of magnetic induction represented by each student from the learning progress category in each type is shown in Table 1.

Table 1. The progress of students' thinking on the concept of magnetic induction from the type of progress based on the learning stages.

Students' argument at the stage			Type/Students' code
Early	Intermediate	Final	
There is a magnetic flow of energy or magnetism	Power/power/magnet energy has a limit	Magnetic forces can flow on other magnetic objects	Consistent with more scientific concepts (3)
		There is electricity to the other clamp	Changed more randomly (24)
There is a similar temperature to the clamp and is connected to the magnet	Magnets have temporary magnetic properties	The existence of a magnetic field, and clamps attached to the magnet is a temporary magnet	Changed more progressive (5)
The strongest magnetic attraction is in objects that are directly in contact with magnets	Power/power/magnet energy has a limit	Magnetic forces can flow on other magnetic objects	Changed more progressive (8)
The strongest	Magnets can attract	The existence of a	Changed more

magnetic attraction is in objects that are directly in contact with magnets	metal objects	magnetic field, and clamps attached to the magnet is a temporary magnet	progressive (7)
	Magnetic attraction only for conductor objects	There is an effect of magnetic poles that can pull magnetic objects	Consistent with basic experience (17)

Table 1 above shows that the movement of students' ideas at each stage of learning as a sample for the categories of student learning progress on the concept of magnetic induction has a more diverse path or pattern toward a more scientific understanding. Suppose that students 5, 7 and 8 have a Type II advancement category (changed more progressively), but the pattern of idea change varies from one to the other.

The results of interviews on students 5 say that the phenomenon of a sticky paper clip on the magnet can attract another clamp when a paper clip is not a magnet because the magnet attaches the same appeal to another paper clip, and if the appeal is not enough then the clamp is not interested anymore. Different things expressed by students 24 with the type of learning progress "change more random" saying that "there is a difference between the flow of magnets with electricity, if the electricity is energy flowing with the conductor, while the magnet does not need a carrier (conductor)"

4. Conclusion

Facilitating the gradual thinking of the students on the concept of magnetic induction through the corresponding new phenomena can increase the number of students who have a change of thinking in the more progressive categories that observed from the mental model changes it has. In addition, through facilitation also obtained data show that not all students have a consistent and progressive thinking, but there are also students who think more randomly.

Acknowledgments

Acknowledgments especially the authors conveyed to my promoter and co-promoter during study at the Universitas Pendidikan Indonesia, which has contributed the most of constructive ways of thinking to an educational research context in science, especially in primary science education.

References

[1] M Başer and Ö Geban 2007 Effect of instruction based on conceptual change activities on students' understanding of static electricity concepts *Res. Sci. Technol. Educ.* **25** (2) 243–267

[2] A M Criado and A García-Carmona 2010 Prospective Teachers' Difficulties in Interpreting Elementary Phenomena of Electrostatic Interactions: Indicators of the status of their intuitive ideas *Int. J. Sci. Educ.* **32** 769–805

[3] J Park 2001 Analysis of students' processes of confirmation and falsification of their prior ideas about electrostatics *Int. J. Sci. Educ.* **23** 1219–1236

[4] R Hamid, A Widodo and W Sopandi 2017 Students' Conceptual Change in Electricity *Adv. Soc. Sci. Educ. Humanit. Res.* **57** 48–52

[5] P S Skosana and R B Monyai 2013 The Teacher as a Catalytic Agent in the Implementation of the Curriculum *Int. J. Humanit. Soc. Sci. Invent.* **2** (9) 90–96

[6] N M Siew 2013 Exploring Primary Science Teachers ' Creativity and Attitudes through Responses to Creative Questions in University Physics Lessons *Br. J. Educ. Soc. Behav. Sci.* **3** (1) 93–108

[7] I Marõ Â A Greca and M A Moreira 2000 Mental models, conceptual models, and modelling 1," *Int. J. Sci. Educ* **22** (1) 1–11

[8] M A K & A Y Emen 2013 Mental Models of the High School Students Related To the Contraction of Matter *Int. J. Educ. Res. Technol.* **4** 1–5

[9] M T H Chi 2008 Three Types of Conceptual Change: Belief Revision, Mental Model Transformation, and Categorical Shift in *Handbook of research on conceptual change* pp. 61–82
[10] Shavelson R J and Kurpius 2012 Reflexion on learning progression in *Learning Progression in Science: Current challenges and future direction* (A. W. Gotwals, Ed.) pp. 13–26
[11] W Harlen 2010 *Principles and big ideas of science education*
[12] G W Fulmer 2013 Constraints on Conceptual Change: How Elementary Teachers' Attitudes and Understanding of Conceptual Change Relate to Changes in Students' Conceptions *J. Sci. Teacher Educ.* **24** 7 1219–1236
[13] J W Pellegrino 2009 The Design of an Assessment System for the Race to the Top: A Learning Sciences Perspective on Issues of Growth and Measurement *Explor. Semin. Meas. Challenges Within Race to Top Agenda* 1–37
[14] M Wilson and C Carstensen 2007 Assessment to improve learning in mathematics: The BEAR assessment system *Assess. Math. Profic.* **53** 311–332
[15] John W. Creswell 2011 *Educational Research: Planning, Conducting, and Evaluating Quantitative and ...* (John W. Creswell - Google Books)
[16] L E A Cohen 2007 R*esearch methods in education*
[17] J Wilcox and L R Richey 2012 Using concrete activities to confront misconceptions about magnetism in the primary grades *Sci. Child.* **50** (2) 62–67
[18] H Kucukozer and A Bostan 2010 Ideas of kindergarten students on the day-night cycles, the seasons and the moon phases *J. Theory Pract. Educ.* **6** (2) 267–280
[19] I Azaiza, V Bar, Y Awad and M Khalil 2012 Pupils' Explanations of Natural Phenomena and Their Relationship to Electricity *Creat. Educ.* 3 (8) 1354–1365
[20] M A Kurnaz 2013 Mental Models of Pre-School Children Regarding the Sun , Earth and Moon *Int. J. Soc. Sci.* **7** (1) 136–143
[21] G Harman, Z Aksan and D Celikler 2015 Mental models with influence the attitudes of science students towards recyling *Int. J. Suitanable Green Energy* **4** (1) 6–11

Construction of Open-Ended Problems for Assessing Elementary Student Mathematical Connection Ability on Plane Geometry

R Y Agustini[1*], D Suryadi [2] and A Jupri [2]

[1]Pendidikan Dasar Sekolah Pascasarjana Universitas Pendidikan Indonesia, Bandung, Indonesia
[2]Departemen Pendidikan Matematika Universitas Pendidikan Indonesia, Bandung, Indonesia

[*]riniyulia178@gmail.com

Abstract. The aim of this study is to construct open-ended problems about plane geometry for measuring mathematical connection ability. And to determine the profile of mathematical connection ability. The instrument test is 4 geometry problems. The problems is assessing four ability in mathematical connection at elementary school. The subject of this research is 50 students 6th grade of one elementary school in Bandung. The instrument is repairing base on suggestions from the four expert validation. The result validity of instrument show that problem no. 1 is 0,579 with categories enough validity, problem no. 2 is 0,809, no. 3 is 0,765, and no. 4 is 0,790 with categories high validity. The validation test show that instrument is a good quality. The reliability of instrument show 0,762 with categories high reliability. The distinguishing capacity for instrument no. 1 is 0,2857 with categories enough, no. 2 is 0,8036 and no. 3 is 0,7679 with categories very good, no. 4 is 0,4643 with categories good. The difficulty level of question for no. 1, 2, 3 is moderate, and no. 4 is difficult. The profile of mathematical connection ability of 6th grade elementary students on plane geometry is low. Base on result and analysis, the conclusion is that open-ended problem can be used to assessing elementary student mathematical connection ability.

1. Introduction

One of the six principles fundamental to high-quality mathematics education is assessment. The term *assessment* is defined as "the process of gathering evidence about a student's knowledge of, ability to use, and disposition toward mathematics and of making inferences from that evidence for a variety of purposes"[1]. Note that "gathering evidence" is not the same as giving a test or quiz. Assessment can and should happen every day as an integral part of instruction. Assessment should support the learning of important mathematics and furnish useful information to both teachers and students.

The Assessment Principle stresses two main ideas: (1) Assessment should enhance students learning, and (2) assessment is a valuable tool for making instructional decisions [2]. It means assessment should not merely be done to students; rather, it should also be done for students, to guide and enhance their learning. Assessment that includes ongoing observation and student interaction encourages students to articulate and, thus, clarify their ideas. Assessment is important role in learning process until after learning ending. Whereas assessment is the most important in the evaluation of learning. It due to evaluation of student learning is done by the teacher to monitor the process and the progress and improvement of learning outcomes on an ongoing basis. This indicates that evaluation is feedback for what learning goes well or not.

Mathematical connection is one of five process standards refer to the mathematical processes through which students should acquire and use mathematical knowledge. The connections standard has two parts. First, it is important to connect within and among mathematical ideas. Second, mathematics should be connected to the real world and to other disciplines [3]. Students should see that mathematics plays a significant role in science, language arts, and social studies. This suggests that mathematics should frequently be integrated with other discipline areas and that applications of mathematics should be explored in real world contexts.

Mathematical connection supports students to comprehend a concept substantially and assists them to improve their understanding and it's helps students provide a mathematical model that illustrates the relationship among concepts, data, and situation [4]. In fact, many find the questions of connection still become a problem for student as especially for elementary student. It show in some research said student have a low of mathematical connections ability [5]. This means that evaluations need to be considered.

The type of problem used in learning proces with open-ended approach is not routine and open problem. Open process means that the task type has several correct ways. Open end products mean that the task type has multiple answer possibilities. Finally, open ways to develop means that when the students have solved their previous problems, they can solve new problems by changing the condition of the previous problem [6]. Using the open-ended approach, students is working individually or in groups, are expected to apply their own unique methodology to solve given problems. These problems are so designed, that there may be more than one correct answer or there may be more than one way of arriving at an answer, thus they are able to challenge students at various levels of cognitive development [7].

To construct open-ended problem we can develop from the given problem. It is easy for students to start mathematical activity from the given open ended problems. It is also so suitable for teachers to investigate how their formulated open-ended problems have been engaged in by the students [8]. One more advantage of open-ended problems is that every student, no matter if he/she is highly capable or struggles with mathematics, can try and find his own solutions to the problems depending on his own scope and level of abilities [9].

The study of geometric shapes and their properties is an essential component of a comprehensive elementary mathematics curriculum [10]. So it's important for elementary student to learn about geometry. Geometry is rich in concepts, problem-solving experiences, and applications [11]. Learn about geometry not just learn about formula, but we can learn others subject to solving problems of geometry. So by using open ended problems on geometry we can give student an experience to solving problems and to applied their knowledge about plane geometry.

Based on some preconception, we try to make a constructing an instrument test for measuring students mathematical connection ability using open-ended problems.

2. Experimental Method

The research method was used quantitative method. Quantitative research was an approach for testing objective theories by examining the relationship among variables. These variables, in turn, could be measured, typically on instruments, so that numbered data can be analyzed using statistical procedures [10]. The final written report had a set structure consisting of introduction, literature and theory, methods, results, and discussion. Like qualitative researchers, those who engage in this form of inquiry have assumptions about testing theories deductively, building in protections against bias, controlling for alternative explanations, and being able to generalize and replicate the findings.

The purpose in this study was construction question and determine the profile of mathematical connection ability elementary students. Type of question was open ended problem in plane geometry. The sequential exploratory strategy was used in this research. The first, collection and analysis data. Two steps for this stage were literature study and expert validity. Instrument test for measuring connection ability in geometry were data obtained. Then data were analysed. Analysis was done for gotten a good instrument test. The second, collection and analysis data were done by statistical procedures. The steps for this stage were reliability, distinguishing capacity, and difficulty level of questions. Then these data were analysed. The last, we were getting full data and were interpreted.

The population of this study was student of 6th grade from one of elementary school in Bandung. Then sample based on population is two of four class with sampling technique used was purposive sampling class. The number of student were 50 as participant. Generally, several steps were needed to construct instrument test was include: 1) Made test specification, such as determine the purpose of test, blue print of test, shape of test, and determine length of test, 2) Wrote question of test, 3) Checked question of test, 4) Done try out of test, 5) Analyzed every question of test, 6) Repaired question of test, 7) Remaked test, 8) Done test, and 9) Interpreted result test [11].

3. Result and Discussion

The purpose construction of instrument test was determine student's connection ability about geometry at elementary school. So that geometry was focus of material in this study. Based on this purpose, the shape of test in this research was open-ended problem. After made test specification and wrote blue print of test, open-ended problems consist of 4 problems, with specification one problem for every indicator in mathematical connection ability. The open-ended problems in plane geometry which contain indicator in mathematical connection ability are presented in Table 1.

Table 1. The open-ended problems in Mathematical Connection Ability

Mathematical Connection Ability	Indicator	Open-Ended Problem
To connect inter-topics in mathematics that connect inter-concept or principle in the same topic	Desribe connection inter-topics in mathematics	Mr. Amir is a gardener. He will make a parallelogram garden with 18 m^2 area. a. Determine the measurement of Mr. Amir garden? Make it in table! b. Determine the bigest around of Mr. Amir garden? Make it in table!
Connection between topics in mathematics that connect one material and other materials in mathematics	Determine solving way that connect one and other materials in mathematics	A rectangle tablecloth look like the picture beside. It's have a 4: 5 proportion between long and wide. 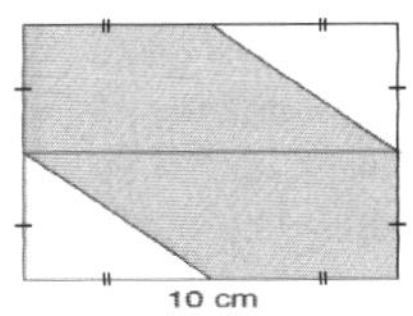 a. Determine the long and wide of the tablecloth? b. Determine the area of the tablecloth? c. Determine the shaded in area of the tablecloth?
Connection between mathematics materials and other science	Solve the problem which connect mathematics materials and other science	Mr. Ali have a rectangle garden look like the picture beside. He will plant guava at trapezoid area and papaya at the shaded in area. 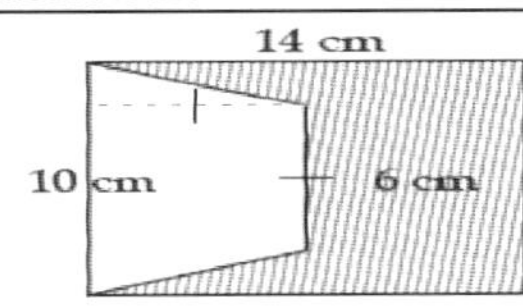 a. Count the whole area of Mr. Ali garden? b. Count the area of guava garden? c. Count the area of papaya garden?
Connection between mathematics and everyday life which can be found by students	Checking the effective way to solve problem with others procedure	A flower park look like picture beside. The park will have fence. a. Count the around of the park? b. Count the area of the park with 2 different way? 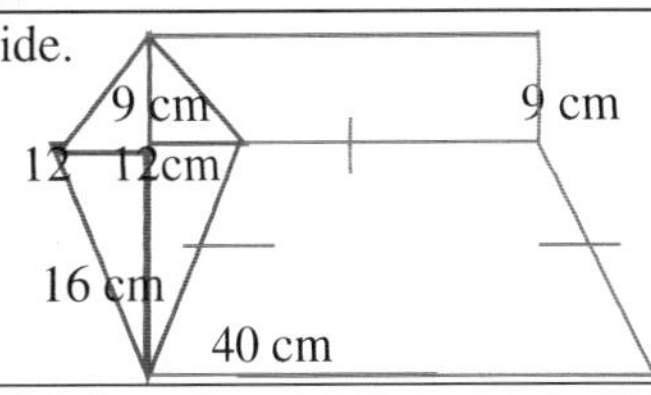

Once the open ended problem was made, the assessment then validated by the expert. Validity was a measure that indicates the level of validity or validity of an instrument [12]. Validity can be analyzed by asking expert opinion (expert judgment). Expert judgment was done by asking for ratings of four experts in accordance with the scope of the investigation to ensure that the instruments was made in accordance with the aspects to be measured in the study. The four persons who have experts give feedback and suggestions on instruments had made in terms of material and suitability indicators with questions that are made at follow in Table 2.

Table 2. Feedback and suggestion from expert validation

No.	Expert Validation	Aspect	Feedback and suggestions
1	Expert Validation I	Material	• The problem must be able to assess connection ability • The problem must contain geometry which is open ended problem
2	Expert Validation II	Compliance material with indicators	• The problem must contain open problem in plane geometry • The material must be show the mathematical connection in plane geometry • The student must have hands on activities so they make conclusion about the formula in plane geometry
3	Expert Validation III	Languange and writing	The picture must be clear and proportion so the student can get information by the picture
4	Expert Validation IV	Languange and writing	The languange and instruction must be clear so the student known must to do

Based on suggestions from the four expert validation, the instrument repaired with the feedback that had been given. After the repair was completed, the instrument then assessed from the aspect of content, technical, and construct. Assessment conducted by two teachers who are already teaching at 6th grade in elementary school. The result validation show that instrument is a good quality.

After validation expert made, the next step was done validity testing. Instrument validity test performed by the test technique that is by trying out the instrument in a different class. Validity is degree to which evidence and theory support the interpretation of test scores entailed by proposed uses of tests. It is our interpretations of the scores that are either valid or invalid. We might interpret this score as representing how much each student has learned relative to other students [13]. The equation used to determine the validity of the instrument at Equation 1.

$$r_{xy} = \frac{N\sum XY - (\sum X)(\sum Y)}{\sqrt{\{N\sum X^2 - (\sum X)^2\}\{N\sum Y^2 - (\sum Y)^2\}}}$$

After we determine validity, the results of calculating of validity was interpreted in a category are presented in Table 3.

Table 3. Interpretation validity in categories

Limited	Category
$0{,}80 < r_{xy} \leq 1{,}00$	Very high
$0{,}60 < r_{xy} \leq 0{,}80$	High
$0{,}40 < r_{xy} \leq 0{,}60$	Enough
$0{,}20 < r_{xy} \leq 0{,}40$	Low
$0{,}00 < r_{xy} \leq 0{,}20$	Very Low

IOP Conf. Series: Journal of Physics: Conf. Series **895** (2017) 012148 doi:10.1088/1742-6596/895/1/012148

The analysis process to determine the validity of the instrument was used SPSS 21. The result validity of instrument showed that problem no. 1 was 0,579 with category enough validity, problem no. 2 was 0,809, no. 3 was 0,765, and no. 4 was 0,790 with category high validity.

After validation expert made, the next step was done reliability testing. Instrument reliability test performed by the test technique that is by trying out the instrument in a different class. Reliability is measured by the correlation coefficient of result test. When the positive and significant correlation coefficient of the instrument has been declared reliable [14]. The equation used to determine the reliability of the instrument at Equation 2.

$$r_{11} = \frac{2r_{\frac{1}{2}\frac{1}{2}}}{(1 + 2r_{\frac{1}{2}\frac{1}{2}})}$$

Having calculated the coefficient of reliability, this value is interpreted in a category at follow in Table 4.

Table 4. Interpretation reliability in categories

Limited	Category
$0{,}80 < r_{11} \leq 1{,}00$	Very high
$0{,}60 < r_{11} \leq 0{,}80$	High
$0{,}40 < r_{11} \leq 0{,}60$	Enough
$0{,}20 < r_{11} \leq 0{,}40$	Low
$0{,}00 < r_{11} \leq 0{,}20$	Very Low

The analysis process to determine the reliability of the instrument was used SPSS 21. The result reliability of instrument showed that 0,762 with category high reliability.

After reliability tests was conducted, the distinguishing capacity intended to determine the extent of open ended problem could differentiate high-ability students with low-ability students [15]. The equation used to determine the distinguishing capacity of the instrument at Equation 3.

$$DP = \frac{JBA - JBB}{JSA}$$

After we determine distinguishing capacity, the results of calculating of distinguishing capacity was interpreted in a category at follow in Table 5.

Table 5. Interpretation distinguishing capacity in clasification

DC Value	Clasification
$0{,}00 < DC \leq 0{,}20$	Bad
$0{,}20 < DC \leq 0{,}40$	Enough
$0{,}40 < DC \leq 0{,}70$	Good
$0{,}70 < DC \leq 1{,}00$	Very Good

Based on the results of calculating, distinguishing capacity for problem no. 1 was enough, problem no. 4 was good, and problem no. 2, 3 was very good.

After distinguishing capacity was conducted, then we calculating the level of convenience. The level of convenience is a number that indicates something about the difficult and the easy [16]. Convenience items is an overall proportion of students who answered correctly on the item was. To calculate the level of ease of each items used Equation 4.

$$P = \frac{\Sigma B}{N}$$

After calculating, we interpreting in a category at follow in Table 6.

Table 6. Interpretation the level of convenience in categories

Value	Category
$0{,}00 < P \leq 0{,}30$	Difficult
$0{,}30 < P \leq 0{,}70$	Moderate
$0{,}70 < P \leq 1{,}00$	Easy

The results of calculating found that category difficult for question no. 4, and then 3 for category moderate for no. 1, 2, 3. After analysis of all questions, 4 questions were made.

The ability in mathematical connection at elementary school can be represented by four indicator as follows; (1) to connect inter-topics in mathematics that connect inter-concept or principle in the same topic, (2) connection between topics in mathematics that connect one material and other materials in mathematics, (3) connection between mathematics materials and other science, (4) connection between mathematics and everyday life which can be found by students. The profile of elementary student's mathematical connection ability on geometry can be seen from the answers and the obtained scores on open ended problems. The four mathematical connection ability are presented in Table 7, 8, 9, and 10.

Table 7. Ability profile in connect inter-topics in mathematics

Interval Value	Frequency	Percentage (%)	Category
$75< t \leq 100$	1	2	Very good
$58.33< t \leq 75$	6	12	Good
$41.67< t \leq 58.33$	21	42	Enough
$25< t \leq 41.67$	16	32	Low
$0< t \leq 25$	6	12	Very low
The average value	40.00		Low

Table 7 shows that the ability profile in connect inter-topics in mathematics is low. Only one student who have very good category. And 6 students who have very low category. The student still have some difficulty to determine parallelogram measurement with certain area. And to determine the bigest around of parallelogram with certain area. It's shows that the students ability to connect between area, measurement, and around concept on plane geometry is still low.

Table 8. Ability profile in connection between topics in mathematics

Interval Value	Frequency	Percentage (%)	Category
$75< t \leq 100$	10	20	Very good
$58.33< t \leq 75$	2	4	Good
$41.67< t \leq 58.33$	8	16	Enough
$25< t \leq 41.67$	7	14	Low
$0< t \leq 25$	23	46	Very low
The average value	34.50		Low

Table 8 shows that the ability profile in connection between topics in mathematics that connect one material and other materials in mathematics is low. Ten students have very good category. And 23 students have very low category. The student still have some difficulty to determine area from the proportion of long and wide that given. And determine the shaded areafrom picture given. It's show that the students ability to make connection between proportion with plane geometry topics is low.

Table 9. Ability profile in connection between mathematics materials and other science

Interval Value	Frequency	Percentage (%)	Category
$75< t \leq 100$	14	28	Very good
$58.33< t \leq 75$	3	6	Good
$41.67< t \leq 58.33$	3	6	Enough
$25< t \leq 41.67$	9	18	Low
$0< t \leq 25$	21	42	Very low
The average value	40.00		Low

Table 9 shows that the ability in connection between mathematics materials and other science is low. 14 students have very good category. And 21 students have very low category. The students still have some difficulty to count whole area and certain area from picture given. The picture are show trapezoid in rectangle. It's show that students ability to make connection between plane geometry topics with garden area context is low.

Tabel 10. Ability profile in connection between mathematics and everyday life

Interval Value	Frequency	Percentage (%)	Category
$75< t \leq 100$	2	4	Very good
$58.33< t \leq 75$	1	2	Good
$41.67< t \leq 58.33$	7	14	Enough
$25< t \leq 41.67$	17	34	Low
$0< t \leq 25$	23	46	Very low
The average value	21.00		Low

Table 10 shows that the ability in connection between mathematics and everyday life which can be found by students is low. Only two student have very good category. And 23 students have very low category. The students still have some difficulty to count the around and area from picture given. Student can see the picture as kite and two trapezoid, or trapezoid, rectangle and two triangle. It's need spatial ability to solving the problem. It's show that student ability to make connection between logical ability and spatial ability is still low.

4. Conclusion

The open ended problem on plane geometry can used to assessing mathematical connection ability at elementary student. The mathematical connection ability profile of 6 th grade students in elementary school on plane geometry is low. To construct open ended problem at plane geometry, we must to consider what student already about that subject so we can give them new experiences by solving the open-ended problems.

Acknowledgments

My thank to the school principal who has given permission data retrieval, teachers and student 6th grade who actively participated in this study.

References

[1] NCTM 1995 *Assessment Standards for School Mathematics* (USA: NCTM Inc) p 3
[2] NCTM 2000 *Principles and Standards for School Mathematics* (USA: NCTM Inc) p 22
[3] Van de Walle J A 2013 *Elementary and Middle School Mathematics: Teaching Developmentally* (USA: Pearson Education) p 4
[4] Hendriana H et. al. 2014 Mathematical connection ability and self-confidence (an experiment on junior high school students through contextual teaching and learning with mathematical manipulative) *International Journal of Education* **8** (1)

[5] Saminanto and Kartono 2015 Analysis of mathematical connection ability in linear equation with one variable based on connectivity theory *International Journal of Education and Research* 3 (4)
[6] Fattah A et. al 2016 Open-ended approach: an effort in cultivating students' mathematical creative thinking ability and self-esteem in mathematics *Journal on Mathematics Education* **7** (1) pp 11-20
[7] Munroe L 2015 The open-ended approach framework *European Journal of Educational Research* **4** (3) pp 97 - 104
[8] Inprasitha M 2006 Open-Ended Approach and Teacher Education *Tsukuba Journal of Educational Study in Mathematics* 25
[9] Mihajlović and Dejić 2015 Using open-ended problems and problem posing activities in elementary mathematics classroom *9TH International MCG Conference*
[10] Musser G L et al 2008 *Mathematics for Elementary Teachers A Contemporary Approach. 8th ed.* USA: John Wiley & Sons, Inc.
[11] Leicester M and Taylor D 2010 *Critical Thinking Across The Curriculum Developing Critical Thinking Skills, Literacy and Philosophy in The Primary Classroom.* UK : Mc Graw-Hill Education
[12] Creswell 2013 *Research Design : Qualitative Quantitative and Mixed Methods Approaches* (USA: SAGE Publications Inc)
[13] Mardapi D 2012 *Pengukuran Penilaian dan Evaluasi Pendidikan* (Yogyakarta: Nuha Medika) pp 53-55
[14] Arikunto S 2013 *Prosedur Penelitian Suatu Pendekatan Praktik* (Jakarta: Rineka Cipta)
[15] Gall M et. al. 2003 *Educational Research An Introduction. Seventh Edition* (USA: Ablongman)
[16] Susetyo B 2014 *Statistika untuk Analisis Data Penelitian* (Bandung : Refika Aditama)
[17] Suherman 2003 *Evaluasi Pembelajaran Matematika* (Bandung: UPI) p 160
[18] Riduwan 2012 *Dasar-dasar Statistika* (Bandung : Alfa Beta) p 183

AUTHOR INDEX

AUTHOR INDEX